ESSAYS ON FOURIER ANALYSIS IN HONOR OF ELIAS M. STEIN

Princeton Mathematical Series

EDITORS: LUIS A. CAFFARELLI, JOHN N. MATHER, *and* ELIAS M. STEIN

ESSAYS ON FOURIER ANALYSIS IN HONOR OF ELIAS M. STEIN

EDITED BY

Charles Fefferman, Robert Fefferman
and Stephen Wainger

PRINCETON UNIVERSITY PRESS
PRINCETON, NEW JERSEY

Library of Congress Cataloging-in-Publication Data

Essays on Fourier Analysis in honor of Elias M. Stein / edited by Charles
Fefferman, Robert Fefferman, and Stephen Wainger.

      p. cm.—(Princeton mathematical series ; 42)
      Proceedings of the Princeton Conference in Harmonic
      Analysis, held May 13–17, 1991.
      Includes bibliographical references.
      ISBN: 978-0-691-60365-0
      1. Fourier analysis—Congresses.   I. Stein, Elias M., 1931–.
II. Fefferman, Charles, 1949–.   III. Fefferman, Robert, 1951–. IV.
Wainger, Stephen, 1936–.   V. Series.
     QA403.5.P76 1993
     515′.2433—dc20   92-43054
                                    CIP

10 9 8 7 6 5 4 3 2 1

CONTENTS

INTRODUCTION

This is the proceedings of the Princeton Conference in Harmonic Analysis, which was held from May 13 through May 17, 1991. This conference was a very special event because it celebrated Elias M. Stein's extraordinary impact on mathematics, on the occasion of his sixtieth birthday. Through his brilliant research, books, lectures, and teaching, E. M. Stein has changed the way central problems in analysis are approached. The articles presented here are a clear reflection of this and a tribute to his stellar contribution.

We would like to take this opportunity to express our gratitude to those who made the conference possible. We thank the National Science Foundation for funding and the Princeton Mathematics Department for both its hospitality and its financial support. It is a pleasure to acknowledge the contributions of M. Christ and M. Machedon, who provided us with financial assistance through their PYI grants. Thanks go to J. Ryff for his many helpful suggestions and his encouragement, and to Scott Kenney and the staff of the Princeton Mathematics Department for the great amount of work they did to make things run smoothly. Finally, we wish to thank the speakers and the many participants in the conference for their energy and enthusiasm, which made possible its success.

Charles Fefferman
Robert Fefferman
Stephen Wainger

ESSAYS ON FOURIER ANALYSIS IN HONOR OF ELIAS M. STEIN

1

Selected Theorems by Eli Stein

Charles Fefferman

INTRODUCTION

The purpose of this survey article is to give the general reader some idea of the scope and originality of Eli Stein's contributions to analysis. His work deals with representation theory, classical Fourier analysis, and partial differential equations. He was the first to appreciate the interplay among these subjects, and to perceive the fundamental insights in each field arising from that interplay. No one else really understands all three fields; therefore, no one else could have done the work I am about to describe. However, deep understanding of three fields of mathematics is by no means sufficient to lead to Stein's main ideas. Rather, at crucial points, Stein has shown extraordinary originality, without which no amount of work or knowledge could have succeeded. Also, large parts of Stein's work (e.g., the fundamental papers [26], [38], [41], [44], [59] on complex analysis in tube domains) don't fit any simple one-paragraph description such as the one above.

It follows that no single mathematician is competent to present an adequate survey of Stein's work. As I attempt the task, I am keenly aware that many of Stein's papers are incomprehensible to me, while others were of critical importance to my own work. Inevitably, therefore, my survey is biased, as any reader will see. Fortunately, S. Gindikin provided me with a layman's explanation of Stein's contributions to representation theory, thus keeping the bias (I hope) within reason. I am grateful to Gindikin for his help, and also to Y. Sagher for a valuable suggestion.

For purposes of this article, representation theory deals with the construction and classification of the irreducible unitary representations of a semisimple Lie group. Classical Fourier analysis starts with the L^p-boundedness of two fundamental operators, the maximal function

$$f^*(x) = \sup_{h>0} \frac{1}{2h} \int_{x-h}^{x+h} |f(y)| \, dy,$$

and the Hilbert transform

$$Hf(x) = \lim_{\epsilon \to 0+} \frac{1}{\pi} \int_{|x-y|>\epsilon} \frac{f(x)dy}{x-y}.$$

Finally, we shall be concerned with those problems in partial differential equations that come from several complex variables.

COMPLEX INTERPOLATION

Let us begin with Stein's work on interpolation of operators. As background, we state and prove a classical result, namely the

M. Riesz Convexity Theorem. *Suppose X, Y are measure spaces, and suppose T is an operator that carries functions on X to functions on Y. Assume T is bounded from $L^{p_0}(X)$ to $L^{r_0}(Y)$, and from $L^{p_1}(X)$ to $L^{r_1}(Y)$. (Here, $p_0, p_1, r_0, r_1 \in [1, \infty]$.) Then T is bounded from $L^p(X)$ to $L^r(Y)$ for $\frac{1}{p} = \frac{t}{p_1} + \frac{(1-t)}{p_0}, \frac{1}{r} = \frac{t}{r_1} + \frac{(1-t)}{r_0}, 0 \le t \le 1$.*

The Riesz Convexity Theorem says that the points $(\frac{1}{p}, \frac{1}{r})$ for which T is bounded from L^p to L^r form a convex region in the plane. A standard application is the Hausdorff–Young inequality: We take T to be the Fourier transform on $\mathbb{R}^n$, and note that T is obviously bounded from L^1 to L^∞, and from L^2 to L^2. Therefore, T is bounded from L^p to the dual class $L^{p'}$ for $1 \le p \le 2$.

The idea of the proof of the Riesz Convexity Theorem is to estimate $\int_Y (Tf) \cdot g$ for $f \in L^p$ and $g \in L^{r'}$. Say $f = Fe^{i\phi}$ and $g = Ge^{i\psi}$ with $F, G \ge 0$ and ϕ, ψ real. Then we can define analytic families of functions f_z, g_z by setting $f_z = F^{az+b}e^{i\phi}$, $g_z = G^{cz+d}e^{i\psi}$, for real a, b, c, d to be picked in a moment.

Define

(1) $$\Phi(z) = \int_Y (Tf_z)g_z.$$

Evidently, Φ is an analytic function of z.

For the correct choice of a, b, c, d we have

(2) $|f_z|^{p_0} = |f|^p$ and $|g_z|^{r_o'} = |g|^{r'}$ when $\text{Re } z = 0$;

(3) $|f_z|^{p_1} = |f|^p$ and $|g_z|^{r_1'} = |g|^{r'}$ when $\text{Re } z = 1$;

(4) $f_z = f$ and $g_z = g$ when $z = t$.

From (2) we see that $\| f_z \|_{L^{p_0}}, \| g_z \|_{L_0^{r'}} \leq C$ for Re $z = 0$. So the definition (1) and the assumption $T : L^{p_0} \to L^{r_0}$ show that

(5)
$$|\Phi(z)| \leq C' \qquad \text{for} \quad \text{Re } z = 0.$$

Similarly, (3) and the assumption $T : L^{p_1} \to L^{r_1}$ imply

(6)
$$|\Phi(z)| \leq C' \qquad \text{for} \quad \text{Re } z = 1.$$

Since Φ is analytic, (5) and (6) imply $|\Phi(z)| \leq C'$ for $0 \leq$ Re $z \leq 1$ by the maximum principle for a strip. In particular, $|\Phi(t)| \leq C'$. In view of (4), this means that $| \int_Y (Tf)g | \leq C'$, with C' determined by $\| f \|_{L^p}$ and $\| g \|_{L^{r'}}$.

Thus, T is bounded from L^p to L^r, and the proof of the Riesz Convexity Theorem is complete.

This proof had been well-known for over a decade, when Stein discovered an amazingly simple way to extend its usefulness by an order of magnitude. He realized that an ingenious argument by Hirschman [H] on certain multiplier operators on $L^p(\mathbb{R}^n)$ could be viewed as a Riesz Convexity Theorem for analytic families of operators. Here is the result.

Stein Interpolation Theorem. *Assume T_z is an operator depending analytically on z in the strip $0 \leq$ Re $z \leq 1$. Suppose T_z is bounded from L^{p_0} to L^{r_0} when Re $z = 0$, and from L^{p_1} to L^{r_1} when Re $z = 1$. Then T_t is bounded from L^p to L^r, where $\frac{1}{p} = \frac{t}{p_1} + \frac{(1-t)}{p_0}, \frac{1}{r} = \frac{t}{r_1} + \frac{(1-t)}{r_0}$ and $0 \leq t \leq 1$.*

Remarkably, the proof of the theorem comes from that of the Riesz Convexity Theorem by adding a single letter of the alphabet. Instead of taking $\Phi(z) = \int_Y (Tf_z)g_z$ as in (1), we set $\Phi(z) = \int_Y (T_z f_z)g_z$. The proof of the Riesz Convexity Theorem then applies with no further changes.

Stein's Interpolation Theorem is an essential tool that permeates modern Fourier analysis. Let me just give a single application here, to illustrate what it can do. The example concerns Cesaro summability of multiple Fourier integrals.

We define an operator $T_{\alpha R}$ on functions on $\mathbb{R}^n$ by setting

$$\widehat{T_{\alpha R} f}(\xi) = \left(1 - \frac{|\xi|^2}{R^2} \right)_+^\alpha \hat{f}(\xi).$$

Then

(7)
$$\| T_{\alpha R} f \|_{L^p(\mathbb{R}^n)} \leq C_{\alpha p} \| f \|_{L^p(\mathbb{R}^n)}, \qquad \text{if} \quad \left| \frac{1}{p} - \frac{1}{2} \right| < \frac{\alpha}{n-1}.$$

This follows immediately from the Stein Interpolation Theorem. We let α play the role of the complex parameter z, and we interpolate between the elementary cases $p = 1$ and $p = 2$. Inequality (7), due to Stein, was the first non-trivial progress on spherical summation of multiple Fourier series.

REPRESENTATION THEORY I

Our next topic is the Kunze-Stein phenomenon, which links the Stein Interpolation Theorem to representations of Lie groups. For simplicity we restrict attention to $G = SL(2, \mathbb{R})$, and begin by reviewing elementary Fourier analysis on G. The irreducible unitary representations of G are as follows:

The *principal series*, parametrized by a sign $\sigma = \pm 1$ and a real parameter t;
The *discrete series*, parametrized by a sign $\sigma = \pm 1$ and an integer $k \geq 0$; and
The *complementary series*, parametrized by a real number $t \in (0, 1)$.

We don't need the full description of these representations here.

The irreducible representations of G give rise to a Fourier transform. If f is a function on G, and U is an irreducible unitary representation of G, then we define

$$\hat{f}(U) = \int_G f(g)U_g dg,$$

where dg denotes Haar measure on the group. Thus, $\hat{f}$ is an operator-valued function defined on the set of irreducible unitary representations of G. As in the Euclidean case, we can analyze convolutions in terms of the Fourier transform. In fact,

(8) $$\widehat{f * g} = \hat{f} \cdot \hat{g}$$

as operators. Moreover, there is a Plancherel formula for G, which asserts that

$$\|f\|^2_{L^2(G)} = \int \|\hat{f}(U)\|^2_{\text{Hilbert-Schmidt}} d\mu(U)$$

for a measure μ (the Plancherel measure). The Plancherel measure for G is known, but we don't need it here. However, we note that the complementary series has measure zero for the Plancherel measure.

These are, of course, the analogues of familiar results in the elementary Fourier analysis of $\mathbb{R}^n$. Kunze and Stein discovered a fundamental new phenomenon in Fourier analysis on G that has no analogue on $\mathbb{R}^n$. Their result is as follows.

Theorem (Kunze–Stein Phenomenon). *There exists a uniformly bounded representation $U_{\sigma,\tau}$ of G, parametrized by a sign $\sigma = \pm 1$ and a complex number τ in a strip Ω, with the following properties.*
(A) *The $U_{\sigma,\tau}$ all act on the same Hilbert space H.*

(B) *For fixed* $\sigma = \pm 1$, $g \in G$, *and* ξ, $\eta \in H$, *the matrix element* $\langle (U_{\sigma,\tau})_g \xi, \eta \rangle$ *is an analytic function of* $\tau \in \Omega$.

(C) *The* $U_{\sigma,\tau}$ *for* Re $\tau = \frac{1}{2}$ *are equivalent to the representations of the principal series.*

(D) *The* $U_{+1,\tau}$ *for suitable* τ *are equivalent to the representations of the complementary series.*

(See [14] for the precise statement and proof, as well as Ehrenpreis-Mautner [EM] for related results.)

The Kunze-Stein Theorem suggests that analysis on G resembles a fictional version of classical Fourier analysis in which the basic exponential $\xi \longmapsto \exp(i\xi \cdot x)$ is a bounded analytic function on a strip $|\operatorname{Im} \xi| \leq C$, uniformly for all x.

As an immediate consequence of the Kunze-Stein Theorem, we can give an analytic continuation of the Fourier transform for G. In fact, we set $\hat{f}(\sigma, \tau) = \int_G f(g)(U_{\sigma,\tau})_g \, dg$ for $\sigma = \pm 1$, $\tau \in \Omega$.

Thus, $f \in L^1(G)$ implies $\hat{f}(\sigma, \cdot)$ analytic and bounded on Ω. So we have continued analytically the restriction of $\hat{f}$ to the principal series. It is as if the Fourier transform of an L^1 function on $(-\infty, \infty)$ were automatically analytic in a strip. If $f \in L^2(G)$, then $\hat{f}(\sigma, \tau)$ is still defined on the line $\{\operatorname{Re} \tau = \frac{1}{2}\}$, by virtue of the Plancherel formula and part (C) of the Kunze-Stein Theorem. Interpolating between $L^1(G)$ and $L^2(G)$ using the Stein Interpolation Theorem, we see that $f \in L^p(G)$ $(1 \leq p < 2)$ implies $\hat{f}(\sigma, \cdot)$ analytic and satisfying an $L^{p'}$-inequality on a strip Ω_p. As p increases from 1 to 2, the strip Ω_p shrinks from Ω to the line $\{\operatorname{Re} \tau = \frac{1}{2}\}$. Thus we obtain the following results.

Corollary 1. *If* $f \in L^p(G)$ $(1 \leq p < 2)$, *then* $\hat{f}$ *is bounded almost everywhere with respect to the Plancherel measure.*

Corollary 2. *For* $1 \leq p < 2$ *we have the convolution inequality* $\|f * g\|_{L^2(G)} \leq C_p \|f\|_{L^p(G)} \|g\|_{L^2(G)}$.

To check Corollary 1, we look separately at the principal series, the discrete series, and the complementary series. For the principal series, we use the $L^{p'}$-inequality established above for the analytic function $\tau \longmapsto \hat{f}(\sigma, \tau)$ on the strip Ω_p. Since an $L^{p'}$-function analytic on a strip Ω_p is clearly bounded on an interior line $\{\operatorname{Re} \tau = \frac{1}{2}\}$, it follows at once that $\hat{f}$ is bounded on the principal series. Regarding the discrete series $U_{\sigma,k}$ we note that

$$(9) \qquad \left(\sum_{\sigma,k} \mu_{\sigma,k} \| \hat{f}(U_{\sigma,k}) \|^{p'} \right)^{1/p'} \leq \|f\|_{L^p(G)}$$

for suitable weights $\mu_{\sigma,k}$ and for $1 \leq p \leq 2$. The weights $\mu_{\sigma,k}$ amount to the Plancherel measure on the discrete series, and (9) is proved by a trivial interpolation, just like the standard Hausdorff-Young inequality. The boundedness of the $\|\hat{f}(U_{\sigma,k})\|$ is immediate from (9). Thus the Fourier transform $\hat{f}$ is bounded on both the principal series and the discrete series, for $f \in L^p(G)$ $(1 \leq p < 2)$. The complementary series has measure zero with respect to the Plancherel measure, so the proof of Corollary 1 is complete. Corollary 2 follows trivially from Corollary 1, the Plancherel formula, and the elementary formula (8).

This proof of Corollary 2 poses a significant challenge. Presumably, the Corollary holds because the geometry of G at infinity is so different from that of Euclidean space. For example, the volume of the ball of radius R in G grows exponentially as $R \to \infty$. This must have a profound impact on the way mass piles up when we take convolutions on G. On the other hand, the statement of Corollary 2 clearly has nothing to do with cancellation; proving the Corollary for two arbitrary functions f, g is the same as proving it for $|f|$ and $|g|$. When we go back over the proof of Corollary 2, we see cancellation used crucially, e.g., in the Plancherel formula for G; but there is no explicit mention of the geometry of G at infinity. Clearly there is still much that we do not understand regarding convolutions on G.

The Kunze-Stein phenomenon carries over to other semisimple groups, with profound consequences for representation theory. We will continue this discussion later in the article. Now, however, we turn our attention to classical Fourier analysis.

CURVATURE AND THE FOURIER TRANSFORM

One of the most fascinating themes in Fourier analysis in the last two decades has been the connection between the Fourier transform and curvature. Stein has been the most important contributor to this set of ideas. To illustrate, I will pick out two of his results. The first is a "restriction theorem," i.e., a result on the restriction $\hat{f}|_\Gamma$ of the Fourier transform of a function $f \in L^p(\mathbb{R}^n)$ to a set Γ of measure zero. If $p > 1$, then the standard inequality $\hat{f} \in L^{p'}(\mathbb{R}^n)$ suggests that $\hat{f}$ should not even be well-defined on Γ, since Γ has measure zero. Indeed, if Γ is (say) the x-axis in the plane $\mathbb{R}^2$, then we can easily find functions $f(x_1, x_2) = \varphi(x_1)\psi(x_2) \in L^p(\mathbb{R}^2)$ for which $\hat{f}|_\Gamma$ is infinite everywhere. Fourier transforms of $f \in L^p(\mathbb{R}^2)$ clearly cannot be restricted to straight lines. Stein proved that the situation changes drastically when Γ is curved. His result is as follows.

Stein's Restriction Theorem. *Suppose Γ is the unit circle, $1 \leq p < \frac{8}{7}$, and $f \in C_0^\infty(\mathbb{R}^2)$. Then we have the a priori inequality $\|\hat{f}|_\Gamma\|_{L^2} \leq C_p \|f\|_{L^p(\mathbb{R}^2)}$, with C_p depending only on p.*

Using this a priori inequality, we can trivially pass from the dense subspace C_0^∞ to define the operator $f \longmapsto \hat{f}|_\Gamma$ for all $f \in L^p(\mathbb{R}^2)$. Thus, the Fourier transform of $f \in L^p$ ($p < \frac{8}{7}$) may be restricted to the unit circle.

Improvements and generalizations were soon proven by other analysts, but it was Stein who first demonstrated the phenomenon of restriction of Fourier transforms.

Stein's proof of his restriction theorem is amazingly simple. If μ denotes uniform measure on the circle $\Gamma \subset \mathbb{R}^2$, then for $f \in C_0^\infty(\mathbb{R}^2)$ we have

$$(10) \qquad \int_\Gamma |\hat{f}|^2 = \int_{\mathbb{R}^2} (\hat{f}\mu)\overline{(\hat{f})} = \langle f * \hat{\mu}, f \rangle \leq \|f\|_{L^p}\|f * \hat{\mu}\|_{L^{p'}}.$$

The Fourier transform $\hat{\mu}(\xi)$ is a Bessel function. It decays like $|\xi|^{-\frac{1}{2}}$ at infinity, a fact intimately connected with the curvature of the circle. In particular, $\hat{\mu} \in L^q$ for $4 < q \leq \infty$, and therefore $\|f * \hat{\mu}\|_{L^{p'}} \leq C_p\|f\|_{L^p}$ for $1 \leq p < \frac{8}{7}$, by the usual elementary estimates for convolutions. Putting this estimate back into (10), we see that $\int_\Gamma |\hat{f}|^2 \leq C_p\|f\|_{L^p}^2$, which proves Stein's Restriction Theorem. The Stein Restriction Theorem means a lot to me personally, and has strongly influenced my own work in Fourier analysis.

The second result of Stein's relating the Fourier transform to curvature concerns the differentiation of integrals on $\mathbb{R}^n$.

Theorem. *Suppose $f \in L^p(\mathbb{R}^n)$ with $n \geq 3$ and $p > \frac{n}{n-1}$. For $x \in \mathbb{R}^n$ and $r > 0$, let $F(x, r)$ denote the average of f on the sphere of radius r centered at x. Then $\lim_{r \to 0} F(x, r) = f(x)$ almost everywhere.*

The point is that unlike the standard Lebesgue Theorem, we are averaging f over a small sphere instead of a small ball. As in the restriction theorem, we are seemingly in trouble because the sphere has measure zero in $\mathbb{R}^n$, but the curvature of the sphere saves the day. This theorem is obviously closely connected to the smoothness of solutions of the wave equation.

The proof of the above differentiation theorem relies on an

Elementary Tauberian Theorem. *Suppose that $\lim_{R \to 0} \frac{1}{R} \int_0^R F(r)dr$ exists and $\int_0^\infty r \left| \frac{dF}{dr} \right|^2 dr < \infty$. Then $\lim_{R \to 0} F(R)$ exists, and equals $\lim_{R \to 0} \frac{1}{R} \int_0^R F(r)dr$.*

This result had long been used, e.g., to pass from Cesaro averages of Fourier series to partial sums. (See Zygmund [Z].) On more than one occasion, Stein has shown the surprising power hidden in the elementary Tauberian Theorem. Here we apply it to $F(x, r)$ for a fixed x. In fact, we have $F(x, r) = \int f(x + ry)d\mu(y)$, with μ equal to normalized surface measure on the unit sphere, so that the Fourier transforms of F and f are related by $\hat{F}(\xi, r) = \hat{f}(\xi)\hat{\mu}(r\xi)$ for each fixed r.

Therefore, assuming $f \in L^2$ for simplicity, we obtain

$$\int_{\mathbb{R}^n} \left(\int_0^\infty r \left| \frac{\partial}{\partial r} F(x, r) \right|^2 dr \right) dx = \int_0^\infty r \left[\int_{\mathbb{R}^n} \left| \frac{\partial}{\partial r} F(x, r) \right|^2 dx \right] dr$$

$$= \int_0^\infty r \left[\int_{\mathbb{R}^n} \left| \frac{\partial}{\partial r} \widehat{F}(\xi, r) \right|^2 d\xi \right] dr = \int_{\mathbb{R}^n} \int_0^\infty r \left| \frac{\partial}{\partial r} \hat{\mu}(r\xi) \right|^2 |\hat{f}(\xi)|^2 dr \, d\xi$$

$$= \int_{\mathbb{R}^n} \left\{ \int_0^\infty r \left| \frac{\partial}{\partial r} \hat{\mu}(r\xi) \right|^2 dr \right\} |\hat{f}(\xi)|^2 d\xi = (\text{const.}) \int_{\mathbb{R}^n} \left| \hat{f}(\xi) \right|^2 d\xi < \infty.$$

(Here we make crucial use of curvature, which causes $\hat{\mu}$ to decay at infinity, so that the integral in curly brackets converges.) It follows that $\int_0^\infty r \left| \frac{\partial}{\partial r} F(x, r) \right|^2 dr < \infty$ for almost every $x \in \mathbb{R}^n$. On the other hand, $\frac{1}{R} \int_0^R F(x, r) dr$ is easily seen to be the convolution of f with a standard approximate identity. Hence the usual Lebesgue differentiation theorem shows that $\lim_{R \to 0} \frac{1}{R} \int_0^R F(x, r) dr = f(x)$ for almost every x.

So for almost all $x \in \mathbb{R}^n$, the function $F(x, r)$ satisfies the hypotheses of the elementary Tauberian theorem. Consequently,

$$\lim_{r \to 0} F(x, r) = \lim_{R \to 0} \frac{1}{R} \int_0^R F(x, r) dr = f(x)$$

almost everywhere, proving Stein's differentiation theorem for $f \in L^2(\mathbb{R}^n)$.

To prove the full result for $f \in L^p(\mathbb{R}^n)$, $p > \frac{n}{n-1}$, we repeat the above argument with surface measure μ replaced by an even more singular distribution on $\mathbb{R}^n$. Thus we obtain a stronger conclusion than asserted, when $f \in L^2$. On the other hand, for $f \in L^{1+\varepsilon}$ we have a weaker result than that of Stein, namely Lebesgue's differentiation theorem. Interpolating between L^2 and $L^{1+\varepsilon}$, one obtains the Stein differentiation theorem.

The two results we picked out here are only a sample of the work of Stein and others on curvature and the Fourier transform. For instance, J. Bourgain has dramatic results on both the restriction problem and spherical averages.

We refer the reader to Stein's address at the Berkeley congress [128] for a survey of the field.

H^p-SPACES

Another essential part of Fourier analysis is the theory of H^p-spaces. Stein transformed the subject twice, once in a joint paper with Guido Weiss, and again in a joint paper with me. Let us start by recalling how the subject looked before Stein's work. The classical theory deals with analytic functions $F(z)$ on the unit disc. Recall that F belongs to H^p $(0 < p < \infty)$ if the norm $\|F\|_{H^p} \equiv \lim_{r \to 1-} (\int_0^{2\pi} |F(re^{i\theta})|^p d\theta)^{\frac{1}{p}}$ is finite.

The classical H^p-spaces serve two main purposes. First, they provide growth conditions under which an analytic function tends to boundary values on the unit circle. Secondly, H^p serves as a substitute for L^p to allow basic theorems on Fourier series to extend from $1 < p < \infty$ to all $p > 0$. To prove theorems about $F \in H^p$, the main tool is the Blaschke product

$$(11) \qquad B(z) = \Pi_\nu e^{i\theta_\nu} \frac{z_\nu - z}{1 - \bar{z}_\nu z},$$

where $\{z_\nu\}$ are the zeroes of the analytic function F in the disc, and θ_ν are suitable phases. The point is that $B(z)$ has the same zeroes as F, yet it has absolute value 1 on the unit circle. We illustrate the role of the Blaschke product by sketching the proof of the Hardy-Littlewood maximal theorem for H^p. The maximal theorem says that $\|F^*\|_{L^p} \leq C_p \|F\|_{H^p}$ for $0 < p < \infty$, where $F^*(\theta) = \sup_{z \in \Gamma(\theta)} |F(z)|$, and $\Gamma(\theta)$ is the convex hull of $e^{i\theta}$ and the circle of radius $\frac{1}{2}$ about the origin.

This basic result is closely connected to the pointwise convergence of $F(z)$ as $z \in \Gamma(\theta)$ tends to $e^{i\theta}$. To prove the maximal theorem, we argue as follows.

First suppose $p > 1$. Then we don't need analyticity of F. We can merely assume that F is harmonic, and deduce the maximal theorem from real variables. In fact, it is easy to show that F arises as the Poisson integral of an L^p function f on the unit circle. The maximal theorem for f, a standard theorem of real variables, says that $\|Mf\|_{L^p} \leq C_p \|f\|_{L^p}$, where $Mf(\theta) = \sup_{h>0} (\frac{1}{2h} \int_{\theta-h}^{\theta+h} |f(t)| dt)$. It is quite simple to show that $F^*(\theta) < CMf(\theta)$. Therefore $\|F^*\|_{L^p} \leq C\|Mf\|_{L^p} < C'\|F\|_{H^p}$, and the maximal theorem is proven for H^p $(p > 1)$.

If $p \leq 1$, then the problem is more subtle, and we need to use analyticity of $F(z)$. Assume for a moment that F has no zeroes in the unit disc. Then for $0 < q < p$, we can define a single-valued branch of $(F(z))^q$, which will belong to $H^{\frac{p}{q}}$ since $F \in H^p$. Since $\tilde{p} \equiv \frac{p}{q} > 1$, the maximal theorem for $H^{\tilde{p}}$ is already known. Hence, $\max_{z \in \Gamma(\theta)} |(F(z))^q| \in L^{\frac{p}{q}}$, with norm

$$\int_{-\pi}^{\pi} \left(\max_{z \in \Gamma(\theta)} |(F(z))^q| \right)^{\frac{p}{q}} d\theta \leq C_{p,q} \|F^q\|_{H^{\frac{p}{q}}}^{\frac{p}{q}} = C_{p,q} \|F\|_{H^p}^p.$$

That is, $\|F^*\|_{L^p} \leq C_{p,q} \|F\|_{H^p}$, proving the maximal theorem for functions without zeroes.

To finish the proof, we must deal with the zeroes of an $F \in H^p$ $(p \leq 1)$. We bring in the Blaschke product $B(z)$, as in (11). Since $B(z)$ and $F(z)$ have the same zeroes and since $|B(z)| = 1$ on the unit circle, we can write $F(z) = G(z)B(z)$ with G analytic, and $|G(z)| = |F(z)|$ on the unit circle. Thus, $\|G\|_{H^p} = \|F\|_{H^p}$.

Inside the circle, G has no zeroes and $|B(z)| \leq 1$. Hence $|F| \leq |G|$, so

$$\| \max_{z \in \Gamma(\theta)} |F(z)| \|_{L^p} \leq \| \max_{z \in \Gamma(\theta)} |G(z)| \|_{L^p} \leq C_p \|G\|_{H^p} = C_p \|F\|_{H^p},$$

by the maximal theorem for functions without zeroes. The proof of the maximal theorem is complete. (We have glossed over difficulties that should not enter an expository paper.)

Classically H^p theory works only in one complex variable, so it is useful only for Fourier analysis in one real variable. Attempts to generalize H^p to several complex variables ran into a lot of trouble, because the zeroes of an analytic function $F(z_1 \ldots z_n) \in H^p$ form a variety V with growth conditions. Certainly V is much more complicated than the discrete set of zeroes $\{z_v\}$ in the disc. There is no satisfactory substitute for the Blaschke product. For a long time, this blocked all attempts to extend the deeper properties of H^p to several variables.

Stein and Weiss [13] realized that several complex variables was the wrong generalization of H^p for purposes of Fourier analysis. They kept clearly in mind what H^p spaces are supposed to do, and they kept an unprejudiced view of how to achieve it. They found a version of H^p theory that works in several variables.

The idea of Stein and Weiss was very simple. They viewed the real and imaginary parts of an analytic function on the disc as the gradient of a harmonic function. In several variables, the gradient of a harmonic function is a system $\vec{u} = (u_1, u_2, \ldots, u_n)$ of functions on $\mathbb{R}^n$ that satisfies the Stein-Weiss Cauchy-Riemann equations

$$(12) \qquad \frac{\partial u_j}{\partial x_k} = \frac{\partial u_k}{\partial x_j}, \qquad \sum_k \frac{\partial u_k}{\partial x_k} = 0.$$

In place of the Blaschke product, Stein and Weiss used the following simple observation. If $\vec{u} = (u_1 \ldots u_n)$ satisfies (12), then $|\vec{u}|^p = (u_1^2 + u_2^2 + \cdots + u_n^2)^{\frac{p}{2}}$ is subharmonic for $p > \frac{n-2}{n-1}$. We sketch the simple proof of this fact, then explain how an H^p theory can be founded on it.

To see that $|\vec{u}|^p$ is subharmonic, we first suppose $|\vec{u}| \neq 0$ and calculate $\triangle(|\vec{u}|^p)$ in coordinates that diagonalize the symmetric matrix $\left(\frac{\partial u_j}{\partial x_k}\right)$ at a given point. The result is

$$(13) \qquad \triangle(|\vec{u}|^p) = p|\vec{u}|^{p-2}\{|\vec{w}|^2|\vec{u}|^2 - (2 - p)|\vec{v}|^2\},$$

with $w_k = \frac{\partial u_k}{\partial x_k}$ and $v_k = u_k w_k$.

Since $\sum_{k=1}^n w_k = 0$ by the Cauchy-Riemann equations, we have

$$|w_k|^2 = \left| \sum_{j \neq k} w_j \right|^2 \leq (n-1) \sum_{j \neq k} |w_j|^2 = (n-1)|\vec{w}|^2 - (n-1)|w_k|^2;$$

i.e., $|w_k|^2 \leq \frac{n-1}{n} |\vec{w}|^2$. Hence $|\vec{v}|^2 \leq \left(\max_k |w_k|^2\right)|\vec{u}|^2 \leq \left(\frac{n-1}{n}\right)|\vec{w}|^2 |\vec{u}|^2$, so the expression in curly brackets in (13) is non-negative for $p \geq \frac{n-2}{n-1}$, and $|\vec{u}|^p$ is subharmonic.

So far, we know that $|\vec{u}|^p$ is subharmonic where it isn't equal to zero. Hence for $0 < r < r(x)$ we have

$$(14) \qquad\qquad |\vec{u}(x)|^p \leq Av_{|y-x|=r}|\vec{u}(y)|^p,$$

provided $|\vec{u}(x)| \neq 0$. However, (14) is obvious when $|\vec{u}(x)| = 0$, so it holds for any x. That is, $|\vec{u}|^p$ is a subharmonic function for $p \geq \frac{n-2}{n-1}$, as asserted.

Now let us see how to build an H^p theory for Cauchy-Riemann systems, based on subharmonicity of $|\vec{u}|^p$. To study functions on $\mathbb{R}^{n-1}$ $(n \geq 2)$, we regard $\mathbb{R}^{n-1}$ as the boundary of $\mathbb{R}^n_+ = \{(x_1 \ldots x_n)|x_n > 0\}$, and we define $H^p(\mathbb{R}^n_+)$ as the space of all Cauchy-Riemann systems $(u_1, u_2, \ldots, u_n)$ for which the norm

$$\|\vec{u}\|^p_{H^p} = \sup_{t>0} \int_{\mathbb{R}^{n-1}} |\vec{u}(x_1 \ldots x_{n-1}, t)|^p dx_1 \ldots dx_{n-1}$$

is finite. For $n = 2$ this definition agrees with the usual H^p spaces for the upper half-plane.

Next we show how the Hardy-Littlewood maximal theorem extends from the disc to $\mathbb{R}^n_+$.

Define the maximal function $M(\vec{u})(x) = \sup_{|y-x|<t} |\vec{u}(y, t)|$ for $x \in \mathbb{R}^{n-1}$. Then for $\vec{u} \in H^p(\mathbb{R}^n_+)$, $\frac{n-2}{n-1} < p$, we have $M(\vec{u}) \in L^p(\mathbb{R}^{n-1})$ with norm $\int_{\mathbb{R}^{n-1}} \left(M(\vec{u})\right)^p dx \leq C\|\vec{u}\|^p_{H^p}$.

As in the classical case, the proof proceeds by reducing the problem to the maximal theorem for L^p $(p > 1)$. For small $h > 0$, the function $F_h(x, t) = |\vec{u}(x, t + h)|^{\frac{n-2}{n-1}}$ $(x \in \mathbb{R}^{n-1}, t \geq 0)$ is subharmonic on $\mathbb{R}^n_+$ and continuous up to the boundary. Therefore,

$$(15) \qquad\qquad F_h(x, t) \leq \text{P.I.}(f_h),$$

where P.I. is the Poisson integral and $f_h(x) = F_h(x, 0) = |\vec{u}(x, h)|^{\frac{n-2}{n-1}}$. By definition of the H^p-norm, we have

$$(16) \qquad \int_{\mathbb{R}^{n-1}} |f_h(x)|^{\tilde{p}} dx \leq \|\vec{u}\|^p_{H^p}, \qquad \text{with} \quad \tilde{p} = \left(\frac{n-1}{n-2}\right) p > 1.$$

On the other hand, since the Poisson integral arises by convolving with an approximate identity, one shows easily that

$$(17) \qquad\qquad \sup_{|y-x|<t} \text{P.I.}(f_h)(y, t) \leq Cf_h^*(x)$$

with

$$f_h^*(x) = \sup_{r>0} r^{-(n-1)} \int_{|x-y|<r} |f_h(y)| \, dy \quad (x \in \mathbb{R}^{n-1}).$$

The standard maximal theorem of real variables gives

$$\int_{\mathbb{R}^{n-1}} (f_h^*)^{\tilde{p}} \leq C_p \int_{\mathbb{R}^{n-1}} |f_h|^{\tilde{p}},$$

since $\tilde{p} > 1$. Hence (15), (16), and (17) show that

$$\int_{x \in \mathbb{R}^{n-1}} \left(\sup_{|y-x|<t} F_h(y, t) \right)^{\tilde{p}} dy \leq C_p \|\vec{u}\|_{H^p}^p, \text{ i.e.}$$

(18)
$$\int_{x \in \mathbb{R}^{n-1}} \left(\sup_{|y-x|+h<t} |\vec{u}(y, t)| \right)^p dy \leq C_p \|\vec{u}\|_{H^p}^p.$$

The constant C_p is independent of h, so we can take the limit of (18) as $h \to 0$ to obtain the maximal theorem for H^p. The point is that subharmonicity of $|\vec{u}|^{\frac{n-2}{n-1}}$ substitutes for the Blaschke product in this argument.

Stein and Weiss go on in [13] to obtain n-dimensional analogues of the classical theorems on existence of boundary values of H^p functions. They also extend to $\mathbb{R}_+^n$ the classical F and M Riesz theorem on absolute continuity of H^1 boundary values. They begin the program of using $H^p(\mathbb{R}_+^n)$ in place of $L^p(\mathbb{R}^{n-1})$, to extend the basic results of Fourier analysis to $p = 1$ and below. We have seen how they deal with the maximal function. They prove also an H^p-version of the Sobolev theorem.

It is natural to try to get below $p = \frac{n-2}{n-1}$, and this can be done by studying higher gradients of harmonic functions in place of (12). See Calderón-Zygmund [CZ].

A joint paper [60] by Stein and me completed the task of developing basic Fourier analysis in the setting of the H^p-spaces. In particular, we showed in [60] that singular integral operators are bounded on $H^p(\mathbb{R}_+^n)$ for $0 < p < \infty$. We proved this by finding a good viewpoint, and we found our viewpoint by repeatedly changing the definition of H^p. With each new definition, the function space H^p remained the same, but it became clearer to us what was going on. Finally we arrived at a definition of H^p with the following excellent properties. First of all, it was easy to prove that the new definition of H^p was equivalent to the Stein-Weiss definition and its extensions below $p = \frac{n-2}{n-1}$. Secondly, the basic theorems of Fourier analysis, which seemed very hard to prove from the original definition of $H^p(\mathbb{R}_+^n)$, became nearly obvious in terms of the new definition. Let me retrace the steps in [60].

Burkholder-Gundy-Silverstein [BGS] had shown that an analytic function $F = u + iv$ on the disc belongs to H^p ($0 < p < \infty$) if and only if the maximal function $u^*(\theta) = \sup_{z \in \Gamma(\theta)} |u(z)|$ belongs to L^p (Unit Circle). Thus, H^p can be defined purely in terms of harmonic functions u, without recourse to the harmonic conjugate v. Stein and I showed in [60] that the same thing happens in n dimensions. That is, a Cauchy-Riemann system $(u_1, u_2, \ldots u_n)$ on $\mathbb{R}^n_+$ belongs to the Stein-Weiss H^p space $\left(p > \frac{n-2}{n-1}\right)$ if and only if the maximal function $u^*(x) = \sup_{|y-x|<t} |u_n(y, t)|$ belongs to $L^p(\mathbb{R}^{n-1})$. (Here, the nth function u_n plays a special role because $\mathbb{R}^n_+$ is defined by $\{x_n > 0\}$.) Hence, H^p may be viewed as a space of harmonic functions $u(x, t)$ on $\mathbb{R}^n_+$. The result extends below $p = \frac{n-2}{n-1}$ if we pass to higher gradients of harmonic functions.

The next step is to view H^p as a space of distributions f on the boundary $\mathbb{R}^{n-1}$. Any reasonable harmonic function $u(x, t)$ arises as the Poisson integral of a distribution f, so that $u(x, t) = \varphi_t * f(x)$, $\varphi_t =$ Poisson kernel. Thus, it is natural to say that $f \in H^p(\mathbb{R}^{n-1})$ if the maximal function

$$(19) \qquad f^*(x) \equiv \sup_{|y-x|<t} |\varphi_t * f(y)|$$

belongs to L^p. Stein and I found in [60] that this definition is independent of the choice of the approximate identity φ_t, and that the "grand maximal function"

$$(20) \qquad \mathcal{M}f(x) = \sup_{\{\varphi_t\}\in\mathcal{A}} \sup_{|y-x|<t} |\varphi_t * f(y)|$$

belongs to L^p, provided $f \in H^p$. Here $\mathcal{A}$ is a neighborhood of the origin in a suitable space of approximate identities. Thus, $f \in H^p$ if and only if $f^* \in L^p$ for some reasonable approximate identity. Equivalently, $f \in H^p$ if and only if the grand maximal function $\mathcal{M}f$ belongs to L^p. The proofs of these various equivalencies are not hard at all.

We have arrived at the good definition of H^p mentioned above.

To transplant basic Fourier analysis from L^p $(1 < p < \infty)$ to H^p $(0 < p < \infty)$, there is a simple algorithm. Take Calderón-Zygmund theory, and replace every application of the standard maximal theorem by an appeal to the grand maximal function. Only small changes are needed, and we omit the details here. Our paper [60] also contains the duality of H^1 and BMO. Before leaving [60], let me mention an application of H^p theory to L^p-estimates. If σ denotes uniform surface measure on the unit sphere in $\mathbb{R}^n$, then $f \longmapsto \left(\frac{\partial}{\partial x}\right)^\alpha \sigma * f$ is bounded on $L^p(\mathbb{R}^n)$, provided $n \geq 3$ and $\left|\frac{1}{p} - \frac{1}{2}\right| \leq \frac{1}{2} - \frac{\alpha}{n-1}$. Clearly, this result gives information on solutions to the wave equation.

The proof uses complex interpolation involving the analytic family of operators

$$T_\alpha : f \longmapsto (-\triangle)^{\frac{\alpha}{2}} \sigma * f \qquad (\alpha \text{ complex}),$$

as is clear to anyone familiar with the Stein interpolation theorem. The trouble here is that $(-\triangle)^{\frac{\alpha}{2}}$ fails to be bounded on L^1 when α is imaginary. This makes it impossible to prove the sharp result $\left(\left|\frac{1}{p} - \frac{1}{2}\right| = \frac{1}{2} - \frac{\alpha}{n-1}\right)$ using

L^p alone. To overcome the difficulty, we use H^1 in place of L^1 in the interpolation argument. Imaginary powers of the Laplacian are singular integrals, which we know to be bounded on H^1. To show that complex interpolation works on H^1, we combined the duality of H^1 and BMO with the auxiliary function $f^{\#}(x) = \sup_{Q \ni x} \frac{1}{|Q|} \int_Q |f(y) - (\text{mean}_Q f)| dy$. We refer the reader to [60] for an explanation of how to use $f^{\#}$, and for other applications.

Since [60], Stein has done a lot more on H^p, both in "higher rank" settings, and in contexts related to partial differential equations.

REPRESENTATION THEORY II

Next we return to representation theory. We explain briefly how the Kunze-Stein construction extends from $SL(2, \mathbb{R})$ to more general semisimple Lie groups, with profound consequences for representation theory. The results we discuss are contained in the series of papers by Kunze-Stein [20], [22], [33], [63], Stein [35], [48], [70], and Knapp-Stein [43], [46], [50], [53], [58], [66], [73], [93], [97]. Let G be a semisimple Lie group, and let U^{π} be the unitary principal series representations of G, or one of its degenerate variants. The U^{π} all act on a common Hilbert space, whose inner product we denote by $\langle \xi, \eta \rangle$. We needn't write down U^{π} here, nor even specify the parameters on which it depends. A finite group W, the Weyl group, acts on the parameters π in such a way that the representations U^{π} and $U^{w\pi}$ are unitarily equivalent for $w \in W$. Thus there is an *intertwining operator* $A(w, \pi)$ so that

$$(21) \quad A(w, \pi)U_g^{w\pi} = U_g^{\pi}A(w, \pi) \qquad \text{for} \quad g \in G, \ w \in W, \text{ and for all } \pi.$$

If U^{π} is irreducible (which happens for most π), then $A(w, \pi)$ is uniquely determined by (21) up to multiplication by an arbitrary scalar $a(w, \pi)$. The crucial idea is as follows. If the $A(w, \pi)$ are correctly normalized (by the correct choice of $a(w, \pi)$), then $A(w, \pi)$ continues analytically to complex parameter values π. Moreover, for certain complex (w, π), the quadratic form

$$(22) \quad \big((\xi, \eta)\big)_{w,\pi} = \langle (\text{TRIVIAL FACTOR})A(w, \pi)\xi, \eta \rangle$$

is positive definite.

In addition, the representation U^{π} (defined for complex π by a trivial analytic continuation) is unitary with respect to the inner product (22). Thus, starting with the principal series, we have constructed a new series of unitary representations of G. These new representations generalize the complementary series for $SL(2, \mathbb{R})$. Applications of this basic construction are as follows.

(1) Starting with the unitary principal series, one obtains understanding of the previously discovered complementary series, and construction of new ones, e.g., on $Sp(4, \mathbb{C})$. Thus, Stein exposed a gap in a supposedly complete list of complementary series representations of $Sp(4, \mathbb{C})$ [GN]. See [33].

(2) Starting with a degenerate unitary principal series, Stein constructed new irreducible unitary representations of $SL(n, \mathbb{C})$, in startling contradiction to the standard, supposedly complete list [GN] of irreducible unitary representations of that group. Much later, when the complete list of representations of $SL(n, \mathbb{C})$ was given correctly, the representations constructed by Stein played an important role.

(3) The analysis of intertwining operators required to carry out analytic continuation also determines which exceptional values of π lead to reducible principal series representations. For example, such reducible principal series representations exist already for $SL(n, \mathbb{R})$, again contradicting what was "known". See Knapp-Stein [53].

A very recent result of Sahi and Stein [139] also fits into the same philosophy. In fact, Speh's representation can also be constructed by a more complicated variant of the analytic continuation defining the complementary series. Speh's representation plays an important role in the classification of the irreducible unitary representations of $SL(2n, \mathbb{R})$.

The main point of Stein's work in representation theory is thus to analyze the intertwining operators $A(w, \pi)$. In the simplest non-trivial case, $A(w, \pi)$ is a singular integral operator on a nilpotent group N. That is, $A(w, \pi)$ has the form

$$(23) \qquad Tf(x) = \int_N K(xy^{-1})f(y)dy,$$

with $K(y)$ smooth away from the identity and homogeneous of the critical degree with respect to "dilations"

$$(\delta_t)_{t>0} : N \longrightarrow N.$$

In (23), dy denotes Haar measure on N. We know from the classical case $N = \mathbb{R}^1$ that (23) is a bounded operator only when the convolution kernel $K(y)$ satisfies a cancellation condition. Hence we assume $\int_{B_1 \backslash B_0} K(y)dy = 0$, where the B_i are dilates ($B_i = \delta_{t_i}(B)$) of a fixed neighborhood of the identity in N.

It is crucial to show that such singular integrals are bounded on $L^2(N)$, generalizing the elementary L^2-boundedness of the Hilbert transform.

COTLAR-STEIN LEMMA

In principle, L^2-boundedness of the translation-invariant operator (23) should be read off from the representation theory of N. In practice, representation theory provides a necessary and sufficient condition for L^2-boundedness that no one knows how to check. This fundamental analytic difficulty might have proved fatal to the study of intertwining operators. Fortunately, Stein was working simultaneously on a seemingly unrelated question, and made a discovery that saved the day. Originally motivated by desire to get a simple proof of Calderón's theorem on commutator integrals [Ca], Stein proved a simple, powerful lemma in functional analysis. His contribution was to generalize to the critically important non-commutative case the remarkable lemma of Cotlar [Co]. The Cotlar-Stein lemma turned out to be the perfect tool to prove L^2-boundedness of singular integrals on nilpotent groups. In fact, it quickly became a basic, standard tool in analysis. We will now explain the Cotlar-Stein lemma, and give its amazingly simple proof. Then we will return to its application to singular integrals on nilpotent groups.

The Cotlar-Stein lemma deals with a sum $T = \sum_\nu T_\nu$ of operators on a Hilbert space. The idea is that if the T_ν are almost orthogonal, like projections onto the various coordinate axes, then the sum T will have norm no larger than $\max_\nu \|T_\nu\|$. The precise statement is as follows.

Cotlar-Stein Lemma. *Suppose $T = \sum_{k=1}^M T_k$ is a sum of operators on Hilbert space. Assume $\|T_j^* T_k\| \leq a(j-k)$ and $\|T_j T_k^*\| \leq a(j-k)$. Then $\|T\| \leq \sum_{-M}^M \sqrt{a(j)}$.*

Proof. $\|T\| = (\|TT^*\|)^{\frac{1}{2}}$, so

$$\|T\|^{2s} \leq \|(TT^*)^s\| = \left\| \sum_{j_1 \ldots j_{2s}=1}^M T_{j_1} T_{j_2}^* \cdots T_{j_{2s-1}} T_{j_{2s}}^* \right\|$$

(24)

$$\leq \sum_{j_1 \ldots j_{2s}=1}^M \| T_{j_1} T_{j_2}^* \cdots T_{j_{2s-1}} T_{j_{2s}}^* \|.$$

We can estimate the summand in two different ways.

Writing $T_{j_1} T_{j_2}^* \cdots T_{j_{2s-1}} T_{j_{2s}}^* = (T_{j_1} T_{j_2}^*)(T_{j_3} T_{j_4}^*) \cdots (T_{j_{2s-1}} T_{j_{2s}}^*)$, we get

(25) $\| T_{j_1} T_{j_2}^* \cdots T_{j_{2s-1}} T_{j_{2s}}^* \| \leq a(j_1 - j_2) a(j_3 - j_4) \cdots a(j_{2s-1} - j_{2s}).$

On the other hand, writing

$$T_{j_1} T_{j_2}^* \cdots T_{j_{2s-1}} T_{j_{2s}}^* = T_{j_1} (T_{j_2}^* T_{j_3})(T_{j_4}^* T_{j_5}) \cdots (T_{j_{2s-2}}^* T_{j_{2s-1}}) T_{j_{2s}}^*,$$

we see that

(26) $\| T_{j_1} T_{j_2}^* \cdots T_{j_{2s-1}} T_{j_{2s}}^* \|$

$$\leq \left(\max_j \|T_j\| \right)^2 a(j_2 - j_3) a(j_4 - j_5) \cdots a(j_{2s-2} - j_{2s-1}).$$

Taking the geometric mean of (25), (26) and putting the result into (24), we conclude that

$$\|T\|^{2s} \leq \sum_{j_1 \cdots j_{2s}=1}^{M} \left(\max_j \|T_j\| \right) \sqrt{a(j_1 - j_2)} \sqrt{a(j_2 - j_3)} \cdots \sqrt{a(j_{2s-1} - j_{2s})}$$

$$\leq \left(\max_j \|T_j\| \cdot M \right) \left(\sum_\ell \sqrt{a(\ell)} \right)^{2s-1}$$

Thus, $\|T\| \leq \left(\max_j \|T_j\| \cdot M \right)^{\frac{1}{2s}} \cdot \left(\sum_\ell \sqrt{a(\ell)} \right)^{\frac{2s-1}{2s}}$. Letting $s \to \infty$, we obtain the conclusion of the Cotlar-Stein Lemma. ■

To apply the Cotlar-Stein lemma to singular integral operators, take a partition of unity $1 = \sum_\nu \varphi_\nu(x)$ on N, so that each φ_ν is a dilate of a fixed C_0^∞ function that vanishes in a neighborhood of the origin. Then $T : f \to K * f$ may be decomposed into a sum $T = \sum_\nu T_\nu$, with $T_\nu : f \to (\varphi_\nu K) * f$. The hypotheses of the Cotlar-Stein lemma are verified trivially, and the boundedness of singular integral operators follows. The L^2-boundedness of singular integrals on nilpotent groups is the *Knapp-Stein Theorem*.

Almost immediately after this work, the Cotlar-Stein lemma became the standard method to prove L^2 boundedness of operators. Today one knows more, e.g., the $T(1)$ theorem of David and Journé. Still it is fair to say that the Cotlar-Stein lemma remains the most important tool for L^2-boundedness.

Singular integrals on nilpotent groups were soon applied by Stein in a context seemingly far from representation theory.

$\bar{\partial}$-PROBLEMS

We prepare to discuss Stein's work on the $\bar{\partial}$-problems of several complex variables and related questions. Let us begin with the state of the subject before Stein's contributions. Suppose we are given a domain $D \subset \mathbb{C}^n$ with smooth boundary. If we try to construct analytic functions on D with given singularities at the boundary, then we are led naturally to the following problems.

I. Given a $(0, 1)$ form $\alpha = \sum_{k=1}^{n} f_k \overline{dz_k}$ on D, find a function u on D that solves $\bar{\partial}u = \alpha$, where $\bar{\partial}u = \sum_{k=1}^{n} \frac{\partial u}{\partial \bar{z}_k} \overline{dz_k}$. Naturally, this is possible only if α satisfies the consistency condition $\bar{\partial}\alpha = 0$, i.e. $\frac{\partial}{\partial \bar{z}_k} f_j = \frac{\partial}{\partial \bar{z}_j} f_k$. Moreover, u is determined only modulo addition of an arbitrary analytic function on D. To make u unique, we demand that u be orthogonal to analytic functions in $L^2(D)$.

II. There is a simple analogue of the $\bar{\partial}$-operator for functions defined only on the boundary ∂D. In local coordinates, we can easily find $(n - 1)$ linearly independent complex vector fields $L_1 \cdots L_{n-1}$ of type $(0, 1)$ (i.e., $L_j = a_{j1} \frac{\partial}{\partial \bar{z}_1} + a_{j2} \frac{\partial}{\partial \bar{z}_2} + \cdots + a_{jn} \frac{\partial}{\partial \bar{z}_n}$ for smooth, complex-valued a_{jk}) whose real and imaginary parts are all tangent to ∂D. The restriction u of an analytic function to ∂D clearly satisfies $\bar{\partial}_b u = 0$, where in local coordinates $\bar{\partial}_b u = (L_1 u, L_2 u, \ldots, L_{n-1} u)$. The boundary analogue of the $\bar{\partial}$-problem (I) is the inhomogeneous $\bar{\partial}_b$-equation $\bar{\partial}_b u = \alpha$. Again, this is possible only if α satisfies a consistency condition $\bar{\partial}_b \alpha = 0$, and we impose the side condition that u be orthogonal to analytic functions in $L^2(\partial D)$.

Just as analytic functions of one variable are related to harmonic functions, so the first-order systems (I) and (II) are related to second-order equations $\square$ and $\square_b$, the $\bar{\partial}$-Neumann and Kohn Laplacians. Both fall outside the scope of standard elliptic theory. Even for the simplest domains D, they posed a fundamental challenge to workers in partial differential equations. More specifically, $\square$ is simply the Laplacian in the interior of D, but it is subject to non-elliptic boundary conditions. On the other hand, $\square_b$ is a non-elliptic system of partial differential operators on ∂D, with no boundary conditions (since ∂D has no boundary). Modulo lower-order terms (which, however, are important), $\square_b$ is the scalar operator $\mathcal{L} = \sum_{k=1}^{n-1}(X_k^2 + Y_k^2)$, where X_k and Y_k are the real and imaginary parts of the basic complex vector fields L_k. At a given point in ∂D, the X_k and Y_k are linearly independent, but they don't span the tangent space of ∂D. This poses the danger that $\mathcal{L}$ will behave like a partial Laplacian such as $\Delta' = \frac{\partial^2}{\partial x^2} + \frac{\partial^2}{\partial y^2}$ acting on function $u(x, y, z)$. The equation $\Delta' u = f$ is very bad. For instance, we can take $u(x, y, z)$ to depend on z alone, so that $\Delta' u = 0$ with u arbitrarily rough. Fortunately, $\mathcal{L}$ is more like the full Laplacian than like Δ', because the X_k and Y_k together with their commutators $[X_k, Y_k]$ span the tangent space of ∂D for suitable D. Thus, $\mathcal{L}$ is a well-behaved operator, thanks to the intervention of commutators of vector fields.

It was Kohn in the 1960's who proved the basic C^∞ regularity theorems for $\square$, $\square_b$, $\bar{\partial}$ and $\bar{\partial}_b$ on strongly pseudoconvex domains (the simplest case). His proofs were based on subelliptic estimates such as $\langle \square_b w, w \rangle \geq c \|w\|_{(\varepsilon)}^2 - C \|w\|^2$, and brought to light the importance of commutators. Hörmander proved a celebrated theorem on C^∞ regularity of operators,

$$L = \sum_{j=1}^N X_j^2 + X_0,$$

where $X_0, X_1, \ldots, X_N$ are smooth, real vector fields which, together with their repeated commutators, span the tangent space at every point.

If we allow X_0 to be a complex vector field, then we get a very hard problem that is not adequately understood to this day, except in very special cases.

Stein made a fundamental change in the study of the the $\bar{\partial}$-problems by bringing in constructive methods. Today, thanks to the work of Stein with several collaborators, we know how to write down explicit solutions to the $\bar{\partial}$-problems modulo negligible errors on strongly pseudoconvex domains. Starting from these explicit solutions, it is then possible to prove sharp regularity theorems. Thus, the $\bar{\partial}$ equations on strongly pseudoconvex domains are understood completely. It is a major open problem to achieve comparable understanding of weakly pseudoconvex domains.

Now let us see how Stein and his co-workers were able to crack the strongly pseudoconvex case. We begin with the work of Folland and Stein [67]. The simplest example of a strongly pseudoconvex domain is the unit ball. Just as the disc is equivalent to the half-plane, the ball is equivalent to the Siegel domain $D_{\text{Siegel}} = \{(z, w) \in \mathbb{C}^{n-1} \times \mathbb{C} \mid \text{Im } w > |z|^2\}$. Its boundary $H = \partial D_{\text{Siegel}}$ has an important symmetry group, including the following.

(a) *Translations* $(z, w) \longmapsto (z, w) \cdot (z', w') \equiv (z + z', w + w' + 2iz \cdot \bar{z}')$ for $(z', w') \in H$;

(b) *Dilations* $\delta_t : (z, w) \longmapsto (tz, t^2 w)$ for $t > 0$;

(c) *Rotations* $(z, w) \longmapsto (Uz, w)$ for unitary $(n - 1) \times (n - 1)$ matrices U.

The multiplication law in (a) makes H into a nilpotent Lie group, the Heisenberg group. Translation-invariance of the Siegel domain allows us to pick the basic complex vector-fields $L_1 \cdots L_{n-1}$ to be translation-invariant on H. After we make a suitable choice of metric, the operators $\mathcal{L}$ and $\Box_b$ become translation- and rotation-invariant, and homogeneous with respect to the dilations δ_t. Therefore, the solution[1] of $\Box_b w = \alpha$ should have the form of a convolution $w = K * \alpha$ on the Heisenberg group. The convolution kernel K is homogeneous with respect to the dilations δ_t and invariant under rotations. Also, since K is a fundamental solution, it satisfies $\Box_b K = 0$ away from the origin. This reduces to an elementary ODE after we take the dilation- and rotation-invariance into account. Hence one can easily find K explicitly and thus solve the $\Box_b$-equation for the Siegel domain. To derive sharp regularity theorems for $\Box_b$, we combine the explicit fundamental solution with the Knapp-Stein theorem on singular integrals on the Heisenberg group. For instance, if $\Box_b w \in L^2$, then $L_j L_k w$, $\bar{L}_j L_k w$, $L_j \bar{L}_k w$, and $\bar{L}_j \bar{L}_k w$ all belong to L^2. To see this, we write

$$\Box_b w = \alpha, \quad w = K * \alpha, \quad L_j L_k w = (L_j L_k K) * \alpha,$$

---

[1]Kohn's work showed that $\Box_b w = \alpha$ has a solution if we are in complex dimension > 2. In two complex dimensions, $\Box_b w = \alpha$ has no solution for most α. We assume dimension > 2 here.

and note that $L_j L_k K$ has the critical homogeneity and integral 0. Thus $L_j L_k K$ is a singular integral kernel in the sense of Knapp and Stein, and it follows that $\|L_j L_k w\| \leq C\|\alpha\|$. For the first time, nilpotent Lie groups have entered into the study of $\bar{\partial}$-problems.

Folland and Stein viewed their results on the Heisenberg group not as ends in themselves, but rather as a tool to understand general strongly pseudoconvex CR manifolds. A CR-manifold M is a generalization of the boundary of a smooth domain $D \subset \mathbb{C}^n$. For simplicity we will take $M = \partial D$ here. The key idea is that near any point w in a strongly pseudoconvex M, the CR structure for M is very nearly equivalent to that of the Heisenberg group H via a change of coordinates $\Theta_w : M \to H$. More precisely, Θ_w carries w to the origin, and it carries the CR-structure on M to a CR-structure on H that agrees with the usual one at the origin. Therefore, if $w = K * \alpha$ is our known solution of $\square_b w = \alpha$ on the Heisenberg group, then it is natural to try

$$(27) \qquad w(z) = \int_M K(\Theta_w(z))\alpha(w)dw$$

as an approximate solution of $\square_b w = \alpha$ on M. (Since w and α are sections of bundles, one has to explain carefully what (27) really means.) If we apply $\square_b$ to the w defined by (27), then we find that

$$(28) \qquad \square_b w = \alpha - \mathcal{E}\alpha,$$

where $\mathcal{E}$ is a sort of Heisenberg version of $(-\triangle)^{-\frac{1}{2}}$. In particular, $\mathcal{E}$ gains smoothness, so that $(I - \mathcal{E})^{-1}$ can be constructed modulo infinitely smoothing operators by means of a Neumann series. Therefore (27) and (28) show that the full solution of $\square_b w = \alpha$ is given (modulo infinitely smoothing errors) by

$$(29) \qquad w(z) = \sum_k \int_M K(\Theta_w(z))(\mathcal{E}^k\alpha)(w)dw,$$

from which one can deduce sharp estimates to understand completely $\square_b^{-1}$ on M.

The process is analogous to the standard method of "freezing coefficients" to solve variable-coefficient elliptic differential equations. Let us see how the sharp results are stated. As on the Heisenberg group, there are smooth, complex vector fields L_k that span the tangent vectors of type $(0, 1)$ locally. Let X_j be the real and imaginary parts of the L_k. In terms of the X_j we define "non-Euclidean" versions of standard geometric and analytic concepts. Thus, the non-Euclidean ball $\mathbb{B}(z, \rho)$ may be defined as an ellipsoid with principal axes of length $\sim \rho$ in the codimension 1 hyperplane spanned by the X_j, and length $\sim \rho^2$ perpendicular to that hyperplane. In terms of $\mathbb{B}(z, \rho)$, the non-Euclidean Lipschitz spaces $\Gamma_\alpha(M)$

are defined as the set of functions u for which $|u(z) - u(w)| < C\rho^\alpha$ for $w \in \mathbb{B}(z, \rho)$. (Here, $0 < \alpha < 1$. There is a natural extension to all $\alpha > 0$.) The non-Euclidean Sobolev spaces $S_{m,p}(M)$ consist of all distributions u for which all $X_{j_1} X_{j_2} \cdots X_{j_s} u \in L^p(M)$ for $0 \leq s \leq m$.

Then the sharp results on $\square_b$ are as follows. If $\square_b w = \alpha$ and $\alpha \in S_{m,p}(M)$, then $w \in S_{m+2,p}(M)$ for $m \geq 0$, $1 < p < \infty$. If $\square_b w = \alpha$ and $\alpha \in \Gamma_\alpha(M)$, then $X_j X_k w \in \Gamma_\alpha(M)$ for $0 < \alpha < 1$ (say). For additional sharp estimates, and for comparisons between the non-Euclidean and standard function spaces, we refer the reader to [67].

To prove their sharp results, Folland and Stein developed the theory of singular integral operators in a non-Euclidean context. The Cotlar-Stein lemma proves the crucial results on L^2-boundedness of singular integrals. Additional difficulties arise from the non-commutativity of the Heisenberg group. In particular, standard singular integrals or pseudodifferential operators commute modulo lower-order errors, but non-Euclidean operators are far from commuting. This makes more difficult the passage from L^p estimates to the Sobolev spaces $S_{m,p}(M)$.

Before we continue with Stein's work on $\bar{\partial}$, let me explain the remarkable paper of Rothschild-Stein [72]. It extends the Folland-Stein results and viewpoint to general Hörmander operators $\mathcal{L} = \sum_{j=1}^N X_j^2 + X_0$. Actually, [72] deals with systems whose second-order part is $\sum_j X_j^2$, but for simplicity we restrict attention here to $\mathcal{L}$. In explaining the proofs, we simplify even further by supposing $X_0 = 0$. The goal of the Rothschild-Stein paper is to use nilpotent groups to write down an explicit parametrix for $\mathcal{L}$ and prove sharp estimates for solutions of $\mathcal{L}u = f$. This ambitious hope is seemingly dashed at once by elementary examples. For instance, take $\mathcal{L} = X_1^2 + X_2^2$ with

$$(30) \qquad X_1 = \frac{\partial}{\partial x}, \qquad X_2 = x \frac{\partial}{\partial y} \qquad \text{on} \quad \mathbb{R}^2.$$

Then X_1 and $[X_1, X_2]$ span the tangent space, yet $\mathcal{L}$ clearly cannot be approximated by translation-invariant operators on a nilpotent Lie group in the sense of Folland-Stein. The trouble is that $\mathcal{L}$ changes character completely from one point to another. Away from the y-axis $\{x = 0\}$, $\mathcal{L}$ is elliptic, so the only natural nilpotent group we can reasonably use is $\mathbb{R}^2$. On the y-axis, $\mathcal{L}$ degenerates, and evidently cannot be approximated by a translation-invariant operator on $\mathbb{R}^2$. The problem is so obviously fatal, and its solution by Rothschild and Stein so simple and natural, that [72] must be regarded as a gem. Here is the idea:

Suppose we add an extra variable t and "lift" X_1 and X_2 in (30) to vector fields

$$(31) \qquad \tilde{X}_1 = \frac{\partial}{\partial x}, \qquad \tilde{X}_2 = x \frac{\partial}{\partial y} + \frac{\partial}{\partial t} \qquad \text{on} \quad \mathbb{R}^3.$$

Then the Hörmander operator $\widetilde{\mathcal{L}} = \widetilde{X}_1^2 + \widetilde{X}_2^2$ looks the same at every point of $\mathbb{R}^3$, and may be readily understood in terms of nilpotent groups as in Folland-Stein [67]. In particular, one can essentially write down a fundamental solution and prove sharp estimates for $\widetilde{\mathcal{L}}^{-1}$. On the other hand, $\widetilde{\mathcal{L}}$ reduces to $\mathcal{L}$ when acting on functions $u(x, y, t)$ that do not depend on t. Hence, sharp results on $\widetilde{\mathcal{L}}u = f$ imply sharp results on $\mathcal{L}u = f$.

Thus we have the Rothschild-Stein program: First, add new variables and lift the given vector fields $X_1 \cdots X_N$ to new vector fields $\widetilde{X}_1 \cdots \widetilde{X}_N$ whose underlying structure does not vary from point to point. Next, approximate $\widetilde{\mathcal{L}} = \sum_1^N \widetilde{X}_j^2$ by a translation-invariant operator $\widehat{\mathcal{L}} = \sum_1^N Y_j^2$ on a nilpotent Lie group $\mathcal{N}$. Then analyze the fundamental solution of $\widehat{\mathcal{L}}$, and use it to write down an approximate fundamental solution for $\widetilde{\mathcal{L}}$. From the approximate solution, derive sharp estimates for solutions of $\widetilde{\mathcal{L}}u = f$. Finally, descend to the original equation $\mathcal{L}u = f$ by restricting attention to functions u, f that do not depend on the extra variables.

To carry out the first part of their program, Rothschild and Stein prove the following result.

Theorem A. *Let $X_1 \cdots X_N$ be smooth vector fields on a neighborhood of the origin in $\mathbb{R}^n$. Assume that the X_j and their commutators $[\,[\,[X_{j_1}, X_{j_2}], X_{j_3}] \cdots, X_{j_s}]$ of order up to r span the tangent space at the origin. Then we can find smooth vector fields $\widetilde{X}_1 \cdots \widetilde{X}_N$ on a neighborhood $\widetilde{U}$ of the origin in $\mathbb{R}^{n+m}$ with the following properties.*

(a) *The $\widetilde{X}_j$ and their commutators up to order r are linearly independent at each point of $\widetilde{U}$, except for the linear relations that follow formally from the antisymmetry of the bracket and the Jacobi identity.*

(b) *The $\widetilde{X}_j$ and their commutators up to order r span the tangent space of $\widetilde{U}$.*

(c) *Acting on functions on $\mathbb{R}^{n+m}$ that do not depend on the last m coordinates, the $\widetilde{X}_j$ reduce to the given X_j.*

Next we need a nilpotent Lie group $\mathcal{N}$ appropriate to the vector fields $\widetilde{X}_1 \cdots \widetilde{X}_N$. The natural one is the free nilpotent group $\mathcal{N}_{Nr}$ of step r on N generators. Its Lie algebra is generated by $Y_1 \cdots Y_N$ whose Lie brackets of order higher than r vanish, but whose brackets of order $\leq r$ are linearly independent, except for relations forced by antisymmetry of brackets and the Jacobi identity. We regard the Y_j as translation-invariant vector fields on $\mathcal{N}_{Nr}$. It is convenient to pick a basis $\{Y_\alpha\}_{\alpha \in A}$ for the Lie algebra of $\mathcal{N}_{Nr}$, consisting of $Y_1 \cdots Y_N$ and some of their commutators.

On $\mathcal{N}_{Nr}$ we form the Hörmander operator $\widehat{\mathcal{L}} = \sum_1^N Y_j^2$. Then $\widehat{\mathcal{L}}$ is translation-invariant and homogeneous under the natural dilations on $\mathcal{N}_{Nr}$. Hence $\widehat{\mathcal{L}}^{-1}$ is given by convolution on $\mathcal{N}_{Nr}$ with a homogeneous kernel $K(\cdot)$ having a weak singularity at the origin. Hypoellipticity of $\widehat{\mathcal{L}}$ shows that K is smooth away from the origin. Thus we understand the equation $\widehat{\mathcal{L}}u = f$ very well.

We want to use $\widehat{\mathcal{L}}$ to approximate $\widetilde{\mathcal{L}}$ at each point $y \in \widetilde{U}$. To do so, we have to identify a neighborhood of y in $\widetilde{U}$ with a neighborhood of the origin in $\mathcal{N}_{Nr}$. This has to be done just right, or else $\widehat{\mathcal{L}}$ will fail to approximate $\widetilde{\mathcal{L}}$. The idea is to use exponential coordinates on both $\widetilde{U}$ and $\mathcal{N}_{Nr}$. Thus, if $x = \exp(\sum_{\alpha \in A} t_\alpha Y_\alpha)$ (identity) $\in \mathcal{N}_{Nr}$, then we use $(t_\alpha)_{\alpha \in A}$ as coordinates for x. Similarly, let $(\widetilde{X}_\alpha)_{\alpha \in A}$ be the commutators of $\widetilde{X}_1 \cdots \widetilde{X}_N$ analogous to the Y_α, and let $y \in \widetilde{U}$ be given. Then given a nearby point $x = \exp(\sum_{\alpha \in A} t_\alpha \widetilde{X}_\alpha)y \in \widetilde{U}$, we use $(t_\alpha)_{\alpha \in A}$ as coordinates for x.

Now we can identify $\widetilde{U}$ with a neighborhood of the identity in $\mathcal{N}_{Nr}$, simply by identifying points with the same coordinates. Denote the identification by $\Theta_y : \widetilde{U} \longrightarrow \mathcal{N}_{Nr}$, and note that $\Theta_y(y) = $ identity.

In view of the identification Θ_y, the operators $\widehat{\mathcal{L}}$ and $\widetilde{\mathcal{L}}$ live on the same space. The next step is to see that they are approximately equal. To formulate this, we need some bookkeeping on the nilpotent group $\mathcal{N}_{Nr}$. Let $\{\delta_t\}_{t>0}$ be the natural dilations on $\mathcal{N}_{Nr}$. If $\varphi \in C_0^\infty(\mathcal{N}_{Nr})$, then write φ_t for the function $x \longmapsto \varphi(\delta_t x)$. When φ is fixed and t is large, then φ_t is supported in a tiny neighborhood of the identity. Let $\mathcal{D}$ be a differential operator acting on functions on $\mathcal{N}_{Nr}$. We say that $\mathcal{D}$ has "degree" at most k if for each $\varphi \in C_0^\infty(\mathcal{N}_{Nr})$ we have $|\mathcal{D}(\varphi_t)| = O(t^k)$ for large, positive t. According to this definition, $Y_1, \cdots, Y_N$ have degree 1 while $[Y_j, Y_k]$ has degree 2, and the degree of $a(x)\,[Y_j, Y_k]$ depends on the behavior of $a(x)$ near the identity. Now we can say in what sense $\widetilde{\mathcal{L}}$ and $\widehat{\mathcal{L}}$ are approximately equal. The crucial result is as follows.

Theorem B. *Under the map* Θ_y^{-1}, *the vector field* $\widetilde{X}_j$ *pulls back to* $Y_j + Z_{y,j}$, *where* $Z_{y,j}$ *is a vector field on* $\mathcal{N}_{Nr}$ *of "degree"* ≤ 0.

Using Theorem B and the map Θ_y, we can produce a parametrix for $\widetilde{\mathcal{L}}$ and prove that it works. In fact, we take

$$(32) \qquad \widetilde{K}(x, y) = K(\Theta_y x),$$

where K is the fundamental solution of $\widehat{\mathcal{L}}$. For fixed y, we want to know that

$$(33) \qquad \widetilde{\mathcal{L}}\widetilde{K}(x, y) = \delta_y(x) + \mathcal{E}(x, y),$$

where $\delta_y(\cdot)$ is the Dirac delta-function and $\mathcal{E}(x, y)$ has only a weak singularity at $x = y$. To prove this, we use Θ_y to pull back to $\mathcal{N}_{Nr}$. Recall that $\widetilde{\mathcal{L}} = \sum_1^N \widetilde{X}_j^2$ while $\widehat{\mathcal{L}} = \sum_1^N Y_j^2$. Hence by Theorem B, $\widetilde{\mathcal{L}}$ pulls back to an operator of the form $\widehat{\mathcal{L}} + \mathcal{D}_y$, with $\mathcal{D}_y$ having "degree" at most 1. Therefore (33) reduces to proving that

(34) $$(\widehat{\mathcal{L}} + \mathcal{D}_y)K(x) = \delta_{\mathrm{id.}}(x) + \widehat{\mathcal{E}}(x),$$

where $\widehat{\mathcal{E}}$ has only a weak singularity at the identity. Since $\widehat{\mathcal{L}}K(x) = \delta_{\mathrm{id.}}(x)$, (34) means simply that $\mathcal{D}_y K(x)$ has only a weak singularity at the identity. However, this is obvious from the smoothness and homogeneity of $K(x)$, and from the fact that $\mathcal{D}_y$ has degree ≤ 1. Thus, $\widetilde{K}(x, y)$ is an approximate fundamental solution for $\widetilde{\mathcal{L}}$.

From the explicit fundamental solution for the lifted operator $\widetilde{\mathcal{L}}$, one can "descend" to deal with the original Hörmander operator $\mathcal{L}$ in two different ways.

a. Prove sharp estimates for the lifted problem, then specialize to the case of functions that don't depend on the extra variables.
b. Integrate out the extra variables from the fundamental solution for $\widetilde{\mathcal{L}}$, to obtain a fundamental solution for $\mathcal{L}$.

Rothschild and Stein used the first approach. They succeeded in proving the estimate

(35) $$\|X_0 u\|_{L^p(U)} + \|X_j X_k u\|_{L^p(U)} \leq C_p \left\|\left(\sum_{j=1}^N X_j^2 + X_0\right) u\right\|_{L^p(V)} +$$

$$C_p \|u\|_{L^p(V)} \qquad \text{for} \quad 1 < p < \infty \text{ and } U \subset\subset V.$$

This is the most natural and the sharpest estimate for Hörmander operators. It was new even for $p = 2$. Rothschild and Stein also proved sharp estimates in spaces analogous to the Γ_α and $S_{m,p}$ of Folland-Stein [67], as well as in standard Lipschitz and Sobolev spaces. We omit the details, but we point out that commuting derivatives past a general Hörmander operator here requires additional ideas.

Later, Nagel, Stein, and Wainger [119] returned to the second approach ("b" above) and were able to estimate the fundamental solution of a general Hörmander operator. This work overcomes substantial problems.

In fact, once we descend from the lifted problem to the original equation, we again face the difficulty that Hörmander operators cannot be modelled directly on nilpotent Lie groups. So it isn't even clear how to state a theorem on the fundamental solution of a Hörmander operator. Nagel, Stein and Wainger [119] realized that a family of non-Euclidean "balls" $B_{\mathcal{L}}(x, \rho)$ associated to the Hörmander operator $\mathcal{L}$ plays the basic role. They defined the $B_{\mathcal{L}}(x, \rho)$ and proved their essential properties. In particular, they saw that the family of balls survives the projection from the lifted problem back to the original equation, even though the nilpotent Lie group structure is destroyed. Non-Euclidean balls had already played an important part in Folland-Stein [67]. However, it was simple in [67] to guess the correct family of balls. For general Hörmander operators $\mathcal{L}$ the problem of defining and controlling non-Euclidean balls is much more subtle. Closely related results appear also in [FKP], [FS].

Let us look first at a nilpotent group such as $\mathcal{N}_{Nr}$, with its family of dilations $\{\delta_t\}_{t>0}$. Then the correct family of non-Euclidean balls $B_{\mathcal{N}_{Nr}}(x, \rho)$ is essentially dictated by translation and dilation-invariance, starting with a more or less arbitrary harmless "unit ball" $B_{\mathcal{N}_{Nr}}$ (identity, 1). Recall that the fundamental solution for $\widehat{\mathcal{L}} = \sum_1^N Y_j^2$ on $\mathcal{N}_{Nr}$ is given by a kernel $K(x)$ homogeneous with respect to the δ_t. Estimates that capture the size and smoothness of $K(x)$ may be phrased entirely in terms of the non-Euclidean balls $B_{\mathcal{N}_{Nr}}(x, \rho)$. In fact, the basic estimate is as follows.

$$
(36) \qquad |Y_{j_1} Y_{j_2} \cdots Y_{j_m} K(x)| \le \frac{C_m \rho^{2-m}}{\text{vol } B_{\mathcal{N}_{Nr}}(0, \rho)}
$$

$$
\text{for } x \in B_{\mathcal{N}_{Nr}}(0, \rho) \smallsetminus B_{\mathcal{N}_{Nr}}\left(0, \frac{\rho}{2}\right) \text{ and } m \ge 0.
$$

Next we associate non-Euclidean balls to a general Hörmander operator. For simplicity, take $\mathcal{L} = \sum_1^N X_j^2$ as in our discussion of Rothschild-Stein [72]. One definition of the balls $B_{\mathcal{L}}(x, \rho)$ involves a moving particle that starts at x and travels along the integral curve of X_{j_1} for time t_1. From its new position x' the particle then travels along the integral curve of X_{j_2} for time t_2. Repeating the process finitely many times, we can move the particle from its initial position x to a final position y in a total time $t = t_1 + \cdots + t_m$. The ball $B_{\mathcal{L}}(x, \rho)$ consists of all y that can be reached in this way in time $t < \rho$. For instance, if $\mathcal{L}$ is elliptic, then $B_{\mathcal{L}}(x, \rho)$ is essentially the ordinary (Euclidean) ball about x of radius ρ. If we take $\widehat{\mathcal{L}} = \sum_1^N Y_j^2$ on $\mathcal{N}_{N,r}$, then the balls $B_{\widehat{\mathcal{L}}}(x, \rho)$ behave naturally under translations and dilations; hence they are essentially the same as the $B_{\mathcal{N}_{N,r}}(x, \rho)$ appearing in (36). Nagel-Wainger-Stein analyzed the relations between $B_{\widehat{\mathcal{L}}}(x, \rho)$, $B_{\widetilde{\mathcal{L}}}(x, \rho)$ and $B_{\mathcal{L}}(x, \rho)$ for an arbitrary Hörmander operator $\mathcal{L}$. (Here $\widetilde{\mathcal{L}}$ and $\widehat{\mathcal{L}}$ are as in our previous discussion of Rothschild-Stein.) This allowed them to integrate out the extra variables in the fundamental solution of $\widetilde{\mathcal{L}}$, to derive the following sharp estimates from (36).

Theorem. *Suppose $X_1 \cdots X_N$ and their repeated commutators span the tangent space. Also, suppose we are in dimension greater than 2. Then the solution of $(\sum_1^N X_j^2)u = f$ is given by $u(x) = \int K(x, y) f(y) dy$ with*

$$
|X_{j_1} \cdots X_{j_m} K(x, y)| \le \frac{C_m \rho^{2-m}}{(\text{vol } B_{\mathcal{L}}(y, \rho))} \qquad \text{for} \quad x \in B_{\mathcal{L}}(y, \rho) \smallsetminus B_{\mathcal{L}}\left(y, \frac{\rho}{2}\right)
$$

$$
\text{and } m \ge 0.
$$

Here the X_{j_i} act either in the x- or the y-variable.

Let us return from Hörmander operators to the $\overline{\partial}$-problems on strongly pseu-doconvex domains $D \subset \mathbb{C}^n$. Greiner and Stein derived sharp estimates for the Neumann Laplacian $\Box w = \alpha$ in their book [78]. This problem is hard, because two different families of balls play an important role. On the one hand, the standard (Euclidean) balls arise here, because $\Box$ is simply the Laplacian in the interior of D. On the other hand, non-Euclidean balls (as in Folland-Stein [67]) arise on ∂D, because they are adapted to the non-elliptic boundary conditions for $\Box$. Thus, any understanding of $\Box$ requires notions that are natural with respect to either family of balls. A key notion is that of an *allowable vector field* on $\overline{D}$. We say that a smooth vector field X is allowable if its restriction to the boundary ∂D lies in the span of the complex vector fields $L_1 \cdots L_{n-1}, \overline{L}_1 \cdots \overline{L}_{n-1}$. Here we have retained the notation of our earlier discussion of $\overline{\partial}$-problems. At an interior point, an allowable vector field may point in any direction, but at a boundary point it must be in the natural codimension-one subspace of the tangent space of ∂D. Allowable vector fields are well-suited both to the Euclidean and the Heisenberg balls that control $\Box$. The sharp estimates of Greiner-Stein are as follows.

Theorem. *Suppose $\Box w = \alpha$ on a strictly pseudoconvex domain $D \subset \mathbb{C}^n$. If α belongs to the Sobolev space L_k^p, then w belongs to L_{k+1}^p ($1 < p < \infty$). Moreover, if X and Y are allowable vector fields, then XYw belongs to L_k^p. Also, $\overline{L}w$ belongs to L_{k+1}^p if $\overline{L}$ is a smooth complex vector field of type $(0, 1)$. Similarly, if α belongs to the Lipschitz space $\mathrm{Lip}(\beta)$ ($0 < \beta < 1$), then the gradient of w belongs to $\mathrm{Lip}(\beta)$ as well. Also the gradient of $\overline{L}w$ belongs to $\mathrm{Lip}(\beta)$ if L is a smooth complex vector field of type $(0, 1)$; and XYw belongs to $\mathrm{Lip}(\beta)$ for X and Y allowable vector fields.*

These results for allowable vector fields were new even for L^2. We sketch the proof.

Suppose $\Box w = \alpha$. Ignoring the boundary conditions for a moment, we have $\triangle w = \alpha$ in D, so

$$(37) \qquad\qquad w = G\alpha + \text{P.I.}(\widetilde{w})$$

where $\widetilde{w}$ is defined on ∂D, and G, P.I. denote the standard Green's operator and Poisson integral, respectively. The trouble with (37) is that we know nothing about $\widetilde{w}$ so far. The next step is to bring in the boundary condition for $\Box w = \alpha$. According to Calderón's work on general boundary-value problems, (37) satisfies the $\overline{\partial}$-Neumann boundary conditions if and only if

$$(38) \qquad\qquad A\widetilde{w} = \{B(G\alpha)\} |_{\partial D}$$

for a certain differential operator B on D, and a certain pseudodifferential operator A on ∂D. Both A and B can be determined explicitly from routine computation.

Greiner and Stein [78] derive sharp regularity theorems for the pseudodifferential equation $A\widetilde{w} = g$, and then apply those results to (38) in order to understand $\widetilde{w}$ in terms of α. Once they know sharp regularity theorems for $\widetilde{w}$, formula (37) gives the behavior of w.

Let us sketch how Greiner-Stein analyzed $A\widetilde{w} = g$. This is really a system of n pseudodifferential equations for n unknown functions ($n = \dim \mathbb{C}^n$). In a suitable frame, one component of the system decouples from the rest of the problem (modulo negligible errors) and leads to a trivial (elliptic) pseudodifferential equation. The non-trivial part of the problem is a first-order system of $(n-1)$ pseudodifferential operators for $(n-1)$ unknowns, which we write as

$$(39) \qquad \Box_+ w^{\#} = \alpha^{\#}.$$

Here $\alpha^{\#}$ consists of the non-trivial components of $\{B(G\alpha)\}\,|_{\partial D}$, $w^{\#}$ is the unknown, and $\Box_+$ may be computed explicitly.

Greiner and Stein reduce (39) to the study of the Kohn-Laplacian $\Box_b$. In fact, they produce a matrix $\Box_-$ of first-order pseudodifferential operators similar to $\Box_+$, and then show that $\Box_-\Box_+ = \Box_b$ modulo negligible errors.[2] Applying $\Box_-$ to (39) yields

$$(40) \qquad \Box_b w^{\#} = \Box_- \alpha^{\#} + \text{negligible}.$$

From Folland-Stein [67] one knows an explicit integral operator K that inverts $\Box_b$ modulo negligible errors. Therefore,

$$(41) \qquad w^{\#} = K\Box_- \alpha^{\#} + \text{negligible}.$$

Equations (37) and (41) express w in terms of α as a composition of various explicit operators, including: the Poisson integral; restriction to the boundary; $\Box_-$; K; G. Because the basic notion of allowable vector fields is well-behaved with respect to both the natural families of balls for $\Box w = \alpha$, one can follow the effect of each of these very different operators on the relevant function spaces without losing information. To carry this out is a big job. We refer the reader to [78] for the rest of the story.

There have been important recent developments in the Stein program for several complex variables. In particular, we refer the reader to Phong's paper in this volume for a discussion of singular Radon transforms; and to Nagel-Rosay-Stein-Wainger

---

[2] This procedure requires significant changes in two complex variables, since then $\Box_b$ isn't invertible.

[131], D.-C. Chang-Nagel-Stein [132], and [McN], [Chr], [FK] for the solution of the $\bar{\partial}$-problems on weakly pseudoconvex domains of finite type in $\mathbb{C}^2$.

Particularly in several complex variables are we able to see in retrospect the fundamental interconnections among classical analysis, representation theory, and partial differential equations, which Stein was the first to perceive.

I hope this article has conveyed to the reader the order of magnitude of Stein's work. However, let me stress that it is only a selection, picking out results which I could understand and easily explain. Stein has made deep contributions to many other topics, e.g.,

Limits of sequences of operators
Extension of Littlewood-Paley Theory from the disc to $\mathbb{R}^n$
Differentiability of functions on sets of positive measure
Fourier analysis on $\mathbb{R}^N$ when $N \to \infty$
Function theory on tube domains
Analysis of diffusion semigroups
Pseudodifferential calculus for subelliptic problems.

The list continues to grow.

Princeton University

BIBLIOGRAPHY OF E. M. STEIN

1. "Interpolation of linear operators." *Trans. Amer. Math. Soc.* **83** (1956), 482–492.
2. "Functions of exponential type." *Ann. of Math.* **65** (1957), 582–592.
3. "Interpolation in polynomial classes and Markoff's inequality." *Duke Math. J.* **24** (1957), 467–476.
4. "Note on singular integrals." *Proc. Amer. Math. Soc.* **8** (1957), 250–254.
5. (with G. Weiss) "On the interpolation of analytic families of operators action on H^p spaces." *Tohoku Math. J.* **9** (1957), 318–339.
6. (with E.H. Ortrow) "A generalization of lemmas of Marcinkiewicz and Fine with applications to singular integrals." *Annuli Scula Normale Superiore Pisa* **11** (1957), 117–135.
7. "A maximal function with applications to Fourier series." *Ann. of Math.* **68** (1958), 584–603.
8. (with G. Weiss) "Fractional integrals on n-dimensional Euclidean space." *J. Math. Mech.* **77** (1958), 503–514.
9. (with G. Weiss) "Interpolation of operators with change of measures." *Trans. Amer. Math. Soc.* **87** (1958), 159–172.
10. "Localization and summability of multiple Fourier series." *Acta Math.* **100** (1958), 93–147.
11. "On the functions of Littlewood-Paley, Lusin, Marcinkiewicz." *Trans. Amer. Math. Soc.* **88** (1958), 430–466.

12. (with G. Weiss) "An extension of a theorem of Marcinkiewicz and some of its applications." *J. Math. Mech.* **8** (1959), 263–284.

13. (with G. Weiss) "On the theory of harmonic functions of several variables I, The theory of H^p spaces." *Acta Math.* **103** (1960), 25–62.

14. (with R. A. Kunze) "Uniformly bounded representations and harmonic analysis of the 2×2 real unimodular group." *Amer. J. Math.* **82** (1960), 1–62.

15. "The characterization of functions arising as potentials." *Bull. Amer. Math. Soc.* **67**(1961), 102–104; II, **68** (1962), 577–582.

16. "On some functions of Littlewood-Paley and Zygmund." *Bull. Amer. Math. Soc.* **67** (1961), 99–101.

17. "On limits of sequences of operators." *Ann. of Math.* **74** (1961), 140–170.

18. "On the theory of harmonic functions of several variables II. Behavior near the boundary." *Acta Math.* **106** (1961), 137–174.

19. "On certain exponential sums arising in multiple Fourier series." *Ann. of Math.* **73** (1961), 87–109.

20. (with R. A. Kunze) "Analytic continuation of the principal series." *Bull. Amer. Math. Soc.* **67** (1961), 543–546.

21. "On the maximal ergodic theorem." *Proc. Nat. Acad. Sci.* **47** (1961), 1894–1897.

22. (with R. A. Kunze) "Uniformly bounded representations II. Analytic continuation of the principal series of representations of the $n \times n$ complex unimodular groups." *Amer. J. Math.* **83** (1961), 723–786.

23. (with A. Zygmund) "Smoothness and differentiability of functions." *Ann. Univ. Sci. Budapest, Sectio Math., III-IV* (1960–61), 295–307.

24. "Conjugate harmonic functions in several variables." Proceedings of the International Congress of Mathematicians, Djursholm-Linden, Instut Mittag-Leffler (1963), 414–420.

25. (with A. Zygmund) "On the differentiability of functions." *Studia Math.* **23** (1964), 248–283.

26. (with G. and M. Weiss) H^p-classes of holomorphic functions in tube domains." *Proc. Nat. Acad. Sci.* **52** (1964), 1035–1039.

27. (with B. Muckenhoupt) "Classical expansions and their relations to conjugate functions." *Trans. Amer. Math. Soc.* **118** (1965), 17–92.

28. "Note on the boundary of holomorphic functions." *Ann. of Math.* **82** (1965), 351–353.

29. (with S. Wainger) "Analytic properties of expansions and some variants of Parseval-Plancherel formulas." *Arkiv. Math., Band 5* **37** (1965), 553–567.

30. (with A. Zygmund) "On the fractional differentiability of functions." *London Math. Soc. Proc.* **34A** (1965), 249–264.

31. "Classes H^2, multiplicateurs, et fonctions de Littlewood-Paley." *Comptes Rendues Acad. Sci. Paris* **263** (1966), 716–719; 780–781; also **264** (1967), 107–108.

32. (with R. Kunze) "Uniformly bounded representations III. Intertwining operators." *Amer. J. Math.* **89** (1967), 385–442.

33. "Singular integrals, harmonic functions and differentiability properties of functions of several variables." *Proc. Symp. Pure Math.* **10** (1967), 316–335.

34. "Analysis in matrix spaces and some new representations of SL(N, C)." *Ann. of Math.* **86** (1967), 461–490.

35. (with A. Zygmund) "Boundedness of translation invariant operators in Hölder spaces and L^p spaces." *Ann. of Math.* **85** (1967), 337–349.

36. "Harmonic functions and Fatou's theorem." In Proceeding of the C.I.M.E. Summer Course on Homogeneous Bounded Domains, Cremonese, 1968.

37. (with A. Koranyi) "Fatou's theorem for generalized halfplanes." *Annali di Pisa* **22** (1968), 107–112.

38. (with G. Weiss) "Generalizations of the Cauchy-Riemann equations and representations of the rotation group." *Amer. J. Math.* **90** (1968), 163–196.

39. (with A. Grossman and G. Loupias) "An algebra of pseudodifferential operators and quantum mechanics in phase space." *Ann. Inst. Fourier, Grenoble* **18** (1968), 343–368.

40. (with N. J. Weiss) "Convergence of Poisson integrals for bounded symmetric domains." *Proc. Nat. Acad. Sci.* **60** (1968), 1160–1162.

41. "Note on the class $L \log L$." *Studia Math.* **32** (1969), 305–310.

42. (with A. W. Knapp) "Singular integrals and the principal series." *Proc. Nat. Acad. Sci.* **63** (1969), 281–284.

43. (with N. J. Weiss) "On the convergence of Poisson integrals." *Trans. Amer. Math. Soc.* **140** (1969), 35–54.

44. *Singular integrals and differentiability properties of functions.* Princeton Mathematical Series, 30. Princeton University Press, 1970.

45. (with A. W. Knapp) "The existence of complementary series." In *Problems in Analysis.* Princeton University Press, 1970.

46. *Topics in harmonic analysis related to the Littlewood-Paley theory.* Annals of Mathematics Studies, 103. Princeton University Press, 1970.

47. "Analytic continuation of group representations." *Adv. Math.* **4** (1970), 172–207.

48. "Boundary values of holomorphic functions." *Bull. Amer. Math. Soc.* **76** (1970), 1292–1296.

49. (with A. W. Knapp) "Singular integrals and the principal series II." *Proc. Nat. Acad. Sci.* **66** (1970), 13–17.

50. (with S. Wainger) "The estimating of an integral arising in multipier transformations." *Studia Math.* **35** (1970), 101–104.

51. (with G. Weiss) *Introduction to Fourier analysis on Euclidean spaces.* Princeton University Press, 1971.

52. (with A. Knapp) "Intertwining operators for semi-simple groups." *Ann. of Math.* **93** (1971), 489–578.

53. (with C. Fefferman) "Some maximal inequalities." *Amer. J. Math.* **93** (1971), 107–115.

54. "L^p boundedness of certain convolution operators." *Bull. Amer. Math. Soc.* **77** (1971), 404–405.

55. "Some problems in harmonic analysis suggested by symmetric spaces and semi-simple groups." *Proceedings of the International Congress of Mathematicians*, Paris: Gauthier-Villers **1** (1971), 173–189.

56. "Boundary behavior of holomorphic functions of several complex variables." Princeton Mathematical Notes. Princeton University Press, 1972.

57. (with A. Knapp) *Irreducibility theorems for the principal series.* (Conference on Harmonic Analysis, Maryland) Lecture Notes in Mathematics, No. 266. Springer Verlag, 1972.

58. (with A. Koranyi) "H^2 spaces of generalized half-planes." *Studia Math.* **XLIV** (1972), 379–388.

59. (with C. Fefferman) "H^p spaces of several variables." *Acta Math.* **129** (1972), 137–193.

60. "Singular integrals and estimates for the Cauchy-Riemann equations." *Bull. Amer. Math. Soc.* **79** (1973), 440–445.

61. "Singular integrals related to nilpotent groups and $\bar{\partial}$-estimates." *Proc. Symp. Pure Math.* **26** (1973), 363–367.

62. (with R. Kunze) "Uniformly bounded representations IV. Analytic continuation of the principal series for complex classical groups of types B_n, C_n, D_n." *Adv. Math.* **11** (1973), 1–71.

63. (with G. B. Folland) "Parametrices and estimates for the $\bar{\partial}_b$ complex on strongly pseudoconvex boundaries." *Bull. Amer. Math. Soc.* **80** (1974), 253–258.

64. (with J. L. Clerc) "L^p multipliers for non-compact symmetric spaces." *Proc. Nat. Acad. Sci.* **71** (1974), 3911–3912.

65. (with A. Knapp) "Singular integrals and the principal series III." *Proc. Nat. Acad. Sci.* **71** (1974), 4622–4624.

66. (with G. B. Folland) "Estimates for the $\bar{\partial}_b$ complex and analysis on the Heisenberg group." *Comm. Pure and Appl. Math.* **27** (1974), 429–522.

67. "Singular integrals, old and new." In Colloquium Lectures of the 79th Summer Meeting of the American Mathematical Society, August 18–22, 1975. American Mathematical Society, 1975.

68. "Necessary and sufficient conditions for the solvability of the Lewy equation." *Proc. Nat. Acad. Sci.* **72** (1975), 3287–3289.

69. "Singular integrals and the principal series IV." *Proc. Nat. Acad. Sci.* **72** (1975), 2459–2461.

70. "Singular integral operators and nilpotent groups." In Proceedings of the C.I.M.E., Differential Operators on Manifolds. Edizioni, Cremonese, 1975: 148–206.

71. (with L. P. Rothschild) "Hypoelliptic differential operators and nilpotent groups." *Acta Math.* **137** (1976), 247–320.

72. (with A. W. Knapp) "Intertwining operators for SL(n, r)." *Studies in Math. Physics.* E. Lieb, B. Simon and A. Wightman, eds. Princeton University Press, 1976: 239–267.

73. (with S. Wainger) "Maximal functions associated to smooth curves." *Proc. Nat. Acad. Sci.* **73** (1976), 4295–4296.

74. "Maximal functions: Homogeneous curves." *Proc. Nat. Acad. Sci.* **73** (1976), 2176–2177.

75. "Maximal functions: Poisson integrals on symmetric spaces." *Proc. Nat. Acad. Sci.* **73** (1976), 2547–2549.

76. "Maximal functions: Spherical means." *Proc. Nat. Acad. Sci.* **73** (1976), 2174–2175.

77. (with P. Greiner) "Estimates for the $\bar{\partial}$-Neumann problem." Mathematical Notes 19. Princeton University Press, 1977.

78. (with D. H. Phong) "Estimates for the Bergman and Szegö projections." *Duke Math. J.* **44** (1977), 695–704.

79. (with N. Kerzman) "The Szegö kernels in terms of Cauchy-Fontappie kernels." *Duke Math. J.* **45** (1978), 197–224.

80. (with N. Kerzman) "The Cauchy kernels, the Szegö kernel and the Riemann mapping function." *Math. Ann.* **236** (1978), 85–93.

81. (with A. Nagel and S. Wainger) "Differentiation in lacunary direction." *Proc. Nat. Acad. Sci.* **73** (1978), 1060–1062.

82. (with A. Nagel) "A new class of pseudo-differential operators." *Proc. Nat. Acad. Sci.* **73** (1978), 582–585.

83. (with N. Kerzman) "The Szegö kernel in terms of the Cauchy-Fontappie kernels." In Proceedings of the Conference on Several Complex Variables, Cortona, 1977. 1978.

84. (with P. Greiner) "On the solvability of some differential operators of the type $\Box_b$." In Proceedings of the Conference on Several Complex Variables, Cortona, 1977. 1978.

85. (with S. Wainger) "Problems in harmonic analysis related to curvature." *Bull. Amer. Math. Soc.* **84** (1978), 1239–1295.

86. (with R. Grundy) "H^p theory for the poly-disc." *Proc. Nat. Acad. Sci.* **76** (1979), 1026–1029.

87. "Some problems in harmonic analysis." *Proc. Symp. Pure and Appl. Math.* **35** (1979), Part I, 3–20.

88. (with A. Nagel and S. Wainger) "Hilbert transforms and maximal functions related to variable curves." *Proc. Symp. Pure and Appl. Math.* **35** (1979), Part I., 95–98.

89. (with A. Nagel) "Some new classes of pseudo-differential operators." *Proc. Symp. Pure and Appl. Math.* **35** (1979), Part II, 159–170.

90. "A variant of the area integral." *Bull. Sci. Math.* **103** (1979), 446–461.

91. (with A. Nagel) "Lectures on pseudo-differential operators: Regularity theorems and applications to non-elliptic problems." Mathematical Notes 24. Princeton University Press, 1979.

92. (with A. Knapp) "Intertwining operators for semi-simple groups II." *Invent. Math.* **60** (1980), 9–84.

93. "The differentiability of functions in $\mathbb{R}^n$." *Ann. of Math.* **113** (1981), 383–385.

94. "Compositions of pseudo-differential operators." In Proceedings of Journées Equations aux derivées partielles, Saint-Jean de Monts, Juin 1981, Sociéte Math. de France, Conférence #5, 1–6.

95. (with A. Nagel and S. Wainger) "Boundary behavior of functions holomorphic in domains of finite type." *Proc. Nat. Acad. Sci.* **78** (1981), 6596–6599.

96. (with A. Knapp) "Some new intertwining operators for semi-simple groups." In *Non-commutative harmonic analysis on Lie groups*, Colloq. Marseille-Luminy, 1981. Lecture Notes in Mathematics, no. 880. Springer Verlag, 1981.

97. (with M. H. Taibleson and G. Weiss) "Weak type estimates for maximal operators on certain H^p classes." *Rendiconti Circ. mat. Pelermo*, Suppl. n. **1** (1981), 81–97.

98. (with D. H. Phong) "Some further classes of pseudo-differential and singular integral operators arising in boundary value problems, I, Composition of operators." *Amer. J. Math.* **104** (1982), 141–172.

99. (with D. Geller) "Singular convolution operators on the Heisenberg group." *Bull. Amer. Math. Soc.* **6**(1982), 99–103.

100. (with R. Fefferman) "Singular integrals in product spaces." *Adv. Math.* **45** (1982), 117–143.

101. (with G. B. Folland) "Hardy spaces on homogeneous groups." Mathematical Notes 28. Princeton University Press, 1982.

102. "The development of square functions in the work of A. Zygmund." *Bull. Amer. Math. Soc.* **7** (1982).

103. (with D. M. Oberlin) "Mapping properties of the Radon transform." *Indiana Univ. Math. J.* **31** (1982), 641–650.

104. "An example on the Heisenberg group related to the Lewy operator." *Invent. Math.* **69** (1982), 209–216.

105. (with R. Fefferman, R. Gundy, and M. Silverstein) "Inequalities for ratios of functionals of harmonic functions." *Proc. Nat. Acad. Sci.* **79** (1982), 7958–7960.

106. (with D. H. Phong) "Singular integrals with kernels of mixed homogeneites." (Conference in Harmonic Analysis in honor of Antoni Zygmund, Chicago, 1981), W. Beckner, A. Calderón, R. Fefferman, P. Jones, eds. Wadsworth, 1983.

107. "Some results in harmonic analysis in $\mathbb{R}^n$, for $n \to \infty$." *Bull. Amer. Math. Soc.* **9** (1983), 71–73.

108. "An H^1 function with non-summable Fourier expansion." In *Proceedings of the Conference in Harmonic Analysis, Cortona, Italy, 1982*. Lecture Notes in Mathematics, no. 992. Springer Verlag, 1983.

109. (with R. R. Coifman and Y. Meyer) "Un nouvel espace fonctionel adapté a l'étude des opérateurs définis pour des intégrales singulières." In *Proceedings of the Conference in Harmonic Analysis, Cortona, Italy, 1982*. Lecture Notes in Mathematics, no. 992. Springer Verlag, 1983.

110. "Boundary behaviour of harmonic functions on symmetric spaces: Maximal estimates for Poisson integrals." *Invent. Math.* **74** (1983), 63–83.

111. (with J. O. Stromberg) "Behavior of maximal functions in $\mathbb{R}^n$ for large n." *Arkiv f. Math.* **21** (1983), 259–269.

112. (with D. H. Phong) "Singular integrals related to the Radon transform and boundary value problems." *Proc. Nat. Acad. Sci.* **80** (1983), 7697–7701.

113. (with D. Geller) "Estimates for singular convolution operators on the Heisenberg group." *Math. Ann.* **267** (1984), 1–15.

114. (with A. Nagel) "On certain maximal functions and approach regions." *Adv. Math.* **54** (1984), 83–106.

115. (with R. R. Coifman and Y. Meyer) "Some new function spaces and their applications to harmonic analysis." *J. Funct. Anal.* **62** (1985), 304–335.

116. "Three variations on the theme of maximal functions." (Proceedings of the Seminar on Fourier Analysis, El Escorial, 1983.) *Recent Progress in Fourier Analysis*. I. Peral and J. L. Rubiode Francia, eds.

117. Appendix to the paper "Unique continuation. . . ." *Ann. of Math.* **121** (1985), 489–494.

118. (with A. Nagel and S. Wainger) "Balls and metrics defined by vector fields I: Basic properties." *Acta Math.* **155** (1985), 103–147.

119. (with C. Sogge) "Averages of functions over hypersurfaces." *Invent. Math.* **82** (1985), 543–556.

120. "Oscillatory integrals in Fourier analysis." In *Beijing lectures on Harmonic analysis*. Annals of Mathematics Studies, 112. Princeton University Press, 1986.

121. (with D. H. Phong) "Hilbert integrals, singular integrals and Radon transforms II." *Invent. Math.* **86** (1986), 75–113.

122. (with F. Ricci) "Oscillatory singular integrals and harmonic analysis on nilpotent groups." *Proc. Nat. Acad. Sci.* **83** (1986), 1–3.

123. (with F. Ricci) "Homogeneous distributions on spaces of Hermitian matricies." *Jour. Reine Angw. Math.* **368** (1986), 142–164.

124. (with D. H. Phong) "Hilbert integrals, singular integrals and Radon transforms I." *Acta Math.* **157** (1986), 99–157.

125. (with C. D. Sogge) "Averages over hypersurfaces: II." *Invent. Math.* **86** (1986), 233–242.

126. (with M. Christ) "A remark on singular Calderón-Zygmund theory." *Proc. Amer. Math. Soc.* **99, 1** (1987), 71–75.

127. "Problems in harmonic analysis related to curvature and oscillatory integrals." Proc. Int. Congress of Math., Berkeley **1** (1987), 196–221.

128. (with F. Ricci) "Harmonic analysis on nilpotent groups and singular integrals I." *J. Funct. Anal.* **73** (1987), 179–194.

129. (with F. Ricci) "Harmonic analysis on nilpotent groups and singular integrals II." *J. Funct. Anal.* **78** (1988), 56–84.

130. (with A. Nagel, J. P. Rosay and S. Wainger) "Estimates for the Bergman and Szegö kernels in certain weakly pseudo-convex domains." *Bull. Amer. Math. Soc.* **18** (1988), 55–59.

131. (with A. Nagel and D. C. Chang) "Estimates for the $\bar{\partial}$-Neumann problem for pseudo-convex domains in $\mathbb{C}^2$ of finite type." *Proc. Nat. Acad. Sci.* **85** (1988), 8771-8774.

132. (with A. Nagel, J. P. Rosay, and S. Wainger) "Estimates for the Bergman and Szegö kernels in $\mathbb{C}^2$." *Ann. of Math.* **128** (1989), 113–149.

133. (with D. H. Phong) "Singular Radon transforms and oscillatory integrals." *Duke Math. J.* **58** (1989), 347–369.

134. (with F. Ricci) "Harmonic analysis on nilpotent groups and singular integrals III." *J. Funct. Anal.* **86** (1989), 360–389.

135. (with A. Nagel and F. Ricci) "Fundamental solutions and harmonic analysis on nilpotent groups." *Bull. Amer. Math. Soc.* **23** (1990), 139–143.

136. (with A. Nagel and F. Ricci) "Harmonic analysis and fundamental solutions on nilpotent Lie groups in Analysis and P.D.E." A collection of papers dedicated to Mischa Cotlar. Marcel Decker, 1990.

137. (with C. D. Sogge) "Averages over hypersurfaces, smoothness of generalized Radon transforms." *J. d' Anal. Math.* **54** (1990), 165–188.

138. (with S. Sahi) "Analysis in matrix space and Speh's representations." *Invent. Math.* **101** (1990), 373–393.

139. (with S. Wainger) "Discrete analogues of singular Radon transforms." *Bull. Amer. Math. Soc.* **23** (1990), 537–544.

140. (with D. H. Phong) "Radon transforms and torsion." *Duke Math. J.* (Int. Math. Res. Notices) #4, (1991), 44–60.

141. (with A. Seeger and C. Sogge) "Regularity properties of Fourier integral operators." *Ann. of Math.* **134** (1991), 231–251.

142. (with J. Stein) "Stock price distributions with stochastic volatility: an analytic approach." *Rev. Fin. Stud.* **4** (1991), 727–752.

143. (with D. C. Chang and S. Krantz) "Hardy spaces and elliptic boundary value problems." In the Madison Symposium on Complex Analysis, *Contemp. Math.*, **137** (1992), 119–131.

OTHER REFERENCES

[BGS] D. Burkholder, R. Gundy, and M. Silverstein. "A maximal function characterization of the class H^p." *Trans. Amer. Math. Soc.* **157** (1971): 137–153.

[Ca] A. P. Calderón. "Commutators of singular integral operators." *Proc. Nat. Acad. Sci.* **53** (1965): 1092–1099.

[Chr] M. Christ. "On the $\overline{\partial}_b$-equation and Szegö projection on a CR manifold." In *Proceedings, El Escorial Conference on Harmonic Analysis 1987*. Lecture Notes in Mathematics, no. 1384. Springer Verlag, 1987.

[Co] M. Cotlar. "A unified theory of Hilbert transforms and ergodic theory." *Rev. Mat. Cuyana I* (1955): 105–167.

[CZ] A. P. Calderón and A. Zygmund. "On higher gradients of harmonic functions." *Studia Math.* **26** (1964): 211–226.

[EM] L. Ehrenpreis and F. Mautner. "Uniformly bounded representations of groups." *Proc. Nat. Acad. Sci.* **41** (1955): 231–233.

[FK] C. Fefferman and J. J. Kohn, "Estimates of kernels on three-dimensional CR manifolds." *Rev. Mat. Iber.* 4, no. 3 (1988): 355–405.

[FKP] C. Fefferman, J. J. Kohn, and D. Phong. "Subelliptic eigenvalue problems." In *Proceedings, Conference in Honor of Antoni Zygmund*. Wadsworth, 1981.

[FS] C. Fefferman and A. Sanchez-Calle, "Fundamental solutions for second order subelliptic operators." *Ann. of Math.* **124** (1986): 247–272.

[GN] I. M. Gelfand and M. A. Neumark. *Unitäre Darstellungen der Klassischen Gruppen*. Akademie Verlag, 1957.

[H] I. I. Hirschman, Jr."Multiplier transformations I." *Duke Math. J.* **26** (1956): 222–242;"Multiplier transformations II." *Duke Math. J.* **28** (1961): 45–56.

[McN] J. McNeal, "Boundary behavior of the Bergman kernel function in $\mathbb{C}^2$." *Duke Math. J.* **58** (1989): 499–512.

[Z] A. Zygmund, *Trigonometric Series*. Cambridge University Press, 1959.

2

Geometric Inequalities
in Fourier Analysis

*William Beckner**

1 INTRODUCTION

Geometric ideas occur in almost every aspect of Fourier analysis. Beginning with the symmetry structure of the domain and the product structure of the operator, geometric concepts control deep facts about the Fourier transform. The symmetry structure of a Riemannian manifold defines not only the natural objects of analysis for the domain such as the Laplace-Beltrami operator, Green's function, global transforms, and boundary operators, but also determines intrinsic ways to compare the "size" of such objects as measured by the classical function spaces. The geometric structure of a manifold is manifest in the character of analytic operator and variational inequalities. Such estimates are the building-blocks of "everyday analysis."

Convolution is a natural object viewed both as an averaging process for a translation-invariant measure and as the dual operator to the Fourier transform. Fractional integration is even more natural arising in the context of Green's functions and potential theory, restriction phenomena for the Fourier transform, intertwining operators for representations of the Lorentz groups and correlation functions in conformal field theory and statistical mechanics. A framework for the analysis of convolution inequalities on a manifold is developed here, especially in terms of (1) conformal invariance; (2) geometric symmetrization and equimeasurable rearrangement of functions; (3) complex symmetry structure on Lie groups; and (4) embedding of geometric and probabilistic information in exact constants

*This research was supported in part by the National Science Foundation.

for variational problems. Several themes from E. M. Stein's work have influenced this program, including his treatment of Riesz transforms, spherical harmonics, and the Hardy-Littlewood-Sobolev theorem on fractional integration; his viewpoint on Lie groups and boundary manifolds including $SL(2, \mathbb{R})$ and the Heisenberg group; and his emphasis on integral transforms and boundary behavior in several complex variables. Philosophically the roots for the overall approach go back to Hardy and Littlewood. In addition, a strong connection can be drawn to problems in physics, including the Bargmann-Fock analysis of the hydrogen atom, ideas of Bargmann, Fock, and Segal on quantum field theory, and Polyakov's quantum string theory. Three important calculations for sharp inequalities due to J. Moser, E. Onofri, and E. H. Lieb directly motivate much of this program. The interaction of ideas and methods from differential geometry, Fourier analysis, and quantum string theory has led to a richer understanding of the role of algebraic invariance and geometric structure in analysis on manifolds.

2 CLASSICAL INEQUALITIES

The basic convolution inequality on a unimodular Lie group is Young's inequality

$$(f * g)(x) = \int_G f(xy^{-1})g(y) \, dy$$

(1)
$$\|f * g\|_{L^r(G)} \leq \|f\|_{L^p(G)} \|g\|_{L^q(G)}$$

with $1/p + 1/q - 1 = 1/r$, $1 \leq p, q, r \leq \infty$. Using product structure and radial symmetry, one can show that on the Euclidean space $\mathbb{R}^n$ this inequality can be improved ([8],[16]) with the extremal result holding only for gaussian functions

(2)
$$\|f * g\|_{L^r(\mathbb{R}^n)} \leq (A_p A_q A_{r'})^n \|f\|_{L^p(\mathbb{R}^n)} \|g\|_{L^q(\mathbb{R}^n)}$$

with

$$A_p = \left[p^{1/p}/p'^{1/p'} \right]^{1/2}$$

and primes always denoting dual exponents, $1/p + 1/p' = 1$. It is relatively easy to see that this inequality extends to include weak Lorentz classes and in fact corresponds to the Hardy-Littlewood-Sobolev theorem for fractional integration ([68])

(3)
$$\left\| |x|^{-\lambda} * f \right\|_{L^r(\mathbb{R}^n)} \leq C \|f\|_{L^p(\mathbb{R}^n)}$$

with $\lambda = n/q$ for $1 < q < \infty$ (r, p, q related as above). The function $|x|^{-\lambda}$ is characteristic for the Lorentz class $L_{q,\infty}(\mathbb{R}^n)$ so this inequality has the equivalent form

(4)
$$\|f * g\|_{L^r(\mathbb{R}^n)} \leq C \|g\|_{L_{q,\infty}(\mathbb{R}^n)} \|f\|_{L^p(\mathbb{R}^n)}.$$

The linking step between these inequalities involves a symmetrization argument of Riesz-Sobolev type.

Riesz-Sobolev Lemma.

(5)
$$\left| \int_{\mathbb{R}^n \times \mathbb{R}^n} f(x)g(y)h(x-y)\, dx\, dy \right| \leq \int_{\mathbb{R}^n \times \mathbb{R}^n} f^*(x)g^*(y)h^*(x-y)\, dx\, dy.$$

Here $*$ denotes the equimeasurable radial decreasing rearrangement applied to the modulus of a function.

This technical result is central to the analysis of positive convolution operators and Sobolev inequalities ([70]), and is geometric in nature being equivalent to the Brunn-Minkowski inequality. The symmetry structure intrinsic to these inequalities includes: translation invariance, rotational symmetry, dilation invariance, and Euclidean product structure.

It is useful to take into account this product structure even though it does not provide good control of constants. Using the relation between arithmetic and geometric means, the one-dimensional Hardy-Littlewood-Sobolev inequality controls not only fractional integration on classical n-dimensional Riemannian manifolds but also on nilpotent Lie groups. Moreover, the Riesz-Sobolev symmetrization lemma can be applied on a nilpotent Lie group to remove the non-abelian structure and give the following extension for the inequalities that appear above.

Theorem 1. *Let G be a nilpotent Lie group of dimension n, homogeneous dimension m with $|x|$ denoting the canonical distance on G. Then for $1/p + 1/q - 1 = 1/r, 1 \leq p, q, r \leq \infty$*

(6)
$$\|f * g\|_{L^r(G)} \leq (A_p A_q A_{r'})^n \|f\|_{L^p(G)} \|g\|_{L^q(G)}$$
$$A_p = [p^{1/p}/p'^{1/p'}]^{1/2}$$

and for $\lambda = m/q, 1 < q < \infty$

(7)
$$\|f * g\|_{L^r(G)} \leq C \|f\|_{L^p(G)} \|g\|_{L_{q,\infty}(G)}$$

(8)
$$\left\| |x|^{-\lambda} * g \right\|_{L^r(G)} \leq C \|f\|_{L^p(G)}.$$

No extremal functions exist for inequality (6) when the constant is less than one.

Interpolation arguments are the standard method used to prove weak Lorentz class inequalities, but dilation invariance and the Riesz-Sobolev lemma provide a more elementary reduction to the one-dimensional form of Young's inequality on the real line. To illustrate this point consider the inequality

$$(9) \qquad \left| \int_{\mathbb{R}^n \times \mathbb{R}^n} f(x)g(y)|x - y|^{-\lambda} \, dx \, dy \right| \le C \|f\|_{L^p(\mathbb{R}^n)} \|g\|_{L^q(\mathbb{R}^n)}$$

for $\lambda = n(1/p' + 1/q') < n$ and $1 < p, q < \infty$. Using symmetrization the problem of verifying this inequality is reduced to non-negative radial decreasing functions which satisfy the estimates

$$|x|^{n/p} f(x) \le C\|f\|_p, \qquad |y|^{n/p} g(y) \le C\|g\|_q.$$

For the most part, the symbol C will denote a generic constant. Since the problem is one-variable, dilation invariance provides a formulation as a convolution inequality on the line. Set $\tilde{u}(t) = |x|^{n/p} f(x), x = \exp t$ and $\tilde{v}(s) = |y|^{n/q} g(y), y = \exp s$; then the inequality above becomes

$$\int_{\mathbb{R} \times \mathbb{R}} \tilde{u}(t)\tilde{v}(s)\psi(t - s) \, ds \, dt \le C \|\tilde{u}\|_{L^p(\mathbb{R})} \|\tilde{v}\|_{L^q(\mathbb{R})},$$

but under the conditions $\psi \in L^1 \cap L_{n/\lambda,\infty}$, $\|\tilde{u}\|_r \le C\|f\|_{L^p(\mathbb{R}^n)}$, $p \le r \le \infty$ and $\|\tilde{v}\|_r \le C\|g\|_{L^q(\mathbb{R}^n)}$, $q \le r \le \infty$. The estimate now follows immediately from Young's inequality (1).

E. H. Lieb recognized ([48]) that when the map

$$f \to |x|^{-\lambda} * f$$

is a map from a space to its dual, then two additional types of symmetry are evident which can be used to calculate the best constant for the inequality

$$(10) \qquad \left| \int_{\mathbb{R}^n \times \mathbb{R}^n} f(x)g(y)|x - y|^{-\lambda} \, dx \, dy \right| \le C \|f\|_{L^p(\mathbb{R}^n)} \|g\|_{L^p(\mathbb{R}^n)}$$

with $\lambda = 2n/p'$ and $1 < p < 2$. There is a quadratic functional symmetry which makes the operator positive-definite self-adjoint so that one may take $f = g$. More importantly, the inequality is conformally invariant. Suppose τ is a conformal transformation and let J denote the modulus of the Jacobian determinant for this change of variables. Then under the transformation

$$x \to \tau x, \qquad f \to \tilde{f}(x) = f(\tau x) J(\tau, x)^{1/p}$$

the functional inequality (10) is invariant. This application of conformal invariance to convolution problems has a rich history in mathematical physics. It was used by Bargmann and Fock to give a group-theoretical analysis of the spectrum of the hydrogen atom and more recently by Onofri ([57]) to study the variation of the zeta-function determinant of the Laplacian in two dimensions under conformal deformation. The critical issue here is the algebraic invariance of the metric. In addition, conformal invariance implies that equivalent forms of inequality (10) exist on any domain conformally equivalent to the plane $\mathbb{R}^n$, including the sphere S^n and the two-sheeted hyperboloid $\mathbb{H}^n$.

Using the Riesz-Sobolev lemma twice, one can easily observe that extremal functions exist for inequality (10). Symmetrization on $\mathbb{R}^n$ reduces the problem to one variable. Then the dilation structure provides a one-dimensional inequality which can be put on the multiplicative group $\mathbb{R}_+$. That is, let $u(t) = |x|^{n/p} f(x) \geq 0, t = |x|, v(s) = |y|^{n/p} g(y) \geq 0, s = |y|$ and

$$\psi(t) = \int_{S^{n-1}} [t + 1/t - 2\xi \cdot \eta]^{-\lambda/2} \, d\xi;$$

now inequality (10) becomes

(11) $$\int_{\mathbb{R}_+ \times \mathbb{R}_+} u(t)v(s)\psi(t/s) \frac{dt}{t} \frac{ds}{s} \leq C_p \|u\|_{L^p(\mathbb{R}_+)} \|v\|_{L^p(\mathbb{R}_+)}.$$

But the kernel ψ is symmetric decreasing away from the origin $t = 1$ so one can apply the Riesz-Sobolev lemma a second time to insure that u and v are radial decreasing functions on $\mathbb{R}_+$ and as observed above they must also be uniformly bounded. Now choose a sequence $u_n = v_n$ with $\|u_n\|_p = 1$ such that

$$\int u_n(t)u_n(s)\psi(s/t) \frac{dt}{t} \frac{ds}{s} \rightarrow C_p$$

where C_p is the best constant in (11). Since the u_n's are decreasing functions, one can use the Helly selection principle to choose a subsequence that converges almost everywhere to a function $u \in L^p(\mathbb{R}_+)$. By Fatou's lemma $\|u\|_p \leq 1$. But

$$u_n(t) \leq C[1 + |\ln t|]^{-1/p} = \alpha(t)$$

and $\alpha(t)\alpha(s)\psi(t/s) \in L^1(\mathbb{R}_+ \times \mathbb{R}_+)$ so applying Lebesgue's dominated convergence theorem gives

$$\int u_{n_k}(t)u_{n_k}(s)\psi(t/s) \frac{dt}{t} \frac{ds}{s} \rightarrow \int u(t)u(s)\psi(t/s) \frac{dt}{t} \frac{ds}{s} = C_p.$$

Since C_p is the best constant, $\|u\|_p = 1$ and u must be an extremal function for inequality (11) which then provides an extremal function for the fractional integral inequality (10).

The different symmetries of the conformal structure determine the form of the extremal functions. This is clearly realized in terms of the conformal transformation (stereographic projection) from the plane $\mathbb{R}^n$ to the sphere S^n where the Jacobian determinant is proportional to $(1 + |x|^2)^{-n}$. Since the sphere is a compact manifold, one expects constants to be extremal on that domain. This would then imply that on $\mathbb{R}^n$ the extremal functions are given by the function $A(1 + |x|^2)^{-n/p}$ up to conformal automorphism.

To obtain this result, one makes use of the fact that the different kernel functions studied here are all strictly decreasing functions of a single variable. This means that whenever a symmetrization is effected, there must either be some positive increase in the value of the functional, or the unsymmetrized functions are translates of symmetric decreasing functions on the domain. The technical tool needed here is a variation of a symmetrization lemma of Baernstein and Taylor ([5],[9]). The first application of this style of argument for such functions is due to Lieb ([47]).

Symmetrization Lemma (after Baernstein and Taylor). *Let M denote $\mathbb{R}^n$, S^n, or $\mathbb{H}^n$ (real hyperbolic space) with symmetric decreasing rearrangement defined in terms of geodesic distance. Let K be a monotone decreasing function and $d(x, y)$ denote the distance between the points x and y in M. Then*

$$\int_{M \times M} f(x) K[d(x, y)] g(y) \, dx \, dy \le \int_{M \times M} f^*(x) K[d(x, y)] g^*(y) \, dx \, dy$$

where f, g are non-negative measurable functions with f^*, g^* denoting their respective equimeasurable geodesically decreasing rearrangements on M, dx denotes the measure invariant under the group action on the symmetric space M and the integrand on the left is in $L^1(M \times M)$. If K is strictly decreasing, then the above inequality is strict unless $f(x) = f^*(\tau x)$ and $g(x) = g^*(\tau x)$ almost everywhere for τx a "translate" of x.

Using the conformal equivalence between $\mathbb{R}^n$ and S^n-{pole} given by

$$\xi = \left(\frac{1 - |x|}{1 + |x|^2}, \frac{2x}{1 + |x|^2} \right), \qquad \eta = \left(\frac{1 - |y|^2}{1 + |y|^2}, \frac{2y}{1 + |y|^2} \right)$$

and setting

$$F(\xi) = (1 + |x|^2)^{n/p} f(x), \qquad G(\eta) = (1 + |y|^2)^{n/p} g(y)$$

an equivalent fractional integral inequality to (10) is obtained for the sphere S^n.

(12) $\qquad \left| \int_{S^n \times S^n} F(\xi) G(\eta) |\xi - \eta|^{-\lambda} \, d\xi \, d\eta \right| \le B_p \|F\|_{L^p(S^n)} \|G\|_{L^p(S^n)}.$

Now consider an extremal function for the $\mathbb{R}^n$ fractional integral inequality (10) which also has the inversion symmetry given by symmetrizing on the $\mathbb{R}_+$ inequality (11). If one now transforms this extremal function to the setting of the sphere, one finds that the inversion symmetry gives a function symmetric with respect to hemispheres. But applying the Baernstein and Taylor lemma would produce a positive increase to the left-hand side of inequality (12). This contradicts the fact that the function is extremal, so on the sphere this function must be radial decreasing away from some pole. But constants are the only measurable functions symmetric with respect to hemispheres and decreasing from a pole. Hence conformal factors given in terms of the Jacobian determinant are the only extremal functions for these conformally invariant fractional integral problems. The essential point of this argument is that symmetrization on $\mathbb{R}_+$ picks out a specific extremal which "breaks the S^n symmetry" unless the equivalent function on S^n is constant almost everywhere.

In summary, the analysis of the fractional integral inequality as a dual-space mapping depends on the invariance of the functional under the action of the conformal group and the strict monotonicity of the kernel (as a function of one variable). In studying the variation of the zeta-function determinant of the Laplacian under conformal deformation of metric on S^2, Onofri ([57]) had realized that this conformal group action was a natural technique for the solution of geometric variational problems. Carlen and Loss ([19]) later observed that the two techniques described above, conformal action and geometric symmetrization, could be put together to solve the variational problem for the Hardy-Littlewood-Sobolev inequality (10) on a single domain since in a generic sense a conformal transformation will break the geodesic symmetry. This was quite a useful remark since for some problems the range of symmetrization techniques is more extensive in non-compact domains. A second example where the competing symmetries of different geometries determine the extremal functions for a variational problem occurs in the author's paper ([9]). There radial symmetry on $\mathbb{R}^n$ is played off against the infinite-dimensional spherical symmetry of gaussian measure to provide an elementary proof of Nelson's inequality ([55]). That idea was an outgrowth of earlier work with D. Jerison on optimal information, gaussian symmetry, and a problem of H. Chernoff.

The first immediate application of the sharp Hardy-Littlewood-Sobolev inequality is that one obtains the classical sharp Sobolev inequalities for L^2 control of the gradient and the Dirichlet form for harmonic extension from boundary-values ([9],[10]). Transforming these inequalities to the sphere and taking the infinite-dimensional limit also gives the logarithmic Sobolev inequality of Gross for gaussian measure. Such arguments suggest that considerable geometric and probabilistic information is contained in sharp fractional integral inequalities. It is also apparent that the most workable domain for developing the full consequences of inequality (10) may be the n-dimensional sphere. In this setting spectral data can be visualized in the broadest sense. A good example to illustrate this point is the spectrum of the hydrogen atom where the degeneracy was originally puzzling on the Euclidean side but readily transparent for the equivalent spectral problem for the three-dimensional sphere.

Using stereographic projection and the Funk-Hecke formula, the sharp Hardy-Littlewood-Sobolev inequality (10) gives a conformally invariant fractional integral inequality on the sphere and an equivalent multiplier inequality expressed in terms of spherical harmonics.

Theorem 2. *For $F, G \in L^p(S^n)$ with $1 \le p < 2$ and $\lambda = 2n/p'$*

(13)
$$\left| \int_{S^n \times S^n} F(\xi) |\xi - \eta|^{-\lambda} G(\eta) \, d\xi \, d\eta \right| \le B_p \|F\|_{L^p(S^n)} \|G\|_{L^p(S^n)}$$

$$B_p = \int_{S^n} |\xi - \eta|^{-\lambda} \, d\eta = 2^{-\lambda} \frac{\Gamma(n)\Gamma(\frac{n-\lambda}{2})}{\Gamma(\frac{n}{2})\Gamma(n - \frac{\lambda}{2})}.$$

For $F = \sum Y_k$ and $1 \le p \le 2$

(14)
$$\sum_{k=0}^{\infty} \gamma_k \int_{S^n} |Y_k|^2 \, d\xi \le \left[\|F\|_{L^p(S^n)} \right]^2$$

(15)
$$\gamma_k = \frac{(\frac{\lambda}{2}) \cdots (\frac{\lambda}{2} + k - 1)}{(n - \frac{\lambda}{2}) \cdots (n - \frac{\lambda}{2} + k - 1)} = \frac{\Gamma(\frac{n}{p})\Gamma(\frac{n}{p'} + k)}{\Gamma(\frac{n}{p'})\Gamma(\frac{n}{p} + k)}.$$

Equality is attained if $1 < p < 2$ only for functions of the form

$$F = G, \qquad F_\zeta(\xi) = A|1 - \zeta \cdot \xi|^{-n/p}, \qquad |\zeta| < 1.$$

The sharp value B_p here corresponds to

$$C = \pi^{\lambda/2} \frac{\Gamma(\frac{n-\lambda}{2})}{\Gamma(n - \frac{\lambda}{2})} \left[\frac{\Gamma(\frac{n}{2})}{\Gamma(n)} \right]^{-1 + \frac{\lambda}{n}}$$

for inequality (10). The correspondence between these inequalities comes from considering the map from $\mathbb{R}^n$ to $S^n - \{(0, 0, \ldots, -1)\}$ with the "north pole" corresponding to $(0, 0, \ldots, 1)$:

$$\xi = (u, s) = \left(\frac{2x}{1 + |x|^2}, \frac{1 - |x|^2}{1 + |x|^2} \right) \qquad x \in \mathbb{R}^n, \ u \in \mathbb{R}^n, \ -1 < s \le 1.$$

The inverse map is $x = u/(1 + s)$ and the change of measure is given by

$$d\xi = \pi^{-n/2} \left[\Gamma(n)/\Gamma(n/2) \right] (1 + |x|^2)^{-n} \, dx$$

where $d\xi$ is normalized surface measure on S^n. Let $x, y \in \mathbb{R}^n$ with ξ, η the corresponding points on S^n and set $F(\xi) = (1 + |x|^2)^{n/p} f(x)$ and $G(\eta) = (1 + |y|^2)^{n/p} g(y)$. The algebraic invariance of the metric is expressed by

$$|x - y| = \frac{1}{2} |\xi - \eta| \left[(1 + x^2)(1 + y^2) \right]^{1/2}.$$

The equivalence between (13) and (14) follows from the fact that the kernel $|\xi - \eta|^{-\lambda}$ defines a positive-definite self-adjoint operator that commutes with rotations.

Since such operators arise as intertwining operators for representations of the Lorentz groups, it is natural to see the relation with the Selberg point-pair product on symmetric spaces. The Funk-Hecke formula is used to find the multiplier action in terms of spherical harmonics that is used for (14).

Funk-Hecke Formula. *For K an integral function on $[-1, 1]$ and Y_k a spherical harmonic of degree k on S^n, then*

$$\int_{S^n} K(\xi \cdot \eta) Y_k(\eta) \, d\eta = \lambda_k Y_k(\xi)$$

$$\lambda_k = c_n \frac{\Gamma(k + 1)}{\Gamma(k + n - 1)} \int_{-1}^{1} K(\omega) C_k^{(n-1)/2}(\omega)(1 - \omega^2)^{\frac{n}{2} - 1} \, d\omega$$

where $C_k^\nu(\omega)$ is the Gegenbauer polynomial defined by

$$(1 - 2\omega t + t^2)^{-\nu} = \sum C_k^\nu(\omega) t^k$$

and $c_n = \pi^{-1} 2^{n-2} \Gamma(\frac{n+1}{2}) \Gamma(\frac{n-1}{2})$.

Results in this paper are generally obtained as a priori inequalities for a smooth class of functions and then the full estimates follow by taking limits. Logarithmic integrals are in general indeterminate, but the form of the estimates occurring here will imply that all logarithmic integrals are well-defined.

It is particularly useful to study the inequalities in Theorem 2 under variation of the parameters p and n, especially in terms of end-point information for p and asymptotic limits for n. A wealth of geometric and probabilistic information is contained in this fractional integral inequality including Nelson's hypercontractive estimates for gaussian measure, hypercontractive estimates for the Poisson and heat semigroups on the sphere, logarithmic Sobolev inequalities, Moser-Trudinger inequalities in higher dimensions, Carleson-Chang inequalities, entropy estimates for logarithmic kernels and extremal properties for the zeta-function determinant of the conformal Laplacian and square of the Dirac operator in two and four dimensions under conformal deformation of metric. A nice feature of this structure is that one often captures the intrinsic nature of a problem by writing the inequality in a form where the operator norm is one. For end-point information, the idea is to find a value of the parameter where equality is attained in the functional inequality for a large class of functions and then to take a limit of difference quotients so that one obtains a "differentiated inequality" from the original inequality. Here one can consider this process at $p = 2$ for general functions and at $p = 1$ for non-negative functions. In modern analysis this technique was utilized by L. Gross in a striking manner to show that Nelson's hypercontractive semigroup estimates were equivalent to a logarithmic Sobolev inequality.

The first issue treated here is the asymptotic limit for large n. Observe that

$$(16) \qquad (p-1)^k \leq \frac{\Gamma(\frac{n}{p})\Gamma(\frac{n}{p'}+k)}{\Gamma(\frac{n}{p'})\Gamma(\frac{n}{p}+k)} \xrightarrow[n\to\infty]{} (p-1)^k.$$

Now one has a choice of limits, both of which give Nelson's inequality for gaussian measure. If F is restricted to be a function of the polar angle, then the spherical harmonics are Gegenbauer polynomials which go over in an appropriate limit to Hermite polynomials. The normalization used here is defined by

$$\exp\left(-\frac{1}{2}t^2 + tx\right) = \sum_{k=0}^{\infty} \frac{t^k}{k!} H_k(x)$$

corresponding to the one-dimensional gaussian measure

$$d\mu(x) = (2\pi)^{-1/2}\exp(-x^2/2)\,dx.$$

One could also rescale the inequality to be on a sphere of radius $\sqrt{n}$ and take the infinite-dimensional Poincaré limit[1] to obtain an inequality on $\mathbb{R}^\infty$ with gaussian measure ([9],[50]). For $|\omega| \leq 1$ consider the operator defined on Hermite polynomials

$$T_\omega : H_k \to \omega^k H_k$$

and its product extension to higher dimensions. T_ω may be represented as an integral operator

$$(T_\omega g)(x) = \int T_\omega(x, y)g(y)\,d\mu(y)$$

using the Mehler kernel

$$T_\omega(x, y) = (1 - \omega^2)^{-1/2}\exp\left\{-\frac{\omega^2(x^2+y^2)}{2(1-\omega^2)} + \frac{\omega xy}{1-\omega^2}\right\}$$

and in arbitrary dimension by the semigroup operator

$$T_\omega = e^{-tN}, \qquad N = -\Delta + x\cdot\nabla, \qquad \omega = e^{-t}$$

where N corresponds to the "number operator" in Fock space.

Corollary 1 (Nelson's inequality). *For $d\mu$ denoting the product gaussian measure and real ω with $|\omega| \leq \sqrt{p-1}$ for $1 \leq p \leq 2$*

$$(17) \qquad \|T_\omega g\|_{L^2(d\mu)} \leq \|g\|_{L^p(d\mu)}.$$

---

[1]Though this limit is often attributed to Poincaré, Dan Stroock has told me that it was used earlier by Mehler.

Using the left-hand side of (16), the analogous estimate is obtained for the Poisson semigroup (see [11]). For $0 < r < 1$ the Poisson kernel on the sphere S^n is defined by

$$P_r(\xi, \eta) = \frac{1 - r^2}{|\xi - r\eta|^{n+1}}$$

with

$$(P_r F)(\xi) = \int_{S^n} P_r(\xi, \eta) F(\eta) \, d\eta.$$

The action of this kernel on spherical harmonics is given by

$$(P_r F)(\xi) = \sum_{k=0}^{\infty} r^k Y_k(\xi)$$

if $F = \sum Y_k$.

Corollary 2. *For $0 < r < \sqrt{p - 1}$ and $1 \leq p \leq 2$*

(18) $\|P_r F\|_{L^2(S^n)} \leq \|F\|_{L^p(S^n)}.$

The most useful limit that can be applied in the context of Theorem 2 is the classical differentiation argument that extracts "end-point information" from the parameter range of the mapping. One starts with a functional inequality depending on a parameter p, say $\Phi_p(f) \geq 0$ for which equality is attained for a class of functions f at the parameter value p_0, i.e., $\Phi_{p_0}(f) \equiv 0$. Then one can differentiate the inequality at the value p_0 to produce a new functional inequality

(19) $\Phi'_{p_0}(f) = \lim_{p \to p_0} \frac{\Phi_p(f) - \Phi_{p_0}(f)}{p - p_0} = \lim_{p \to p_0} \frac{\Phi_p(f)}{p - p_0} \geq 0$

if $p \geq p_0$ and the reverse inequality if $p \leq p_0$. The inequalities of Theorem 2 provide "end-point information" at (L^1, L^∞) and at L^2. The first case (L^1, L^∞) will be treated in Section 4 where this gives a sharp Moser-Trudinger inequality for the n-dimensional sphere and further geometric information about zeta-function determinants. In a rough sense, (L^1, L^∞) is the "geometric end-point" while L^2 is the "probabilistic end-point" in that information is obtained which can be applied in an infinite-dimensional setting.

Inequality (14) becomes an equality for all functions when $p = 2$ so the differentiation argument given by (19) can be used for this end-point.

Theorem 3. *For $F \in L^2(S^n)$ with $\int |F|^2 \, d\xi = 1$ and $F = \sum Y_k$,*

(20) $\int |F|^2 \ln |F| \, d\xi \leq \sum_{k=1}^{\infty} \Delta_k(n) \int_{S^n} |Y_k|^2 \, d\xi$

$$\Delta_k(n) = \frac{n}{2} \sum_{\ell=0}^{k-1} \frac{1}{\frac{n}{2} + \ell} \quad k \geq 1.$$

Note that $\Delta_1(n) = 1$, $\Delta_k(n) < k$ if $k \geq 2$ and asymptotically $\Delta_k(n) \simeq (n/2) \ln k$ for large k. The logarithmic character of this factor $\Delta_k(n)$ is not surprising if one thinks about the relation of this problem to fractional integration on the circle. Now using these estimates a more useful logarithmic inequality follows.

Theorem 4. *For $F \in L^2(S^n)$ with $\int |F|^2\, d\xi = 1$ and u denoting the harmonic extension of F to the interior of the unit ball in $\mathbb{R}^{n+1}$ and $c_n = 2\pi^{(n+1)/2}/\Gamma(\frac{n+1}{2})$*

(21)
$$\int |F|^2 \ln |F|\, d\xi \leq \sum_{k=1}^{\infty} k \int_{S^n} |Y_k|^2\, d\xi$$

$$= \frac{1}{c_n} \int_{|x|\leq 1} |\nabla u|^2\, dx \leq \frac{1}{n} \int_{S^n} |\nabla F|^2\, d\xi.$$

This logarithmic Sobolev inequality now gives by a standard argument ([9]) that both the Poisson and heat semigroups are hypercontractive on the n-dimensional sphere. This result was first realized by Weissler ([76]) in one dimension using a direct combinatorial argument and by Janson ([43]) in two dimensions using a weighted convolution of Bernoulli trials. The logarithmic Sobolev inequality for the heat semigroup was first obtained by Mueller and Weissler ([53]) and then by Bakry and Emery ([6]) using "iterated gradients" and the Bochner-Lichnerowicz-Weitzenböck formula.

Theorem 5. *The Poisson semigroup defines a contraction mapping from $L^p(S^n)$ to $L^q(S^n)$ with $1 \leq p \leq q \leq \infty$ and r real if and only if $|r| \leq [(p-1)/(q-1)]^{1/2}$:*

(22)
$$\|P_r F\|_{L^q(S^n)} \leq \|F\|_{L^p(S^n)}.$$

The possibility of proving such operator inequalities was first suggested to the author by E. M. Stein. Though the role of logarithmic Sobolev inequalities has been closely associated with arguments of Segal, Nelson, Federbush, and Gross in mathematical physics, an early application of a logarithmic inequality of this type relating entropy to smoothness occurs in the Calderón-Zygmund theory of singular integrals ([18]).

3 MULTILINEAR CORRELATION INEQUALITIES

The interplay of Fourier analysis, differential geometry, and conformal field theory has marked several interesting directions in contemporary analysis. As one pushes to develop more fully the analysis and geometry of higher-dimensional spaces, both for the classical geometries and for Lie groups, the intrinsic operators that characterize basic geometric questions display greater complexity in terms of both multilinear and tensor structure. The natural resource to exploit here is the symmetry and algebraic invariance of the operators, especially in terms of such groups as the conformal group or the symmetric group. Conformal field theory and statistical mechanics constitute a useful laboratory producing exact model calculations that provide considerable insight for the development of this program.

A characteristic example that seems likely to be very rich in both analytic and geometric consequences is the conformally invariant multilinear fractional integral. This operator arises naturally in two different contexts: 1) as an invariant operator-valued m-point correlation function in conformal field theory (see [26] for example) and 2) in terms of restriction phenomena for the Fourier transform ([23]). In addition, it is an intertwining operator for the Lorentz group, and the operator norms are related to higher-dimensional Selberg integrals. In his thesis, M. Christ showed that an end-point estimate for controlling the restriction of the Fourier transform of an $L^p(\mathbb{R}^n)$ function to a curve in $\mathbb{R}^n$ was guaranteed by a sharp fractional integral estimate. The foundation of this analysis rested on methods developed by Fefferman and Stein ([29]) and Carleson and Sjölin ([21]). A basic tool in the analysis of such estimates is the generalization of the Riesz-Sobolev rearrangement lemma by Brascamp, Lieb, and Luttinger ([17]).

Rearrangement Lemma (Brascamp, Lieb, and Luttinger).

(23)
$$\left| \int_{\mathbb{R}^n \times \cdots \times \mathbb{R}^n} \prod_{k=1}^{N} f_k \left(\sum_{\ell=1}^{m} a_{k\ell} x_\ell \right) dx_1 \ldots dx_m \right|$$

$$\leq \int_{\mathbb{R}^n \times \cdots \times \mathbb{R}^n} \prod_{k=1}^{N} f_k^* \left(\sum_{\ell=1}^{m} a_{k\ell} x_\ell \right) dx_1 \ldots dx_m$$

where $\{a_{k\ell}\}$ are real numbers and $x_k \in \mathbb{R}^n, k = 1, \ldots, m$.

Using the framework established in Section 2, four equivalent theorems are set down here that give the conformally invariant multilinear Hardy-Littlewood-Sobolev fractional integral inequality in different settings.

Theorem 6. *For functions* $f_k \in L^{p_k}(\mathbb{R}^n), k = 1, \ldots, N$ *and* $p_k > 1, \sum \frac{1}{p_k} > 1$
(24)
$$\left| \int \prod_{k=1}^{N} f_k(x_k) \prod_{1 \leq i < j \leq N} |x_i - x_j|^{-\gamma_{ij}} dx_1 \ldots, dx_N \right| \leq C_{p,\gamma,n,N} \prod_{k=1}^{N} \| f_k \|_{L^{p_k}(\mathbb{R}^n)}$$

where $x_k \in \mathbb{R}^n$, *p denotes* $\{p_k\}$, γ *denotes* $\{\gamma_{ij}\}$, $n > \gamma_{ij} = \gamma_{ji} \geq 0$,

$$\sum_{i \neq k} \gamma_{ik} = \frac{2n}{p_k'}, \qquad \sum_{i < j} \gamma_{ij} = n \sum \frac{1}{p_k'} < (N-1)n$$

with p_k *and* p_k' *dual exponents. The best constant* $C_{p,\gamma,n,N}$ *is attained for the extremal functions* $f_k(x) = A(1 + |x|^2)^{-n/p_k}$ *up to conformal automorphism.*

Because this problem is conformally invariant, there is an equivalent inequality on the sphere S^n.

Theorem 7. *For functions* $F_k \in L^{p_k}(S^n)$, $k = 1, \ldots, N$ *and* $p_k > 1$, $\sum \frac{1}{p_k} > 1$

$$(25) \quad \left| \int \prod_{k=1}^N F_k(\xi_k) \prod_{i < j} |\xi_i - \xi_j|^{-\gamma_{ij}} \, d\xi_1 \ldots d\xi_N \right| \leq B_{p,\gamma,n,N} \prod_{k=1}^N \|F_k\|_{L^{p_k}(S^n)}$$

where $\xi_k \in S^n$, $d\xi_k$ *denotes normalized surface measure,* $n > \gamma_{ij} = \gamma_{ji} \geq 0$,

$$\sum_{i \neq k} \gamma_{ik} = \frac{2n}{p_k'}, \qquad \sum_{i < j} \gamma_{ij} < (N-1)n$$

with p_k *and* p_k' *dual exponents. The best constant* $B_{p,\gamma,n,N}$ *is attained only for the extremal functions* $F_k(\xi) = A(1 - \zeta \cdot \xi)^{-n/p_k}$ *where* $\zeta \in \mathbb{R}^{n+1}$ *with* $|\zeta| < 1$.

The two constants are related by the equation
(26)

$$B_{p,\gamma,n,N} = \left[\frac{\Gamma(n)}{(4\pi)^{n/2}\Gamma(n/2)} \right]^{\sum \frac{1}{p_k'}} C_{p,\gamma,n,N} = \int \prod_{i < j} |\xi_i - \xi_j|^{-\gamma_{ij}} \, d\xi_1 \ldots d\xi_N.$$

This formula makes the connection to the Selberg integral and the representation theory of the symmetric group readily apparent. By taking asymptotic limits of inequality (25) one obtains a multilinear version of Nelson's inequality which was first developed in the author's thesis from its equivalence to a generalized sharp Young's inequality. If the γ_{ij} are taken to be constant, then the one-dimensional form of inequality (24) (without sharp constants) was used by M. Christ ([23]) in obtaining optimal estimates for the restriction of the Fourier transform to curves in $\mathbb{R}^N$.

In addition to the flat and spherical geometric pictures for this inequality, the conformal invariance of the problem provides a realization in terms of real hyperbolic geometry. Suppose $\mathbb{H}^n$ denotes the unit two-sheeted hyperboloid in $\mathbb{R}^{n+1}$

$$\mathbb{H}^n = \left\{ P = (p_0, \bar{p}) \in \mathbb{R} \times \mathbb{R}^n : p_0^2 - |\bar{p}|^2 = 1 \right\}$$

where $\mathbb{H}^n_+$ denotes the upper hyperboloid with $p_0 \geq 1$, and $\mathbb{H}^n_-$ denotes the lower hyperboloid with $p_0 \leq -1$. $\mathbb{H}^n$ is then a homogeneous space for the Lorentz group $O(1, n)$. Consider the map from $\mathbb{R}^n - \{|x| = 1\}$ to $\mathbb{H}^n$ given by

$$P = \left(\frac{1 + |x|^2}{1 - |x|^2}, \frac{2x}{1 - |x|^2} \right)$$

$$|x - y|^2 = \frac{|1 - |x|^2| \, |1 - |y|^2|}{2} |pq - 1|$$

where (only in the context of hyperbolic geometry) $pq = p_0 q_0 - \bar{p}\bar{q}$ and

$$dv(p) = 2\delta(1 - p^2) \, dp = 2\delta(1 + \bar{p}^2 - p_0^2) \, dp_0 \, d\bar{p} = 2^n |1 - |x|^2|^{-n} \, dx$$

is an $O(1, n)$ invariant measure on $\mathbb{H}^n$. Then the multilinear Hardy-Littlewood-Sobolev inequality (24) has an equivalent formulation on the homogeneous manifold $\mathbb{H}^n$.

Theorem 8. *For functions* $G_k \in L^{p_k}(\mathbb{H}^n)$, $k = 1, \ldots, N$ *and* $p_k > 1$, $\sum \frac{1}{p_k} > 1$

(27)
$$\left| \int \prod_{k=1}^n G_k(q_k) \prod_{i<j} |q_i q_j - 1|^{-\gamma_{ij}/2} \, dv(q_1) \ldots dv(q_N) \right|$$

$$\leq E_{p,\gamma,n,N} \prod_{k=1}^N \|G\|_{L^{p_k}(\mathbb{H}^n)}$$

where $q_k \in \mathbb{H}^n, n > \gamma_{ij} = \gamma_{ji} \geq 0$

$$\sum_{i \neq k} \gamma_{ik} = \frac{2n}{p_k'}, \qquad \sum_{i<j} \gamma_{ij} < n(N - 1)$$

with p_k *and* p_k' *dual exponents. The best constant* $E_{p,\gamma,n,N}$ *is attained for the extremal functions* $G_k(q) = A[1 + \bar{q}^2]^{-n/(2p_k)}$ *up to conformal automorphism.*

Inequality (27) results from using the stereographic projection map from the plane $\mathbb{R}^n$ to the two-sheeted hyperboloid $\mathbb{H}^n$ described above, making the function change $G_k(q) = |1 - |x|^2|^{n/p_k} f_k(x)$ in (24) and observing that

$$E_{p,\gamma,n,N} = (\sqrt{2})^{n \sum \frac{1}{p_k}} C_{p,\gamma,n,N}.$$

The relation between the parameters γ, n, and p is determined by dilation invariance and the requirement for a conformal structure. To check the upper bound on the size of γ, consider each f_k to be the characteristic function of a ball centered at the origin. Using the Euclidean product structure of the kernel, one could show boundedness for multilinear fractional integrals by induction on dimension and an interpolation argument. A direct proof of Theorem 6 can be given by playing off the conformal invariance which breaks the geodesic symmetry and symmetrization which improves the inequality by the Rearrangement Lemma. In fact, this lemma allows an equivalent formulation as a one-variable problem in terms of the dilation structure. Apply the lemma to (24) and let $y_k = \ln |x_k|$ with $h_k(y_k) = \exp(ny_k/p_k)f_k(y_k)$.

Theorem 9. *For functions* $h_k \in L^{p_k}(\mathbb{R}), k = 1, \ldots, N$ *and* $p_k > 1, \sum \frac{1}{p_k} > 1$

$$\left| \int \prod_{k=1}^{N} h_k(y_k) \left\{ \prod_{i<j} [2\cosh(y_i - y_j) - 2 + |\eta_i - \eta_j|^2]^{-\gamma_{ij}/2} d\eta_1 \ldots d\eta_N \right\} \right.$$

(28) $\left. dy_1 \ldots dy_N \right|$

$$\leq A_{p,\gamma,n,N} \prod_{k=1}^{N} \|h_k\|_{L^{p_k}(\mathbb{R})}.$$

where $y_k \in \mathbb{R}, \eta_k \in S^{n-1}, n > \gamma_{ij} = \gamma_{ji} \geq 0,$

$$\sum_{i \neq k} \gamma_{ik} = \frac{2n}{p_k'}, \qquad \sum_{i<j} \gamma_{ij} < n(N-1).$$

The constants in Theorems 6 and 9 are related by

$$A_{p,\gamma,n,N} = \left[\frac{\Gamma(n/2)}{2\pi^{n/2}} \right]^{\sum \frac{1}{p_k'}} C_{p,\gamma,n,N}.$$

Fractional integral estimates always have equivalent formulations in terms of weak Lorentz spaces when rearrangement arguments are accessible, so Theorem 6 would, in fact, give sharp norms for such functional inequalities.

The interplay here between conformal invariance and tensor structure provides considerable insight into higher-order differential operators. Moreover, these theorems allow one to study conformally invariant operators acting between non-dual L^p spaces. In addition, algebraic invariance associated to the symmetric group is central to understanding the geometric aspects of this work. The recognition that restriction phenomena would be closely associated to the geometric analysis of sharp constants arose in work with A. Carbery.

4 MOSER-TRUDINGER INEQUALITY

One of the most interesting aspects of the Hardy-Littlewood-Sobolev theorem is its relation to limiting Sobolev inequalities and geometric variational problems for conformal deformation. In view of the role played by sharp Sobolev constants in the arguments of Aubin, Obata, and Schoen for the Yamabe problem, this development is quite natural. But it was the impetus of quantum string theory and the focus on computing functional determinants that emphasized the critical inequality of Moser-Trudinger-Onofri

$$(29) \qquad \ln \int_{S^2} e^F \, d\xi \leq \int_{S^2} F \, d\xi + \frac{1}{4} \int_{S^2} |\nabla F|^2 \, d\xi.$$

Here equality is attained only for functions of the form

$$F(\xi) = -2 \ln |1 - \zeta \cdot \xi| + C, \qquad |\zeta| < 1.$$

There are, in fact, two sharp constants here and two different geometric variational problems for conformal deformation: to characterize gaussian curvature and to extremize the functional determinant of the Laplacian. The first constant is the factor one-fourth which was obtained by Moser using geometric symmetrization. Motivated by Polyakov's string theory models, Onofri determined the sharp normalization constant using conformal invariance. The effort to fully understand this inequality led to several incisive results, including: 1) a geometric derivation of (29) and its analogue for the circle by Osgood, Phillips, and Sarnak ([59]) in the context of computing functional determinants on Riemann surfaces; 2) the Carleson-Chang theorem for the disc ([20]); 3) the explicit calculations of zeta-function determinants on four-manifolds by Branson and Ørsted ([15]); and 4) the study of the linearized Adams inequality on the four-sphere by Branson, Chang, and Yang ([14]).

Not only is inequality (29) and the corresponding Carleson-Chang inequality for non-negative functions with zero-boundary value on the unit disc in $\mathbb{R}^2$

$$(30) \quad \ln \left[\frac{1}{\pi} \int_{|x| \leq 1} e^{2f} \, dx \right] + \left[\frac{1}{\pi} \int_{|x| \leq 1} e^{2f} \, dx \right]^{-1} \leq 1 + \frac{1}{4\pi} \int_{|x| \leq 1} |\nabla f|^2 \, dx$$

a consequence of the sharp Hardy-Littlewood-Sobolev inequality, but one can find by the same techniques a family of higher-dimensional Moser-Trudinger inequalities of considerable geometric importance. A striking result is that under conformal deformation with fixed volume the zeta-function determinant of the conformal Laplacian is extremized by the standard metric up to conformal automorphism (see [10],[14],[15]). The essential feature of the analysis is to combine conformal invariance and geometric symmetrization, while exploiting the paradigm that "fractional integration controls Sobolev estimates."

Theorem 10. *For a real-valued function F defined on the sphere S^n with an expansion in spherical harmonics $F = \sum_{k=0}^{\infty} Y_k$ an exponential-class a priori* *inequality of Moser-Trudinger type holds*

$$(31) \quad \ln \int_{S^n} e^{F(\xi)} \, d\xi \leq \int_{S^n} F(\xi) \, d\xi + \frac{1}{2n} \sum_{k=1}^{\infty} c_k(n) \int_{S^n} |Y_k|^2 \, d\xi$$

$$c_k(n) = \frac{\Gamma(n+k)}{\Gamma(n)\Gamma(k)} = \frac{1}{\Gamma(n)} k(k+1) \cdots (k+n-1).$$

Using the spectral representation for the Laplacian, this inequality can be written in terms of a positive-definite conformally invariant operator $P_n(-\Delta)$ which is a differential operator for even n and a pseudo-differential operator of boundary type for odd n.

$$(32) \quad \ln \int_{S^n} e^{F(\xi)} \, d\xi \leq \int_{S^n} F(\xi) \, d\xi + \frac{1}{2n!} \int_{S^n} F(P_n F) \, d\xi$$

$$P_n(-\Delta) = \prod_{\ell=0}^{\frac{n-2}{2}} [-\Delta + \ell(n-1-\ell)] \, n \text{ even}$$

$$= \left[-\Delta + \left(\frac{n-1}{2} \right)^2 \right]^{1/2} \prod_{\ell=0}^{\frac{n-3}{2}} [-\Delta + \ell(n-1-\ell)] \, n \text{ odd}.$$

This inequality is invariant under the conformal transformation

$$F(\xi) \longrightarrow F(\tau\xi) + \ln J(\tau, \xi)$$

with τ an element of the conformal group of S^n and J the modulus of the corresponding Jacobian determinant. Equality in the above inequalities is attained only for functions of the form

$$F_\zeta(\xi) = -n \ln |1 - \zeta \cdot \xi| + C \quad \xi \in S^n$$

where $\zeta \in \mathbf{B}^{n+1} = \{x \in \mathbb{R}^{n+1} : |x| < 1\}$. Here $d\xi$ corresponds to normalized surface measure on S^n.

There are three parts to the proof of the theorem: 1) the sharp inequality, 2) conformal invariance of the functional inequality, and 3) equality being attained only for conformal factors. The first two parts are readily treated on the "differential operator side" using the differentiation argument discussed previously. Determining conditions for equality is more natural from the "fractional integral side" using a method that goes back to Sobolev for n-dimensional integrals. Invert the "differential operator" by a "fractional integral" and analyze an equivalent sharp inequality in the dual space setting. In fact, one could do the entire problem in terms of fractional integrals, but the applications to geometric analysis suggest that one should understand as much of the problem as possible on the differential operator side. For the exponential class the dual space setting is the Orlicz class $L \ln L \cap L^1$ since the exponential and the logarithm are complementary Young's functions. The dual space result is contained in the following theorem.

Theorem 11. *Suppose F and G are non-negative functions on the sphere S^n with $\int F \, d\xi = \int G \, d\xi = 1$. Then*

$$(33) \qquad -n \int_{S^n \times S^n} F(\xi) \ln |\xi - \eta|^2 G(n) \, d\xi \, d\eta$$

$$\leq -n \int_{S^n} \ln |\xi - \eta|^2 \, d\xi + \int_{S^n} F \ln F \, d\xi + \int_{S^n} G \ln G \, d\xi$$

and equality is attained only for functions of the form

$$F = G, \qquad F_\zeta(\xi) = A|1 - \zeta \cdot \xi|^{-n}, \qquad |\zeta| < 1.$$

$$-n \int_{S^n} \ln |\xi - \eta|^2 \, d\xi = -n[\ln 4 + \psi(n/2) - \psi(n)] \quad \psi = (\ln \Gamma)'.$$

For $F = G = 1 + \sum_{k=1}^{\infty} Z_k$ inequality (4) is equivalent to

$$(34) \qquad n \sum_{k=1}^{\infty} \frac{\Gamma(n)\Gamma(k)}{\Gamma(n+k)} \int_{S^n} |Z_k|^2 \, d\xi \leq 2 \int_{S^n} G \ln G \, d\xi$$

which by duality inverts inequality (3).

These two theorems are directly equivalent by duality, but either can be obtained by a differentiation argument ([10]) applied to the Hardy-Littlewood-Sobolev inequality (12) transferred to the sphere. They are the "end-point information" at (L^1, L^∞) for this inequality, just as the sharp logarithmic Sobolev inequality (20) for the sphere is the "end-point information" at L^2. Conformal invariance suggests that the Green's function for the operator will play a critical role. This is already implicit in the work of Carleson and Chang, though the equivalence of inequalities (29) and (30) is not so apparent. Especially in conformal field theory, the Green's function is often at the forefront of the analysis.[2] The combination of these ideas about the Green's function and nonlinear differential equations from analysis, differential geometry, quantum field theory, and statistical mechanics suggests that much remains to be discovered about the implications of these results, especially in terms of geometric structure.

---

[2] After the author proved these theorems, two preprints appeared that touched on similar themes. Using a mean-field calculation in statistical mechanics, M. Kiessling recognized the relation between the logarithmic potential and the Moser-Trudinger inequality in two dimensions. Subsequently E. Carlen and M. Loss developed a proof of the conditions for equality in much the same spirit as the argument given here.

Both Theorems 10 and 11 are consequences of Theorem 2 using end-point differentiation arguments at $p = 1$. But a quick calculation of inequality (31) is obtained by inverting inequality (14) for $q = p' \geq 2$ and taking an appropriate limit for $q \to \infty$. For $G = \sum Y_k$

$$
\text{(35)} \qquad \left[\|G\|_{L^q(S^n)}\right]^2 \leq \sum_{k=0}^{\infty} \frac{1}{\gamma_k} \int_{S^n} |Y_k|^2 \, d\xi = \int_{S^n} G^*(TG) \, d\xi
$$

$$
= \sum_{k=0}^{\infty} \frac{\Gamma(n/q)\Gamma(n/q' + k)}{\Gamma(n/q')\Gamma(n/q + k)} \int_{S^n} |Y_k|^2 \, d\xi
$$

where T is now a positive-definite self-adjoint operator. For a bounded real-valued function F, set $G = 1 + (1/q)F$ and apply the limit $q \to \infty$ to

$$
\int_{S^n} \left|1 + \frac{1}{q} F\right|^q d\xi
$$

$$
\leq \left[1 + \frac{2}{q} \int_{S^n} F \, d\xi + \frac{1}{q^2} \sum_{k=0}^{\infty} \frac{\Gamma(n/q)\Gamma(n/q' + k)}{\Gamma(n/q')\Gamma(n/q + k)} \int_{S^n} |Y_k|^2 \, d\xi\right]^{q/2}
$$

which immediately gives inequality (31). To rephrase this step as an end-point differentiation argument, set $\theta = 1/q$ and rewrite this inequality as

$$
\text{(36)} \qquad \left[\|1 + \theta F\|_{L^{1/\theta}(S^n)}\right]^2 \leq 1 + 2\theta \int_{S^n} F \, d\xi + \theta^2 \int_{S^n} F(TF) \, d\xi.
$$

There is no loss of generality here since all inequalities are a priori and (36) is completely equivalent to the original fractional integral inequality (13) on the sphere. The operator T depends on θ but θT is stable at $\theta = 0$, namely $\lim_{\theta \to 0} \theta T = \frac{1}{n!} P_n$ on functions orthogonal to constants where P_n is the operator occurring in Theorem 10. Now inequality (36) is an identity at $\theta = 0$ for all bounded F so that this functional inequality has a positive derivative at $\theta = 0$, namely

$$
\text{(37)} \qquad 2 \int_{S^n} F \, d\xi + \frac{1}{n!} \int_{S^n} F(P_n F) \, d\xi - 2 \ln \int_{S^n} e^F \, d\xi \geq 0.
$$

But this is simply inequality (32).

The condition on boundedness of F is easily removed by taking limits. Conformal invariance can also be obtained by either a limiting argument from the conformal invariance of (14), directly from the form of inequality (33), or by an explicit analysis of the operators $P_n(-\Delta)$. The fourth-order conformally invariant operator

$$
-\Delta(-\Delta + 2)
$$

was observed by Paneitz ([60]) to be important in the interplay between the conformal and gauge groups for Maxwell's equations.

To see the conformal invariance of inequality (37) directly from the calculation of the above limit process, observe that inequality (36) is invariant under the conformal transformation

$$\left[1 + \theta F(\xi)\right] \rightarrow \left[1 + \theta F(\tau \xi)\right] J^{\theta}(\tau, \xi)$$

where τ is a conformal automorphism of S^n and $J(\tau, \xi)$ is the modulus of the corresponding Jacobian determinant. In the limit $\theta \rightarrow 0$, this is equivalent to replacing F in inequality (36) by $\widetilde{F}(\xi) = F(\tau \xi) + \ln J(\tau, \xi)$ so that

$$2 \int_{S^n} F \, d\xi + \frac{1}{n!} \int_{S^n} F(P_n F) \, d\xi - 2 \ln \int_{S^n} e^F \, d\xi$$

$$= 2 \int_{S^n} \widetilde{F} \, d\xi + \frac{1}{n!} \int_{S^n} \widetilde{F}(P_n \widetilde{F}) \, d\xi - 2 \ln \int_{S^n} e^{\widetilde{F}} \, d\xi$$

and in fact the form

$$2 \int_{S^n} F \, d\xi + \frac{1}{n!} \int_{S^n} F(P_n F) \, d\xi$$

is invariant under the transformation $F \rightarrow \widetilde{F}$.

Since constant functions are extremal for inequality (37), conformal invariance shows that the functions

$$F_{\zeta}(\xi) = -n \ln |1 - \zeta \cdot \xi| + C$$

for $\xi \in S^n$ and $\zeta \in \mathbb{R}^{n+1}, |\zeta| < 1$ are also extremal. It remains to show that these conformal factors are the only extremals. The classical method to obtain extremal functions is to apply symmetrization arguments and, for the case of a differential operator, to invert the operator and study a fractional integral in a dual setting. The idea goes back to Sobolev for the problem simply of showing boundedness and is exactly the step used to show that the multiplier inequality (35) is an equality only for functions of the form $G_{\zeta}(\xi) = A|1 - \zeta \cdot \xi|^{-n/q}$ with ζ a point in the open unit ball in $\mathbb{R}^{n+1}$. Thus to determine the extremal functions in Theorem 10, it suffices to determine the extremal functions for Theorem 11.

The entropy inequality (33) can be directly obtained using an end-point differentiation argument at $p = 1$ from the fractional integral inequality on the sphere (13). For non-negative functions F and G having integral one

$$-n \int_{S^n} \ln |\xi - \eta|^2 \, d\xi + \int_{S^n} F \ln F \, d\xi + \int_{S^n} G \ln G \, d\xi$$

$$\geq -n \int_{S^n} F(\xi) \ln |\xi - \eta|^2 G(\eta) \, d\xi \, d\eta.$$

The logarithmic kernel is a monotone strictly decreasing function of the distance so one can apply the Symmetrization Lemma from section 2. The argument is usually made for positive kernels, but the sign of the kernel plays no role as long as it is a monotone decreasing function of the distance between two points and the other functions are nonnegative. This symmetrization lemma shows that an extremal function for the inequalities in Theorem 11 must be radial decreasing with respect to distance on the sphere from some pole. On every domain which is conformally equivalent to the sphere S^n, one can find an equivalent inequality written in terms of a strictly decreasing integral kernel. Moreover, one can find an equivalent one-dimensional inequality on $\mathbb{R}_+$ which captures the "dilation character" of the inequality. The same symmetrization argument can be applied there ($\mathbb{R}_+ \cong \mathbb{R}$ with dilation on $\mathbb{R}_+$ going over to the translation on $\mathbb{R}$) forcing the extremal function to be symmetric with respect to distance on $\mathbb{R}_+$ which can then be traced back to its equivalent formulation on S^n where this inversion symmetry means that the extremal function must be symmetric with respect to a hemisphere. Since constants are the only measurable function which are both radial decreasing away from a pole and symmetric with respect to a hemisphere, the only extremal functions for inequality (33) are the constants and the corresponding Jacobian factors for conformal transformations

$$G_\zeta(\xi) = A|1 - \zeta \cdot \xi|^{-n}$$

where ζ is a point in the interior of the unit ball in $\mathbb{R}^{n+1}$ and A is a positive normalization constant. Since (33) is equivalent to (31), this means that the only extremals for the n-dimensional Moser-Trudinger inequality are of the form

$$F_\zeta(\xi) = -n \ln |1 - \zeta \cdot \xi| + C$$

with $|\zeta| < 1$. This is the same method that was used above in section 2. The point is that symmetrization on $\mathbb{R}_+$ picks out a specific extremal which "breaks the S^n symmetry" unless the equivalent function on S^n is constant almost everywhere. A second approach applies an observation of Carlen and Loss ([19]) that in a generic sense "conformal transformations break symmetry" and allows the problem to be done on a single domain. Suppose one has an extremal function for inequality (33) which is radial decreasing away from a pole. Perform a conformal transformation

$$G(\xi) \to G(\tau\xi)J(\tau, \xi) = \widetilde{G}(\xi)$$

where τ is not simply a rotation; by the conformal invariance of the inequality, this will give a new extremal function which cannot be radial decreasing away from some pole unless the original extremal function was a conformal factor. But then the symmetrization argument would force some increase in the integral

$$-n \int_{S^n \times S^n} \widetilde{G}(\xi) \ln |\xi - \eta|^2 \widetilde{G}(\eta) \, d\xi \, d\eta$$

so, in fact, the only extremals are conformal factors.

The dual setting a priori inequalities used above are collected in the following theorem.

Theorem 12 (Entropy Inequalities). *Suppose f and g are non-negative $L \ln L$ functions on $\mathbb{R}^n$ with $\int_{\mathbb{R}^n} f \, dx = \int_{\mathbb{R}^n} g \, dx = 1$. Then*

(38)
$$-n \int_{\mathbb{R}^n \times \mathbb{R}^n} f(x) \ln |x - y|^2 g(y) \, dx \, dy \leq C_n + \int_{\mathbb{R}^n} f \ln f \, dx + \int_{\mathbb{R}^n} g \ln g \, dx$$

$$C_n = n \ln \pi + n[\psi(n) - \psi(n/2)] - 2 \ln[\Gamma(n)/\Gamma(n/2)]$$

where $\psi(z) = \frac{d}{dz} \ln \Gamma(z)$ and extremals are given up to conformal automorphism by

$$f(x) = g(x) = A(1 + |x|^2)^{-n}.$$

Suppose u and v are non-negative $L \ln L$ functions on $\mathbb{R}_+$ with $\int_0^\infty u(t) \frac{dt}{t} = \int_0^\infty v(t) \frac{dt}{t} = 1$ and

$$\varphi(t) = -n \int_{S^{n-1}} \ln[t + 1/t - 2\xi \cdot \eta] \, d\xi.$$

Then

(39)
$$\int_{\mathbb{R}_+ \times \mathbb{R}_+} u(t) \varphi(t/s) v(s) \frac{dt}{t} \frac{ds}{s} \leq D_n + \int_{\mathbb{R}_+} u \ln u \frac{dt}{t} + \int_{\mathbb{R}_+} v \ln v \frac{dt}{t}$$

$$D_n = \left\{ 4\pi^n / [\Gamma(n/2)]^2 \right\} C_n$$

and extremals are given up to translation on $\mathbb{R}_+$ by

$$u(t) = v(t) = A(t + 1/t)^{-n}.$$

Since the kernel here has variable sign, one must check that the conditions on the functions guarantee that the inequalities are well-defined (that is, there is no cancellation of infinities). Inequalities (38) and (39) follow by straightforward differentiation arguments from the Hardy-Littlewood-Sobolev inequalities for smooth functions with compact support on the domains. Using the observation that the kernel is integrable over any compact neighborhood of the identity, then the inequality for smooth functions insures by a limiting argument that the integral of the positive part of the integrand over the product space must be finite. Hence, inequalities (28) and (29) are well-defined under the condition that the functions should be both in L^1 and $L \ln L$. It is useful to emphasize that the functional inequalities in Theorems 11 and 12 are directly equivalent for smooth functions with compact support using conformal transformations, but the class $L \ln L \cap L^1$ is not preserved by these transformations. However, the class of extremal functions is guaranteed to be preserved.

As explained at the beginning of Section 2, fractional integral inequalities correspond to extending Young's inequality for convolution of functions on $\mathbb{R}^n$ to include a weak Lorentz class. Inequality (38) has a corresponding interpretation here. Consider the class of measurable functions K on $\mathbb{R}^n$ such that

$$\exp[K/2] \in L_{1,\infty}(\mathbb{R}^n)$$

which is equivalent to

$$K^*(x) \leq -n \ln |x|^2 + C$$

where K^* denotes the equimeasurable radial decreasing rearrangement of K. The smallest value of C defines a size $\widetilde{C}_K$ for this class. Since the sets $\{x : K(x) > \alpha\}$ have finite measure, the following theorem is obtained using the Riesz-Sobolev Lemma and Theorem 12.

Theorem 13. *Let K satisfy the property $\exp[K/2] \in L_{1,\infty}(\mathbb{R}^n)$ and f and g be non-negative $L \ln L$ functions on $\mathbb{R}^n$ with integral equal to one. Then*

$$(40) \qquad \int_{\mathbb{R}^n} f(x)(K * g)x)\, dx \leq \widetilde{C}_K + C_n + \int_{\mathbb{R}^n} f \ln f \, dx + \int_{\mathbb{R}^n} g \ln g \, dx$$

$$C_n = n \ln \pi + n\big[\psi(n) - \psi(n/2)\big] - 2 \ln\big[\Gamma(n)/\Gamma(n/2)\big].$$

In addition, the Hardy-Littlewood-Sobolev inequality provides an extension of the classical sharp Sobolev inequalities in the context of dimensions one and two and more generally an interpolating result for the operators $P_n(-\Delta)$ of Theorem 10 in higher dimensions. These inequalities are not conformally invariant but do contain the higher-dimensional Moser-Trudinger inequality as a limiting case.

Theorem 14.

$$(41) \qquad \left(\int_{S^n} |F(\xi)|^q \, d\xi\right)^{2/q} \leq \frac{q-2}{n} \int_{S^n} |\nabla F|^2 \, d\xi + \int_{S^n} |F|^2 \, d\xi$$

for $2 \leq q < \infty$ if $n = 1$ or 2 and $2 \leq q \leq \frac{2n}{n-2}$ if $n \geq 3$;

$$(42) \qquad \left(\int_{S^n} |F(\xi)|^q \, d\xi\right)^{2/q} \leq \frac{q-2}{c_n} \int_{|x| \leq 1} |\nabla u|^2 \, dx + \int_{S^n} |F|^2 \, d\xi$$

for $2 \leq q < \infty$ if $n = 1$ and $2 \leq q \leq \frac{2n}{n-1}$ if $n \geq 2$ where u is the harmonic extension of F to the interior of the unit ball in $\mathbb{R}^{n+1}$ and $c_n = 2\pi^{(n+1)/2}/\Gamma((n+1)/2)$.

Theorem 15. *On S^n for $2 \le q < \infty$*

$$(43) \qquad \left(\int_{S^n} |F|^q \, d\xi \right)^{2/q} \le \frac{q-2}{n!} \int_{S^n} F\left[P_n(-\Delta) F \right] d\xi + \int_{S^n} |F|^2 \, d\xi.$$

The proofs of these two theorems depend on convexity properties of the spectral data for the Hardy-Littlewood-Sobolev fractional integral on the sphere ([10]). An interesting geometric P.D.E. proof of the classical part of (42) was given by Escobar ([27],[28]) using the character of an Einstein metric.

Interest in this problem goes back to some remarks made by J. Moser at a lunch organized by Narasimhan at Chicago in the late 1970's and a talk given by S.-Y.A. Chang at Arcata in the summer of 1985. But the real impetus for seeing the possibility of these theorems came from the observation that a special case of (41),

$$(44) \qquad \left(\int_{S^2} |F|^4 \, d\xi \right)^{1/2} \le \int_{S^2} |\nabla F|^2 \, d\xi + \int_{S^2} |F|^2 \, d\xi,$$

was a consequence of $SL(2, \mathbb{R})$ invariance for sharp Sobolev inequalities on the Heisenberg group. An unusual feature of these problems is that the algebraic invariance may not always be evident as with the Carleson-Chang inequality (30) or the Heisenberg inequality (44). But, in fact, this indicates a deep connection with geometric structure. The sharp form of the Carleson-Chang inequality and its equivalence with the Moser-Trudinger-Onofri inequality on the two-sphere is part of recent work by S.-Y.A. Chang and the author. It is now evident that there is a strong structural relation determined by conformal invariance and the Green's function which connects the Moser-Trudinger inequalities of Theorem 10, the work of Carleson and Chang ([20]), the Adams inequalities ([1]), and the entropy estimates of Theorems 11 and 12. This point will be discussed in detail in a forthcoming paper.

5 ZETA-FUNCTION DETERMINANTS

The interaction of ideas from differential geometry, representation theory, and string theory has focused attention on conformally invariant operators, their relation to conformal deformation and Einstein metrics, and especially the calculation of functional determinants. An elegant example coming out of Polyakov's string theory ([57],[58],[64]) was the computation that under conformal deformation of metric on the two-sphere $ds^2 = e^F \, ds_0^2$

$$\det(-\widetilde{\Delta}) / \det(-\Delta) = e^{(-1/3)S(F)}$$

$$S(F) = \frac{1}{4} \int_{S^2} |\nabla F|^2 \, d\xi + \int_{S^2} F \, d\xi - \ln \int_{S^2} e^F \, d\xi$$

so that the sharp Moser-Trudinger inequality (29) implies that the determinant of the Laplacian on S^2 is maximized by the standard metric. Here, $\widetilde{\Delta}$ denotes the Laplace-Beltrami operator for the deformed metric. But a classical result already occurs in this context as Widom pointed out that the one-dimensional Moser-Trudinger inequality (31) is part of Szegö's limit theorem for determinants of Toeplitz operators ([75],[35],[79]). Branson and Ørsted ([15]) calculated explicit formulas for functional determinants in four dimensions, including the conformal Laplacian and the square of the Dirac operator. Branson, Chang, and Yang ([14]) recently studied the general character of such operators under conformal deformation on a four-manifold and the relation to the linearized Adams exponential-class Sobolev inequality for second-order operators on S^4. Using the Adams inequality, they were able to show the existence of extremals for the variational problem. S.-Y.A. Chang and the author together recognized that the sharp S^4 inequality from Theorem 10 would determine the extremals for the determinant of the conformal Laplacian. Under conformal deformation of metric with fixed volume, this determinant is extremized by the standard metric up to conformal automorphism.

Theorem (Branson and Ørsted). *Consider a conformal deformation $ds^2 = e^{F/2} ds_0^2$ on the sphere S^4. Let A denote either the conformal Laplacian or the square of the Dirac operator while $\widetilde{A}$ denotes the same operator under conformal deformation. Then*

(45)
$$\det \widetilde{A}/\det A = e^{-\beta_1 S_1(F) - \beta_2 S_2(F)}$$

$$S_1(F) = \frac{1}{48} \int_{S^4} F\big[-\Delta(-\Delta + 2)F\big]\, d\xi + \int_{S^4} F\, d\xi - \ln \int_{S^4} e^F\, d\xi$$

$$S_2(F) = \int_{S^4} e^{-F/2}(\Delta e^{F/4})^2\, d\xi - \frac{1}{4} \int_{S^4} |\nabla F|^2\, d\xi.$$

Relating the values given in Branson and Ørsted ([15]) to the normalization used here, $\beta_1 = \beta_2 = -180$ for the conformal Laplacian and $7\beta_1 = 22\beta_2 = 77/180$ for the square of the Dirac operator. Note that the notation here differs from that in ([14]) and ([15]). $S_1(F)$ is non-negative by Theorem 1 and $S_2(F)$ is non-negative by the following lemma. Both functionals are conformally invariant and vanish only for functions of the form

$$F_\zeta(\xi) = -4 \ln |1 - \zeta \cdot \xi| + C, \quad |\zeta| < 1.$$

Lemma 1.

(46)
$$\int_{S^4} e^{-2w}(\Delta e^w)^2\, d\xi \geq 4 \int_{S^4} |\nabla w|^2\, d\xi.$$

This lemma was proved by the author using "iterated gradients" and the Bochner-Lichnerowicz-Weitzenböck formula and by P. Yang using the sharp Sobolev inequality. Yang's method clearly demonstrates the conformal invariance of the functional and extends easily to include higher-order differential operators related to the Hardy-Littlewood-Sobolev inequality.

Lemma 2. *Suppose T is a positive-definite linear operator with $T1 = 0$ and for some $q > 2$*

$$\left[\|f\|_{L^q(S^n)}\right]^2 \le \frac{1}{c} \int_{S^n} f(Tf)\, d\xi + \int_{S^n} |f|^2\, d\xi;$$

then for $f \ge 0$

(47)
$$c \le \left\| \frac{1}{f} Tf + c \right\|_{S^{q/(q-2)}(S^n)}.$$

From the Hardy-Littlewood-Sobolev inequality for $f = \sum Y_k$

$$\left[\|f\|_{L^q(S^n)}\right]^2 \le \sum_{k=0}^{\infty} \frac{\Gamma(\frac{n}{q})\Gamma(\frac{n}{q'} + k)}{\Gamma(\frac{n}{q'})\Gamma(\frac{n}{q} + k)} \int |Y_k|^2\, d\xi$$

together with the lemma, one can obtain a family of conformally invariant inequalities for higher-order differential operators.

Theorem 16. *On S^n let $\alpha = n/2 - n/q$ for $q > 2$ and define*

$$B = \left[-\Delta + \left(\frac{n-1}{2} \right)^2 \right]^{1/2} \qquad D_{2\alpha} = \frac{\Gamma(B + \frac{1}{2} + \alpha)}{\Gamma(B + \frac{1}{2} - \alpha)};$$

then

(48)
$$\left[\|F\|_{L^q(S^n)}\right]^2 \le \frac{\Gamma(\frac{n}{2} - \alpha)}{\Gamma(\frac{n}{2} + \alpha)} \int_{S^n} F(D_{2\alpha}F)\, d\xi$$

and for $f > 0$

(49)
$$\frac{\Gamma(\frac{n}{2} + \alpha)}{\Gamma(\frac{n}{2} - \alpha)} \le \left\| \frac{1}{f} D_{2\alpha} f \right\|_{L^{(n/2\alpha)}(S^n)}.$$

Equality holds only for functions of the form

$$f(\xi) = A|1 - \zeta \cdot \xi|^{\alpha - \frac{n}{2}}, \qquad |\zeta| < 1.$$

This inequality seems likely to be most useful in the cases where α is a positive integer and $n/2\alpha$ is an even integer ($D_{2\alpha}$ is a differential operator). The more explicit realization of the Hardy-Littlewood-Sobolev inequality in terms of intertwining operators for $SO(n + 1, 1)$ extends a suggestion of T. Branson.[3]

6 LIE GROUPS

The principal theme of the program of geometric analysis outlined here is to understand the patterns of symmetry that exist for a problem. As the central questions in analysis involve increasingly and more fully higher-order differential operators, multilinear tensor analysis, non-homogeneous dilations, and surface singularities, "hard analysis" on Lie groups and symmetric spaces becomes more important. Techniques developed in this program have already made contributions in terms of Lie groups that arise naturally in real and complex analysis: namely $SL(2, \mathbb{R})$, the Lorentz groups, the Heisenberg group, and quaternionic groups.

To illustrate this general approach, some results about $SL(2, \mathbb{R})$ are described. Three intrinsic features of $SL(2, \mathbb{R})$ are its natural action on the upper half-plane, the exponential growth of geodesic balls and the role of $SL(2, \mathbb{R})/SO(2)$ as the "dilation structure" of the Heisenberg group $\mathbb{H}_n$. The hyperbolic structure extending the natural $SL(2, \mathbb{R})$ action on a domain is at the heart of the inequalities described in the previous section. From representation theory Kunze and Stein recognized a surprising extension of Young's inequality for $SL(2, \mathbb{R})$. This phenomenon reflected the exponential character of Haar measure on this group.

Kunze-Stein Phenomenon. *For* $G = SL(2, \mathbb{R})$ *and* $1 \le p < 2$

(50) $$\|f * g\|_{L^2(G)} \le A_p \|f\|_{L^p(G)} \|g\|_{L^2(G)}.$$

Stein extended this theorem to the complex classical groups and conjectured that it should hold for semi-simple Lie groups with finite center. In addition, for the case where f is bi-invariant, he gave a simple proof using Harish-Chandra's estimates for the decay of spherical functions. Cowling later gave a complete proof of the conjecture using representation theory. But symmetrization and the rearrangement lemma of Brascamp, Lieb, and Luttinger provide an immediate reduction for $SL(2, \mathbb{R})$ to the case where f is bi-invariant.

Theorem 17. *For* $G = SL(2, \mathbb{R})$

(51) $$\left| \int_G h(x)(f * g)(x) \, dx \right| \le \int_G h_\#(x)(f_\# * g_\#)(x) \, dx.$$

---

[3] A research announcement describing the results of Sections 4 and 5 appears in "Moser-Trudinger Inequality in Higher Dimensions," *Int. Math. Res. Not.* (1991), 83–91.

*# denotes an equimeasurable rearrangement of the absolute value of a function
on G such that for a Cartan decomposition $x = (p, k)$, the rearranged function
is radial decreasing in the p variable. Here $p \in M = SL(2, \mathbb{R})/SO(2)$ and
$k \in K = SO(2)$.*

This theorem allows the Kunze-Stein inequality (50) to be reduced to functions
radial in the p variable and constant in the k variable, so, in fact, to bi-invariant
functions. A similar proof works for $SL(2, \mathbb{C})$, the Lorentz groups, and any
appropriate group with abelian nilpotent part.

An interesting feature of analysis on the Heisenberg group and complex analysis
in several variables is the lack of reflection symmetry. In part, this is simply a consequence of the two-parameter dilation structure given by $M = SL(2, \mathbb{R})/SO(2)$.
Using the sharp Hardy-Littlewood-Sobolev inequality and analysis of fractional
integration on the Heisenberg group, a one-parameter family of Sobolev inequalities is obtained for M. To set notation, let $w = x + iy \in M$ for $y > 0$;
$dm = y^{-2} \, dx \, dy$ is left-invariant measure on M, the gradient is $D = y\nabla$, and
the invariant distance function is

$$d(w, w') = \frac{|w - w'|}{2\sqrt{yy'}}.$$

Theorem 18. *For $1 \leq p < 2$, $1/p + 1/p' = 1$ and $\lambda = 2(n + 1)/p'$ for
integral n*

$$(52) \left| \int_{M \times M} F(w)G(w')\psi_\lambda[d(w, w')] \, dm(w) \, dm(w') \right| \leq A_p \|F\|_{L^p(M)} \|G\|_{L^p(M)}$$

$$\psi_\lambda(u) = (1 + u^2)^{-\lambda/2} F\left(\lambda/2, \lambda/2; n + 1; \frac{1}{1 + u^2}\right).$$

*Extremal functions are given by $[1 + d^2(w, i)]^{-(n+1)/p}$ up to conformal automor-
phisms of M.*

For $q = 2(n + 1)/n$

(53)

$$\|F\|_{L^q(M)} \leq \frac{1}{\sqrt{\pi n}} \left[\frac{4\pi}{n}\right]^{1/q} \left\{\int_M |DF|^2 \, dm + \frac{n}{4}(n - 2) \int_M |F|^2 \, dm\right\}^{1/2}.$$

*For $n > 1$, extremal functions are given by $[1 + d^2(w, i)]^{-n/2}$ up to conformal
automorphisms.*

Further results for the Heisenberg group and nilpotent groups of type-H, $\mathbb{P}^n(\mathbb{C})$
and $SL(2, \mathbb{R})$ will be discussed in forthcoming work. Analysis of convolution
inequalities and complex symmetry structure is a very rich subject in Fourier

Analysis. The foundations for this program were set down by Hardy and Littlewood in the 1920's and in higher dimensions by Sobolev in the 1930's. Modern contributions have been made by many individuals mentioned in this article. The intersection of ideas from analysis, geometry, probability, and physics has made this subject a useful part of contemporary mathematics.

Remarks. I have a special feeling for Eli Stein. His warmth and enthusiasm— that sparkle in his eyes—lit up the hallways of old Fine Hall the day I met him. His lectures on Fourier Analysis that fall were filled with insight and marked the direction of my mathematical work. Both as a friend and as a mathematician, he has given me much inspiration.

Acknowledgments. I want to thank the following individuals for their generous comments: Tom Branson, Tony Carbery, Alice Chang, José Escobar, Charlie Fefferman, David Jerison, Carlos Kenig, Bent Ørsted, Aleksandr Pełczyński, D. H. Phong, Peter Sarnak, and Eli Stein. Portions of this work were completed while visiting the University of Chicago.

University of Texas, Austin

REFERENCES

1. D. R. Adams. "A sharp inequality of J. Moser for higher order derivatives." *Ann. of Math.* **128** (1988), 385–398.
2. O. Alvarez. "Theory of strings with boundary." *Nucl. Phys. B* **216** (1983), 125–184.
3. T. Aubin. "Meilleures constantes dans le théorèm d'inclusion de Sobolev et un théorèm de Fredholm non linéaire pour la transformation conforme de la courbure scalaire." *J. Funct. Anal.* **32** (1979), 148–174.
4. T. Aubin. "Best constants in the Sobolev imbedding theorem: the Yamabe problem." In *Seminar on Differential Geometry* edited by S.-T. Yau, Princeton University Press, 1982.
5. A. Baernstein and B. A. Taylor. "Spherical rearrangements, subharmonic functions, and ∗-functions in *n*-space." *Duke Math. J.* **43** (1976), 245–268.
6. D. Bakry and M. Emery. "Hypercontractivité de semi-groupes de diffusion." *C. R. Acad. Sci. Paris* **299** (1984), 775–778.
7. M. Bander and C. Itzykson. "Group theory and the hydrogen atom." *Rev. Mod. Phys.* **38** (1966), 330–358.
8. W. Beckner. "Inequalities in Fourier analysis." *Ann. of Math.* **102** (1975), 159–182.
9. W. Beckner. "A generalized Poincaré inequality for Gaussian measures." *Proc. Amer. Math. Soc.* **105** (1989), 397–400.
10. W. Beckner. "Moser-Trudinger inequality in higher dimensions." *Int. Math. Res. Not.* (1991), 83–91.

11. W. Beckner. "Sobolev inequalities, the Poisson semigroup and analysis on the sphere S^n." *Proc. Nat. Acad. Sci.* **89** (1992), 4816–4819.

12. W. Beckner. "Sharp Sobolev inequalities on the sphere and the Moser-Trudinger inequality." *Ann. of Math.* **138** (1993), 213–242.

13. A. Beurling. *Etudes sur un problème de majoration.* Thesis, Uppsala, 1933.

14. T. Branson, S.-Y. A. Chang, and P. Yang. "Estimates and extremals for zeta function determinants on four-manifolds.." *Comm. Math. Phys.* **149** (1992), 241–262.

15. T. Branson and B. Ørsted. "Explicit functional determinants in four dimensions." *Proc. Amer. Math. Soc.* **113** (1991), 671–684.

16. H. J. Brascamp and E. H. Lieb. "Best constants in Young's inequality, its converse, and its generalization to more than three functions." *J. Funct. Anal.* **20** (1976), 151–173.

17. H. J. Brascamp, E. H. Lieb, and J. M. Luttinger. "A general rearrangement inequality for multiple integrals." *J. Funct. Anal.* **17** (1974), 227–237.

18. A. P. Calderón M. Weiss, and A. Zygmund. "On the existence of singular integrals." *Proc. Symp. Pure Math.* **10** (1967), 56–73.

19. E. A. Carlen and M. Loss. "Extremals of functionals with competing symmetries." *J. Funct. Anal.* **88** (1990), 437–456.

20. L. Carleson and S.-Y. A. Chang. "On the existence of an extremal function for an inequality of J. Moser." *Bull. Sci. Math.* **110** (1986), 113–127.

21. L. Carleson and P. Sjölin. "Oscillatory integrals and a multiplier problem for the disc." *Studia Math.* **44** (1972), 287–299.

22. S.-Y. A. Chang and D. E. Marshall. "On a sharp inequality concerning the Dirichlet integral." *Amer. J. Math.* **107** (1985), 1015–1033.

23. M. Christ. "On the restriction of the Fourier transform to curves: endpoint results and the degenerate case." *Trans. Amer. Math. Soc.* **287** (1985), 223–238.

24. E. D'Hoker and D. H. Phong. "On determinants of Laplacians on Riemann surfaces." *Comm. Math. Phys.* **104** (1986), 537–545.

25. E. D'Hoker and D. H. Phong. "The geometry of string perturbation theory." *Rev. Mod. Phys.* **60** (1988), 917–1065.

26. Vl. S. Dotsenko. "Lectures on conformal field theory." *Advanced Studies in Pure Mathematics* **16** (1988) "Conformal Field Theory and Solvable Lattice Models," 123–170.

27. J. F. Escobar. "Sharp constant in a Sobolev trace inequality." *Indiana Math. J.* **37** (1988), 687–698.

28. J. F. Escobar. "Uniqueness theorems on conformal deformation of metrics, Sobolev inequalities and an eigenvalue estimate." *Comm. Pure Appl. Math.* **43** (1990), 857–883.

29. C. Fefferman. "Inequalities for strongly singular convolution operators." *Acta Math.* **124** (1970), 9–36.

30. C. Fefferman. "Monge-Ampere equations, the Bergman kernel and geometry of pseudoconvex domains." *Ann. of Math.* **103** (1976), 395–416.

31. C. Fefferman. "Parabolic invariant theory in complex analysis." *Adv. Math.* **31** (1979), 131–262.

32. C. Fefferman and C. R. Graham. "Conformal invariants" in "Élie Cartan et les Mathématiques d'aujourd'hui." *Astérisque* (1985), 95–116.

33. G. B. Folland and E. M. Stein. "Estimates for the $\bar{\partial}_b$ complex and analysis on the Heisenberg group." *Comm. Pure Appl. Math.* **27** (1974), 429–522.

34. G. B. Folland and E. M. Stein. *Hardy Spaces on Homogeneous Groups*. Princeton University Press, 1982.
35. U. Grenander and G. Szegö. *Toeplitz Forms and Their Applications*. University of California Press, 1958.
36. L. Gross. "Logarithmic Sobolev inequalities." *Amer. J. Math.* **97** (1975), 1061–1083.
37. G. H. Hardy, J. E. Littlewood, and G. Pólya. *Inequalities*. Cambridge University Press, 1934.
38. S. Helgason. *Differential Geometry, Lie Groups and Symmetric Spaces*. Academic Press, 1978.
39. S. Helgason. *Groups and Geometric Analysis*. Academic Press, 1984.
40. L. Hörmander. "Oscillatory integrals and multipliers on FL^p." *Arkiv. Math.* **11** (1973), 1–11.
41. C. Itzykson. "Remarks on boson commutation rules." *Comm. Math. Phys.* **4** (1967), 92–122.
42. H. P. Jakobsen and M. Vergne. "Wave and Dirac operators and representations of the conformal group." *J. Funct. Anal.* **24** (1977), 52–106.
43. S. Janson. "On hypercontractivity for multipliers on orthogonal polynomials." *Arkiv. Math.* **21** (1983), 97–110.
44. D. Jerison and J. M. Lee. "A subelliptic nonlinear eigenvalue problem on CR manifolds." *Contemp. Math.* **27** (1984), 57–63.
45. D. Jerison and J. M. Lee. "Extremals for the Sobolev inequality on the Heisenberg group and the CR Yamabe problem." *J. Amer. Math. Soc.* **1** (1988), 1–13.
46. S. Kobayashi and K. Nomizu. *Foundations of Differential Geometry*. John Wiley, 1969.
47. E. H. Lieb. "Existence and uniqueness of the minimizing solution of Choquard's nonlinear equation." *Stud. Appl. Math.* **57** (1977), 93–105.
48. E. H. Lieb. "Sharp constants in the Hardy-Littlewood-Sobolev and related inequalities." *Ann. of Math.* **118** (1983), 349–374.
49. P. L. Lions. "The concentration-compactness principle in the calculus of variations II. The limit case." *Rev. Mat. Iber.* **1** (1985), 45–121.
50. H. P. McKean. "Geometry of differential space." *Ann. Prob.* **1** (1973), 197–206.
51. J. Moser. "A sharp form of an inequality by N. Trudinger." *Indiana Math. J.* **20** (1971), 1077–1092.
52. J. Moser. On a nonlinear problem in differential geometry. In *Dynamical Systems*, edited by M. M. Peixoto. Academic Press, 1973.
53. C. E. Mueller and F. B. Weissler. "Hypercontractivity for the heat semigroup for ultraspherical polynomials and on the *n*-sphere." *J. Funct. Anal.* **48** (1982), 252–283.
54. A. Nagel and E. M. Stein. *Lectures on Pseudo-differential Operators*. Princeton University Press, 1979.
55. E. Nelson. "The free Markoff field." *J. Funct. Anal.* **12** (1973), 211–227.
56. M. Obata. "The conjectures on conformal transformations of Riemannian manifolds." *J. Diff. Geom.* **6** (1971), 247–258.
57. E. Onofri. "On the positivity of the effective action in a theory of random surfaces." *Comm. Math. Phys.* **86** (1982), 321–326.
58. E. Onofri and M. Virasoro. "On a formulation of Polyakov's string theory with regular classical solutions." *Nuc. Phys. B* **201** (1982), 159–175.
59. B. Osgood, R. Phillips, and P. Sarnak. "Extremals of determinants of Laplacians." *J. Funct. Anal.* **80** (1988), 148–211.

60. S. Paneitz. "A quartic conformally covariant differential operator for arbitrary pseudo-Riemannian manifolds." Unpublished preprint, 1983.

61. D. H. Phong. "Complex geometry and string theory." In *Proceedings of the Geometry Festival, July 1990*, edited by R. Greene and S.-T. Yau.

62. J. Polchinski. "Evaluation of the one loop string path integral." *Comm. Math. Phys.* **104** (1986), 37–47.

63. G. Pólya and G. Szegö. *Isoperimetric Inequalities in Mathematical Physics*. Princeton University Press, 1951.

64. A. M. Polyakov. "Quantum geometry of bosonic strings." *Phys. Lett.* B **103** (1981), 207–210.

65. R. Schoen. "Conformal deformation of a Riemannian metric to constant scalar curvature." *J. Diff. Geom.* **20** (1984), 479–495.

66. J. Schwinger. "Coulomb Green's function." *J. Math. Phys.* **5** (1964), 1606–1608.

67. A. Selberg. "Harmonic analysis and discontinuous groups in weakly symmetric Riemannian spaces with applications to Dirichlet series." *J. Indian Math. Soc.* **20** (1956), 47–87.

68. S. L. Sobolev. "On a theorem of functional analysis." *Mat. Sb. (N.S.)* **4** No. 46 (1938), 471–497.

69. E. M. Stein. "Analytic continuation of group representations." *Adv. Math.* **4** (1970), 172–207.

70. E. M. Stein. *Singular Integrals and Differentiability Properties of Functions*. Princeton University Press, 1970.

71. E. M. Stein. *Some problems in harmonic analysis suggested by symmetric spaces and semi-simple groups*. Actes Congrès Intern. Math. (Nice, 1970), Tome 1, 173–189.

72. E. M. Stein. *Topics in Harmonic Analysis Related to the Littlewood-Paley Theory*. Princeton University Press, 1970.

73. E. M. Stein. *Boundary Behavior of Holomorphic Functions of Several Complex Variables*. Princeton University Press, 1972.

74. E. M. Stein and G. Weiss. *Introduction to Fourier Analysis on Euclidean Spaces*. Princeton University Press, 1971.

75. G. Szegö. *On certain hermitian forms associated with the Fourier series of a positive function*. Festskrift Marcel Riesz (Lund, 1952), 228–238.

76. G. Talenti. "Best constant in Sobolev inequality." *Ann. Mat. Pura Appl.* **110** (1976), 353–372.

77. N. Th. Varopoulos. "Sobolev inequalities on Lie groups and symmetric spaces." *J. Funct. Anal.* **86** (1989), 19–40.

78. F. B. Weissler. "Logarithmic Sobolev inequalities and hypercontractive estimates on the circle." *J. Funct. Anal.* **37** (1980), 218–234.

79. H. Widom. "On an inequality of Osgood, Phillips and Sarnak." *Proc. Amer. Math. Soc.* **102** (1988), 773–774.

80. A. Zygmund. *Trigonometric Series*. 2d ed. Cambridge University Press, 1959.

3

Representing Measures for Holomorphic Functions on Type 2 Wedges

Al Boggess and Alexander Nagel

1 INTRODUCTION

A striking fact about holomorphic function theory in several variables is that, under suitable convexity hypotheses on a domain $\Omega \subset C^n$ with $n > 1$, if $z_0 \in \Omega$ is "close" to the boundary $\partial\Omega$, then there are representing measures for z_0 whose support on $\partial\Omega$ is contained in a set which is concentrated near the orthogonal projection of z_0 onto the boundary. This is false for more general domains in C^n and false for harmonic functions on domains in R^n.

For a simple example of this phenomenon, let $\Omega = \{(z, w); Re(w) > |z|^2\}$. We will denote the boundary of Ω by $\Sigma = \{Re(w) = |z|^2\}$ which can be identified with the Heisenberg group. We wish to represent the value of a holomorphic function F on Ω at the point $(0, r) \in \Omega$, $r > 0$, by integrating F against a suitable measure on Σ. To do this, we let $\phi \in C_0^\infty(C)$ be a radial function with support in the unit disc and whose integral over C is one. Let

$$\phi_r(w) = \frac{4}{r^2} \phi \left(\frac{2(w - r)}{r} \right).$$

The function ϕ_r has support in the disc centered at r with radius $\frac{r}{2}$ and the integral of ϕ_r over C is one. The mean value property for holomorphic functions shows that

$$F(0, r) = \int_{w \in C} F(0, w)\phi_r(w)dxdy$$

where we have written $w = x + \iota y$. The support of ϕ_r is contained in the square $\{\frac{r}{2} \le x \le \frac{3r}{2}, |y| \le \frac{r}{2}\}$. Therefore, $\sqrt{x}$ is well defined on the support of ϕ_r. Using the mean value property of F in the first variable, we obtain

$$F(0, r) = \frac{1}{2\pi} \int_{w \in C} \int_0^{2\pi} F(\sqrt{x}e^{i\theta}, x + \iota y)\phi_r(x + \iota y)\, d\theta dxdy.$$

Now, we change variables by letting $x = t^2$ and then we let $z = te^{i\theta}$. We obtain

$$F(0, r) = \frac{1}{\pi} \int \int_{z \in C} \int_{y \in R} F(z, |z|^2 + \iota y)\phi_r(|z|^2 + \iota y)d\lambda(z)dy$$

where $d\lambda(z)$ denotes Lebesgue measure on the complex plane. The map $(z, y) \mapsto (z, |z|^2 + \iota y)$ for $z \in C$ and $y \in R$ is a parameterization for Σ and $d\lambda(z)dy$ is comparable to surface measure on Σ, denoted by $d\sigma$. Therefore, we obtain

(1) $$F(0, r) = \int_{\Sigma} F(\zeta)K_r(\zeta)d\sigma(\zeta)$$

where $K_r(\cdot)$ is a smooth function. Note that the support of $K_r(\zeta)$ is contained in a ball centered at the origin of radius $C\sqrt{r}$ (where C is a uniform constant). Thus, K_r is our desired local representing measure for holomorphic functions on the Heisenberg group (near the origin).

The same argument with inequalities replacing equalities implies that if u is a non-negative plurisubharmonic function on Ω which is continuous up to Σ, then

(2) $$u(0, r) \le \int_{\Sigma} u(\zeta)K_r(\zeta)d\sigma(\zeta).$$

The key idea in the above analysis is to slice Ω with analytic discs whose boundary is contained in Σ (in the above case, the analytic disc is the map $\zeta \mapsto (z\zeta, |z|^2 + iy)$ for ζ in the unit disc in C). Then we average our holomorphic F (or subaverage our plurisubharmonic u) over the boundary of the discs. A similar slicing argument can be used for a more general convex or strictly pseudoconvex domain and a representation analogous to (1) and an estimate analogous to (2) can be derived. The case of weakly pseudoconvex domains of finite type presents additional difficulties since the boundary cannot always be locally convexified, but here, too, one can obtain local representing measures by imbedding suitable analytic discs (see [BDN]).

In this paper, we generalize these ideas to the case of "wedge domains" with an "edge" given by a CR submanifold $M \subset C^n$ of real codimension greater than 1. Our goal is to find local representing measures on the edge for points in the wedge near the edge. We restrict our attention to the case where the edge of the wedge has type 2, which is the analogue of the case of a domain with an edge given by a strictly pseudoconvex boundary (as discussed above). We have been greatly influenced in our work by E. M. Stein's seminal observation that a strictly pseudoconvex boundary can be modeled at each point by a nilpotent Lie group, the Heisenberg group, and that the boundary behavior of holomorphic functions in strictly pseudoconvex domains is intimately connected with the approximating group structure on the boundary. In this paper, we carefully analyze a "model case" for our wedge domains where the edge is a nilpotent Lie group of step 2 (analogous to the Heisenberg group). This is done in section 4 of this paper. In section 2, we state our theorem on representing measures for more general wedge domains, and in section 3, we state an application to the boundary behavior of H^p functions defined on wedge domains.

The details of the results given in sections 2 and 3 will appear in a later paper. The results in section 2 for general wedges with type 2 edges are obtained by a three-stage process which is again inspired by the work of Stein and his collaborators (see for example [FS] and [RS]): (i) we pass from the original object of study to a "free" object by adding appropriate variables; (ii) we solve the problem on the freed object by approximating it suitably by the model case; (iii) we return to the original object by integrating out the extra variables. The results of section 3 are related to recent work of Rosay [R].

2 THE STATEMENT OF THE GENERAL THEOREM

In order to state our theorem, we first recall some basic facts about CR manifolds. For more details, see [B]. Let M be a C^∞ submanifold of C^n of real codimension d. For $p \in M$, let T_pM denote the real tangent space to M at p. Then the maximal complex subspace of T_pM is $T_p^C = T_pM \cap J(T_pM)$, where J is the complex structure map given by multiplication by $\iota = \sqrt{-1}$. M is called a *CR submanifold*, with *CR dimension* m if for all $p \in M$, $dim_C T_p^C M = m$. M is called *generic* if for all $p \in M$, $T_pM + J(T_pM) = C^n$. If M is a generic CR submanifold of real codimension d and CR dimension m, it follows that $n = m + d$.

For $p \in M$, let Y_pM denote the orthogonal complement to $T_p^C M$ in T_pM, and let N_pM be the orthogonal complement of T_pM in T_pC^n (i.e., N_pM is the normal space of M at p). We have

$$T_pM = T_p^C \oplus Y_pM$$

$$T_pC^n = T_pM \oplus N_pM$$

where all the direct sum decompositions are orthogonal. Let $\pi_p : T_pC^n \mapsto N_pM$ be the orthogonal projection. Let $H_p^{1,0}(M)$ denote the subspace of the complexified tangent space to M at p spanned by tangent vectors of type $(1, 0)$. If $Z \in H^{1,0}(M)$, then the vector field $\frac{1}{2\iota}[Z, \overline{Z}]$ is a real tangent vector field.

Definition 2.1. *The Levi form is the well defined quadratic mapping* $\mathcal{L}_p$: $H_p^{1,0}(M) \mapsto N_p M$ *given by*

$$\mathcal{L}_p(Z) = -\pi_p(J(\frac{1}{2i}[Z, \overline{Z}]_p)).$$

where $Z \in H^{1,0}(M)$ *is any vector field extension of* Z_p. *The convex hull of the image of* $\mathcal{L}_p$ *is a cone in* $N_p M$ *and is denoted by* Γ_p.

Definition 2.2. *Let* $M \subset C^n$ *be a generic CR submanifold. Then* M *is of type 2 at* $p \in M$ *if and only if* Γ_p *has nonempty interior in* $N_p M$.

This definition is equivalent to the condition that the real and imaginary parts of the tangential Cauchy-Riemann vector fields, together with their commutators, span the real tangent space to M at p.

For any subset $K \subset N_p M$ and any $\epsilon > 0$, let $K_\epsilon = \{z \in N_p M \mid z \in K$, and $|z| < \epsilon\}$. If γ_1 and γ_2 are two cones in $N_p M$, we say that γ_1 *is smaller than* γ_2 and write $\gamma_1 < \gamma_2$ if $\overline{\gamma_1} \cap S$ is a compact subset of the interior of γ_2 where S is the unit sphere in $N_p M$. We now describe the type of wedge domains that we will study.

Definition 2.3. *Let* $M \subset C^n$ *be a generic CR submanifold of type 2. An open set* $\Omega \subset C^n$ *is a domain with edge* M *if:*
(1) $M \subset \overline{\Omega}$;
(2) *For each* $p \in M$ *and for each cone* $\gamma < \Gamma_p$ *there exists an open set* $\omega \subset M$ *containing* p *and an* $\epsilon > 0$ *such that*

$$\omega + \gamma_\epsilon \subset \Omega.$$

Property 2) roughly states that near a point $p \in M$, Ω locally contains translates of M in directions strictly interior to the cone Γ_p.

The set $\omega + \gamma_\epsilon$ is parameterized in a natural way by $\omega \times \gamma_\epsilon$. After shrinking ϵ if necessary, we can consider the orthogonal projection Π of $\omega + \gamma_\epsilon$ onto M so that for $z \in \omega + \gamma_\epsilon$,

$$z - \Pi(z) \in N_{\Pi(z)} M.$$

We write

$$r(z) = |z - \Pi(z)|$$

where the absolute value denotes the length of a vector in $N_{\Pi(z)} M$.

We shall need to use the nonisotropic pseudometric and corresponding nonisotropic balls on M induced by the ambient complex structure on C^n. We only summarize the construction, which can be carried out for any CR submanifold M

of finite type (see [NSW] for more details). Suppose $L_1 = X_1 + \iota Y_1, \ldots, L_m = X_m + \iota Y_m$ is a basis for $H^{1,0}(M)$ on an open set $\omega \subset M$.

Definition 2.4. *For $p \in \omega$ and $\delta > 0$, let*

$$B(p, \delta) = \left\{ \exp\left[\sum_{j=1}^{m} \alpha_j X_j + \beta_j Y_j + \gamma T \right](p); \quad \alpha_j, \beta_j, \gamma \in R \right.$$

$$\left. with \; |\alpha_j|, |\beta_j| < \delta, \; |\gamma| < \delta^2 \right\},$$

where T is a fixed vector field transverse to $L_1, \ldots, L_m$.

There also exists a pseudo-metric $D : M \times M \mapsto R$ so that

$$B(p, \delta) = \{ \zeta \in M; D(p, \zeta) < \delta \}.$$

These balls have Euclidean dimension δ in the m complex tangent space directions of M at p and so D behaves roughly Euclidean in these directions. In the d totally real tangent directions, these balls have Euclidean dimension roughly δ^2 (since M has type 2) and so D behaves roughly like the square root of the Euclidean distance in the totally real directions.

We can now state our main result on the existence of a local integral representation formula for holomorphic and plurisubharmonic functions in domains with type 2 edges.

Theorem 1. *Let $U \subset C^n$ be an open set and let $M \subset U$ be a generic CR submanifold of type 2. Let Ω be a domain with edge M. Let $p \in M$, let $\gamma < \Gamma_p$, and let $L_1, \ldots, L_m$ be a basis for the space $H^{1,0}(M)$ near p. Then there exist a neighborhood $\omega \subset M$ of p, a constant $\epsilon > 0$, a constant $C < \infty$, and a C^∞ function*

$$K : \{ \omega + \gamma_\epsilon \} \times M \mapsto [0, \infty)$$

with the following properties:
(1) For $z \in \omega + \gamma_\epsilon$ fixed, the function $K_z(\zeta) = K(z, \zeta)$ has compact support in $B(\Pi(z), C\sqrt{r(z)})$.
(2) For every noncommuting polynomial $P(L, \overline{L})$ of degree $k \geq 0$ in the vector fields $L_1, \ldots, L_m, \overline{L}_1, \ldots, \overline{L}_m$, there is a constant C_P so that

$$|P(L, \overline{L})(K_z)(\zeta)| \leq C_P r(z)^{-k/2} |B(\Pi(z), \sqrt{r(z)})|^{-1}.$$

(3) *If* F *is a function continuous on the closure of the set* Ω *and holomorphic on* Ω, *then for* $z \in \omega + \gamma$,

$$F(z) = \int_M K(z, \zeta) F(\zeta) \, d\sigma(\zeta)$$

where $d\sigma$ *is the surface area measure on* M.

(4) *If* u *is a non-negative function which is continuous on the closure of the set* Ω *and plurisubharmonic on* Ω, *then for* $z \in \omega + \gamma$,

$$u(z) \leq \int_M K(z, \zeta) u(\zeta) \, d\sigma(\zeta)$$

where $d\sigma$ *is the surface area measure on* M.

It is easy to see that the kernel for the Heisenberg group constructed in the introduction satisfies the properties listed in the above theorem. The function $(z, y) \mapsto \phi_r(|z|^2 + iy)$ appearing in that kernel has support in the rectangle $\{|z| \leq C\sqrt{r}, |y| \leq Cr\}$ for a suitable uniform constant C. This set is comparable to the projection of the nonisotropic ball $B(0, C\sqrt{r})$ onto the tangent space of Σ at the origin. In addition, from the definition of ϕ_r we obtain the estimate

$$|\phi_r(w)| \leq \frac{C}{r^2} \approx \frac{C}{|B(0, \sqrt{r})|}$$

where $| \cdot |$ denotes Lebesgue (surface) measure.

In section 4, we shall prove theorem 1 for a model case for wedge domains, which is a generalization of the Heisenberg group.

3 BOUNDARY BEHAVIOR OF H^p FUNCTIONS

As an application, Theorem 1 can be used to study the boundary behavior of H^p functions defined on a wedge relative to a type 2 edge. We begin with the definition of this class of functions.

Definition 3.1. *Let* $\Omega \subset C^n$ *be a domain with edge* M. *Then for* $0 < p < \infty$, $H^p_{loc}(\Omega, M)$ *is the space of holomorphic functions on* Ω *such that for every* $q \in M$ *and every* $\gamma < \Gamma_q$ *there exist* $\epsilon > 0$ *and a neighborhood* $\omega \subset M$ *of* q *so that*

$$\sup_{z \in \gamma_\epsilon} \int_{\zeta \in \omega} |F(\zeta + z)|^p \, d\sigma(\zeta) < +\infty.$$

If $p = \infty$ *we require that*

$$\sup_{z \in \gamma_\epsilon + \omega} |F(z)| < +\infty.$$

Next we define certain nontangential approach regions in domains with edges.

Definition 3.2. *Let $\Omega \subset C^n$ be a domain with edge M. Assume that M is a generic CR submanifold of type 2. Let $q_0 \in M$; let $\gamma < \Gamma_{q_0}$; and let $\alpha > 0$. By the definition of a domain with an edge, there is a neighborhood ω of q_0 in M and an $\epsilon > 0$ so that $\omega + \gamma_\epsilon \subset \Omega$. Let $q \in \omega$. Then define*

$$\mathcal{A}(\gamma, \alpha, q) = \{z \in \omega + \gamma_\epsilon; \, D(\Pi(z), q)^2 \leq \alpha \, \text{dist}(z, M)\}.$$

Recall that Π is the projection from Ω to M which is defined at least near M, and $D(p, q)$ is the nonisotropic distance between p and q. Also *dist* refers to the Euclidean distance. Since D is roughly Euclidean in the complex tangential directions, the approach regions $\mathcal{A}$ allow quadratic approach to M along these directions. Since D^2 is roughly Euclidean in the totally real directions, these regions allow only nontangential approach in these directions. Thus the regions $\mathcal{A}(\gamma, \alpha, q)$ are the analogues of admissible approach regions for strictly pseudoconvex domains.

It is easy to see that if Ω is a domain with edge M, and if $F \in H^p_{loc}(\Omega, M)$, then F has polynomial growth locally near M. To be precise, given $p > 0$, and given a point $q \in M$ and a compact cone $\gamma < \Gamma_q$, there is a real number s, a neighborhood $\omega \subset M$ of q and an $\epsilon > 0$ so that for every $F \in H^p_{loc}(\Omega, M)$ there exists a constant C so that

$$|F(z)| \leq C \, \text{dist}(z, M)^{-s}$$

for $z \in \omega + \gamma_\epsilon$. This polynomial growth in turn implies that every $F \in H^p_{loc}(\Omega, M)$ has a distributional limit F^* along M in the sense that if $\varphi \in C_0^\infty(M)$ then

$$\lim_{\epsilon \mapsto 0} \int_M F(\zeta + \epsilon z)\varphi(\zeta) \, d\sigma(\zeta) = F^*(\varphi).$$

Our main results deal with the existence of pointwise and dominated limits, rather than distributional limits, and we obtain a partial characterization of the boundary value distributions.

Theorem 2. *Suppose M is a type 2 CR submanifold of C^n and suppose Ω is a domain with edge M. Let $f \in H^p_{loc}(\Omega, M), 0 < p \leq \infty$. For almost all $q \in M$ the following holds: Given $\gamma < \Gamma_q$ and $\alpha > 0$, then*

$$\lim_{\substack{z \mapsto q \\ z \in \mathcal{A}(\gamma, \alpha, q)}} f(z) \quad \text{exists.}$$

This limit defines an element of $L^p_{loc}(M)$. Conversely, if $1 \leq p \leq \infty$ and $f \in L^p_{loc}(M)$ is a CR distribution, then f has a unique holomorphic extension $F \in H^p_{loc}(\Omega, M)$ for some open set Ω with edge M.

The key ingredient of the proof of this theorem is the following maximal function estimate for plurisubharmonic functions which is derived from the estimates given in Theorem 1 in section 2. Let u be non-negative and plurisubharmonic on the wedge Ω and continuous up to M; then

$$\sup_{z \in A(\gamma, \alpha, q)} u(z) \leq C_{\gamma, \alpha} \mathcal{M}(u)(q)$$

where $\mathcal{M}$ denotes the maximal function associated to our family of nonisotropic balls and $C_{\gamma, \alpha}$ is a constant which depends only on the cone γ and the aperture α. This maximal function estimate (with $u = |F|^{p/2}$) is used in the proof of Theorem 2 in much the same way as the analogous maximal function estimates are used in the classical analysis of the boundary values of H^p functions on domains in the complex plane.

4 THE MODEL CASE

In this section, we derive the reproducing kernel for Theorem 1 of section 2 for the following model case. We let $M_m \cong C^{m^2}$ denote the complex vector space of $m \times m$ complex matrices, and let H_m denote the real vector subspace of $m \times m$ Hermitian matrices. For any $r \times s$ complex matrix A, let A^* denote the conjugate transpose $s \times r$ matrix. For $W \in M_m$, set

$$\mathcal{R}(W) = \frac{1}{2}(W + W^*);$$

$$\mathcal{I}(W) = \frac{1}{2\iota}(W - W^*).$$

Then for $W \in M_m$, $\mathcal{R}(W)$ and $\mathcal{I}(W)$ are Hermitian; $W = \mathcal{R}(W) + \iota\mathcal{I}(W)$; and $M_m = H_m \oplus \iota H_m$. This exhibits H_m as a maximal totally real subspace of M_m.

We shall view elements of C^m as $m \times 1$ complex matrices, and so if

$$z = \begin{bmatrix} z_1 \\ \vdots \\ z_m \end{bmatrix} \quad \text{then} \quad z^* = [\bar{z}_1, \ldots, \bar{z}_m].$$

Define a quadratic form $Q_m : C^m \mapsto H_m$ given by

$$Q_m(z) = zz^* = \begin{bmatrix} z_1\bar{z}_1 & z_1\bar{z}_2 & \cdots & z_1\bar{z}_m \\ z_2\bar{z}_1 & z_2\bar{z}_2 & \cdots & z_2\bar{z}_m \\ \vdots & \vdots & \ddots & \vdots \\ z_m\bar{z}_1 & z_m\bar{z}_2 & \cdots & z_m\bar{z}_m \end{bmatrix}.$$

Definition 4.1. *Define* $\rho : C^m \times M_m \mapsto H_m$ *by setting*

$$\rho(z, Z) = \mathcal{R}(Z) - Q_m(z).$$

Then set

$$\Sigma_m = \{(z; Z) \in C^m \times M_m; \mathcal{R}(Z) = zz^*\}$$
$$= \{(z; Z) \in C^m \times M_m; \rho(z; Z) = 0\}$$

and

$$\Omega_m = \{(z; Z) \in C^m \times M_m; \mathcal{R}(Z) > zz^*\}$$
$$= \{(z; Z) \in C^m \times M_m; \rho(z; Z) > 0\}$$

Here, $\rho(z; Z) > 0$ means that the matrix $\rho(z; Z)$ is positive definite.

It is easy to check that Σ_m is a generic CR submanifold of $C^m \times M_m \cong C^{m+m^2}$ of type 2, and that Ω_m is a domain with edge Σ_m. We often identify Σ_m with $C^m \times H_m$ via the correspondence

$$C^m \times H_m \ni (z, Y) \leftrightarrow (z; zz^* + \iota Y) \in \Sigma_m.$$

A typical point in Ω_m is of the form $(z; Z)$ where $Z = zz^* + X + \iota Y$ with $X, Y \in H_m$, and $X > 0$. There is a natural projection $\Pi : \Omega_m \mapsto \Sigma_m$ given by

$$\Pi\big((z; zz^* + X + \iota Y)\big) = (z; zz^* + \iota Y).$$

We are interested in the existence of analytic discs in Ω_m with boundary in Σ_m. Every analytic disc in $C^m \times M_m$ is a continuous map $A = (Z; W) : \overline{D} \mapsto C^m \times M_m$ which is holomorphic on D. We write

$$Z(\zeta) = \begin{bmatrix} Z_1(\zeta) \\ \vdots \\ Z_m(\zeta) \end{bmatrix},$$

where each $Z_j(\cdot)$ is a (scalar) holomorphic function. $W(\cdot)$ is an $m \times m$ matrix-valued holomorphic function. Such an analytic disc maps the boundary of the unit disc D into Σ_m if and only if

$$\mathcal{R}(W(e^{\iota\theta})) = Z(e^{\iota\theta})Z(e^{\iota\theta})^*$$

for $0 \leq \theta \leq 2\pi$.

We shall introduce the notation $A_0(\zeta)$ for the special analytic disc

$$A_0(\zeta) = (Z_0(\zeta); W_0(\zeta))$$

where

$$Z_0(\zeta) = \begin{bmatrix} \zeta \\ \zeta^2 \\ \vdots \\ \zeta^m \end{bmatrix},$$

and

$$W_0(\zeta) = \begin{bmatrix} 1 & 0 & 0 & \cdots & 0 \\ 2\zeta & 1 & 0 & \cdots & 0 \\ 2\zeta^2 & 2\zeta & 1 & \cdots & 0 \\ \vdots & \vdots & \vdots & \ddots & \vdots \\ 2\zeta^{m-1} & 2\zeta^{m-2} & 2\zeta^{m-3} & \cdots & 1 \end{bmatrix}.$$

Note that

$$\mathcal{R}(W_0(\zeta)) = \begin{bmatrix} 1 & \overline{\zeta} & \overline{\zeta}^2 & \cdots & \overline{\zeta}^{m-1} \\ \zeta & 1 & \overline{\zeta} & \cdots & \overline{\zeta}^{m-2} \\ \zeta^2 & \zeta & 1 & \cdots & \overline{\zeta}^{m-3} \\ \vdots & \vdots & \vdots & \ddots & \vdots \\ \zeta^{m-1} & \zeta^{m-2} & \zeta^{m-3} & \cdots & 1 \end{bmatrix}.$$

Also,

$$Q_m(Z_0(\zeta)) = \begin{bmatrix} \zeta\overline{\zeta} & \zeta\overline{\zeta}^2 & \zeta\overline{\zeta}^3 & \cdots & \zeta\overline{\zeta}^m \\ \zeta^2\overline{\zeta} & \zeta^2\overline{\zeta}^2 & \zeta^2\overline{\zeta}^3 & \cdots & \zeta^2\overline{\zeta}^m \\ \zeta^3\overline{\zeta} & \zeta^3\overline{\zeta}^2 & \zeta^3\overline{\zeta}^3 & \cdots & \zeta^3\overline{\zeta}^m \\ \vdots & \vdots & \vdots & \ddots & \vdots \\ \zeta^m\overline{\zeta} & \zeta^m\overline{\zeta}^2 & \zeta^m\overline{\zeta}^3 & \cdots & \zeta^m\overline{\zeta}^m \end{bmatrix}.$$

Thus when $|\zeta| = 1$,

$$\mathcal{R}(W_0(\zeta)) = Q_m(Z_0(\zeta)) = Z_0(\zeta)Z_0(\zeta)^*$$

and so A_0 maps the boundary of the unit disc to Σ_m. If $\xi = (\xi_1, \ldots, \xi_m) \in C^m$, then $\big(\mathcal{R}(W_0(\zeta))\xi, \xi\big)$ is a harmonic function of ζ, while $\big(Q_m(Z_0(\zeta))\xi, \xi\big)$ $= |\sum_{j=1}^m \zeta^j \xi_j|^2$ is a subharmonic function of ζ. It is easy to check that

$((\mathcal{R}(W_0(0)) - Q_m(Z_0(0)))\xi, \xi) = |\xi|^2$ and hence it follows from the minimum principle for superharmonic functions that $\mathcal{R}(W_0(\zeta)) - Q_m(Z(\zeta)) > 0$ for $|\zeta| < 1$. Therefore, the analytic disc A_0 maps the open unit disc to Ω_m.

We now construct a local integral representation formula for Ω_m. Let F be holomorphic on Ω_m, and continuous on $\overline{\Omega_m}$. Let I denote the $m \times m$ identity matrix, so that $(0; I) \in \Omega_m$. Our first goal is to construct a representing measure for the point $(0; I) \in \Omega_m$ with compact support and C^∞ density; i.e., we wish to find a C^∞ function K_0 with compact support, such that

$$F((0; I)) = \int_{\Sigma_m} F(\zeta) K_0(\zeta) \, d\lambda(\zeta)$$

where $d\lambda$ is Lebesgue measure on $\Sigma_m = C^m \times H_m$.

Since F is holomorphic on the subset $\{(0; W) \in C^m \times M_m; \mathcal{R}(W) > 0\}$, we can use the mean value property to conclude that

$$F(0; I) = \int_{M_m} F(0; I + w)\varphi(w) \, dw,$$

where φ is any radial function with compact support in a small neighborhood of $0 \in M_m$ and with total integral equal to one. Write $w = H + \iota Y$ where both $H, Y \in H_m$. For a suitable function φ with compact support we have

$$F(0; I) = \int_{H_m} \int_{H_m} F(0; (I + H) + \iota Y)\varphi(H, Y) \, dH \, dY.$$

Note that the integrand has support in a small neighborhood of the origins of each copy of H_m. Next, we make the change of variables

$$(I + H) = (I + X)^2,$$

that is,

$$H = \Psi(X) = 2X + X^2$$

where $X \in H_m$, and $\Psi : H_m \mapsto H_m$. This mapping is a diffeomorphism from some small neighborhood of the origin to a neighborhood of the origin. Hence there is a smooth function ψ with support near the origin of $H_m \times H_m$ so that

$$F(0; I) = \int_{H_m} \int_{H_m} F(0; (I + X)^2 + \iota Y)\psi(X, Y) \, dX \, dY.$$

Let $U(m)$ denote the group of $m \times m$ unitary matrices, and let dg denote Haar measure on $U(m)$. Since $Z_0(0) = 0$ and $W_0(0) = I$, for every $g \in U(m)$ we have

$$F(0; (I + X)^2 + \iota Y) = F((I + X)gZ_0(0); (I + X)gW_0(0)g^*(I + X) + \iota Y)$$

and so $F(0; I)$ is given by the integral

$$\int_{H_m} \int_{H_m} \int_{U(m)} F((I + X)gZ_0(0);$$

$$(I + X)gW_0(0)g^*(I + X) + \iota Y)\psi(X, Y)\, dg\, dX\, dY.$$

We replace $I + X$ by the variable $X \in H_m$. By averaging over the analytic disc

$$D \ni \zeta \mapsto (XgZ_0(\zeta); XgW_0(\zeta)g^*X + \iota Y)$$

we see that $F(0; I)$ is given by the integral

$$\frac{1}{2\pi} \int_{H_m} \int_{H_m} \int_{U(m)} \int_0^{2\pi} F(XgZ_0(e^{\iota\theta});$$

$$XgW_0(e^{\iota\theta})g^*X + \iota Y)\psi(X - I, Y)d\theta\, dg\, dX\, dY$$

Note that for any $X, Y \in H_m$, $g \in U(m)$ and $\theta \in [0, 2\pi]$, the point

$$\Psi(X, Y, g, \theta) = (XgZ_0(e^{\iota\theta}); XgW_0(e^{\iota\theta})g^*X + \iota Y)$$

belongs to Σ_m. We now study this mapping $\Psi : H_m \times H_m \times U(m) \times [0, 2\pi] \mapsto \Sigma_m$. We have

Lemma 4.2. *For each fixed $g_0 \in U(m)$ and $\theta_0 \in [0, 2\pi]$, the mapping Ψ has maximal rank at the point $(I, 0, g_0, \theta_0)$.*

Proof. After we identify Σ_m with $C^m \times H_m$, the mapping Ψ becomes

$$\Psi(X, Y, g, \theta) = (XgZ_0(e^{\iota\theta}), Y) \in C^m \times H_m.$$

Thus it suffices to show that the mapping

$$(X, g, \theta) \mapsto Xg \begin{bmatrix} e^{\iota\theta} \\ e^{2\iota\theta} \\ \vdots \\ e^{m\iota\theta} \end{bmatrix} \in C^m$$

has rank m at the point (I, g_0, θ_0). This is easily shown by restricting X to diagonal matrices and g to diagonal multiples of g_0. ∎

Using this lemma, we can integrate out the extra variables in the integral formula for $F(0; I)$ and obtain the following.

Lemma 4.3. *There is a non-negative function $K_0 \in C_0^\infty(\Sigma_m)$ such that:*

(1) *For every function* F *continuous on* $\overline{\Omega_m}$ *and holomorphic on* Ω,

$$F(0; I) = \int_{\Sigma_m} F(\xi) K_0(\xi)\, d\lambda(\xi)$$

$$= \int_{C^m} \int_{H_m} F(z, Y) K_0(z, Y)\, dY\, dz;$$

(2) *For every non-negative function* u *continuous on* $\overline{\Omega_m}$ *and plurisubharmonic on* Ω,

$$u(0; I) \leq \int_{\Sigma_m} u(\xi) K_0(\xi)\, d\lambda(\xi)$$

$$= \int_{C^m} \int_{H_m} u(z, Y) K_0(z, Y)\, dY\, dz.$$

Our discussion so far has dealt with representing the value of a holomorphic function, but if the function is plurisubharmonic, all equalities are replaced by inequalities and estimate (2) can be established. The representation formula (1) and the estimate (2) are generalizations of (1) and (2) given in section 1 (with $r = 1$) for the Heisenberg group.

We can obtain a representation formula for other points in Ω_m by making use of the group structure and dilations on Σ_m. Let g be an element of $U(m)$. Both Σ_m and Ω_m are invariant under the rescaling map

$$(z; Z) \mapsto (gz; gZg^*).$$

In addition for $(z_0; Z_0) \in \Sigma_m$, the map

$$(z; Z) \mapsto (z + z_0; 2zz_0^* + Z + Z_0)$$

also preserves Σ_m and Ω_m. (This is the map induced by translation under the group structure of Σ_m.) Using lemma 4.3 and the translation map and rescaling map, we obtain the following representation formula for holomorphic functions at an arbitrary point in Ω_m. To state the result, note that an arbitrary point in Ω_m can be written as $(z_0; Z_0) = (z_0; z_0 z_0^* + X_0 + \imath Y_0) \in \Omega_m$ where X_0 is positive definite (and therefore, X_0 has a unique positive square root $X_0^{1/2} \in H_m$).

Theorem 3. *There is a non-negative,* C^∞ *function* $K : \Omega_m \times \Sigma_m \mapsto [0, \infty)$ *such that if*

$$(z_0; Z_0) = (z_0; z_0 z_0^* + X_0 + \imath Y_0) \in \Omega_m$$

then:

(1) *For every function F continuous on $\overline{\Omega_m}$ and holomorphic on Ω_m,*

$$F(z_0; Z_0) = \int_{C^m} \int_{H_m} F(z, Y)K\big((z_0; Z_0), (z, Y)\big) \, dY \, dz.$$

(2) *For every non-negative function u continuous on $\overline{\Omega_m}$ and plurisubharmonic on Ω_m,*

$$u(z_0; Z_0) \leq \int_{C^m} \int_{H_m} u(z, Y)K\big((z_0; Z_0), (z, Y)\big) \, dY \, dz.$$

Moreover,

$$K\big((z_0; z_0 z_0^* + X_0 + \iota Y_0), (z, Y)\big)$$
$$= \det(X_0)^{-m-1} K_0(X_0^{-1/2}(z - z_0), X_0^{-1/2}(Y - Y_0 - 2zz_0^*)X_0^{-1/2}).$$

If $z = (z_0; Z_0) = (z_0; z_0 z_0^* + X_0 + \iota Y_0)$ is a point in Ω_m, then the quantity $r(z) = \|X_0\|$ is proportional to the distance from z to Σ_m. Moreover, in a fixed compact subcone, $\det(X_0)$ is proportional to $\|X_0\|^m$. In view of the fact that K_0 is a bounded function, the above formula implies the following estimate for K:

$$|K(z_0, z)| \leq \frac{C}{r(z)^{m^2+m}}.$$

The CR (complex) dimension of Σ_m is m and the real dimension of the totally real tangent space of Σ $(= \dim_R H_m)$ is m^2. Therefore, the right side of the above estimate is proportional to $1/|B(z_0, \sqrt{r(z)})|$. This establishes the $k = 0$ part of the estimate on K given in theorem 1. The estimate for $k \geq 1$ can be obtained by differentiating the above expression for K by holomorphic tangent vector fields.

Texas A & M University

University of Wisconsin, Madison

REFERENCES

[BDN] A. Boggess, R. Dwilewicz, and A. Nagel. "The hull of holomorphy of a nonisotropic ball in a real hypersurface of finite type." *Trans. Amer. Math. Soc.* **323** (1991), 209–232.

[FS] G. Folland and E. M. Stein. "Estimates for the $\overline{\partial}_b$ complex and analysis on the Heisenberg group." *Comm. Pure Appl. Math.* **27** (1974), 429–522.

[NSW] A. Nagel, E. M. Stein, and S. Wainger. "Balls and metrics defined by vector fields I: basic properties." *Acta Math.* **155** (1985), 103–147.

[R] J. Rosay. "On the radial maximal function and the Hardy-Littlewood maximal function in wedges." Preprint.

[RS] L. Rothschild and E. M. Stein. "Hypoelliptic differential operators and nilpotent groups." *Acta Math.* **137** (1976), 247–340.

4

Some New Estimates on Oscillatory Integrals

Jean Bourgain

1 INTRODUCTION

This paper is mainly a summary of recent work of the author on the subject of the title, and most of its content relates to [B1] and [B2]. The main objects underlying the problems discussed here are certain operators expressed by an oscillatory integral, roughly of the form

$$(1.1) \qquad Tf(x) = \int e^{i\varphi(x,y)} f(y) dy$$

where φ will be a smooth function of x, y (this phase function φ will be more specified later). Our aim will be to understand the mapping properties of T, namely the pairs of (L^{p_1}, L^{p_2})-spaces where T acts as a bounded operator, and to estimate its norm. This question is presently only very partially understood, even in the simplest cases. In this report, we specify the problem to the following themes:

(i) Restriction and extension problems;
(ii) The Bochner-Riesz summation operators;
(iii) The (more general) oscillatory integrals considered by Hörmander [Hör].

We do not intend to give a survey of these topics but, instead, to comment on some progress made in [B1], [B2]. We also want to mention the paper [St1], which the reader may like to consult for more background and related material.

(i)   The model cases (in the compact setting) are the operators

$$(1.2) \qquad\qquad f \longmapsto \hat{f}\,\Big|_{S_{d-1}} \qquad \text{(restriction)}$$

and

(1.3) $\mu \in M(S_{d-1}) \longmapsto \hat{\mu}$ (extension).

Here $S_{d-1} = S$ the unit sphere $\{x \in \mathbb{R}^d \mid \sum_1^d x_i^2 = 1\}$ in $\mathbb{R}^d$ with its invariant measure $\sigma = \sigma_{d-1}$, and the measure μ on S considered in (1.3) is assumed absolutely continuous with respect to σ. The operators (1.2) and (1.3) are of course formally dual to each other.

The main relevant geometrical feature of the sphere is that it is a compact smooth manifold with nonvanishing curvature. Considering locally such a hypersurface as a graph, (1.3) may be expressed by an operator of the form (1.1), namely

(1.4) $Tf(x) = \int e^{i(x_1 y_1 + \cdots + x_{d-1} y_{d-1} + x_d \psi(y))} f(y) dy$

where $x \in \mathbb{R}^d$, y is taken in a neighborhood of $0 \in R^{d-1}$, ψ is a smooth function on that neighborhood satisfying

(1.5) $\det \left(\dfrac{\partial^2}{\partial y^2} \psi \right) \neq 0,$

and f is a function of the y-variable corresponding to the derivative $\frac{du}{d\sigma}$.

(ii) The Bochner-Riesz multipliers are defined by the formula

(1.6)
$$\begin{cases} m_\lambda(\xi) = (1 - |\xi|^2)^\lambda & \text{if } |\xi| \leq 1 \\ \quad\quad\; = 0 & \text{if } |\xi| > 1 \end{cases}$$

with $\lambda > 0$. We are interested in the corresponding Fourier multiplication operator

(1.7) $f \longmapsto (\hat{f} \cdot m_\lambda)^\vee,$

more precisely, when it acts as a bounded operator on $L^p(\mathbb{R}^d)$. Thus the operator is given by a convolution with kernel

(1.8) $\hat{m}_\lambda(x) \sim \dfrac{e^{i|x|}}{|x|^{(d+1)/2+\lambda}}.$

One has obtained estimates using various methods, either dealing with the Fourier-multiplier definition (1.7) or the convolution operator

(1.9) $Tf(x) = \int f(y) \dfrac{e^{i|x-y|}}{|x - y|^{(d+1)/2+\lambda}} \, dy.$

Considering the form (1.9), restricting the variables x, y to disjoint balls at distance $\sim N > 1$ and rescaling, one gets operators

(1.10)
$$T_N f(x) = \int f(y) e^{iN|x-y|} a(x, y) dy$$

where a is a smooth function on a neighborhood of $(0, 0)$ in $\mathbb{R}^d \times \mathbb{R}^d$ vanishing on a neighborhood of the diagonal. The approach of [C-S] and [Hör] consists of first passing from this $\mathbb{R}^d - \mathbb{R}^d$ problem to an $\mathbb{R}^{d-1} - \mathbb{R}^d$ problem, by fixing one coordinate of the y-variable, say $y_d = 1$, keeping x, $(y_1, \ldots, y_{d-1})$ in a neighborhood of $0 \in \mathbb{R}^d$ (resp. $\mathbb{R}^{d-1}$). The phase function $|x - y|$ becomes

(1.11)
$$\left[(x_1 - y_1)^2 + \cdots + (x_{d-1} - y_{d-1})^2 + (x_d - 1)^2 \right]^{1/2}$$

Making an asymptotic expansion of (1.11), eliminating the purely x and y terms and performing changes of variables in x and y separately, one is led to a phase function of the form (1.4), but with the presence of non-linear x-terms $0\left(|x|^2|y|^2\right)$. This generalization of (1.4) gives precisely the operations considered in [Hör].

(iii) In [Hör], one studies the behavior of operators

(1.12)
$$T_N f(x) = \int e^{iN\varphi(x,y)} a(x, y) f(y) dy$$

where $a \in C_0^\infty(\mathbb{R}^{2d-1})$ and $\varphi \in C^\infty(\mathbb{R}^{2d-1})$ is a real valued function satisfying the conditions for $(x, y) \in \operatorname{supp} a$:

(1.13)
$$\operatorname{rank} \partial^2 \varphi / \partial x \partial y = d - 1$$

(1.14) $\dfrac{\partial}{\partial y} \left\langle \dfrac{\partial \varphi}{\partial x}, t \right\rangle = 0, 0 \neq t \in \mathbb{R}^d \implies \det \left(\dfrac{\partial^2}{\partial y^2} \left\langle \dfrac{\partial \varphi}{\partial x}, t \right\rangle \right) \neq 0,$

condition (1.14) meaning that the map $y \mapsto \left\langle \frac{\partial \varphi}{\partial x}, t \right\rangle$ has only non-degenerate critical points. After coordinate changes, φ may be given the form
(1.15)
$$\varphi(x, y) = x_1 y_1 + \cdots + x_{d-1} y_{d-1} + x_d \langle Ay, y \rangle + O(|x| |y|(|x|^2 + |y|^2))$$

where A is a symmetric matrix with $\det A \neq 0$. The number $N > 1$ is a parameter and one seeks uniform estimates of the form

(1.16)
$$\|T_N f\|_q < C_\varphi N^{-d/q} \|f\|_r$$

for certain pairs (q, r). Those pairs are specified in [Hör] for $d = 2$ and the higher-dimensional setting is proposed as a question. Thus the setting (iii) generalizes both (i) and (ii).

It is conjectured that m_λ defines a bounded Fourier multiplier on $L^p(\mathbb{R}^d)$ iff $\lambda > 0$ and

(1.17)
$$\frac{2d}{d+1+2\lambda} < p < \frac{2d}{d-1-2\lambda}.$$

We assume $\lambda \leq \frac{d-1}{2}$ and $p \neq 2$ (trivial case). The condition $\lambda > 0$ is necessary because of C. Fefferman's solution of the ball multiplier problem [F1]. It follows from (1.8) that the range (1.17) is optimal.

It is conjectured that (1.2) maps boundedly from $L^{q'}(\mathbb{R}^d)$ to $L^{p'}(S_{d-1})$ or, more generally, (1.4) fulfills an inequality

(1.18)
$$\|Tf\|_{L^q(\mathbb{R}^d)} \leq C\|f\|_{L^r(S_{d-1})}$$

for

(1.19)
$$q > \frac{2d}{d-1} \quad \text{and} \quad \frac{d+1}{(d-1)q} + \frac{1}{r} \leq 1.$$

The condition (1.19) is the best possible one may expect (recall that $\hat{\sigma}_{d-1} \sim \frac{1}{|x|^{(d-1)/2}}$ for $|x| \to \infty$).

One considers the same range (1.19) for the inequality (1.16) (proposed for $d > 2$ as a question in [Hör]).

In dimension $d = 2$, problems (i), (ii), (iii) are essentially completely understood and the optimal results stated above are known to be valid. Thus the multipliers m_λ, $\lambda > 0$ act boundedly (as a Fourier multiplier) on $L^p(\mathbb{R}^2)$ for $p \vee p' < 4$ and (1.16), (1.18) hold for (r, q) interpolated between the pairs (2, 6); (4, 4) (not allowing (4, 4)). See [St1] for a more complete bibliography. Apparently, the only results (preceding [B1]) in dimension $d \leq 3$ were those derived from a purely "L^2-method." Thus (1.18) was shown for $r = 2, q > \frac{2d}{d+1}$ by P. Tomas [T] and E. Stein (see [St1]) if $r = 2, q \geq \frac{2d}{d+1}$. Also (1.16) is shown in [St1] to be valid in this case. Following [F2], one may then obtain the boundedness of (1.7) under the condition (1.17), provided $p^* = p \vee p' \geq \frac{2(d+1)}{d-1}$. The argument is based mainly on an L^2-factorization of the L^p-operator, invoking Parseval's theorem and the L^2-restriction result for spheres of [T] mentioned earlier.

The author obtained in [B1] results beyond the range of the L^2-techniques. It is shown, for instance, that if $\mu \in M(S_{d-1})$, $\frac{d\mu}{d\sigma} \in L^\infty(S)$, then the Fourier transform $\hat{\mu}$ is p-integrable for some $p < \frac{2(d+1)}{d-1}$. Also the Bochner-Riesz conjecture is verified for certain p with $p^* < \frac{2(d+1)}{d-1}$. The approach makes essential use of certain new estimates on higher dimensional Besicovitch-type maximal operators, which will be described later. The role of such geometric objects became clear in [F1], where the existence of the 2-dimensional Kakeya-Besicovitch set was exploited to disprove the ball-multiplier conjecture. A systematic investigation of the geometric structures in order to approach the Fourier Analysis problems mentioned above was proposed in [F2]. The paper [B1] contains the first results along these lines of investigation (for $d > 2$).

Surprisingly, for $d \geq 3$, inequality (1.16) may fail under the conditions (1.19) and hence this approach to the Bochner-Riesz multipliers for $d > 2$ is a bit too general. In fact, it turns out that if $d > 2$ is *odd*, (1.16) may only hold if $q \geq \frac{2(d+1)}{d-1}$, even for $r = \infty$. In [B2], we study the case $d = 3$, showing that for a "generic" phase function φ one has

(1.20)
$$3 \quad < \inf \left\{ q \mid \|T_N f\|_q < C N^{-3/q} \|f\|_\infty \text{ for } N > 1 \right\} < \quad 4$$
$$\| \qquad\qquad\qquad\qquad\qquad\qquad\qquad\qquad\qquad\qquad\qquad\qquad \|$$
$$\frac{2d}{d-1} \qquad\qquad\qquad\qquad\qquad\qquad\qquad\qquad\qquad\qquad \frac{2(d+1)}{d-1}.$$

Here the relevant geometric structures are generalizations of the Besicovitch structures, in the sense that straight lines are replaced by curves Γ_y, described by the equation

(1.21)
$$\frac{\partial \varphi}{\partial y}(x, y) = a(y).$$

For such systems of curves, certain "compression phenomena" may occur, leading to the failure of (1.16) for $r = \infty$ and certain $q > \frac{2d}{d-1}$ (with various degrees of strength).

Thus the non-linear x-terms in φ given by (1.15) play a significant role if $d > 2$.

Very recently the author observed that if d is *even* one may obtain (1.16) when $r = \infty$ for certain $q < \frac{2(d+1)}{d-1}$ whenever φ satisfies the conditions (1.13), (1.14) (this is not the case for d odd, as said previously). The main reason for this is a different behavior of the families of curves Γ_y defined by (1.21), in the sense that they cannot be contained in a set $A \subset \mathbb{R}^d$ of dim $A = \frac{d+1}{2}$ (which is possible if d is odd). At the end of the paper, we give a sketch of the reason for this fact for even d. More precise statements and details will appear elsewhere.

We also give an application of the non-L^2 restriction results in the context of a.e. convergence of solutions of the 2-dimensional Schrödinger equation

(1.22)
$$\begin{cases} \Delta u = i \partial_t u \\ u(x, 0) = f(x) \end{cases}$$

for $t \to 0$, improving on Vega's result (for $d = 2$) (see [B3], [V]).

2 ESTIMATES BASED ON L^2-METHODS

It was shown by Tomas [T] that for $f \in L^p(\mathbb{R}^d)$, $p < \frac{2(d+1)}{d+3}$, the restriction $\hat{f}|_{S_{d-1}}$ yields an L^2-function on the sphere. In [St1], complex interpolation is used to get the endpoint result $p \leq \frac{2(d+1)}{d+3}$. Also (1.16) is proved for $r = 2, q \geq \frac{2(d+1)}{d-1}$ (see [St1], Th. 10). The aim of this section is to make some further comments on those estimates, that will play a role later on (we will not be concerned here with the endpoint behavior). The discussion below will provide further information on distribution of level sets.

Putting $x = \frac{x'}{N}$, $|x'| \ll N$, we replace the operator T_N defined by (1.12) by

$$(2.1) \qquad T'_N f(x') = \int e^{iN\varphi(x'/N, y)} a(y) f(y) dy$$

where a is supported by a neighborhood of 0 in $\mathbb{R}^{d-1}$.

Proposition 2.2. *For $0 < \lambda < 1$*

$$(2.3) \qquad let\ A_\lambda = \{|x| < N\ and\ |T'_N f(x)| > \lambda\}$$

denote the level sets. Assume $\|f\|_2 \leq 1$. Then the following inequality holds for $R > 1$:

$$(2.4) \qquad |A_\lambda| \leq \lambda^{-2}\left[R + \sum_{\substack{R < \rho < N \\ \rho\ \text{dyadic}}} \rho^{-(d-1)/2} \sup_{|z| < N} |A_\lambda \cap B(z, \rho)|\right].$$

Here $|A|$ denotes the measure of A and $B(z, \rho)$ the ball centered at z of radius ρ.

Proof. Denote $A = A_\lambda$ and consider a collection $\{B_\alpha\}$ of subsets of A such that

$$(2.5) \qquad\qquad\qquad\qquad \operatorname{diam} B_\alpha < R$$

$$(2.6) \qquad \operatorname{dist}\{B_\alpha, B_\beta\} > R \qquad \text{for}\quad \alpha \neq \beta$$

$$(2.7) \qquad\qquad\qquad\qquad |\cup_\alpha B_\alpha| \sim |A|.$$

Let χ_α stand for the indicator function of B_α. Denote T'_N by T. From (2.7), one may assume

$$\operatorname{Re} \sum \langle Tf, \chi_\alpha \rangle > c\lambda |A|$$

$$(2.8) \qquad\qquad \left\|\sum_\alpha T^* \chi_\alpha\right\|_2^2 < c\lambda^2 |A|^2.$$

Expand and estimate the left member of (2.8) as

$$(2.9) \quad \sum_{\alpha} \|T^* \chi_\alpha\|_2^2 + \sum_{\alpha \neq \beta} \langle TT^* \chi_\alpha, \chi_\beta \rangle \leq \|T^*\|_{L^2_{(R)}}^2 \cdot \sum |A_\alpha|$$

$$+ \sum_{\alpha \neq \beta} c(\alpha, \beta) |A_\alpha| \cdot |A_\beta|.$$

Here $\|T^*\|_{L^2_{(R)}}$ denotes the norm of T^* acting on L^2-functions of norm ≤ 1 supported by a ball of radius R; $c(\alpha, \beta) = c(\rho)$, $\rho = \text{dist}(A_\alpha, A_\beta)$ where $c(\rho)$ denotes a uniform bound on the kernel $K(x, x')$ of TT^* for $|x - x'| > \rho$. K is given by the oscillatory integral

$$(2.10) \quad K(x, x') = \int e^{iN[\varphi(x/N, y) - \varphi(x'/N, y)]} a(y) dy.$$

From the hypothesis on φ one has the estimate

$$(2.11) \quad |K(x, x')| < \frac{C}{|x - x'|^{(d-1)/2}}$$

(cf. [St1], Prop. 6). Thus

$$(2.12) \quad c(\rho) < c\rho^{-(d-1)/2}.$$

The square of $\|T^*\|_{L^2_{(R)}}$ may be evaluated by the L^2-norm of the operator with kernel of the form

$$(2.13) \quad L(y, y') = \int e^{iN[\varphi(x/N, y) - \varphi(x/N, y')]} b_R(x - z) dx$$

where $z \in \mathbb{R}^d$, $|z| < N$ and b_R denotes a smooth function such that $b_R(x) = 1$ if $|x| < R$, $b_R(x) = 0$ if $|x| > 2R$ and fulfills the obvious derivative estimates. By Schur's lemma, this L^2-norm is bounded by

$$(2.14) \quad \sup_y \int |L(y, y')| a(y') \, dy'.$$

Since $|L(y, y')| < cR^d[1 + R|y - y'|]^{-d}$, (2.14) is at most C.R. So (2.8), (2.9) imply

$$(2.15) \quad \lambda^2 |A|^2 \lesssim R|A| + \sum_{R < \rho < N} \rho^{-(d-1)/2} \sum_{\text{dist}(A_\alpha, A_\beta) \sim \rho} |A_\alpha| |A_\beta|$$

from which (2.4) immediately follows.

If we let in particular in (2.4) $R \sim \lambda^{-4/(d-1)}$, it follows that

$$(2.16) \quad |A_\lambda| \lesssim \lambda^{-(2(d+1))/(d-1)}$$

implying a bound on T acting between L^2 and $L^{p,\infty}(\mathbb{R}^d)$, $p = \frac{2(d+1)}{d-1}$. The preceding shows, moreover, that if $|A_\lambda| \sim \lambda^{-p}$, then also $|A_\lambda \cap B| \sim \lambda^{-p}$ for some ball B of radius $\lambda^{-4/(d-1)}$. This statement is easily seen to be sharp. Denote, for instance, μ to be the measure on S^{d-1} where $\frac{d\mu}{d\sigma} = \frac{1}{|A|^{1/2}} \chi_A$, A the ε-cap centered at the north pole $(0, \ldots, 0, 1)$. Thus $|A| \sim \varepsilon^{d-1}$ and $\hat{\mu}(x) \sim \varepsilon^{(d-1)/2} = \lambda$ in the tube of essential shape $\underbrace{\varepsilon^{-1} \times \cdots \times \varepsilon^{-1}}_{d-1} \times \varepsilon^{-2}$, of measure $\varepsilon^{-(d+1)} = \lambda^{-p}$. The tube is contained in an $\varepsilon^{-2} = \lambda^{-4/(d-1)}$ ball. In [B4], it is shown that for $d = 3$, $\left(p = \frac{2(d+1)}{d-1} = 4\right)$ and $\frac{d\mu}{d\sigma} = \frac{1}{|A|^{1/2}} \chi_A$, A a measurable subset of S_2, the "extremal" case

$$(2.17) \qquad \|\hat{\mu}\|_{L^4(\mathbb{R}^3)} = 0(1)$$

may, essentially speaking, only occur for caps. More precisely, there is the following estimate.   ∎

Proposition 2.18. *For $0 < \delta < 1$ denote by C_δ a collection of δ-caps τ of S_2 forming a covering of bounded multiplicity. Let r denote the exponent $\frac{16}{9}$ and L_τ^r the space $L^r(\tau)$ where τ is endowed with normalized measure. Then for A and μ as above*

$$(2.19) \qquad \|\hat{\mu}\|_4 \le c\left\{ \sum_{\substack{0<\delta<1 \\ \delta \text{ dyadic}}} \delta^4 \sum_{\tau \in C_\delta} \left\| \frac{d\mu}{d\sigma} \right\|_{L_\tau^r}^4 \right\}^{1/4}.$$

Thus, if (2.17) holds, one gets from (2.19) the following distributional property for the set A:

$$(2.20) \qquad \sum_{\substack{0<\delta<1 \\ \delta \text{ dyadic}}} \sup_{\tau \in C_\delta} \frac{|A \cap \tau|}{|A|^{4/5}|\tau|^{1/5}} = 0(1).$$

Hence for some cap τ on S_2

$$(2.21) \qquad |A| \sim |A \cap \tau| \sim |\tau|$$

from which follows the previous claim.

Proposition 2.18 is shown by direct calculation of the L^4-norm. It seems of interest to have an analogous statement in arbitrary dimension $\left(p = \frac{2(d+1)}{d-1}\right)$.

3 BESICOVITCH-TYPE MAXIMAL OPERATORS

Recall that the classical Kakeya-Besicovitch set in $\mathbb{R}^2$ is a measurable set of zero measure containing a line in every direction. As said before, the existence of such sets was exploited in [F1] to disprove the ball multiplier conjecture, i.e., the operator

$$(3.1) \qquad\qquad f \longmapsto \left(\hat{f}|_{B(0,1)}\right)^{\vee}$$

is unbounded on $L^p(\mathbb{R}^d)$, $d > 1$ (except $p = 2$ of course).

Two-dimensional Besicovitch sets are necessarily of full Hausdorff dimension (see [Fal] for more details on this and related facts). This result is unknown in dimension $d \geq 3$ (at least at the time of this writing). A positive answer to this would be implied by the Fourier Analysis conjectures described in the introduction, more specifically, the restriction conjecture:

$$(3.2) \qquad\qquad L^p(\mathbb{R}^d) \to L^1(S_{d-1}) : f \longmapsto \hat{f}|_S$$

is bounded for all $p < \frac{2d}{d+1}$.

In [B1], new information on these Besicovitch structures was obtained, leading to certain (sharp) maximal inequalities that will demonstrate their importance later on.

Next we summarize the main geometric measure theory results from [B1]. Their proof is of a more combinatorial nature.

We will consider averages over tubes in $\mathbb{R}^d$ of unit length and width δ, with direction $\xi \in S_{d-1}$ (see Fig. 1). Given a bounded measurable function on $\mathbb{R}^d$, we define two maximal functions f_δ^*, f_δ^{**} which we refer to as the Kakeya (resp. Nikodym) maximal function. These names are those appearing in [B1]; in other places they may have been used differently.

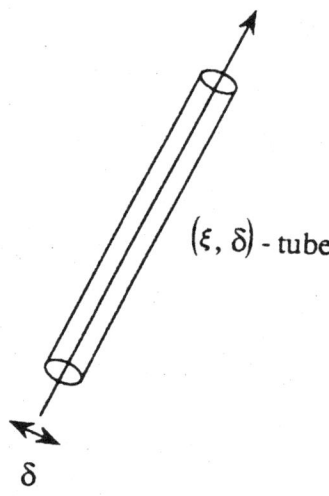

(ξ, δ) - tube

δ

Figure 1.

(i) f_δ^* is defined on the sphere S_{d-1}, letting

(3.3) $$f_\delta^*(\xi) = \sup_\tau \frac{1}{|\tau|} \int_\tau f(x)dx$$

where the supremum is taken over all (ξ, δ)-tubes (fixing ξ and considering different translations). $|\tau|$ denotes the measure $\sim \delta^{d-1}$ of τ.

We call f_δ^* the *Kakeya maximal function* of f for δ eccentricity.

(ii) f_δ^{**} (the *Nikodym maximal function*) is defined on $\mathbb{R}^d$, letting

(3.4) $$f_\delta^{**}(x) = \sup_\tau \frac{1}{|\tau|} \int_\tau f(y)dy$$

where now the supremum is taken over all (ξ, δ)-tubes τ centered at x (thus fixed center x and varying direction ξ).

Considering the case of radial functions and the existence of Besicovitch sets, it seems natural to conjecture the following inequalities

(3.5) $$\|f_\delta^*\|_{L^p(S_{d-1})} \ll \left(\frac{1}{\delta}\right)^{d/p-1+\varepsilon} \|f\|_{L^p(\mathbb{R}^d)}$$

and, similarly,

(3.6) $$\|f_\delta^{**}\|_{L^p(\mathbb{R}^d)} \ll \left(\frac{1}{\delta}\right)^{d/p-1+\varepsilon} \|f\|_{L^p(\mathbb{R}^d)}$$

for $1 \leq p \leq d$. These inequalities are valid for $d = 2$ and shown by Fourier Analysis methods.

In [B1], the following result is obtained.

Proposition 3.7. *For given dimension $d > 2$, (3.5), (3.6) hold provided*

(3.8) $$p \leq p_d$$

where the exponent p_d is given by the recursive formula

(3.9) $$p_2 = 2; \quad p_d\{2p_{d-1} - 1\} = (d + 2)p_{d-1} - d.$$

Thus, in particular,

$$p_2 = 2$$

$$p_3 = \frac{7}{3}$$

$$p_4 = \frac{30}{11}$$

$$p_5 = \frac{155}{49}$$

etc.

Those numbers also yield lower bounds on Hausdorff dimension of Besicovitch sets in corresponding dimension d.

More generally, consider for a k-plane in $\mathbb{R}^d$ the Radon transform

$$(3.10) \qquad F(L) = \int_L f(x)\,dx$$

and the corresponding maximal operator

$$(3.11) \qquad F^*(L) = \sup_{x \in \mathbb{R}^d} F(x + L)$$

defined on the Grassmannian $G(d, k)$.

From Proposition 3.7 one may derive the following corollary, for instance.

Proposition 3.12.

(i) *There is an a priori inequality*

$$(3.13) \qquad \|F^*\|_{L^p_{G(4,2)}} \leq c_p \|f\|_p$$

for $p > 2$ and assuming f supported by the unit ball of $\mathbb{R}^4$.

(ii) *Let A be a measurable subset of $\mathbb{R}^4$ of finite measure. Then for almost all $L \in G(4, 2)$ each translate of L intersects A in a set of finite 2-measure and $\sup_{x \in \mathbb{R}^d} |A \cap (x + L)| < \infty$.*

This result is in the spirit of the work of Oberlin and Stein [O-S]. Following [Fal], call a (d, k)-Besicovitch set a measure zero subset of $\mathbb{R}^d$ containing a translate of every k-plane. Thus Besicovitch's result affirms the existence of $(2, 1)$-Besicovitch sets. It follows from Proposition 3.12 that there are no $(4, 2)$-Besicovitch sets. More generally, one shows in [B1]

Proposition 3.14. *Assume (d, k) fulfill the condition*

$$(3.15) \qquad d \leq 2^{k-1} + k.$$

Then the (d, k)-property holds, i.e., there are no (d, k)-Besicovitch sets.

The proof relies on Proposition 3.7 and Fourier analysis techniques similar to the ones used in [O-S].

4 ESTIMATES ON OSCILLATORY INTEGRALS

We will show in particular how the results of the last two sections may be applied to get new information on the higher-dimensional restriction problem.

Consider more generally φ of the form (1.15) and define for given $N > 1$

(4.1)
$$\psi(x, y) = N\varphi\left(\frac{x}{N}, y\right).$$

Consider the oscillatory integral (2.1)

(4.2)
$$Tf(x) = T_N f(x) = \int e^{i\psi(x,y)} f(y)\,dy$$

(f supported by a neighborhood of 0).

Fix $1 < R \ll N$. Our purpose is to get a new distributional estimate on $Tf|_{B(0,R)}$ by estimating

(4.3)
$$\|Tf\|_{L^p(B(0,R))} \qquad \text{for certain} \quad p < \frac{2(d+1)}{d-1}.$$

This computation is summarized next.

Consider a partition of the y-domain ($=$ neighborhood of 0 in $\mathbb{R}^{d-1}$) in boxes Q_α of size $\frac{1}{\sqrt{R}}$ centered at points y_α. Write for $y \in Q_\alpha$

(4.4)
$$\psi(x, y) = \psi(x, y_\alpha) + \langle \nabla_y \psi(x, y_\alpha), y - y_\alpha \rangle + 0(|y - y_\alpha|^2)$$

where the $0(|y - y_\alpha|^2) = 0\left(\frac{1}{R}\right)$ will be dropped because of the assumption $|x| < R$.

Define the operators

(4.5)
$$T_\alpha f(x) = \int_{Q_\alpha} e^{i\langle \nabla_y \psi(x, y_\alpha), y - y_\alpha \rangle} f(y)\,dy$$

and write

(4.6)
$$Tf(x) = \sum_\alpha e^{i\psi(x, y_\alpha)} T_\alpha f(x).$$

The next point consists in introducing a new variable $z \in B\left(0, \sqrt{R}\right)$. Because of the presence of $y - y_\alpha$, $|y - y_\alpha| < R^{-1/2}$ in the phase function of $T_\alpha f$, one may, roughly speaking, write for $|x| < R^{1/2}$

(4.7)
$$Tf(x + z) \sim \sum_\alpha e^{i\psi(x+z, y_\alpha)} T_\alpha f(x)$$

and estimate (4.3) as
(4.8)
$$\|Tf\|_{L^p(B(0,R))}^p \sim R^{-d/2} \int_{B(0,R)} \left\{ \int_{B(0,R^{1/2})} \left| \sum_\alpha e^{i\psi(x+z, y_\alpha)} T_\alpha f(x) \right|^p dz \right\} dx.$$

Define for fixed x

(4.9)
$$\eta(z, y) = \psi(x + z, y) - \psi(x, y)$$

and the function $g = g(y)$ as

(4.10)
$$g|_{Q_\alpha} = e^{i\psi(x, y_\alpha)} T_\alpha f(x).$$

Because in (4.9) z is restrained to the ball $B(0, R^{1/2})$, one may again roughly set $\eta(z, y) \sim \eta(z, y + u)$ for $|u|$ of size $R^{1/2}$. Hence,

(4.11) $\displaystyle\sum_\alpha e^{i\psi(x+z, y_\alpha)} T_\alpha f(x) = \sum_\alpha e^{i\eta(z, y_\alpha)} g(y_\alpha) \sim R^{(d-1)/2} \int e^{i\eta(z, y)} g(y) dy$

and we have to bound

(4.12)
$$\left\| \int e^{i\eta(z, y)} g(y) dy \right\|_{L^p(B(0, R^{1/2}))}.$$

The exponent p will be taken in the range

(4.13)
$$2 < \frac{2d}{d-1} \leq p \leq p_1 \equiv \frac{2(d+1)}{d-1}$$

to be specified later.

At this point, we just use L^2-theory (see remarks at the end of this chapter) and interpolate between the known $L^2 - L^{p_1}$ result and the $L^2 - L^2$ bound

(4.14)
$$\left\| \int e^{i\eta(z, y)} g(y) dy \right\|_{L^2(B(0, R^{1/2}))} \lesssim R^{1/4} \|g\|_2$$

observed in section 2 (see proof of Proposition 2.2). Hence, putting

(4.15)
$$\frac{1}{p} = \frac{1 - \theta}{2} + \frac{\theta}{p_1},$$

it follows

(4.16)
$$(4.12) \lesssim R^{(1-\theta)/4} \cdot \|g\|_2$$

and thus from (4.11), (4.8) and the definition of the function $g = g_x$

$$\|Tf\|_{L^p(B(0,R))}^p \lesssim R^{-d/2} \cdot R^{((d-1)/2)p} \cdot R^{((1-\theta)/4)p} \cdot \int_{B(0,R)} \|g_x\|_2^p \, dx \lesssim$$

(4.17)
$$R^{-d/2 + (p/4)(d-\theta)} \int_{B(0,R)} \left(\sum_\alpha |T_\alpha f|^2 \right)^{p/2}.$$

Next we will replace suitably

(4.18)
$$|T_\alpha f|^2; \quad T_\alpha f(x) = \int_Q e^{i\langle \nabla_y \psi(x, y_\alpha), y \rangle} f_\alpha(y) dy$$

where

(4.19) $Q = B(0, R^{-1/2}); \quad f_\alpha(y) = f(y_\alpha + y) \qquad \text{for} \quad y \in Q.$

Clearly, one has

(4.20) $|T_\alpha f|^2(x) = \left|\hat{f}_\alpha\right|^2 (\nabla_y \psi(x, y_\alpha)).$

Since supp $f_\alpha \subset Q$, one may write $|\hat{f}_\alpha|^2$ as an average of functions of the form $b\left(\frac{\xi - \cdot}{\sqrt{R}}\right)$ where b is a standard bump function and the average is taken over points ξ, with averaging weight $R^{-(d-1)/2}\|f_\alpha\|_2^2$.

At this point we make the following assumption on f

(4.21) $f \in L^\infty, \quad |f| \leq 1$

improving on the L^2-hypothesis.

Since then $\|f_\alpha\|_2^2 \lesssim R^{-(d-1)/2}$, $|T_\alpha f|^2(x)$ may be recaptured (as a convex combination) from functions of the form

(4.22) $R^{-(d-1)} \cdot b\left(R^{-1/2}(\nabla_y \psi(x, y_\alpha) - \xi_\alpha)\right).$

Thus for some choice of points $\{\xi_\alpha\}$, (4.17) is estimated by

(4.23)

$$R^{-d/2+(p/4)(d-\theta)-(p/2)(d-1)} \cdot \int_{B(0,R)}\left[\sum_\alpha b\left(R^{-1/2}(\nabla_y \psi(x, y_\alpha) - \xi_\alpha)\right)\right]^{p/2} dx.$$

We estimate the $\frac{2}{p}$-power of the integrand in (4.23) by duality. Consider the following related geometric maximal operator

(4.24) $\mathcal{M}_\delta g(y) = \sup_\xi \left\{\delta^{-(d-1)} \int_{\{|\delta^2 \nabla_y \psi(\delta^{-2}x, y) - \xi| < \delta\}} |g(x)| dx\right\}$

assuming $\delta \gg N^{-1/2}$. Making a change of variable $x = Rx'$, (4.23) has the upper bound

(4.25) $R^{d/2+(p/4)(d-\theta)-(p/2)(d-1)}\left[\sum_\alpha R^{-(d-1)/2}(\mathcal{M}_{R^{-1/2}}g)(y_\alpha)\right]^{p/2}$

for some $g \in L^{(p/2)'}(B(0, 1))$ of norm 1.

The sum in (4.25) may be replaced by the y-integral. Hence collecting previous estimates, it follows that if $|f| \leq 1$

(4.26) $\|Tf\|_{L^p(B(0,R))} \lesssim R^{d/(2p)+1/4(d-\theta)-(d-1)/2}\|\mathcal{M}_{R^{-1/2}}\|^{1/2}_{L^{(p/2)'} \to L^1}.$

Finally, substituting the value of θ given by (4.15), (4.26) yields

Proposition 4.27. *For T given by (4.2), $R < N$, one has the bound for $|f| < 1$*

(4.28) $\|Tf\|_{L^p(B(0,R))} \lesssim R^{(3d+1)/(4p)-(3(d-1))/8}\|\mathcal{M}_{R^{-1/2}}\|^{1/2}_{L^{(p/2)'}\to L^1}$

where $2 \le p \le \frac{2(d+1)}{d-1}$ and the maximal operator $\mathcal{M}_\delta$ is given by (4.24).

It remains, of course, to find the estimates on $\mathcal{M}_\delta$, depending on the given phase function φ.

In the case of the restriction problem, φ is linear in x and hence $\psi = \varphi$ in (4.1). The operator $\mathcal{M}_\delta$ defined in (4.24) becomes

(4.29) $\mathcal{M}_\delta g(y) = \sup_\xi \delta^{-(d-1)} \int_{\{|\nabla_y\varphi(x,y)-\xi|<\delta\}} |g(x)|\,dx.$

Letting φ be as in (1.4), i.e.,

(4.30) $\varphi(x, y) = x_1 y_1 + \cdots + x_{d-1} y_{d-1} + x_d \psi(y),$

the gradient equations become

(4.31) $x_i = -\partial_i \psi(y) x_d + \xi_i \quad (i = 1, \ldots, d-1).$

Because of (1.5), the map $y \mapsto \nabla\psi(y)$ yields a diffeomorphism on a neighborhood of $0 \in \mathbb{R}^{d-1}$ and it is therefore easily seen that $\mathcal{M}_\delta$ corresponds to the Kakeya maximal function from Section 3. Thus, applying Proposition 3.7, letting the exponent p from Proposition 4.27 be

$$p = 2p'_d \quad (p_d \text{ defined by (3.9)}),$$

it follows that

(4.32) $\|\mathcal{M}_\delta\|_{L^{(p/2)'}\to L^1} \le \|\mathcal{M}_\delta\|_{(p/2)'} \lesssim \left(\frac{1}{\delta}\right)^{(d/(p/2)')-1}.$

We should observe at this point that p_d given by (3.9) satisfies $p_d > \frac{d+1}{2}$, hence $p < \frac{2(d+1)}{d-1}$. Substitution of (4.32) in (4.28) gives that

(4.33) $\|Tf\|_{L^p(B(0,R))} \ll R^{(d+1)/(4p)-(d-1)/8+\varepsilon}\|f\|_\infty.$

We apply (4.33) in conjunction with Proposition 2.2. Observe that from (4.33) for $|f| < 1$

(4.34) $|A_\lambda \cap B(0, \rho)| \ll \lambda^{-p}\rho^{(d+1)/4-((d-1)/8)p+\varepsilon},$

which yields after substitution in (2.4)

(4.35) $|A_\lambda| \lesssim \lambda^{-2}\left[R + \lambda^{-p}R^{-(d-3)/4-((d-1)/8)p}\right].$

Choosing R optimally, we get the following estimate on the measure of the level set A_λ, assuming $|f| < 1$:

$$(4.36) \qquad |A_\lambda| < \lambda^{-2(2(d+1)+p(d+3))/(2(d+1)+p(d-1))-\varepsilon}, \quad \lambda < 1$$

where $p = 2p'_d$.

Observe that if we let $p = p_1 = \frac{2(d+1)}{d-1}$, we get $|A_\lambda| < \lambda^{-p_1-\varepsilon}$. Hence, since now $p < \frac{2(d+1)}{d-1}$, the exponent in (4.36) will improve. Thus, if we take $\mu \in M(S_{d-1})$ with $\frac{d\mu}{d\sigma} \in L^\infty(\sigma)$ rather than $\frac{d\mu}{d\sigma} \in L^2(\sigma)$, better integrability properties for $\hat{\mu}$ are found. For instance (cf. [B4]),

Proposition 4.37. *Let μ be a measure carried by S_2 with $\frac{d\mu}{d\sigma} \in L^\infty(S_2)$. Then $\hat{\mu} \in L^p(\mathbb{R}^3)$ for $p > \frac{58}{15}$ (< 4). The same conclusion holds assuming $\frac{d\mu}{d\sigma} \in L^p(S^2)$.*

The second part of the statement follows from the Maurey-Nikishin factorization theorem for general operators from L^∞ to L^p ($p > 2$) and considerations of rotational invariance (thus using more specifically the symmetry properties of the sphere).

Remark. In proving Proposition 4.37 by the above method, the use of the crude inequality (4.17) is a weak point in the approach. In the case of the restriction-extension problem, the situation is the following: one considers a system of boxes $\{B_\alpha\}$ approximating a spherical shell and functions f_α, supp $\hat{f}_\alpha \subset B_\alpha$ (see Fig. 2). The square function replacement (4.17) is the first step in the argument and consists in comparing

$$\left\| \sum f_\alpha \right\|_p \qquad \text{with} \qquad \left\| \left(\sum |f_\alpha|^2 \right)^{1/2} \right\|_p.$$

In the 2-dimensional case, we let $p = 4$ and essentially obtain equivalence, because of the geometric properties of the sum sets $B_\alpha + B_\beta$ (this family of

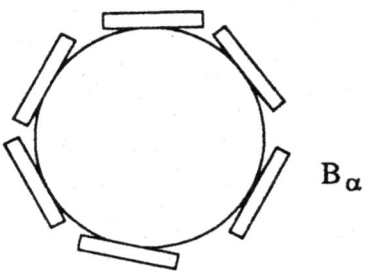

B_α

Figure 2.

sets is of bounded multiplicity). This estimate is more refined than invoking the L^2-extension theorem. It is possible to improve a bit the exponent $\frac{58}{15}$ in Proposition 4.37 by using Proposition 2.18 in developing a substitute for (4.17).

5 APPLICATION TO THE BOCHNER-RIESZ SUMMATION OPERATORS

In [B1], we used a variant of A. Cordoba's 2-dimensional argument (see [Co]), breaking up the Fourier multiplier m_λ in rectangular pieces and applying the new information in the restriction problem, together with an estimate on the Nikodym maximal function described in section 3. We follow here a different approach, namely the Carleson-Sjölin-Hörmander reduction, evaluating (1.9) from estimates on oscillatory integrals and, more precisely, again using Proposition 4.27. In fact, this will give us a better result than proved in [B1], since it will establish the conjecture provided again $p^* = p \vee p' \geq \frac{58}{15}$ for $d = 3$.

Let b be a standard bump function localizing to a neighborhood of 0 in $\mathbb{R}^d$. Using a dyadic partitioning, estimate (1.9) by

$$(5.1) \qquad \sum_{k \geq 0} 2^{-k((d+1)/2+\lambda)} \left\| \int f(y) e^{i|x-y|} b \left(2^{-k}(x-y) \right) dy \right\|_{L^p(dx)}.$$

Fix $N = 2^k$ and rescale by putting $x = Nx'$, $y = Ny'$. The L^p-norm expression then becomes

$$(5.2) \qquad N^{d/p+d} \|S_N g\|_p \qquad \text{where} \quad g(y') = f(Ny')$$

and the operator S_N is given by

$$(5.3) \qquad S_N g(x) = \int g(y) e^{iN|x-y|} b(x-y) dy.$$

Thus, an estimate of the form

$$(5.4) \qquad \|S_N g\|_p \ll N^{-d/p+\varepsilon} \|g\|_p$$

would yield from the preceding a bound on (5.1), provided p fulfills the condition $\frac{d}{p'} < \lambda + \frac{d+1}{2}$. Here p was assumed > 2 and this is thus (1.17). Again by general factorization theory and the rotational invariance, it suffices to get the $L^\infty - L^p$ inequality

$$(5.5) \qquad \|S_N f\|_p \lesssim N^{-d/p} \cdot \|f\|_\infty.$$

As proposed above, we perform the reduction to an $\mathbb{R}^{d-1} - \mathbb{R}^d$ problem, freezing the y_d-variable and obtain the phase function (letting $y_d = 1$)

(5.6)
$$\varphi(x, y) = \left[\sum_{j=1}^{d-1} (x_j - y_j)^2 + (1 - x_d)^2 \right]^{1/2}.$$

Here $y = (y_1, \ldots, y_{d-1})$ and x are taking values in a neighborhood of 0. Let $\delta > N^{-1/2}$, $M = \delta^2 N$ and

(5.7)
$$\eta(x, y) = \left[\sum_{j=1}^{d-1} (x_j - My_j)^2 + (M - x_d)^2 \right]^{1/2}.$$

Rewriting the gradient equations

(5.8)
$$\nabla_y \eta(x, y) = M\xi$$

the regions involved in the definition (4.24) of the maximal function $\mathcal{M}_\delta$ are given by
(5.9)
$$\tau = \tau_{y,\xi} = \left\{ \left| x_j - My_j + \frac{\xi_j}{(1 - |\xi|^2)^{1/2}} (x_d - M) \right| < \delta \ (1 \le j < d) \right\}.$$

(Observe that necessarily $|\xi| = 0(1)$ if we want τ to intersect a neighborhood of 0 [see Fig. 3].)

Thus the regions are neighborhoods of straight lines.

Observe that if we fix a point $\bar{x}$ in a neighborhood of 0 and let $y = (y_1, \ldots, y_{d-1})$ vary, the tubes τ_{y,ξ_y} satisfying $\bar{x} \in \tau_{y,\xi_y}$ will describe all directions. By a straightforward adaptation of the proof of Proposition 5.6 in [B1] on the Nikodym maximal function (to which $\mathcal{M}_\delta$ is obviously related, but with a

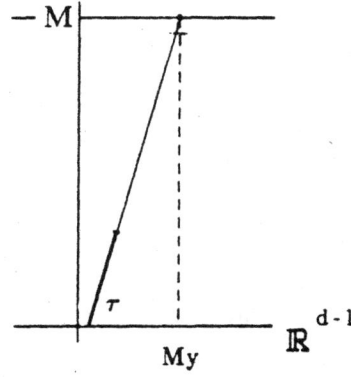

Figure 3.

missing y-dimension), one may obtain the same bounds

(5.10)
$$\|\mathcal{M}_\delta g\|_{L^p(\mathbb{R}^{d-1})} \ll \left(\frac{1}{\delta}\right)^{d/p-1+\varepsilon} \|g\|_p$$

for $p < p_d$, p_d defined by (3.9). Substituting in (4.28) where T is given by (4.2), φ given by (5.6), yields the same estimate (4.34) on the measure $A_\lambda \cap B(0, \rho)$ for the level set A_λ of Tf, $|f| \leq 1$. Hence (4.35), (4.36) are equally valid. Therefore, if

(5.11)
$$p > 2\frac{2(d+1) + q(d+3)}{2(d+1) + q(d-1)}; \quad q = 2p'_d$$

then

(5.12)
$$\left\|\int e^{iN\varphi(x,y)} f(y)a(y)dy\right\|_p \lesssim N^{-d/p}\|f\|_\infty.$$

Integrating over variable y_d (taken here to be 1) leads to the required inequality (5.5). So finally we get

Proposition 5.13. $m_\lambda(\xi) = (1 - |\xi|^2)^\lambda_+$ *defines a bounded multiplier on* $L^p(\mathbb{R}^d)$ *if* p^* *satisfies* (1.17) *and also* (5.11). *In particular, for* $d = 3$, *one gets* $p^* \geq \frac{58}{15}$ *as an extra assumption on* p *satisfying* (1.17).

6 BEHAVIOR IN THE GENERAL CASE

It turns out that in dimension $d \geq 3$, Hörmander's question on the estimates (1.16)

(6.1)
$$\left\|\int e^{iN\varphi(x,y)} f(y)a(x, y)dy\right\|_q \leq CN^{-d/q}\|f\|_r$$

has a negative answer, even if we let $r = \infty$. In fact, there are examples of phase functions φ where the $L^\infty - L^q$ estimate only holds for $q \geq \frac{2(d+1)}{d-1}$ (thus in the L^2-range), if we let d be *odd*.

Define $\varphi(x, y)$ as follows

(6.2)
$$\varphi(x, y) = x_1y_1 + \cdots + x_{d-1}y_{d-1} + 2x_d(y_1y_2 + y_3y_4 + \cdots + y_{d-2}y_{d-1})$$
$$+x_d^2(y_1^2 + y_3^2 + \cdots + y_{d-2}^2).$$

Take

(6.3)
$$f(y) = e^{iN(y_2^2 + y_4^2 + \cdots + y_{d-1}^2)}$$

so that the integral in (6.1) may be written as

(6.4)
$$\int e^{iN\{[(x_d y_1 + y_2)^2 + (x_1 y_1 + x_2 y_2)] + \cdots + [(x_d y_{d-2} + y_{d-1})^2 + (x_{d-2} y_{d-2} + x_{d-1} y_{d-1})]\}}$$

$$a(x, y) dy.$$

For $\delta > 0$, consider the set

(6.5)
$$\Omega_\delta = \bigcap_{j=1}^{(d-1)/2} [|x_{2j} \cdot x_d - x_{2j-1}| < \delta] \cap B(0, 1)$$

of measure

(6.6)
$$|\Omega_\delta| \sim \delta^{(d-1)/2}.$$

Letting $\delta \ll \frac{1}{N}$, the expression between brackets in (6.4) can be replaced by
(6.7)
$$[(x_d y_1 + y_2)^2 + x_2(x_d y_1 + y_2)] + \cdots + [(x_d y_{d-2} + y_{d-1})^2 + x_{d-1}(x_d y_{d-2} + y_{d-1})]$$

for $x \in \Omega_\delta$.

Putting $z_1 = x_d y_1 + y_2$, $z_3 = x_d y_3 + y_4, \ldots, z_{d-2} = x_d y_{d-2} + y_{d-1}$, (6.4) essentially amounts to

(6.8)
$$\prod_{j=1,3,\ldots,d-2} \left(\int e^{iN(z_j^2 + x_{j+1} z_j)} a(z_j) dz_j \right).$$

The individual factors in (6.8) are of size $\sim N^{-1/2}$, hence

(6.9)
$$(6.8) \sim N^{-(d-1)/4}.$$

Thus (6.1) admits the lower bound

(6.10)
$$|\Omega_\delta|^{1/q} \cdot N^{-(d-1)/4} \sim N^{-(d-1)/(2q)-(d-1)/4}$$

restricting x to Ω_δ, using (6.6), (6.9). Inequality (6.1) may thus only hold if

(6.11)
$$\frac{d-1}{2q} + \frac{d-1}{4} \geq \frac{d}{q}$$

thus $q \geq \frac{2(d+1)}{d-1}$.

In particular, for $d = 3$, φ is given by

(6.12)
$$\varphi(x, y) = x_1 y_1 + x_2 y_2 + 2x_3 y_1 y_2 + x_3^2 y_1^2$$

and the gradient equations $\nabla_y \varphi = \xi$ are

(6.13)
$$\begin{cases} x_1 = -2y_2 x_3 - 2y_1 x_3^2 + \xi_1 \\ x_2 = -2y_1 x_3 + \xi_2. \end{cases}$$

Put $\xi = \xi_y = (0, -2y_2)$. For this translation, one thus gets curves $\Gamma_y \subset \mathbb{R}^3$ contained in the 2-dimensional surface $x_1 = x_2 x_3$. This compression phenomenon is responsible for bad behavior of the maximal operators $\mathcal{M}_\delta$ defined by (4.24), in the sense that for $q \geq 2$, $\|\mathcal{M}_\delta\|_{q \to q} \sim \delta^{-1/q}$.

The previous behavior is not generic, however. In fact, one shows in [B2] that, if $d = 3$, in the generic case the best q satisfying (6.1) with $r = \infty$ lies in the open interval $] \frac{2d}{d-1}, \frac{2(d+1)}{d-1} [$ (and depends on φ).

More generally, in the example (6.2), the gradient curves Γ_y are translated in the $\frac{d+1}{2}$-dimensional surface

(6.14) $$x_{2j-1} = x_{2j} \cdot x_d \quad j = 1, 2, \ldots, \frac{d-1}{2}.$$

A compression in a set of $\frac{d+1}{2}$-Hausdorff dimension seems impossible if d is *even*. We give a rough sketch of the argument for $d = 4$.

Proposition 6.15. *Let φ be given by* (1.15) ($d = 4$). *For y in a neighborhood of $0 \in \mathbb{R}^{d-1}$, define*

(6.16) $$\Gamma_{y,\delta} = \left[|\nabla_y \varphi(x, y) - \xi_y| < \delta \right]$$

where the $\xi_y \in \mathbb{R}^{d-1}$ are chosen arbitrarily. Consider the set

(6.17) $$\Omega_\delta = \bigcup_y \Gamma_{y,\delta}.$$

This set has δ-entropy (in the metrical sense) $> \delta^{-((d+1)/2)-\tau}$ for some $\tau = \tau_\varphi > 0$.

The general idea of the proof is as follows. We will first observe that the entropy conclusion is true with $\tau = 0$. Assuming then Ω_δ of δ-entropy $\sim \delta^{-(d+1)/2}$, appearing as an "extremal" case, new geometrical information on the structure of the Γ_y is gained, allowing us to see eventually that the δ-entropy number of Ω_δ needs to be $\gtrsim \delta^{-(d/2)-1}$. The argument produces the conclusion of 6.15, with a $\tau = \tau_\varphi$ which is numerical, provided bounds are imposed on sufficiently many derivatives of φ. To make matters a bit more concrete, take φ of the form

(6.18) $$\varphi(x, y) = \sum_{i=1}^{d-1} x_i y_i - \frac{1}{2} x_d (y_1^2 + \cdots + y_{d-1}^2) + 0(|x|^2 |y|^2 + |x| \, |y|^3)$$

so that $\nabla_y \varphi = \xi$ becomes

(6.19) $$x_i = y_i x_d + 0(|x|^2 |y| + |x| \, |y|^2) + \xi_i \quad (i = 1, \ldots, d-1)$$

describing the curve Γ_y.

(i)   If one specifies a point $\bar{x} \in \Gamma_y$ and lets $x' = x - \bar{x}$, (6.19) clearly becomes

(6.20) $$x_i' = y_i x_d' + 0 \left(|x'| \, |y| (|\bar{x}| + |x'| + |y|) \right) .$$

From (6.20), $|x_i'| \lesssim |y| \, |x_d'|$ $(1 \le i \le d - 1)$. Also, if we fix x_d' away from 0, application of the implicit function theorem yields a diffeomorphic correspondence between $(x_1', \ldots, x_{d-1}')$ and y. For instance, $J_y = x_d' \, Id + o(|x'|)$ and hence is invertible from the preceding (see Fig. 4).

(ii)   Consider in the y-variable a net $\mathcal{E}$ of δ-separated points. Thus $|\mathcal{E}| =$ card$(\mathcal{E}) \sim \delta^{-(d-1)}$. Let $\mathcal{E}_1 \subset \mathcal{E}$ such that the $\Gamma_{y,\delta}$, $y \in \mathcal{E}_1$, have a common point. It follows easily from (i) that

(6.21) $$e_\delta \left(\bigcup_{y \in \mathcal{E}_1} \Gamma_{y,\delta} \right) \gtrsim \delta^{-1} |\mathcal{E}_1|$$

denoting by $e_\delta(\Omega)$ the δ-entropy numbers (= minimal number of δ-balls required for a covering of Ω).

Hence, if we assume Ω_δ defined by (6.17) satisfies

(6.22) $$e_\delta(\Omega_\delta) \lesssim \delta^{-(d+1)/2},$$

it follows that

(6.23) $$|\mathcal{E}_1| \lesssim \delta^{-(d-1)/2}$$

for such family $\mathcal{E}_1$.

(iii) Fix separated values x_d^1, x_d^2 of x_d such that the corresponding hyperplanes H_α: $x_d = x_d^\alpha$ $(\alpha = 1, 2)$ get intersected by all Γ_y in points P_y^α (see Fig. 5). Consider the map $y \mapsto (P_y^1, P_y^2) \in (\Omega_\delta \cap H_1) \times (\Omega_\delta \cap H_2)$.

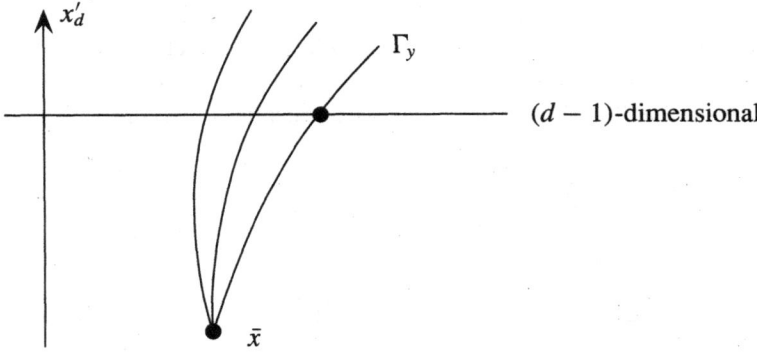

Figure 4.

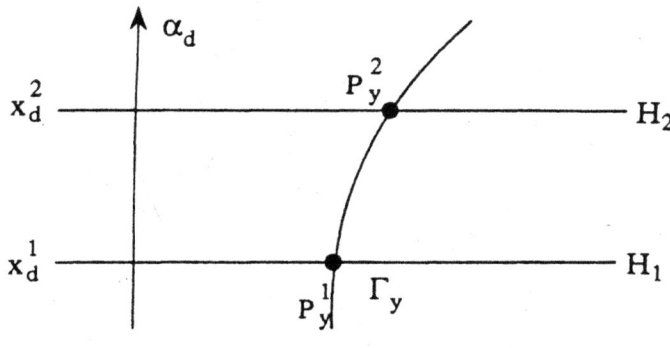

Figure 5.

It follows again from (i) that y may be derived from its image in a Lipschitz way and therefore

(6.24) $\qquad \delta^{-(d-1)} \sim e_\delta(\{y\}) \lesssim e_\delta(\Omega_\delta \cap H_1) \cdot e_\delta(\Omega_\delta \cap H_2)$.

Because of the free choice of the pair (x_d^1, x_d^2), this forces

(6.25) $\qquad\qquad\qquad e_\delta(\Omega_\delta) \gtrsim \delta^{-(d+1)/2}$.

Next, we will assume

(6.26) $\qquad\qquad\qquad e_\delta(\Omega_\delta) \sim \delta^{-(d+1)/2}$

in order to derive more information on the structure of the curves Γ_y, and finally get a contradiction.

(iv) It follows from (iii), assuming (6.26), that a typical $[x_d = \bar{x}_d]$-hyperplane H will satisfy

(6.27) $\qquad\qquad\qquad e_\delta(\Omega_\delta \cap H) \sim \delta^{-(d-1)/2}$.

Fixing H, there is thus a δ-separated system $\mathcal{P}$ of $\sim \delta^{-(d-1)/2}$ points $P \in H$ such that $\Gamma_{y,\delta} \cap \mathcal{P} \neq \emptyset$ for each y. Also, from (ii), each point $P \in \mathcal{P}$ will belong to $\Gamma_{y,\delta}$ for $\lesssim \delta^{-(d-1)/2}$ values of $y \in \mathcal{E}$. This leads to a partition of (a large subset) of $\mathcal{E}$ in sets $\mathcal{E}_P$

(6.28) $\qquad\qquad \mathcal{E} = \bigcup_{P \in \mathcal{P}} \mathcal{E}_P; \quad |\mathcal{E}_P| \sim \delta^{-(d-1)/2}$

and for $P \in \mathcal{P}$ a "bush" $\mathcal{B}_P$ of curves $\{\Gamma_y \mid y \in \mathcal{E}_P\}$, where Γ_y has the equation

(6.29) $\qquad\qquad\qquad \nabla_y \varphi(x, y) = \nabla_y \varphi(P, y)$

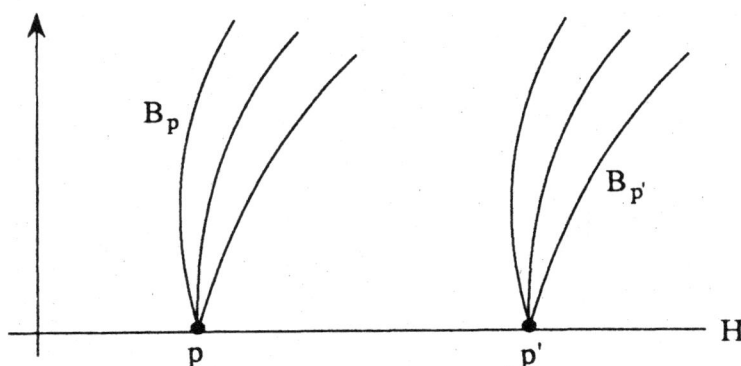

Figure 6.

for $y \in \mathcal{E}_P$. The set Ω_δ is contained in a δ-neighborhood of $\cup_{P \in \mathcal{P}} \mathcal{B}_P$ (see Fig. 6).

(v)   If we fix an individual point P, it follows from (6.28) that

(6.30) $$e_\delta(\mathcal{B}_P) \geq \delta^{-1} \delta^{-(d-1)/2} = \delta^{-(d+1)/2}$$

and hence the δ-neighborhood $\mathcal{B}_P^\delta$ represents a "large" portion of Ω_δ.

(vi)   Let $\delta \ll \varepsilon$ and fix an ε-ball $B(Q, \varepsilon)$ in H. Then

(6.31) $$|\mathcal{P} \cap B(Q, \varepsilon)| \lesssim \left(\frac{\varepsilon}{\delta} \right)^{(d-1)/2}.$$

Indeed, let $\mathcal{P}' = \mathcal{P} \cap B(Q, \varepsilon)$, $|\mathcal{P}'| = K$ and $\mathcal{E}' = \cup_{p \in \mathcal{P}'} \mathcal{E}_P$. Thus by (6.28)

$$|\mathcal{E}'| \sim K \delta^{-(d-1)/2}.$$

Consider an ε-net $\mathcal{I}$ in $\mathcal{E}'$ and let $\mathcal{E}' = \cup_{z \in \mathcal{I}} \mathcal{E}''_z$ be a partitioning of $\mathcal{E}'$ in sets of diameter ε. By construction, if $y \in \mathcal{E}''_z$, the curve Γ_y will be in an 10ε-neighborhood of Γ_z, since the basepoints in H are ε-close and $|y - z| < \varepsilon$. Therefore

(6.32) $$e_\delta \left(\bigcup_{y \in \mathcal{E}'} \Gamma_y \right) \sim \sum_{z \in \mathcal{I}} e_\delta \left(\bigcup_{y \in \mathcal{E}''_z} \Gamma_y \right).$$

Also, the same considerations as in (iii) show that

(6.33) $$e_\delta \left(\bigcup_{y \in \mathcal{E}''_z} \Gamma_y \right) \gtrsim \delta^{-1} |\mathcal{E}''_z|^{1/2}.$$

Since $|\mathcal{E}_z''| < \left(\frac{\varepsilon}{\delta}\right)^{d-1}$, (6.26), (6.32), (6.33) imply that

$$\delta^{-(d+1)/2} \gtrsim \delta^{-1} \sum_{z \in \mathcal{I}} |\mathcal{E}_z''|^{1/2} \gtrsim \delta^{-1} \left(\frac{\varepsilon}{\delta}\right)^{-(d-1)/2} \quad |\mathcal{E}'| \sim \delta^{-1} \varepsilon^{-(d-1)/2} K$$

and hence (6.31) follows.

(vii) It follows from (v), thus (6.26) and (6.30), together with general measure theoretic considerations on set intersections, that

$$(6.34) \qquad e_\delta \left(\mathcal{B}_P^\delta \cap \mathcal{B}_{P'}^\delta \right) \sim \delta^{-(d+1)/2}$$

for $\sim \delta^{-(d-1)}$ pairs (P, P') in the product $\mathcal{P} \times \mathcal{P}$. From (vi), one can, moreover, assume that the points P, P' are far apart. If (6.34) holds, one gets additional information on the curves Γ_y for $y \in \mathcal{E}_P$, since a large subset of $\mathcal{B}_P$ admits a foliation in 2-dimensional surfaces S obtained by fixing some $y' \in \mathcal{E}_{P'}$ and varying Γ_y containing P and intersecting $\Gamma_{y'}$ (see Fig. 7). Based on this consideration, a straightforward construction permits us to replace $\mathcal{B}_P$ by a union of $\sim \delta^{-(d-3)/2}$ surfaces $S \in \mathcal{S}_P$,

$$(6.35) \qquad \mathcal{B}_P = \bigcup_{S \in \mathcal{S}_P} S$$

each S of dimension 2 generated by moving the curve Γ_y described by (6.29) along some other curve, as indicated above.

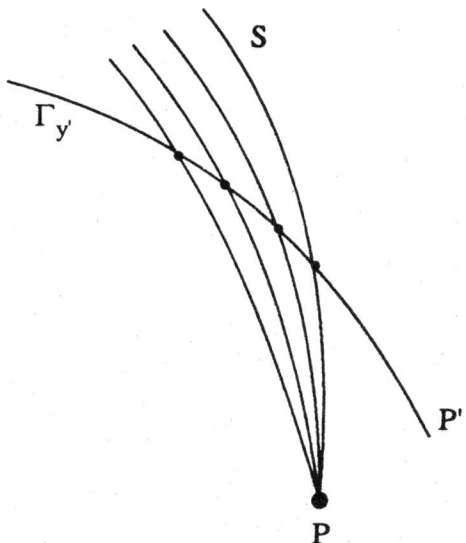

Figure 7.

(viii) Fix a smooth 2-dimensional surface S away from H, a point $\bar{x} \in S$, and let $T_{\bar{x}}$ be the tangent plane of S at $\bar{x}$. Consider the family of curves Γ_y through $\bar{x}$ given by (6.20). By (6.20), the tangent vector $\vec{t}_y = \vec{t} = (t_1, \ldots, t_{d-1}, 1)$ of Γ_y at $\bar{x}$ is described by the equations

$$(6.36) \qquad t_i = y_i + 0(|t|\,|y|(|\bar{x}| + |y|)) \quad i = 1, \ldots, d-1.$$

Hence

$$(6.37) \qquad t_i = y_i + 0(|y|(|\bar{x}| + |y|)).$$

Restricting $\vec{t}$ to lie in $T_{\bar{x}}$, a smooth 1-parameter family of directions is obtained, corresponding to a curve γ_1 in the y-variable and a curve $\gamma_2 = \{\Gamma_y \cap H; y \in \gamma_1\}$ of points in H. Consequently, one may find a neighborhood S' of $\bar{x}$ in S such that if the basepoint P of Γ_y in H lies away from γ_2, Γ_y may only have a "transversal" intersection with S, assuming Γ_y intersects S'.

(ix) From (6.34), (6.35), one may find a point $P \in \mathcal{P}$ such that

$$(6.38) \qquad \sum_{S \in S_P} e_\delta \left(S \cap \mathcal{B}_{P'}^\delta\right) \sim \delta^{-(d+1)/2}$$

for $\sim \delta^{-(d-1)/2}$ points $P' \in \mathcal{P}$. From (6.33), there is some $S \in \mathcal{S}_P$ satisfying

$$(6.39) \qquad e_\delta \left(S \cap \mathcal{B}_{P'}^\delta\right) \sim \delta^{-2}$$

for $\sim \delta^{-(d-1)/2}$ points $P' \in \mathcal{P}$. Identifying S with a point neighborhood S' as constructed in (viii), we get a curve γ in H such that the intersection of S and Γ_y (if any) is transversal, provided its basepoint P' lies away from γ. But from (vi), fixing $\varepsilon \gg \delta$, the number of points $P' \in \mathcal{P}$ at ε-distance from γ is at most

$$(6.40) \qquad \sim \varepsilon^{-1} \left(\frac{\varepsilon}{\delta}\right)^{(d-1)/2} = \varepsilon^{(d-3)/2}\delta^{-(d-1)/2}$$

due to (6.31).

Hence, from the preceding, letting ε be small enough, a point $P' \in \mathcal{P}$ satisfying (6.34) and lying at ε distance from γ is obtained ($d = 4 > 3$). Hence, a large subset of S lies on the Γ_y ($y \in \mathcal{E}_{P'}$)-curves, intersecting, moreover, S in a transversal way (see Fig. 8).

Since the corresponding structure contains a diffeomorph of a neighborhood of 0 in $\mathbb{R}^3$, this implies that

$$(6.41) \qquad e_\delta(\Omega_\delta) \geq e_\delta \left(\mathcal{B}_{P'}^\delta\right) \gtrsim \delta^{-3},$$

contradicting (6.26) for $d = 4$.

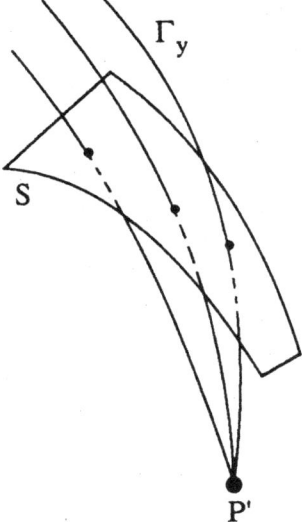

Figure 8.

In the general case, this last construction will yield a representation (6.35) where now the S are 3-dimensional. One repeats the transversality analysis (viii) and gets 2-dimensional exceptional sets. The estimate (6.35) becomes $\varepsilon^{(d-5)/2}\delta^{-(d-1)/2}$, etc., so for $d \geq 6$, 4-dimensional structures are produced. Continuing, one eventually gets

(6.42) $$e_\delta(\Omega_\delta) \gtrsim \delta^{-[(d-1)/2]-2} = \delta^{-d/2-1} \quad (d \text{ is even})$$

contradicting (6.26).

7 AN APPLICATION TO TWO-DIMENSIONAL SCHRÖDINGER OPERATORS

Based on the results from sections 2 and 4 of this paper, we sketch an alternative proof of the following result from [B3].

Proposition 7.1. *Let* $u = u(x, t)$ *be the solution of the Schrödinger equation with initial data* $f \in \mathcal{S}(\mathbb{R}^3)$

(7.2) $$\Delta u = i \frac{\partial}{\partial t} u; u(x, 0) = f(x).$$

Then, assuming f in the Sobolev space $H^s(\mathbb{R}^2)$, for some $s > \rho$ where $\rho = \rho_2 < \frac{1}{2}$, there is almost everywhere convergence

$$\text{(7.3)} \qquad \lim_{t \to 0} u(x, t) = f(x) \quad a.e. \text{ in } x$$

and a control on the maximal operators

$$\text{(7.4)} \qquad u^*(x) = \sup_{|t|>0} |u(x, t)|.$$

The new element is the fact that the ρ-exponent is strictly less than $\frac{1}{2}$. It is known that for dimension $d = 1$, the assumption $f \in H^{1/4}(\mathbb{R})$ ensures (7.3), which is the optimal result (see [Ca], [D-K]). The statement of Proposition 7.1 with $\rho = \frac{1}{2}$ was shown by Vega [V] in arbitrary dimension.

Recall that the solution u of (7.2) is given by the Fourier integral

$$\text{(7.5)} \qquad u(x, t) = \int \hat{f}(\xi) e^{i(<x,\xi> + t|\xi|^2)} d\xi.$$

The statement of Proposition 7.1 will result from an inequality of the form

$$\text{(7.6)} \qquad \left\| \sup_{0<t<1} \left| \int_{|\xi|<N} \hat{f}(\xi) e^{i(<x,\xi> + t|\xi|^2)} d\xi \right| \right\|_{L^2(B(0,1))} \leq N^\rho \left(\int \left| \hat{f}(\xi) \right|^2 d\xi \right)^{1/2}$$

for some $\rho < \frac{1}{2}$.

Choose some $q > 2$ and estimate the left member of (7.6) by appropriate change of variables (rescaling)

$$\text{(7.7)} \qquad N^{2-2/q} \left\| \sup_{0<t<N^2} \left| \int_{|\xi|<1} \hat{f}(N\xi) e^{i(<x,\xi> + t|\xi|^2)} d\xi \right| \right\|_{L^q(dx)}.$$

Using standard considerations, one may then estimate it further by

$$\text{(7.8)} \qquad N^{2(1-1/q)} \left\| \int_{|\xi|<1} \hat{f}(N\xi) e^{i(<x,\xi> + t|\xi|^2)} d\xi \right\|_{L^q(dx\,dt)}.$$

Consider the surface $(\xi, |\xi|^2)$ in $\mathbb{R}^3$, restricting ξ to the unit disc. Since there is obviously curvature and smoothness, the restriction and extension theory applies equally well here as it does for the sphere. Call (p, q) an admissible exponent pair provided

$$\text{(7.9)} \qquad \|\hat{\mu}\|_q \leq C \left\| \frac{d\mu}{d\sigma} \right\|_p$$

with μ a measure carried by the 2-sphere S_2 or the restricted paraboloid $\mathcal{P}$ considered above. The classical L^2-restriction theorem states then that $(2, 4)$ is admissible, and in section 4 admissible pairs were obtained with $q < 4$.

Applying the (p, q) pair in (7.8), one gets a bound

$$(7.10) \quad N^{2(1-1/q)} \left(\int \left| \hat{f}(N\xi) \right|^p d\xi \right)^{1/p} = N^{2(1-1/p-1/q)} \left(\int \left| \hat{f}(\xi) \right|^p d\xi \right)^{1/p}.$$

For $p = 2, q = 4$, that is inequality (7.6) with $\rho = \frac{1}{2}$.

It was shown in section 2 (see discussion following Proposition 2.18) that this (2, 4)-estimate may be improved, unless the density "corresponds" to the indicator function of a cap (as a rough statement). More precisely, if we assume $\hat{f}$ to be of the form

$$(7.11) \qquad\qquad \hat{f}(\xi) = \frac{\chi_\Omega}{|\Omega|^{1/2}} \quad (\Omega \subset B(0, N))$$

then an improvement will be obtained, unless for some square Q one has

$$(7.12) \qquad\qquad |\Omega| \sim |Q| \sim |\Omega \cap Q|.$$

Remark. $\hat{f}$ may always be broken up in level sets. The meaning of $\sim$ actually will allow factors of the form N^δ for some specific δ and so yields an "improvement." That this is the result of the reasoning below I leave to the reader to check.

So assume (7.12) holds. If $|Q| \sim N^2$, apply estimate (7.10) for an admissible pair (p, q) with $q < 4$ gotten from [B1]. Since here

$$(7.13) \qquad\qquad \left(\int \left| \hat{f}(\xi) \right|^p d\xi \right)^{1/p} \sim N^{2/p} N^{-1}$$

we get an estimate of the form

$$(7.14) \qquad\qquad\qquad N^{1-2/q}$$

with $1 - \frac{2}{q} < \frac{1}{2}$.

If $|Q| = N_1^2$, with $N_1 = N^{1-s}$, thus $Q \subset \xi_0 + B(0, N_1)$, we proceed as follows. Write in (7.7)

$$(7.15) \qquad \xi = \xi_0' + \eta; \ |\xi|^2 = |\xi_0'|^2 + 2\langle \xi_0', \eta \rangle + |\eta|^2$$

where $\xi_0' = \frac{\xi_0}{N}$, $|\eta| < \frac{N_1}{N}$.

It is clear from (7.15) that the parameter-values needed to recapture the supremum for $t \in [0, N^2]$ may be taken in a $\frac{N}{N_1}$-net and hence the passage to the t-integral and (7.8) gives a saving factor $\left(\frac{N_1}{N} \right)^{1/q}$. Then continue with the (2, 4)-extension theorem to conclude also that case. If one pursues this argument a bit more explicitly, it easily leads to (7.6) with an exponent $\rho < \frac{1}{2}$.

Institut des Hautes Études Scientifiques, Bures-sur-Yvette, France

REFERENCES

[B1] J. Bourgain. "Besicovitch type maximal operators and applications to Fourier analysis." *Geom. Funct. Anal.* **1:2** (1991), 147–187.

[B2] _____. "L^p-estimates for oscillatory integrals in several variables." *Geom. Funct. Anal.* **1:4** (1991), 321–374.

[B3] _____. "A remark on Schrödinger operators." To appear in *Israel J. Math.*

[B4] _____. *On the restriction and multiplier problem in* $\mathbb{R}^3$. Lecture Notes in Mathematics, no. 1469. Springer Verlag, 1991.

[Ca] L. Carleson. "Some analytical problems related to statistical mechanics." In *Euclidean Harmonic Analysis*. Lecture Notes in Mathematics, no. 779. Springer Verlag, 1979.

[Co] A. Cordoba. "A note on Bochner-Riesz operators." *Duke Math. J.* **46**, N3 (1979), 565–572.

[C-S] L. Carleson and P. Sjölin. "Oscillatory integrals and a multiplier problem for the disc." *Studia Math.* **44** (1972), 287–299.

[D-K] B. Dahlberg and C. Kenig. *A note on almost everywhere behavior of solutions to the Schrödinger equations*. Lecture Notes in Mathematics, no. 908. Springer Verlag, 1982.

[Fal] K. J. Falconer. "The geometry of Fractal sets." *Cambridge Tracts Math.* **85**.

[F1] C. Fefferman. "The multiplier problem for the ball." *Ann. of Math.* **94** (1971), 330–336.

[F2] _____. "A note on spherical summation multipliers." *Israel J. Math.* (1973), 44–52.

[F3] _____. "Inequalities for strongly singular convolution operators." *Acta Math.* **124** (1970), 9–36.

[Hör] L. Hörmander. "Oscillatory integrals and multipliers on FL^p." *Arkiv Math. II* (1973), 1–11.

[St] E. Stein. "Oscillatory integrals in Fourier Analysis." In *Beijing Lectures in Harmonic Analysis*. E. M. Stein, ed. Annals of Mathematics Studies 112. Princeton University Press, 1987.

[T] P. Tomas. "A restriction theorem for the Fourier transform." *Bull. Amer. Math. Soc.* **81** (1975), 477–478.

[V] L. Vega. "The Schrödinger equation: pointwise convergence to the initial data." *Proc. Amer. Math. Soc.* **102** (1988), 875–878.

5

Dilations Associated to Flat Curves in $\mathbb{R}^n$

*Anthony Carbery, James Vance,
Stephen Wainger,* and David Watson*

We would like to give a brief exposition of our recent work on Hilbert transforms and maximal functions along curves in R^n [CVWW]. In particular, we wish to describe a theory of dilations that seems useful in this work. This is an outgrowth of joint work with Mike Christ that dealt with the case $n = 2$ [CCVWW]. We would like to acknowledge the important contributions Mike made to this project in its earlier stage.

We let $\Gamma(t) = (t, \gamma_2(t) \ldots, \gamma_n(t))$ be a smooth curve in R^n with $\Gamma(0) = 0$. For a function f in $C_0^\infty(R^n)$, we define

$$(1) \qquad H_\Gamma f(x) = \int_{-1}^{1} f(x - \Gamma(t)) \frac{dt}{t},$$

and

$$(2) \qquad M_\Gamma f(x) = \sup_{0 < h \le 1} \frac{1}{h} \int_0^h |f(x - \Gamma(t))| dt.$$

We are interested in the problem of obtaining estimates of the form

$$(3) \qquad \|H_\Gamma f\|_{L^p(R^n)} \le A(p, \Gamma) \|f\|_{L^p(R^n)},$$

and

$$(4) \qquad \|M_\Gamma f\|_{L^p(R^n)} \le A(p, \Gamma) \|f\|_{L^p(R^n)}.$$

*Supported, in part, by a grant from the National Science Foundation.

If the estimate (4) holds for some $p < \infty$,

$$\lim_{h \to 0} \frac{1}{h} \int_0^h f(x - \Gamma(t)) dt = f(x) \quad \text{a.e.}$$

for every f locally in L^p.

Positive results for (3) and (4) have been known for a long time under an appropriate "curvature hypothesis" on Γ. The following theorem was due to the efforts of Nagel, Rivière, Stein, and Wainger in the 1970s. See [SW].

Theorem of the 1970s. *Suppose $\Gamma(t)$ satisfies the following curvature condition:*

(5) *For small t, $\Gamma(t)$ lies in the span of the vectors $\Gamma'(0)$, $\Gamma''(0) \ldots$,*

$$\Gamma^{(j)}(0), \ldots$$

Then

$$\|H_\Gamma f\|_{L^p(R^n)} \le A(p, \Gamma) \|f\|_{L^p(R^n)}, \quad 1 < p < \infty,$$

and

$$\|M_\Gamma f\|_{L^p(R^n)} \le A(p, \Gamma) \|f\|_{L^p(R^n)}, \quad 1 < p \le \infty.$$

Examples of curves satisfying the curvature hypothesis (5), are

(6) $$\Gamma(t) = (t, t^{k_2} \ldots, t^{k_n})$$

and

(7) $$\Gamma(t) = (t, t^{k_2} + t^{k_2+\ell_2} \ldots, t^{k_n} + t^{k_n+\ell_n})$$

where the k_j and ℓ_j are positive integers. Examples of curves not satisfying the curvature hypothesis are

(8) $$\Gamma(t) = (t, e^{-1/|t|^{a_2}} \ldots, e^{-1/|t|^{a_n}})$$

where $a_2 \ldots, a_n$ are positive integers.

An important tool used in proving the Theorem of the 1970s was that of a contracting group of linear transformations. A collection of non-singular linear transformations T_λ on R^n defined for $\lambda > 0$ is called a contracting group of linear transformations if

(9) $$T_\lambda T_\mu = T_{\lambda\mu}, \quad \text{for } \lambda > 0 \text{ and } \mu > 0,$$

(10) $$T_1 x = x,$$

for each x in R^n, and

(11) $$\lim_{\lambda \to 0} T_\lambda x = 0,$$

for each x in R^n.

Note that the examples in (6) are "homogeneous" with respect to a dilation group. That is

(12) $$\Gamma(\lambda t) = T_\lambda \Gamma(t)$$

where T_λ is the contracting dilation group defined by

$$T_\lambda(x_1, x_2 \ldots, x_n) = (\lambda x_1, \lambda^{k_2} x_2 \ldots, \lambda^{k_n} x_n).$$

The examples (7) do not satisfy (12), of course, but still for λ positive and small and t small

$$\Gamma(\lambda t) \sim T_\lambda \Gamma(t).$$

(Perhaps more importantly the set of curves

$$\Gamma_\lambda(t) = T_\lambda^{-1} \Gamma(\lambda t), \quad \lambda \leq 1$$

forms a compact family of curves on $[1, 2]$.)

Groups of dilations played three roles in the Theorem of the 1970s. First the dilations allowed one to "normalize" certain estimates. Furthermore, the homogeneity was used as an aid in bounding from below the absolute value of derivatives of $\xi \cdot \Gamma(t)$, for vectors ξ in R^n, so that the lemmas of Van Der Corput might be used. Finally, the dilations provided a variant of the Calderón-Zygmund theory.

Let us first give an illustration of what we mean by "normalizing" estimates. It is not hard to see that the L^2-boundedness of H_Γ is equivalent to the uniform boundedness of

$$m(\xi) = \int_{-1}^{1} \exp(i\xi \cdot \Gamma(t)) \frac{dt}{t},$$

over vectors ξ in R^n. Let

$$m_j(\xi) = \int_{2^{-j} \leq |t| \leq 2 \cdot 2^{-j}} \exp(i\xi \cdot \Gamma(t)) \frac{dt}{t}.$$

One might expect the estimate for $m_j(\xi)$ to be very subtle depending on the curvature of Γ near the origin. In the examples $\Gamma(t) = (t, t^{k_2} \ldots, t^{k_n})$, for instance, one might expect the behavior of $m_j(\xi)$ to depend in a complicated way on $k_2 \ldots, k_n$. However, if $\Gamma(t)$ is homogeneous with respect to a group T_λ, we can argue as follows:

$$m_j(\xi) = \int_{2^{-j} \le |t| \le 2 \cdot 2^{-j}} \exp(i\xi \cdot \Gamma(t)) \frac{dt}{t}$$

$$= \int_{1 \le |t| \le 2} \exp(i\xi \cdot \Gamma(2^{-j}t)) \frac{dt}{t}$$

$$= \int_{1 \le |t| \le 2} \exp(i\xi \cdot T_{2^{-j}}\Gamma(t)) \frac{dt}{t}$$

$$= \int_{1 \le |t| \le 2} \exp(iT_{2^{-j}}^*\xi \cdot \Gamma(t)) \frac{dt}{t}.$$

In the last integral, t stays in a compact interval not containing the origin, and we can expect the curvature of $\Gamma(t)$ to behave in a uniform way as it does in the examples. Thus the various types of behavior that might occur for $m_j(\xi)$ (depending in the examples on $k_2 \ldots, k_n$) is expressed in terms of $T_{2^{-j}}^*\xi$.

Next, the homogeneity of $\Gamma(t)$ implies that the curve $\eta(t) = \Gamma(e^t)$ satisfies the differential equation

(13) $$\eta'(t) = A\eta(t)$$

for an appropriate $n \times n$ matrix A ($\exp(A \log t) = T_t$). The differential equation (13) can be used to obtain estimates from below on derivatives of $\eta(t)$ so that well-known estimates of Van Der Corput can be used to obtain estimates on $m_j(\xi)$. See [SW]. As a matter of fact one obtains the estimate

(14) $$|m_j(\xi)| \le C\|T_{2^{-j}}^*(\xi)\|^{-1/n},$$

which together with the trivial estimate

$$|m_j(\xi)| \le C\|T_{2^{-j}}^*(\xi)\|$$

implies that $m(\xi)$ is bounded.

It turns out that to obtain L^p estimates for $p \ne 2$, one requires an appropriate version of the Calderón-Zygmund theory. It is well known that there is a Calderón-Zygmund theory corresponding to a contracting dilation group. See [R]. In the 1970s L^p estimates were obtained by using the complex interpolation of analytic families of operators due to Stein [S]. For example in studying the operator H_Γ, one introduced the analytic family of operators H_Γ^z defined by

$$\widehat{H_\Gamma^z f}(\xi) = m_z(\xi)\hat{f}(\xi),$$

where

$$m_z(\xi) = \rho^z(\xi) \int |t|^z e^{i\xi \cdot \Gamma(t)} \frac{dt}{t}$$

for an appropriate distance function $\rho(\xi)$. One shows H^z bounded on L^2 for $\text{Re}\, z \leq \epsilon$, for some positive ϵ. Further, one shows that for $\text{Re}\, z < 0$, H^z is convolution with a kernel satisfying the appropriate Calderón-Zygmund condition, and so is bounded in all L^p. Stein's Theorem then implies that H_Γ is bounded in all L^p, $1 < p < \infty$. More recent arguments inspired by work of Christ [C], and Duoandikoextea and Rubio de Francia [DR], employ a Paley-Littlewood argument. In order to employ the argument, one must show that for a smooth function ϕ with compact support vanishing near the origin, the multipliers

(15) $$m_t(\xi) = \Sigma r_j(t)\phi(T^*_{2^{-j}}\xi)$$

(where $r_j(t)$ denotes the usual Rademacher functions) are bounded L^p multipliers. One does this by checking that the inverse Fourier Transform of $m_t(\xi)$ satisfies a Calderón-Zygmund condition.

For a curve such as $\Gamma(t) = (t, t^{k_2} \ldots, t^{k_n})$, the appropriate dilation group

$$T_\lambda(x) = (\lambda x_1, \lambda^{k_2} x_2 \ldots, \lambda^{k_n} x_n)$$

stares one in the face. However, it is not so clear how to choose dilations for a curve such as

$$\Gamma(t) = (t, e^{-1/|t|^{a_1}} \ldots, e^{-1/|t|^{a_n}}).$$

Perhaps a first guess might be to choose

$$T^p_\lambda(x) = (\lambda x_1, \gamma_2(\lambda)x_2 \ldots, \gamma_n(\lambda)x_n)$$

if $\Gamma(t) = (t, \gamma_2(t) \ldots, \gamma_n(t))$ (p is for "provisional"). We shall see that this provisional guess has some problems, but nevertheless we can gain some insight into matters by considering T^p_λ. First of all, we must give up the group property (9). Moreover, it is not even true that $\Gamma_\lambda(t) = T^{-1}_\lambda\Gamma(\lambda t)$ forms a compact family on $[1, 2]$. (For example, $e^{1/\lambda}e^{-1/\lambda t}$ is not uniformly bounded for $1 \leq t \leq 2$.) On the other hand, sums of derivatives of $\Gamma_\lambda(t) \cdot \xi$ do have uniform estimates from below of the type which are useful for the application of Van Der Corput's Lemmas.

We can see that instead of the group property (9), we might hope to have

(16) $$\|T^{-1}_s T_t\| \leq C(t/s)^\epsilon, \quad s \geq t$$

for some positive ϵ. This condition was essentially introduced by Rivière [R]. Something like (16) would seem natural if we want to bound a multiplier like $m_t(\xi)$ defined in (15), for by taking adjoints in (16) we can see that the inequality would imply that the supports of the functions $\phi(T^*_{2^{-j}}\xi)$ would be essentially disjoint. Further, while we do not expect (12) to hold, at least we might expect

(17) $$\Gamma(t) = T_t e$$

for some fixed vector e.

Also normalizing as in the homogeneous case, we see

$$m_j(\xi) = \int_{2^{-j} \le |t| \le 2 \cdot 2^{-j}} \exp(i\xi \cdot \Gamma(t)) \frac{dt}{t}$$

$$= \int_{1 \le |t| \le 2} \exp(i\xi \cdot \Gamma(2^{-j}t)) \frac{dt}{t}$$

$$= \int_{1 \le |t| \le 2} \exp i(\xi \cdot T_{2^{-j}} T_{2^{-j}}^{-1} \Gamma(2^{-j}t)) \frac{dt}{t} .$$

or

(18) $$m_j(\xi) = \int_{1 \le |t| \le 2} \exp(i \cdot T_{2^{-j}}^* \xi \cdot T_{2^{-j}}^{-1} \Gamma(2^{-j}t)) \frac{dt}{t} .$$

Thus if we let

(19) $$\sigma_j(\eta) = \int_{1 \le |t| \le 2} \exp(i\eta \cdot \Gamma_j(t)) \frac{dt}{t} ,$$

where

(20) $$\Gamma_j(t) = T_{2^{-j}}^{-1} \Gamma(2^{-j}t),$$

we seek estimates of the form

(21) $$|\sigma_j(\eta)| \le C \|\eta\|^{-\epsilon}$$

for some $\epsilon > 0$ uniformly in j.

Let us turn now to the reason for discarding T_t^p. First of all, the curve $\Gamma(t) = (t, t \log |t|)$ is homogeneous and the dilation group G_λ is given by the linear transformation

(22) $$T_\lambda(x_1, x_2) = \begin{pmatrix} \lambda & 0 \\ \lambda \ln |\lambda| & \lambda \end{pmatrix} \begin{pmatrix} x_1 \\ x_2 \end{pmatrix} .$$

Thus, if we want our theory to include the homogeneous curve $(t, \gamma(t)) = (t, t \ln |t|)$, we can not define T_λ by the formula

$$T_\lambda(x_1, x_2) = (\lambda x_1, \gamma(\lambda)x_2).$$

Note that the first column of the matrix in (22) is $\Gamma(\lambda)$, and we ask how to interpret the non-zero entry, λ, in the second column.

A quantity that had already played a role in the theory was the function

(23) $$h(t) = t\gamma'(t) - \gamma(t).$$

$(-h(t)$ is the y-intercept of the tangent line to $\Gamma(t) = (t, \gamma(t))$ at $\Gamma(t)$.)

For example we had the following Theorems:

Theorem A [NVWW1]. *Let* $\Gamma(t) = (t, \gamma(t))$. *Assume* $\gamma(t)$ *is convex for* $t > 0$, *and that* $\gamma(t)$ *is odd. Then a necessary and sufficient condition that* H_Γ *be bounded in* $L^2(R^2)$ *is that for some* $C > 1$

(24) $$h(Ct) \geq 2h(t),$$

for $t > 0$.

Concerning the maximal function we have the following result.

Theorem B [NVWW2]. *Suppose* $\gamma(t)$ *is convex and* $h(Ct) \geq 2h(t)$ *for some* $C > 1$. *Then* M_Γ *is bounded in* $L^2(R^2)$.

Now for the curve $\Gamma(t) = (t, t \log |t|)$, $h(t) = t$. Thus for curves $\Gamma(t) = (t, \gamma(t))$ in R^2, we might try

(25) $$T_\lambda = \begin{pmatrix} \lambda & 0 \\ \gamma(\lambda) & h(\lambda) \end{pmatrix}.$$

Moreover, the proof of Theorem B required the use of certain "balls," and those balls can be obtained by applying T_λ to one fixed ball. Of course in the case that $\Gamma(t) = (t, t^k)$, we have been taking

$$T_\lambda = \begin{pmatrix} \lambda & 0 \\ 0 & \lambda^k \end{pmatrix}$$

which is different from the dilations in (25). However, in this case the two families are equivalent in the following sense: Two families of transformations S_λ and T_λ are equivalent if there are balls B_1, B_2, and B_3 such that

$$T_\lambda B_1 \subset S_\lambda B_2 \subset T_\lambda B_3.$$

This means that the kernels treated by the Calderón-Zygmund theory for T_λ and S_λ are the same.

(Of course

$$S_\lambda = \begin{pmatrix} \lambda & 0 \\ \lambda \log |\lambda| & \lambda \end{pmatrix}$$

and

$$T_\lambda = \begin{pmatrix} \lambda & 0 \\ 0 & \lambda \log |\lambda| \end{pmatrix}$$

are not equivalent.) There is another reason for discarding T_λ^p, which we shall come to when we discuss the n-dimensional situation. In any case, the following theorem was obtained by using the dilations

$$T_\lambda = \begin{pmatrix} \lambda & 0 \\ \gamma(\lambda) & h(\lambda) \end{pmatrix}.$$

Theorem C [CCVWW]. *Suppose* $\Gamma(t) = (t, \gamma(t))$ *with* $\gamma(t)$ *convex and that there is an* $\epsilon > 0$ *such that*

(26) $$h'(t) \geq \epsilon \frac{h(t)}{t}.$$

Then

(27) $$\|M_\Gamma f\|_{L^p(R^2)} \leq A(P, \Gamma)\|f\|_{L^2(R^2)}, \quad 1 < p \leq \infty.$$

If in addition $\Gamma(t)$ *is extended to negative t as an odd curve,*

(28) $$\|H_\Gamma f\|_{L^p(R^2)} \leq A(p, \Gamma)\|f\|_{L^p(R^2)}, \quad 1 < p < \infty.$$

The condition (26) is stronger than (24). In fact (24) does not suffice for the positive estimates of Theorem C if $p < 2$ for (27) and for $p \neq 2$ in (28). See [CCVWW] and [C2].

The next questions are what convexity should mean in R^n and how T_λ in (25) should be generalized from R^2 to R^n, $n \geq 3$.

It turns out that there are different directions in which to proceed. One possibility is to proceed inductively. If an estimate is true for a curve $\Gamma(t) = (t, \gamma_2(t) \ldots, \gamma_n(t))$ in R^n, then the same positive estimate holds for the curve $\tilde{\Gamma}(t) = (t, \gamma_2(t) \ldots, \gamma_{n-1}(t))$ in R^{n-1}. Thus one possible avenue is to add assumptions as n increases. This is the approach we shall take. The unfortunate aspect of this approach is that the theory will not be GL(n, R) invariant. In other words if $\Gamma(t)$ satisfies the hypothesis and g is in GL(n, R) the curve $g\Gamma(t)$ will not necessarily satisfy the hypothesis (even though the estimates for H_Γ and M_Γ are equivalent to those for $H_{g\Gamma}$ and $M_{g\Gamma}$). It will be true that our theorems will be invariant under the group of lower triangular matrices with positive diagonal entries and 1 in the upper left-hand corner. An alternative approach which is GL(n, R) invariant is developed in [CVWW]. In that theory the hypotheses are more complicated. The basic difference is that in our theory the condition (16) is proved, while in the GL(n, R) invariant theory (16) is assumed.

Since our theorem is inductive, as explained above, given a curve $\Gamma(t) = (t, \gamma_2(t) \ldots, \gamma_n(t))$ in R^n, it will be natural to consider also the curves

$$\Gamma_r(t) = (t, \gamma_2(t) \ldots, \gamma_r(t)) \quad \text{in } R^r$$

for $2 \le r \le n$. Our basic convexity assumption will be that for $2 \le r \le n$,

$$
(29) \qquad\qquad D_r(t) = \det \begin{pmatrix} \Gamma_r^1(t) \\ \vdots \\ \Gamma_r^{(r)}(t) \end{pmatrix} > 0
$$

for $t > 0$. We are mainly trying to understand what happens if $D_n(t)$ vanishes to infinite order at $t = 0$, so to simplify matters we might assume $D_n(t) \ne 0$ for $t \ne 0$. Now the inductive nature of our hypothesis leads to the assumption that $D_r(t) \ne 0$ for $t \ne 0, 2 \le r \le n$. Finally, by replacing some γ_j by $-\gamma_j$, we arrive at the condition $D_r(t) > 0, t \ne 0, 2 \le r \le n$. (A similar assumption occurs in the theory of differential equations. See [H].)

There are also generalizations of $h(t)$ for curves in R^n. Thus if $\Gamma(t) = (t, \gamma_2(t) \ldots, \gamma_n(t))$ and $\Gamma_2(t) = (t, \gamma_2(t))$, we define $h_2(t)$ to be the $h(t)$ for $\Gamma_2(t)$. To define $h_3(t)$ consider the curve $\Gamma_3(t) = (t, \gamma_2(t), \gamma_3(t))$ in R^3. We consider the tangent line, $L(t)$, to Γ_3 at the point $\Gamma_3(t)$ in R^3. We let $-\tilde{\Gamma}(t)$ be the point of intersection of $L(t)$ with the plane $x_1 = 0$. Then under the hypothesis (29) $\tilde{\Gamma}(t)$ is a convex curve in R^2. We let h_3 be the "h" for $\tilde{\Gamma}(t)$. We define h_r for $r \le n$ in an analogous manner. It turns out that the $h_j(t)$ can be expressed as a ratio of determinants that is easy to calculate asymptotically in examples. As in the case $n = 2$, the $h_j(t)$ were known to have had a role in the theory. This is expressed by the following theorem:

Theorem D. *Suppose $\Gamma(t) = (t, \gamma_2(t) \ldots, \gamma_n(t))$ is convex in the sense of (29) and that $\Gamma(t)$ is odd. Then a necessary and sufficient condition that H_Γ be bounded in L^2 is that for some $C > 1$,*

$$
h_j(Ct) \ge 2h_j(t), \quad t > 0.
$$

for $2 \le j \le n$.

Thus it seems that the generalization of (25) should be

$$
(30) \qquad\qquad T_\lambda = \begin{pmatrix} t & 0 & \cdots & \cdots & 0 \\ \gamma_2(t) & h_2(t) & 0 & \cdots & 0 \\ \gamma_3(t) & ? & h_3(t) & \cdots & 0 \\ \vdots & \vdots & \vdots & \ddots & \vdots \\ \gamma_n(t) & ? & ? & \cdots & h_n(t) \end{pmatrix}.
$$

Looking again at (25) we note that the second column of T_λ is obtained from the first column by applying to the first column a first-order differential operator. Namely, if

$$R\Psi(t) = t^2\left(\frac{\Psi(t)}{t}\right)',$$

R (First column of T_λ) = 2nd column of T_λ. This suggests that we should try to find first-order differential operators $R_1 \ldots, R_{n-1}$ such that

(31) $R_j(j\text{th column of } T_\lambda) = j + 1\text{st column of } T_\lambda.$

It turns out that the convexity assumption (29) allows us to find unique R_j so that (30) and (31) are satisfied with the question marks filled in. If we denote by $T_{\lambda,i,j}$ the i, jth element of the matrix for T_λ, we find

(32) $T_{\lambda,i,1} = \gamma_i(t),$

and

(33) $T_{\lambda,i,j+1} = R_{j+1}T_{\lambda,i,j}$

where

(34) $R_j f(\lambda) = \left(\frac{f(\lambda)}{h_j(\lambda)}\right)' \frac{h_j^2(\lambda)}{h_j'(\lambda)}.$

It turns out then that the hypothesis that $h_j(Ct) > 2h_j(t)$ implies

(35) $\|T_s^{-1}T_t\| \le C\left(\frac{t}{s}\right)^\epsilon, \quad s \ge t,$

for some $\epsilon > 0$. Clearly also

(36) $\Gamma(t) = T_t e,$

where e is the vector $\begin{pmatrix} 1 \\ 0 \\ \vdots \\ 0 \end{pmatrix}.$

The next matter is somehow to relate the dilations to decay estimates of the type (21). We found it convenient to prove a variant of the standard lemmas of Van Der Corput. We consider differential operators of the form

(37) $M_1 f(t) = f'(t)$

and

(38) $M_{j+1}f(t) = \alpha_j(t)\left(\frac{M_j f(t)}{\beta_j(t)}\right)',$

where α_j and β_j are smooth functions. Then we obtain the following theorem.

Theorem 1. *Let J be a bounded interval on R. Assume*

(39) $$\alpha_j(t) > 0 \quad \text{and} \quad \beta_j(t) > 0, \quad 1 \le j \le n - 1$$

for t in J,

(40) $$M_n f(t) \text{ is single-signed for t in J,}$$

(41) $$\beta_1(t) = 1 \text{ and } \alpha_1(t) \le 1/\epsilon \text{ for some positive } \epsilon \text{ and all } t \text{ in } J,$$

(42) $$\sum_{j=1}^{n} |M_j f(t)| \ge \epsilon \quad \text{for } t \text{ in } J,$$

and, finally,

(43) $$\int_x^y \beta_j(t)dt \ge \epsilon(y - x) \max_{x \le t \le y} \alpha_j(t)$$

for $x < y, x, y$ *in* J, *and* $1 \le j \le n - 1$. *Then for* $\lambda \ge 1$,

(44) $$\left| \int_J e^{i\lambda f(t)} dt \right| \le C(\epsilon, n, J) \frac{1}{\lambda^{1/n}}.$$

We apply the proposition with $\alpha_j(t) = \tilde{h}_j(t)$ and $\beta_j(t) = \tilde{h}'_j(t)$ where $\tilde{h}_j$ is the h_j corresponding to the normalized curve

$$\Gamma_j(t) = T_{2^{-j}}^{-1} \Gamma(2^{-j}t).$$

It turns out that to verify the crucial hypothesis (42) it suffices to show

(45) $$\|(\tilde{T}'_\lambda)^{-1}\| \le C(\sigma)$$

for $1 \le \lambda \le \sigma$, where $\tilde{T}_\lambda$ is the dilation matrix formed from Γ_j. (This shows then the way the dilations are used to obtain the decay estimate (21).) Further, (45) can be shown to hold under the assumption $h'_j(t) \ge \epsilon \frac{h_j(t)}{t}$.

The last point concerns the Paley-Littlewood decomposition. We choose a $C^\infty(R^n)$ function $\phi(x)$ such that $\hat{\phi}(\xi) = 1$ for $|\xi| \le 1$ and $\hat{\phi}(\xi) = 0$ for $|\xi| \ge 2$, and we assume that we have a family of non-singular linear transformations A_k satisfying the condition

(46) $$\|A_{k+1}^{-1} A_k\| \le \alpha < 1.$$

(This condition was first introduced by Rivière [R].) We define

(47) $$m_j(\xi) = \hat{\phi}(A_j^* \xi) - \hat{\phi}(A_{j+1}^* \xi),$$

and $S_j f$ by

(48) $$\widehat{S_j f}(\xi) = m_j(\xi) \cdot \hat{f}(\xi).$$

We then obtain the following Littlewood-Paley estimate.

Theorem 2. *If for each integer* k, $\|A_{k+1}^{-1} A_k\| \leq \alpha < 1$, *then*

(49) $$\left\| \left(\sum_j |S_j f(x)|^2 \right)^{1/2} \right\|_{L^p} \leq C_p \|f\|_{L^p}, \quad 1 < p < \infty,$$

and

(50) $$\left\| \sum_j S_j f_j \right\|_{L^p} \leq C_p \left\| \left(\sum |S_j f_j|^2 \right)^{1/2} \right\|_{L^p} \quad 1 < p < \infty.$$

The key step in the proof of Theorem 2 is to show that for $0 \leq t \leq 1$

(51) $$\|U_t f\|_{L^p} \leq C(p)\|f\|_{L^p}, \quad 1 < p < \infty,$$

where

(52) $$U_t f(x) = \sum_j r_j(t) S_j f(x),$$

where again $r_j(t)$ are the standard Rademacher functions. To obtain the estimate (51) we need an appropriate variant of the Calderón-Zygmund theory (which is *not* contained in the theory of spaces of homogeneous type of Coifman and Weiss [CW]).

We denote by B_j, $j \in Z$ the "balls" of the theory. We assume for each $j \in Z$ there is a "ball" B_j. We assume that each B_j is open, convex, balanced and bounded. We then have the following result.

Theorem 3 [CCVWW]. *Assume*
(i) $\cup B_j = R^n$;
(ii) $\cap B_j = \{0\}$;
(iii) $B_{j+1} \subset B_j$; *and*
(iv) *there exists a* $\kappa < \infty$ *so that* $|B_j| \leq \kappa |B_{j+1}|$.
Let T *be a convolution operator bounded on* $L^2(R^n)$, *with convolution kernel* $K(x)$. *Assume for each* $y \in B_j$
(v) $\int_{x \notin 2B_j} |K(x - y) - K(x)| dx \leq A$.
Then for f *in* $C_0^\infty(R^n)$,

$$\|Tf\|_{L^p(R^n)} \leq C\|f\|_{L^p(R^n)}, \quad 1 < p < \infty.$$

The bound C depends only on p, n, A, the L^2 operator norm of T, and the family of balls B_j, but not on f.

The above considerations lead to the following Theorem.

Main Theorem. *Let* $\Gamma(t) = (t, \gamma_2(t) \ldots, \gamma_n(t))$ *be convex in the sense of* (29). *If for some* $\epsilon > 0$,

$$h'_j(t) \geq \epsilon \frac{h_j(t)}{t}, \quad j = 2 \ldots, n,$$

then

$$\|M_\Gamma f\|_{L^p} \leq C_p \|f\|_{L^p}, \quad 1 < p \leq \infty,$$

and if in addition $\Gamma(t)$ *is odd,*

$$\|H_\Gamma f\|_{L^p} \leq C_p \|f\|_{L^p}, \quad 1 < p < \infty.$$

Also, in examples where

$$\Gamma(t) = (t, e^{-1/t^{a_1}} \ldots, e^{1/t^{a_n}})$$

for $t > 0, 0 < a_1 < \cdots < a_n$, it is easy to see that the hypotheses of the Main Theorem are satisfied.

We would like to add a remark pointing out a connection between our work and the theory of uniform asymptotic stability of systems of ordinary differential equations. Suppose one considers the $n \times n$ system of differential equations

(53) $$y'(t) = B(t)y(t)$$

where $y(t)$ is a vector in R^n and $B(t)$ is an $n \times n$ matrix. Let $F(t)$ be a fundamental matrix for the equation (53). That is, the columns of $F(t)$ are solutions of the system and $\det F(t) \neq 0$. Then the system is uniformly asymptotically stable if

$$\|F(t)F^{-1}(s)\| \leq Ce^{-\varepsilon(t-s)}, \quad t \geq s,$$

for some positive ϵ. See [CO]. Thus after a change of the time scale the adjoint of $F(t)$ satisfies the estimate (16). In particular our results imply results on uniform asymptotic stability for systems in which the elements of the coefficient matrix are unbounded. For example, the system

$$z'(t) = B(t)z(t)$$

with

$$B(t) = \begin{pmatrix} 0 & 1 \\ -\dfrac{1}{\sigma(t)} & \dfrac{1 + \sigma(t)}{\sigma(t)} \end{pmatrix}$$

is uniformly asymptotically stable under the hypothesis $0 \leq \sigma(t) \leq C$, for some positive C.

University of Sussex, Brighton, United Kingtom
Wright State University
University of Wisconsin, Madison
Rutgers University, Camden

REFERENCES

[C] M. Christ. Personal communication.

[C2] M. Christ. "Examples of singular maximal functions unbounded in L^p." Preprint.

[CCVWW] A. Carbery, M. Christ, J. Vance, S. Wainger, and D. Watson. "Operators associated to flat plane curves: L^p estimates via dilation methods." *Duke Math. J.* 59 (1989), 675–700.

[CO] W. A. Coppel. *Dichotomies in Stability Theory.* Lecture Notes in Mathematics, no. 629. Springer Verlag, 1978.

[CVWW] A. Carbery, J. Vance, S. Wainger, and D. Watson. "The Hilbert transform and maximal function along flat curves, dilations, and differential equations." To appear.

[CW] R. Coifman and G. Weiss. *Analyse harmonique non commutative sur certains espaces homogènes.* Lecture Notes in Mathematics, no. 242. Springer Verlag, 1971.

[DR] J. Duoandikoextea and J. L. Rubio de Francia. "Maximal and singular integral operators via Fourier transform estimates." *Invent. Math.* 84 (1986), 541–561.

[H] P. Hartman. *Ordinary Differential Equations.* John Wiley, 1964.

[NVWW1] A. Nagel, J. Vance, S. Wainger, and D. Weinberg. "Hilbert transforms for convex curves."*Duke Math. J.* 50 (1983), 735–744.

[NVWW2] A. Nagel, J. Vance, S. Wainger, and D. Weinberg. "Maximal functions for convex curves." *Duke Math. J.* 52 (1985), 715–722.

[NVWW3] A. Nagel, J. Vance, S. Wainger, and D. Weinberg. "Hilbert transform for convex curves in R^n." *Amer. J. of Math.* 108 (1986) 485–504.

[R] N. Rivière. "Singular integrals and multiplier operators." *Arkiv Math.* 9 (1971), 243–278.

[S] E. Stein. "Interpolation of linear operators." *Trans. Amer. Math. Soc.* 83 (1956), 482–492.

[SW] E. Stein and S. Wainger. "Problems in harmonic analysis related to curvature." *Bull. Amer. Math. Soc.* 84 (1978), 1239–1295.

6

Nonexistence of Invariant Analytic Hypoelliptic Differential Operators on Nilpotent Groups of Step Greater than Two

Michael Christ[*]

INTRODUCTION

The development of a large part of the theory of subelliptic partial differential equations has been guided by fundamental papers of Stein and collaborators. Among these are the explicit computation by Greiner and Stein [GS] of the Szegö kernels for certain three-dimensional CR manifolds; the work of Folland and Stein [FS] which developed aspects of harmonic analysis on the Heisenberg group and established its connection with the $\bar{\partial}$ equation on strictly pseudoconvex domains; the proof by Rothschild and Stein [RS] of sharp regularity estimates for sums of squares of vector fields and their forging of the link between these operators and harmonic analysis on graded nilpotent Lie groups; and the estimation by Nagel, Stein, and Wainger [NSW] of the fundamental solutions of such operators in a very precise way. These works have pointed the way to a far-reaching domain of research centered on the confluence of partial differential equations, complex analysis in several variables, harmonic analysis on Euclidean spaces and on nilpotent groups, and real variables.

[*]This research was supported by the National Science Foundation and the Institut des Hautes Études Scientifiques.

The present note serves to announce further progress in this direction. Detailed proofs will appear elsewhere [C5], [C6].

We are concerned with five interrelated themes:

- Analytic hypoellipticity of subelliptic partial differential operators, especially those which are left-invariant and homogeneous on some nilpotent Lie group.
- Hypoellipticity of certain PDE with complex coefficients, in the C^∞ sense.
- Certain one-parameter families of irreducible representations of nilpotent Lie groups and algebras.
- Existence of eigenvalues for certain non-selfadjoint linear operators.
- A scattering problem for ordinary differential equations.

Our investigations have been profoundly influenced by all four of the papers cited above.

Experts should not be misled by our title, in which at least three distortions are perpetrated in the interest of brevity.

1 REVIEW

To begin, we recall some facts concerning constant-coefficient, homogeneous differential operators on $\mathbb{R}^d$. A differential operator L is said to be (C^∞) hypoelliptic in an open set Ω if whenever $\Omega' \subset \Omega$ is open, u is a distribution in $[C_0^\infty(\Omega')]'$, and $Lu \in C^\infty(\Omega')$, then $u \in C^\infty(\Omega')$. L is said to be *analytic* hypoelliptic if $Lu \equiv 0$ in Ω' implies that u is real analytic in Ω'. It is well known that for a homogeneous partial differential operator with constant coefficients in $\mathbb{R}^d$, the following are equivalent:

(1) L is C^∞ hypoelliptic.
(2) L is analytic hypoelliptic.
(3) L is elliptic.

Moreover, if L has constant coefficients but is not necessarily homogeneous, then (2) and (3) remain equivalent.

For a certain very restricted class of nilpotent Lie groups of step 2, including the Heisenberg groups [M], these conditions (1) and (2) remain equivalent for homogeneous, left-invariant differential operators, and are equivalent to a modification of (3), the Rockland criterion; the latter and (1) remain equivalent for all graded nilpotent Lie groups.

We will be concerned in this chapter with the non-validity of the equivalence between (2) and the other conditions, for groups of step three or higher.

In order to motivate later considerations, let us recall how it may be proved that a non-elliptic operator with constant coefficients, in $\mathbb{R}^d$, admits solutions which are not analytic. Let L be such an operator. One approach is to note that if L were analytic hypoelliptic, then for any open sets $\Omega_1 \Subset \Omega_2 \Subset \Omega$, there must exist $C < \infty$ such that for every $u \in C^\infty(\Omega_2)$ satisfying $Lu \equiv 0$, for every multi-index α,

$$(1.1) \qquad \left\| \frac{\partial^\alpha}{\partial x^\alpha} u \right\|_{L^\infty(\Omega_1)} \leq C^{1+|\alpha|} |\alpha|! \, \|u\|_{L^\infty(\Omega_2)}.$$

Assume that L has positive order n, and let P be its symbol. P is a polynomial, which may be considered to be defined on $\mathbb{C}^d$, and then $P(\xi) = 0$ for certain $\xi \in \mathbb{C}^d$. If L is not elliptic, then a sequence $\{\xi_j\} \subset \mathbb{C}^d$ may be chosen such that

$$(1.2) \qquad \begin{cases} P(\xi_j) = 0, \\[2mm] |\xi_j| \to \infty, \\[2mm] \dfrac{|\Im(\xi_j)|}{|\xi_j|} \to 0 \qquad \text{as } j \to \infty. \end{cases}$$

In order to violate the Cauchy estimates (1.1), set

$$f_j(x) = \exp(i\langle x, \xi_j \rangle),$$

so that $Lf_j \equiv 0$. With $\alpha = \alpha(j)$ a multi-index to be determined,

$$\left| \frac{\partial^\alpha}{\partial x^\alpha} f_j(0) \right| = |\xi_j^\alpha|$$

while for any bounded neighborhood Ω of 0,

$$C^{1+|\alpha|} |\alpha|! \, \|f_j\|_{L^\infty(\Omega)} \leq C^{1+|\alpha|} |\alpha|! \, \exp(C_1 |\Im(\xi_j)|)$$

with $C_1 = C_1(\Omega)$ fixed. If each $\xi_j \in \mathbb{R}^d$, then (1.1) is clearly contradicted by fixing any α satisfying $\alpha_j \neq 0$, and exploiting the condition that $|\xi_j| \to \infty$. In the general case, if the imaginary parts are nonzero, then $\exp(C_1 |\Im(\xi_j)|)$ will grow as $j \to \infty$, so that both the left- and right-hand sides of (1.1) grow with j and the situation is more delicate. It can be shown that the third condition in (1.1) still permits the choice of a sequence $\alpha(j)$ such that (1.1) is violated for all finite C, as $j \to \infty$; we shall not reproduce the details. However, for the sequel it is instructive to note that for such a construction to succeed in contradicting the Cauchy estimates, it is permitted that $\|f_j\|_{L^\infty(\Omega)}$ grow as $j \to \infty$, but is equally essential that it not grow too rapidly, in a suitable sense.

From the viewpoint of harmonic analysis, what we have done is to decompose the regular representation of $\mathbb{R}^d$ on $L^2(\mathbb{R}^d)$ into its irreducible unitary components, which are parametrized by $\xi \in \mathbb{R}^d$; to complexify so as to obtain non-unitary representations depending holomorphically on the parameter $\xi \in \mathbb{C}^d$; to associate to each representation π and differential operator L an operator $\pi(L)$ on the Hilbert space $\mathcal{H}_\pi = L^2(\mathbb{R}^0) = \mathbb{C}$; and to exploit those π for which $\pi(L)$ fails to be injective. Ellipticity of L is determined solely by real ξ, that is, by the unitary representations, but the nonunitary ones were vital in the analysis of analytic hypoellipticity. Note that for homogeneous operators, the symbol is a homogeneous polynomial, so that if it is elliptic, then it is nonvanishing in a conic neighborhood of $\mathbb{R}^d$ in $\mathbb{C}^d$, and hence no sequence satisfying (1.2) exists.

A similar situation occurs for homogeneous, left-invariant partial differential operators on graded nilpotent groups. The set of all irreducible unitary representations π may be described quite explicitly [K]. To each such L and each π is associated a differential operator $\pi(L)$ acting on a dense subspace of $\mathcal{H}_\pi = L^2(\mathbb{R}^k)$ for some $k(\pi)$. As conjectured by Rockland [R], and proved by Rothschild and Stein [RS] in the negative direction and by Helffer and Nourrigat [HN1], [HN2] in the positive direction, hypoellipticity of L in the C^∞ category is equivalent to the injectivity of $\pi(L)$ (on its domain) in $\mathcal{H}_\pi$ for every irreducible unitary representation π. This is the modification of (3) mentioned above.

Thus it is natural to expect that the study of analytic hypoellipticity for such operators should be related to the properties of certain non-unitary representations. The subject was first approached from this point of view by Helffer [He]. It proved problematic to determine whether $\pi(L)$ failed to be injective for some π, or not, and that is the problem to be addressed here.

We are able to treat only a very restricted class of groups, for which all the relevant representations satisfy $\mathcal{H}_\pi = L^2(\mathbb{R}^1)$, the simplest case which does not arise for abelian groups. Then $\pi(L)$ is an ordinary differential operator, about which much can be said by elementary means. One of our goals is to develop in detail the analysis of holomorphic one-complex-parameter families of such representations.

Some notation: "$A \sim B$" means that the ratio of these two positive quantities is bounded above and below by positive constants. $\langle z \rangle = 1 + |z|$. The natural logarithm is denoted ln, the real part by $\Re$, and the imaginary part by $\Im$. C^ω denotes the class of real analytic functions.

2 THE MAIN RESULT

$\mathbf{g}$ will denote always a (finite-dimensional) nilpotent Lie algebra, and G a connected Lie group whose Lie algebra is $\mathbf{g}$. Elements of $\mathbf{g}$ will be identified with (real) left-invariant vector fields in a neighborhood of the identity element $0 \in G$ via the exponential map.

Definition 2.1.

(a) *A Lie algebra* $\mathbf{g}$ *is said to be stratified if it admits a vector space decomposition as* $\mathbf{g} = \oplus_{j \geq 1} \mathbf{g}_j$, *with* $[\mathbf{g}_j, \mathbf{g}_k] \subset \mathbf{g}_{j+k}$ *for all* $j, k \in \mathbb{Z}^+$, *such that* $\mathbf{g}_1$ *generates* $\mathbf{g}$ *as a Lie algebra.*

(b) *A stratified Lie algebra* $\mathbf{g}$ *is said to have* k *generators if* $\dim(\mathbf{g}_1) = k$.

(c) *A nilpotent, stratified Lie algebra* $\mathbf{g}$ *is said to be of step* r *if* $\mathbf{g}_{r+1} = \{0\}$ *but* $\mathbf{g}_r \neq \{0\}$.

(d) *If* $\mathbf{g}$ *is stratified, then* $\mathcal{U}_n$ *denotes the linear space of all left-invariant differential operators on* G *which may be represented as homogeneous polynomials of degree* n *in the elements of* $\mathbf{g}_1$.

Definition 2.2. *For any formal polynomial* P *in two non-commuting variables, with complex coefficients,* $\hat{P}$ *denotes the naturally associated polynomial in two complex variables.*

If $\mathbf{g}$ is stratified with two generators, and if a basis $\{X, Y\}$ for $\mathbf{g}_1$ is fixed, then to each such polynomial, homogeneous of some degree n, is associated the element $P(-iX, -iY)$ of $\mathcal{U}_n$. Conversely, any element $\mathcal{L}$ of $\mathcal{U}_n$ may be so expressed; though P will not be uniquely determined, it is easily verified (by consideration of the abelian group whose Lie algebra is $\mathbf{g}/\oplus_{j \geq 2} \mathbf{g}_j$) that $\hat{P}$ is uniquely determined by $\mathcal{L}$.

Definition 2.3. *Let* $\mathbf{g}$ *be a stratified Lie algebra with two generators* $X, Y \in \mathbf{g}_1$.

(a) *A homogeneous polynomial* P *of degree* n *is said to be* generic *if the polynomial* $z \mapsto \hat{P}(z, 1)$ *in one complex variable has degree exactly* n, *and moreover has* n *distinct complex roots.*

(b) $\mathcal{L} \in \mathcal{U}_n$ *is said to be* generic *if it equals* $P(-iX, -iY)$ *for some generic polynomial* P.

It may be verified that the property of being generic is independent of the choice of basis for $\mathbf{g}_1$, so long as the basis itself is suitably generic. Since it is determined by $\hat{P}$, it is an intrinsic property of $\mathcal{L} = P(-iX, -iY)$, independent of the choice of the representing polynomial P.

Our principal result is:

Theorem 2.4. *Let* G *be a Lie group whose Lie algebra* $\mathbf{g}$ *satisfies both:*

- $\mathbf{g}$ *is stratified and has two generators.*

- **g** *is of step greater than or equal to three.*

Then any generic homogeneous, left-invariant differential operator $\mathcal{L}$ on G fails to be analytic hypoelliptic.

Remarks.

1. The proof yields a stronger result in the scale of Gevrey classes. For some open set Ω and real number $s \geq 1$, $G^s(\Omega)$ denotes the class of all C^∞ functions defined on Ω such that, for each compact subset $K \subset \Omega$, there exists $C < \infty$ such that for every multi-index α and every $x \in K$,

$$|\partial^\alpha f(x)| \leq C^{1+|\alpha|}(1 + |\alpha|)^{s|\alpha|}.$$

   For $s = 1$ this is simply the class of real-analytic functions, while for $s > 1$ there are strict inclusions $C^\omega \subset G^s \subset C^\infty$. Then our method establishes existence of a solution to $\mathcal{L}F \equiv 0$, in a neighborhood of 0, such that F fails to belong to G^s for any $s < 3$.

2. The case $\mathcal{L} = X^2 + Y^2$ was treated in the union of the two papers of Helffer [He] and of Pham The Lai and Robert [PR] some time ago, so that Theorem 2.4 is no surprise. However, their argument depends on certain particular features of $X^2 + Y^2$ which are absent for general $\mathcal{L}$.

3. The principal hypothesis is that **g** have step strictly greater than two. There exists a class of nilpotent groups of step two, for which every left-invariant, homogeneous differential operator that is C^∞ hypoelliptic is automatically analytic hypoelliptic [M]. In particular, this occurs for the Heisenberg groups $\mathbb{H}^n$.

4. The hypothesis that **g** have two generators may be weakened to $\dim(\mathbf{g}_2) = 1$, with an appropriate redefinition of genericity. It would be interesting to decide whether the result remains valid without any such hypothesis; this cannot be done by our method as it stands.

5. The hypothesis that $\mathcal{L}$ be generic is made to simplify certain technical aspects of the proof, and might conceivably be eliminated by a more elaborate argument along the same general lines.

6. In order to simplify the exposition slightly we assume henceforth that the roots of $z \mapsto \hat{P}(z, 1)$ have distinct real parts.

3 OUTLINE OF THE PROOF

The first step is to decompose the regular representation of G on $L^2(G)$ into irreducible components, and to focus attention on three particular families of components. For our purpose this amounts simply to separation of variables and may be achieved in the following concrete fashion. Let n_j denote the dimension of

$\mathbf{g}_j$, let $N = \sum n_j$, and let x_{jk} be coordinates in $\mathbb{R}^{n_j}$ for $1 \leq k \leq n_j$. In the coordinates given by the exponential map, G is identified with $\mathbb{R}^N$, and X, Y take the form

$$\sum_{j,k} p_{jk}(x) \frac{\partial}{\partial x_{jk}},$$

where the p_{jk} are polynomials, homogeneous in a natural weighted sense. Moreover, p_{jk} depends only on those variables $x_{i\ell}$ for which $i < j$. Thus the space of all functions on G which depend only on those variables x_{jk} for which $j \leq 3$, is invariant under X and Y, hence under $\mathcal{L}$. It suffices to produce a function in this space which is annihilated by $\mathcal{L}$ in some open set, and which is not analytic. In the same way, we may restrict attention to functions independent of all variables x_{3k} with $k \geq 2$.

Since $\mathbf{g}_1$ and $\mathbf{g}_2$ have dimensions 2 and 1, respectively, our functions now depend on four variables which will be called x, y, s, t. Y takes the form $c_1 \partial_x + c_2 \partial_y + \ell(x, y)\partial_s + p(x, y, s)\partial_t$ where c_i are constants, ℓ is linear, and p is a homogeneous polynomial of degree two in x, y plus a constant times s. X takes a similar form. After composition with an appropriate polynomial diffeomorphism,

$$X = \partial_x \qquad \text{and} \qquad Y = \partial_y - x\partial_s - x^2\partial_t.$$

In effect, we are now working on a particular four-dimensional nilpotent Lie group of step three.

We consider three families of irreducible representations of this group, or more concretely, the action of X, Y on three classes of functions F on $\mathbb{R}^4$:

(I) $\quad F(x, y, s, t) = e^{i\langle \xi, (x,y)\rangle}$ with $\xi \in \mathbb{R}^2$,

(II) $\quad F(x, y, s, t) = f(x)e^{i\lambda s}$ with $0 \neq \lambda \in \mathbb{R}$,

(III) $\quad F(x, y, s, t) = f(x)e^{izy}e^{i\tau t}$ with $0 \neq \tau \in \mathbb{R}$ and $z \in \mathbb{C}$.

The actions of X, Y on these classes of functions generate representations of the Lie algebra which they generate as (possibly unbounded) operators on $\mathbb{C}^1 = L^2(\mathbb{R}^0)$, on $L^2(\mathbb{R}^1)$, and on $L^2(\mathbb{R}^1)$, respectively. X acts as ξ_1, $-i\, d/dx$ and $-i\, d/dx$, and Y acts as ξ_2, $-\lambda s$ and $(z - \tau x^2)$, respectively. The representations (I) factor through $\mathbf{g}/\oplus_{j\geq 2}\mathbf{g}_j$, those of type (II) through $\mathbf{g}/\oplus_{j\geq 3}\mathbf{g}_j$, and those of type (III) through $\mathbf{g}/\oplus_{j\geq 4}\mathbf{g}_j$.

To these representations of $\mathbf{g}$ correspond representations of G. In case (III), when $z \in \mathbb{C}\backslash\mathbb{R}$, these representations are not unitary, and indeed elements of G are represented as unbounded linear operators defined on a certain subspace of $L^2(\mathbb{R})$.

It is a general principle [G], [HN1], [R] that one should study in succession the representations of type (I), then type (II), then type (III) in order to obtain increasingly more refined information about $\mathcal{L}$. For those of type (I), $\mathcal{L}$ acts as $\hat{P}(\xi_1, \xi_2)$;

if $\hat{P}(\xi) = 0$ for some $0 \neq \xi \in \mathbb{R}^2$, then $\mathcal{L}$ annihilates $\exp(ir\langle \xi, (x, y)\rangle)$ for all $r \in \mathbb{R}^+$. By the argument outlined in §1, it follows that $\mathcal{L}$ is not analytic hypoelliptic, nor even C^∞ hypoelliptic. Therefore we may assume henceforth that $\hat{P}$ is an elliptic polynomial.

The following is well known, and expresses the analogue of the last paragraph for the representations of types (II) and (III).

Lemma 3.1.

(i) *If there exists* $f \in L^\infty(\mathbb{R})$, *not identically vanishing, such that* $P(-i\frac{d}{dx}, x)f \equiv 0$, *or such that* $P(-i\frac{d}{dx}, -x)f \equiv 0$, *then there exists a function* F, *defined on an open subset of* G, *such that* $P(-iX, -iY)F \equiv 0$ *but* F *is not* C^∞. *Therefore* $\mathcal{L} = P(-iX, -iY)$ *is not analytic hypoelliptic, nor even* C^∞ *hypoelliptic.*

(ii) *If there exist* $z \in \mathbb{R}$ *and* $0 \neq f \in L^\infty(\mathbb{R})$ *for which* $P(-i\frac{d}{dx}, (z - x^2))f \equiv 0$, *then* $\mathcal{L}$ *is again neither analytic hypoelliptic nor* C^∞ *hypoelliptic.*

(iii) *If there exist* $z \in \mathbb{C}$ *and* $0 \neq f \in L^\infty(\mathbb{R})$ *for which* $P(-i\frac{d}{dx}, (z - x^2))f \equiv 0$, *then* $\mathcal{L}$ *is not analytic hypoelliptic.*

The proof is analogous to that for type (I). As in the argument outlined in §1, some growth restriction on the solution f is essential; in order to contradict the Cauchy estimates one considers the solutions of L

$$f(\tau^{1/3}x)e^{i\tau^{1/3}zy}e^{i\tau t}$$

where τ is a positive parameter which will tend to ∞, and where f, z are as in (iii) and are fixed. Thus the behavior of $f(x)$ for large x comes into play. The condition that f be bounded may be relaxed in case (iii) to $|f(x)| = O(\exp(|x|^{3-\varepsilon}))$ for some $\varepsilon > 0$, but it turns out that (if none of the roots of $\hat{P}$ are purely imaginary) every solution will either decay rapidly to zero as $|x| \to \infty$, or will grow like $\exp(\delta|x|^3)$, that is, too rapidly to lead to the desired negative result. Thus it is irrelevant whether one seeks solutions in the Schwartz class, in L^2, in L^∞, or in some class permitting moderate growth at infinity.

As far as the existence of such bounded solutions for representations of type (II) goes, only the sign of λ is relevant, not its magnitude; this reflects the existence of a group of automorphic dilations of G. A bounded solution f for $\lambda = 1$ leads by an appropriate dilation of the x variable to a bounded solution for any given $\lambda > 0$. Similarly, it is no loss to set $\tau = +1$ in case (III).

Definition 3.2. *$\mathcal{L}$ is said to be nondegenerate if $\hat{P}$ is elliptic and there exist no bounded functions in the nullspaces of* $P(-i\frac{d}{dx}, \pm x)$.

There are certain cases in which a very superficial analysis shows $\mathcal{L}$ to be degenerate. Let

$$\{\gamma_j : 1 \le j \le n\}$$

be the roots of $z \mapsto \hat{P}(z, 1)$, recalling that $\hat{P}$ is assumed to be elliptic.

Proposition 3.3. *Assume* $\mathcal{L} = P(-iX, -iY)$ *to be homogeneous, left-invariant and generic on a nilpotent group whose Lie algebra is stratified with two generators, and assume* $\hat{P}$ *to be elliptic. Then* $\mathcal{L}$ *is degenerate if any of the following occur:*

(1) *The degree of P is odd.*

(2) *Some root γ_j is imaginary.*

(3) *The degree n of P is even and the number of indices j for which $\Re(\gamma_j) > 0$ is not exactly $n/2$.*

The idea of the proof will be indicated later. There is the following consequence:

Corollary 3.4. *Let* $\mathcal{L} = P(-iX, -iY)$ *be a generic homogeneous, left-invariant differential operator on the Heisenberg group* $\mathbb{H}^1$ *of dimension three. If the degree of P is odd, then $\mathcal{L}$ fails to be C^∞ hypoelliptic.*

We assume henceforth not only that $\hat{P}$ is elliptic, but also that none of these conditions are satisfied. These are by no means the only causes of degeneracy: the operators $X^2 + Y^2 + i\alpha[X, Y]$ on $\mathbb{H}^1$ fail to be C^∞ hypoelliptic for a discrete set of real values of the parameter α, while $\hat{P}$ is independent of α. A fundamental issue in our analysis is the difficulty of detecting such more subtle degeneracies.

For our purpose the essential analysis takes place on the family of representations of type (III), which reflect the fact that the group is not of step 2. Let us consider a more general situation by letting $m \in \{2, 3, 4, \ldots\}$ be arbitrary and setting

$$L_z = P(-i\frac{d}{dx}, (z - x^{m-1})).$$

Definition 3.5. $z \in \mathbb{C}$ *is said to be a nonlinear eigenvalue of the operator family* $\{L_z : z \in \mathbb{C}\}$ *if there exists* $f \in L^\infty(\mathbb{R})$, *not identically vanishing, such that* $L_z f = 0$.

In fact, the nonlinear eigenvalues are genuinely eigenvalues for a related problem; see [Ke], [PR]. The set of nonlinear eigenvalues has some structure.

Proposition 3.6. *Assume that P is homogeneous and generic, and that $\hat{P}$ is elliptic and that none of the roots γ_j are imaginary. Let $m \in \{2, 3, 4, \ldots\}$. Then the set* *of nonlinear eigenvalues of $\{L_z\}$ is equal to the set of all zeroes of an entire holomorphic function $\mathcal{W}$, which satisfies*

$$|\mathcal{W}(z)| = O(\exp(C|z|^{m/(m-1)}))$$

for some $C < \infty$. Moreover when $m = 2$, $\mathcal{W}(z) \equiv c_1 \exp(c_2 z^2)$ for some constants c_1, c_2.

In §4 we shall explain how $\mathcal{W}$ is constructed. A class of examples in which it can be computed explicitly is

$$P(-i\frac{d}{dx}, (z - x^{m-1})) = [-\frac{d}{dx} + (z - x^{m-1})] \circ [\frac{d}{dx} + (z - x^{m-1})].$$

Then

$$(3.7) \qquad \mathcal{W}(z) = \begin{cases} \displaystyle\int_{-\infty}^{\infty} e^{2(zs - m^{-1}s^m)}\, ds & \text{for } m \text{ even} \\ \pm 1 & \text{for } m \text{ odd.} \end{cases}$$

On the other hand, if m is even and

$$P(-i\frac{d}{dx}, (z - x^{m-1})) = [\frac{d}{dx} + (z - x^{m-1})] \circ [-\frac{d}{dx} + (z - x^{m-1})],$$

then

$$\mathcal{W}(z) \equiv 0.$$

It was observed in [CG] that the entire function $\int_{-\infty}^{\infty} \exp(2(zs - m^{-1}s^m))\, ds$ has order $m/(m-1)$. Thus at least three behaviors are possible: $\mathcal{W}$ can be identically zero, a nonzero constant, or an entire holomorphic function of order exactly $m/(m-1)$. In general, $\mathcal{W}$ will be of a transcendental nature, as illustrated already by (3.7), and will not be effectively computable.

On the technical level our crucial result is

Proposition 3.8. *Fix an integer $m \geq 3$ and consider $L_z = P(-i\frac{d}{dx}, (z - x^{m-1}))$ where P is generic, homogeneous, and nondegenerate. Then there exists $\delta > 0$ such that*

$$|\mathcal{W}(z)| \geq \delta \exp(\delta|z|^{m/(m-1)}) \qquad as \quad \mathbb{R}^+ \ni z \to \infty.$$

Theorem 2.4 follows at once. If $\mathcal{L}$ is degenerate, then analysis on the level of either type (I) or type (II) representations establishes the existence, on an open subset of G, of a function $F \notin C^\infty$ such that $\mathcal{L}F \equiv 0$; hence $\mathcal{L}$ is hypoelliptic in neither the C^∞ nor the analytic categories. If $\mathcal{L}$ is nondegenerate, then $\mathcal{W}$ is an entire holomorphic function of order exactly $m/(m - 1)$, and since this order is not an integer, $\mathcal{W}$ has a nonempty, in fact infinite, set of zeroes. As discussed earlier, this implies that $\mathcal{L}$ is not analytic hypoelliptic; if any of the zeroes happens to be real, then it is not C^∞ hypoelliptic.

A small extension of the same reasoning establishes

Theorem 3.9. *In* $\mathbb{R}^3$, *with coordinates* (x, y, t), *set*

$$X = \partial_x \qquad and \qquad Y = \partial_y - x^{m-1}\partial_t.$$

Let P *be a generic, homogeneous polynomial. Then for any* $m \in \{3, 4, 5, \ldots\}$, $\mathcal{L} = P(-iX, -iY)$ *is not analytic hypoelliptic.*

Some additional work is required because the type (II) representations do not arise directly as in the previous discussion, but only as a limit of the type (III) representations (with $(z - x^2)$ replaced by $(z - x^{m-1})$), in the fashion described by Helffer and Nourrigat [HN3]. Therefore, in the degenerate case, their results imply only the weaker conclusion that $\mathcal{L}$ fails to be *maximally hypoelliptic*; see [HN3] for the definition. In fact, just as in the group-invariant setting, $\mathcal{L}$ turns out to be hypoelliptic in neither the C^∞ nor the analytic sense.

By two different arguments [PR], [HH], [C2], [C3] the case $L = X^2 + Y^2$ has already been treated. Both used particular properties of this operator not present in general.

4 THE WRONSKIAN $\mathcal{W}$ AND THE SCATTERING PROBLEM

Throughout this section it is assumed that $\hat{P}$ is nondegenerate and generic, and for simplicity of exposition that none of the roots γ_j are imaginary. The principal idea of the proof is to define $\mathcal{W}$ as the determinant of what can be thought of as a scattering matrix. The ordinary differential equation L_z has n linearly independent solutions, and if r denotes the number of indices j for which $\mathfrak{R}(\gamma_j) > 0$ then it happens (see below) that there exist exactly r linearly independent solutions which remain bounded as $x \to +\infty$, and exactly $n - r$ independent solutions which remain bounded as $x \to -\infty$. This happens for every $z \in \mathbb{C}$. Now, the former solutions have an asymptotic behavior which may be understood rather precisely as $x \to +\infty$, and similarly for the latter as $x \to -\infty$. Therefore the only possibility for a globally bounded solution is that these n solutions should become linearly dependent, for some value of z.

The situation is mostly parallel for the level (II) representations, but certain differences have their significance. The number of independent bounded solutions for $P(-i\frac{d}{dx}, -2x)$, as $x \to +\infty$, is again r, but is also r as $x \to -\infty$. For

$P(-i\frac{d}{dx}, 2x)$, the number is $n - r$ in both directions. Therefore if $r > n/2$, one has $2r > n$ solutions, which perforce must be linearly dependent. Hence there exists a bounded solution for $P(-i\frac{d}{dx}, -2x)$. The same goes for $P(-i\frac{d}{dx}, 2x)$ if $r < n/2$. Taking the case where some γ_j has vanishing real part into account, this argument establishes Proposition 3.3.

A precise formulation for the level (III) representations is as follows. Order the roots γ_j so that $\Re(\gamma_k) < \Re(\gamma_j)$ whenever $k < j$. Denote

$$\Phi_j(x) = \gamma_j(zx - \tfrac{1}{3}x^3) + \rho_j \ln(z - x^2)$$

where the ρ_j are certain scalars, and that branch of the logarithm function is chosen whose imaginary part is equal to $i\pi$ on the negative real axis.

Proposition 4.1. *Suppose that $\hat{P}$ is nondegenerate and generic. Then for each j there exists $\rho_j \in \mathbb{C}$ such that for each $z \in \mathbb{C}$, there exists a function $\psi_{zj}^+ \in C^\infty(\mathbb{R})$ satisfying*

$$L_z\psi_{zj}^+ \equiv 0$$

and

$$\psi_{zj}^+(x) = e^{\Phi_j(x)}\big(1 + O(x^{-1})\big) \qquad as \quad x \to +\infty.$$

ψ_{zn}^+ is uniquely determined, while in general each ψ_{zj}^+ is uniquely determined modulo linear combinations of $\{\psi_{zk}^+ : k > j\}$. There exists a constant $C_1 < \infty$ such that for all $z \in \mathbb{C}$, for all $x \geq C_1\langle z\rangle^{1/2}$ and all $0 \leq k < n$,

$$\left| \frac{d^k}{dx^k}\left(\psi_{zj}^+(x) - \gamma_j^{k-1}(z - x^2)^{k-1}e^{\Phi_j(x)}\right)\right| \leq C(|x| + \langle z\rangle^{1/2})^{-1}\left|(z - x^2)^{k-1}e^{\Phi_j(x)}\right|.$$

The ψ_{zj}^+ may be constructed so that each is an entire holomorphic function of z, modulo (for each z) a linear combination of $\{\psi_{zk}^+ : k > j\}$.

There exist also solutions ψ_{zj}^- possessing analogous properties as $x \to -\infty$. There is one formal change in the statement: ψ_{zj}^- is unique modulo linear combinations of the more rapidly decaying solutions $\{\psi_{zk}^- : k < j\}$, and the final clause of the proposition should be modified accordingly. Assume that the number of γ_j with positive real parts is $n/2$.

Definition 4.2.

$$\mathcal{W}(z) = \text{determinant}\left(\frac{d^\ell}{dx^\ell}\psi_{zj}^-(0) : j \leq n/2; \ \frac{d^\ell}{dx^\ell}\psi_{zj}^+(0) : j > n/2; \right)_{0 \leq \ell < n}.$$

The last clause of the proposition implies that $\mathcal{W}$ is an entire holomorphic function. Straightforward estimates yield the upper bound

$$|\mathcal{W}(z)| = O(\exp(C|z|^{3/2})).$$

Comment. The growth estimates for $\mathcal{W}$ are largely artifacts of its normalization, rather than an intrinsic feature of the representations under consideration. In the case $m = 2$, all representations with $z \in \mathbb{R}$ are mutually unitarily equivalent; an intertwining operator is induced by the change of variables $x \mapsto x + z$. Thus one might expect $\mathcal{W}$ to be constant, but in reality it is a Gaussian.

To approach the lower bound of Proposition 3.8, in the case $m = 3$, assume $z \in \mathbb{R}^+$ is large, let λ denote a sufficiently large positive constant, and denote $x_0^{\pm} = \pm z^{1/2}$.

Proposition 4.3. *For all sufficiently large $z \in \mathbb{R}^+$, for all $0 \le \ell < n$, for all $x \ge x_0^+ + \lambda z^{-1/4}$ and all $1 \le j \le n$,*

(4.4)
$$\frac{d^{\ell}}{dx^{\ell}} \psi_{zj}^+(x) = \gamma_j^{\ell-1}(z - x^2)^{\ell-1} e^{\Phi_j(x)}(1 + O(\lambda^{-1})).$$

For $|x - x_0^+| \le \lambda z^{-1/4}$,

(4.5)
$$\sum_{\ell=0}^{n-1} z^{-\ell/4} \left| \frac{d^{\ell}}{dx^{\ell}} \psi_{zj}^+(x) \right| \sim \sum_{\ell=0}^{n-1} z^{-\ell/4} \left| \frac{d^{\ell}}{dx^{\ell}} \psi_{zj}^+(x_0^+ + \lambda z^{-1/4}) \right|,$$

uniformly in z.

In summary, one has excellent control of the ψ_{zj}^+ for $x - x_0^+ \gg z^{-1/4}$, one loses much of this control for $|x - x_0^+| \lesssim z^{-1/4}$, and one has no control for $x \le x_0^+ - \lambda z^{-1/4}$. Thus the situation is analogous to a scattering problem, in which the solutions ψ_{zj}^+ undergo scattering by an obstacle which is localized where $|x - x_0^+| \le \lambda z^{-1/4}$.

The same loss of control occurs for the solutions ψ_{zj}^-, with x_0^+ replaced by x_0^-. Hence the situation is more accurately viewed as one of scattering by two obstacles, widely separated for large z.

Normalize the ψ_{zj}^+ by multiplying by scalars, depending on z, so that the new solutions $\hat{\psi}_{zj}^+$ thus obtained satisfy

$$\sum_{\ell=0}^{n-1} z^{-\ell/4} \left| \frac{d^{\ell}}{dx^{\ell}} \hat{\psi}_{zj}^+(x) \right| \sim 1 \quad \forall |x - x_0^+| \le \lambda z^{-1/4},$$

uniformly in $z \in \mathbb{R}^+$ provided z is sufficiently large.

Define

$$g_{zj}^+(y) = \hat{\psi}_{zj}^+(r^{1/2} + r^{-1/4}y).$$

Lemma 4.6. *As $\mathbb{R}^+ \ni z \to \infty$, the g_{zj}^+ form an equicontinuous family of functions on every compact subset of $\mathbb{R}$. The same holds for all of their derivatives. For any sequence Γ of values of z tending to $+\infty$, if $\{g_{zj}^+ : z \in \Gamma\}$ converge uniformly on compact subsets of $\mathbb{R}$, then the limit is a solution $g_{\infty,j}^+$ of $P(-i\frac{d}{dx}, -2x)$ which does not vanish identically. Moreover, $g_{\infty,j}^+(x)$ decays rapidly to 0 as $x \to +\infty$, for all $j > n/2$.*

There are similar functions g_{zj}^-, and possibly limits $g_{\infty,j}^-$, to which an analogous statement applies.

Let us imagine for a moment that circumstances were simpler, that there was only the single obstacle near x_0^+, that the ψ_{zj}^- satisfied estimates similar to those for the ψ_{zj}^+, for $|x - x_0^+| \le \lambda z^{-1/4}$, and that the g_{zj}^- were redefined accordingly, in terms of the behavior of ψ_{zj}^- near x_0^+. This is actually the situation, except for very minor alterations, when $(z - x^2)$ is replaced by $(z - x^{m-1})$, with m even. There are then two possibilities:

$$(4.7) \qquad |\mathcal{W}(z)| \ge z^{-C_0} \exp\left(\tfrac{2}{3} \sum_j |\Re(\gamma_j)| z^{3/2}\right)$$

as $\mathbb{R}^+ \ni z \to \infty$, for some constant $C_0 \in \mathbb{R}^+$; or as $z \to \infty$ through a sequence for which (4.7) fails for a sufficiently large but fixed C_0, fixing a subsequence for which the g_{zj}^+ and g_{zj}^- all converge to limits $g_{\infty,j}^+$ and $g_{\infty,j}^-$, respectively, the solutions

$$(4.8) \qquad \{g_{\infty,j}^- : j \le n/2\} \cup \{g_{\infty,j}^+ : j > n/2\}$$

are linearly dependent.

In the case (4.8) there exist scalars, not all equal to zero, such that

$$(4.9) \qquad \sum_{j \le n/2} c_j g_{\infty,j}^- \equiv \sum_{j > n/2} c_j g_{\infty,j}^+.$$

Defining

$$f = \sum_{j \le n/2} c_j g_{\infty,j}^-,$$

f is a solution of $P(-i\frac{d}{dx}, -2x)$. It decays rapidly as $x \to -\infty$ because each of the $g_{\infty,j}^-$ does (because $\Re(\gamma_j) < 0$); it also decays rapidly as $x \to +\infty$, because of the alternate representation afforded by (4.9). It follows from the estimates (4.4) at $x_0^+ + \lambda z^{-1/4}$ that the $g_{\infty,j}^+$ must be linearly independent, hence (granting the same for the $g_{\infty,j}^-$ in this hypothetical situation) that f does not vanish identically. Thus we would have Proposition 3.8.

This scheme does not apply directly to $P(-i\frac{d}{dx}, (z - x^2))$ because we know essentially nothing about the $\psi_{zj}^{\pm}$ in the large interval $[x_0^- + \lambda z^{-1/4}, x_0^+ - \lambda z^{-1/4}]$. To overcome this we introduce a third set of solutions ψ_{zj}^0, whose behavior is well understood in exactly this interval. We introduce two additional determinants of the same nature as $\mathcal{W}$, one governing the scattering of the ψ_{zj}^- into ψ_{zj}^0 at x_0^-, the other governing scattering of ψ_{zj}^0 into ψ_{zj}^+ at x_0^+. An inequality relating the three shows that if $\mathcal{W}$ is substantially smaller than the estimate in (4.7), then one of the other two determinants must also be anomalously small. The argument of the preceding paragraph may then be implemented at either of $x_0^{\pm}$, as appropriate. One obtains a bounded solution of $P(-i\frac{d}{dx}, \mp 2x)$. By means of a change of variables, this produces a bounded solution of $P(-i\frac{d}{dx}, \mp x)$.

5 RELATED QUESTIONS

In this section we consider in more detail some issues related to the operator

$$\mathcal{L} = \partial_x^2 + (\partial_y - x^{m-1}\partial_t)^2$$

in $\mathbb{R}^3$, and the ordinary differential operators

$$L_z = -\frac{d^2}{dx^2} + (z - x^{m-1})^2.$$

We assume always that $m \geq 2$ is even. It is useful to introduce

$$L_{\beta,r} = L_z \qquad \text{where} \quad z = re^{i\beta}.$$

Fix m.

Definition 5.1. *A ray* $\mathbb{R}^+ \cdot e^{i\beta}$ *in* $\mathbb{C}$ *is said to be of polynomial growth if there exist* $\sigma \in \mathbb{R}$ *and* $c > 0$ *such that for all sufficiently large* $r \in \mathbb{R}^+$,

(5.2) $$\|L_{\beta,r}f\|_{L^2(\mathbb{R})} \geq cr^{\sigma}\|f\|_{L^2(\mathbb{R})}$$

for all f *in the domain of* $L_{\beta,r}$.

It is permitted that σ be negative. As proved by Pham The Lai and Robert [PR], whenever two rays of polynomial growth can be found such that $|\beta_1 - \beta_2| < \pi\frac{m-1}{m}$, then there exists at least one nonlinear eigenvalue in the sector bounded by the two rays. When m is odd, any ray with $\pi/2 < \beta < 3\pi/2$ is a ray of polynomial growth, as is $\mathbb{R}^+$ (with $\sigma = (m - 2)/2(m - 1)$); this yields sufficiently narrow sectors to guarantee the existence of nonlinear eigenvalues, for odd $m \geq 3$. This is how the failure of $\mathcal{L}$ to be analytic hypoelliptic was first proved for odd m. When m is even, $\mathbb{R}^{\pm}$ are still rays of polynomial growth, with $\sigma = (m - 2)/2(m - 1)$, but it is not apparent that there exist any other such rays.

One may ask (we do so only for even m):

- Where are the nonlinear eigenvalues located?
- Do there exist any rays of polynomial growth, besides $\mathbb{R}^{\pm}$?
- How does $\mathcal{W}$ behave off of the real axis?
- What can be said about C^{∞} hypoellipticity of operators

$$D_{\theta} = \partial_x^2 + (e^{i\theta}\partial_y - x^{m-1}\partial_t)^2,$$

which are sums of squares of non-real vector fields satisfying the condition of Hörmander?

This last question is related to the others; to study C^{∞} hypoellipticity of D_{θ} one should analyze the one-real-parameter family of level (III) representations in which D_{θ} corresponds to $\{L_{\theta,r} : r \in \mathbb{R}\}$ acting on $L^2(\mathbb{R})$.

Some partial answers may be obtained.

Proposition 5.3. *For $m \geq 4$ and even, there are no nonlinear eigenvalues z satisfying $|\arg(z)| < \frac{\pi}{2}\frac{m-1}{m}$ or $|\pi - \arg(z)| < \frac{\pi}{2}\frac{m-1}{m}$.*

This suggests that there may be a corresponding estimate which would affirm that every ray in these two sectors would be a ray of polynomial growth, but this hope is dashed by the next result.

Proposition 5.4. *For any even integer $m \geq 2$, the only rays of polynomial growth are $\mathbb{R}^{\pm}$. More precisely, for any $\beta \in \mathbb{R}$ satisfying $0 < |\beta| < \pi$, there exists $\delta > 0$ such that for every $r \in \mathbb{R}^+$, there exists $f_r \in L^2(\mathbb{R})$ satisfying $\|f_r\|_{L^2} = 1$ but*

$$\|L_{\beta,r}\, f_r\|_{L^2(\mathbb{R})} \leq C \exp(-\delta r^{m/(m-1)}).$$

f_r is constructed by the method of the preceding sections. Solutions $\psi_r^{\pm}$ of $L_{\beta,r}$, which decay rapidly as $x \to \pm\infty$, respectively, are constructed as before. Defining $\hat{\psi}_r^{\pm}$ by multiplying by scalars so that $\hat{\psi}_r^{\pm}$ are suitably normalized on the interval $\{|x - x_0| \leq \lambda r^{-(m-2)/2(m-1)}\}$, where $x_0 = (r\cos(\beta))^{1/(m-1)}$ plays the same role as $x_0^{\pm}$ in the previous analysis, and defining $\hat{\mathcal{W}}$ to be the Wronskian determinant of $\hat{\psi}_r^{\pm}$ so that $|\hat{\mathcal{W}}| \leq C$ uniformly in r, one finds

Proposition 5.5. *The Wronskian of the normalized solutions satisfies*

(5.6) $$|\hat{\mathcal{W}}| \leq C \exp(-\delta r^{m/(m-1)}).$$

We emphasize that the normalizations are carried out so that the $\hat{\psi}_r^{\pm}$ are essentially bounded below by 1 near x_0, and similarly for their derivatives, so (5.6)

means that they are extremely close to being linearly dependent for large r. The functions f_r required in Proposition 5.4 are obtained by pasting $\hat{\psi}_r^{\pm}$ together in the interval $|x - x_0| \leq \lambda r^{-(m-2)/2(m-1)}$ via a partition of unity. The functions $\psi_r^{\pm}$ extend to entire functions of $x \in \mathbb{C}$, their Wronskian is constant, and (5.6) follows by analyzing the equation obtained from $L_{\beta,r}$ by shifting from $x \in \mathbb{R}$ to $x \in \mathbb{R} \cdot e^{i\alpha}$ for a certain $\alpha \neq 0$.

In a sense, the solutions $\psi_r^{\pm}$ constitute the only possible obstruction to the inequality 5.2:

Proposition 5.7. *For each even $m \geq 2$ there exists $\sigma \geq 0$ such that for any β, for any sufficiently large $r \in \mathbb{R}^+$, there exists a subspace V of codimension one in the domain of $L_{\beta,r}$, such that*

$$\|L_{\beta,r} f\|_{L^2(\mathbb{R})} \geq cr^{\sigma}\|f\|_{L^2} \quad \forall f \in V.$$

For $m \geq 4$, σ is strictly positive.

Proposition 5.4 plus standard arguments yield

Proposition 5.8. *For any θ such that $e^{i\theta} \notin \mathbb{R}$, D_{θ} fails to be C^{∞} hypoelliptic at the origin.*

This should be contrasted with the situation for operators which are sums of squares of *real* vector fields satisfying the Hörmander condition. In that case, any small perturbation of the vector fields in the C^k topology, for suitably large k, produces an operator which is still hypoelliptic, even though the perturbation will not in general be of lower order in any sense.

6 FINAL COMMENTS

For homogeneous, constant-coefficient differential operators in $\mathbb{R}^d$, there is an equivalence between ellipticity, C^{∞} hypoellipticity, and analytic hypoellipticity. As presented, our results indicate that this equivalence, with ellipticity replaced, of course, by the Rockland condition that $\pi(L)$ be injective for each irreducible unitary representation π, breaks down for the class of groups in question. However, the analogy with the abelian case goes further, from an alternative point of view.

Consider again the case of $\mathbb{R}^d$, and let L be homogeneous and elliptic, with constant coefficients. Then, although L will be analytic hypoelliptic, still the Cauchy estimates will be violated if the constant $C^{1+|\alpha|}$ is replaced by $C_{\varepsilon} \cdot \varepsilon^{|\alpha|}$ for all sufficiently small ε, no matter how large C_{ε} is chosen to be. This amounts to the fact that while the symbol of L is free of zeroes in a conic neighborhood of $\mathbb{R}^d \subset \mathbb{C}^d$, yet it still does have zeroes.

Consider now the four-dimensional nilpotent group of step 3 which stood at the center of our analysis, and let L be a (generic) homogeneous, left-invariant operator which satisfies the Rockland condition. Then we have proved the existence of certain non-unitary representations π on which $\pi(L)$ is not injective in L^2, an analogue of the fact that every nonconstant polynomial in one complex variable has at least one zero. This means not only that L is not analytic hypoelliptic, but that again, the constant $C^{1+|\alpha|}$ appearing in the Gevrey estimates of order $s = 3$ for solutions of L cannot be replaced by $C_\varepsilon \cdot \varepsilon^{|\alpha|}$ for arbitrarily small ε.

It can be shown that for any such operator on this group which satisfies the Rockland criterion, all solutions of $Lu \equiv 0$ in an open set must belong to the Gevrey class G^3. Thus the situation is quite analogous to that for $\mathbb{R}^d$, provided only that the class G^1 of analytic functions is replaced by G^3.

Substantial progress on related problems has been achieved in the interval between preparation of this manuscript and its publication [C7],[C8]. More detailed information on the nonlinear eigenvalues has been obtained by Ching-Chau Yu in a UCLA PhD dissertation (in preparation).

University of California, Los Angeles

REFERENCES

[C1] M. Christ. "On the $\bar{\partial}$ equation in weighted L^2 norms in $\mathbb{C}^1$." *J. Geom. Anal.* **1** (1991), 193–230.

[C2] _____. "Some non-analytic-hypoelliptic sums of squares of vector fields." *Bull. Amer. Math. Soc.* **116** (1992), 137–140.

[C3] _____. "Certain sums of squares of vector fields fail to be analytic hypoelliptic." *Comm. Part. Diff. Eq.* **16** (1991), 1695–1707.

[C4] _____. "Analytic hypoellipticity breaks down for weakly pseudoconvex Reinhardt domains." *Int. Math. Res. Not.* **1** (1991), 31–40.

[C5] _____. "A family of degenerate differential operators." *J. Geom. Anal.* **3** (1993), 579–597.

[C6] _____. "Analytic hypoellipticity, representations of nilpotent groups, and a nonlinear eigenvalue problem." *Duke Math J.* **72** (1993), 595–639.

[C7] _____. "A necessary condition for analytic hypoellipticity." *Math. Research Letters* **1** (1994), 241–248.

[C8] _____. "The Szegö Projection Need Not Preserve Global Analyticity." Preprint.

[CG] M. Christ and D. Geller. "Counterexamples to analytic hypoellipticity for domains of finite type." *Ann. of Math.* **235** (1992), 551–566.

[CL] E. Coddington and N. Levinson. *Theory of Ordinary Differential Equations.* McGraw-Hill, 1955.

[FS] G. B. Folland and E. M. Stein. "Estimates for the $\bar{\partial}_b$ complex and analysis on the Heisenberg group." *Comm. Pure Appl. Math.* **27** (1974), 429–522.

[G] D. Geller. *Analytic Pseudodifferential Operators for the Heisenberg Group and Local Solvability*. Mathematical Notes 37. Princeton University Press, 1990.

[GS] P. C. Greiner and E. M. Stein. "On the solvability of some differential operators of type $\Box_b$." Several Complex Variables, Proceedings of International Conferences, Cortona, Italy, 1976–77, Scuola Normale Superiore, Pisa, **197**, 106–165.

[He] B. Helffer. "Conditions nécessaires d'hypoanalyticité pour des opérateurs invariants à gauche homogènes sur un groupe nilpotent gradué." *J. Diff. Eq.* **44** (1982), 460–481.

[HH] N. Hanges and A. A. Himonas. "Singular solutions for sums of squares of vector fields." *Comm. Part. Diff. Eq.* **16** (1991), 1503–1511.

[HN1] B. Helffer and J. Nourrigat. "Hypoellipticité pour des groupes nilpotents de rang 3." *Comm. Part. Diff. Eq.* **3** (1978), 643–743.

[HN2] ———. "Caractérisation des opérateurs hypoelliptiques homogènes invariants à gauche sur un groupe nilpotent gradué." *Comm. Part. Diff. Eq.* **4** (1979), 899–958.

[HN3] ———. *Hypoellipticité Maximal pour des Opérateurs Polynômes de Champs de Vecteurs*. Prog. Math. vol. 58. Birkhäuser, 1985.

[K] A. A. Kirillov. "Unitary representations of nilpotent Lie groups." *Russian Math. Surv.* **17** (1962), 53–104.

[Ke] M. V. Keldysh. "On the completeness of the eigenfunctions of classes of non-selfadjoint linear operators." *Russian Math. Surv.* **26** (1971), 15–44.

[M] G. Métivier. "Hypoellipticité analytique sur des groupes nilpotents de rang 2." *Duke Math. J.* **47** (1980), 195–221.

[NSW] A. Nagel, E. M. Stein, and S. Wainger. "Balls and metrics defined by vector fields I: Basic properties." *Acta Math.* **155** (1985), 103–147.

[PR] Pham The Lai and D. Robert. "Sur un problème aux valeurs propres non linéaire." *Israel J. Math.* **36** (1980), 169–186.

[R] C. Rockland. "Hypoellipticity on the Heisenberg group-representation-theoretic criteria." *Trans. Amer. Math. Soc.* **240** (1978), 1–52.

[RS] L. P. Rothschild and E. M. Stein. "Hypoelliptic differential operators and nilpotent groups." *Acta Math.* **137** (1976), 247–320.

7

Opérateurs Bilinéaires
et Renormalisation

R. R. Coifman, S. Dobyinsky, et Y. Meyer

1 INTRODUCTION

Une fois renormalisées, certaines expressions bilinéaires se comportent mieux que prévu et la forme précisée du lemme du div-curl nous permettra d'illustrer cette remarque. Cette forme précisée, conjecturée par P. L. Lions et démontrée dans [2], utilise l'espace $\mathcal{H}^1(\mathbb{R}^n)$ de E. Stein et G. Weiss. Rappelons que l'espace de Hardy $\mathcal{H}^1(\mathbb{R}^n)$ est l'ensemble des fonctions f, appartenant à $L^1(\mathbb{R}^n)$, dont les n transformées de Riesz $R_1(f), \ldots, R_n(f)$ appartiennent également à $L^1(\mathbb{R}^n)$. La norme de f dans $\mathcal{H}^1(\mathbb{R}^n)$ est la somme $\|f\|_1 + \|R_1(f)\|_1 + \cdots + \|R_n(f)\|_1$ et, muni de cette norme, $\mathcal{H}^1(\mathbb{R}^n)$ est un espace de Banach, composé de fonctions d'intégrale nulle. Comme l'ont montré C. Fefferman et E. Stein [4], le dual de $\mathcal{H}^1(\mathbb{R}^n)$ est l'espace BMO de John et Nirenberg. La forme précisée du lemme du div-curl est l'énoncé suivant:

Théorème 1. *Soient* $E(x) = (E_1(x), \ldots, E_n(x))$ *et* $B(x) = (B_1(x), \ldots, B_n(x))$, $x \in \mathbb{R}^n$, *deux champs de vecteurs vérifiant les propriétés suivantes:*

(1.1) $\quad E_j(x) \in L^2(\mathbb{R}^n), \qquad B_j(x) \in L^2(\mathbb{R}^n) \qquad pour \quad 1 \leq j \leq n$

(1.2) $\qquad\qquad \operatorname{div} E(x) = 0, \qquad \operatorname{curl} B(x) = 0.$

Alors $E(x) \cdot B(x) = E_1(x)B_1(x) + \cdots + E_n(x)B_n(x) \in \mathcal{H}^1(\mathbb{R}^n).$

Les dérivées qui interviennent dans (1.2) sont prises au sens des distributions. Chacun des termes $E_j(x)B_j(x)$ du produit scalaire $E(x) \cdot B(x)$ appartient à

$L^1(\mathbb{R}^n)$, sans appartenir à $\mathcal{H}^1(\mathbb{R}^n)$. Si la somme de ces n termes appartient à $\mathcal{H}^1(\mathbb{R}^n)$, ce fait remarquable est dû à la présence de cancellations, provenant des hypothèses (1.2). Ces cancellations seront analysées en appliquant un algorithme de renormalisation à chacun des produits $E_j(x)B_j(x)$. Un algorithme de renormalisation permet d'écrire le produit ponctuel uv entre deux fonctions $u(x)$ et $v(x)$ sous la forme

$$(1.3) \qquad uv = P(u, v) + R(u, v)$$

où l'opérateur bilinéaire P, définissant l'algorithme, satisfait les conditions suivantes:

$$(1.4) \qquad P(u, v) = P(v, u)$$

(1.5) $\quad P(u, v) = 0$ chaque fois que la fonction $u(x)$ est une constante

et où l'opérateur bilinéaire R possède les propriétés suivantes:

(1.6) $\qquad u \in L^2 \quad$ et $\quad v \in L^2 \quad$ impliquent $\quad R(u, v) \in \mathcal{H}^1$

si u_j, $j \geq 1$, et v_j, $j \geq 1$, sont deux suites bornées

(1.7) $\qquad$ en norme L^2 et si $u_j \rightharpoonup u$ $(j \to +\infty)$, $v_j \rightharpoonup v$ $(j \to +\infty)$,

$\qquad$ alors $R(u_j, v_j) \rightharpoonup U(u, v)$ $\quad (j \to +\infty)$.

Nous avons noté $f_j \rightharpoonup f (j \to +\infty)$ la convergence au sens des distributions: $\int f_j(x)\varphi(x)\,dx \to \int f(x)\varphi(x)\,dx$ pour toute fonction de test φ. Dans la conclusion de (1.7), on peut supposer que $\varphi(x)$ appartienne à l'espace VMO qui est la fermeture, pour la norme BMO, de l'espace vectoriel des fonctions continues et nulles à l'infini.

Signalons enfin que $\mathcal{H}^1$ est le dual de VMO.

Voici la signification des deux termes de la décomposition (1.3). Le terme $P(u, v)$ est la partie principale du produit uv. Si au lieu d'appartenir à $L^2(\mathbb{R}^n)$, les deux fonctions u et v appartenaient à $L^p(\mathbb{R}^n)$ et si $p < 2$, alors le produit uv n'aurait plus de sens, en tant que distribution, et il conviendrait de lui soustraire certaines quantités infinies pour retrouver un terme suffisamment oscillant pour être une distribution tempérée. Ces quantités infinies sont présentes dans le terme $P(u, v)$. Le terme $R(u, v)$ appartiendra à l'espace de Hardy $\mathcal{H}^{p/2}$ lorsque u et v appartiennent à L^p et que p est inférieur à 2. *Le terme $R(u, v)$ est donc le produit uv, renormalisé par soustraction de $P(u, v)$.*

Cette renormalisation doit être la plus simple possible: nous ne voulons soustraire que le strict minimum et, la plupart du temps, ne rien soustraire du tout. Si par exemple la fonction $u(x)$ est une constante, le produit entre cette constante et une fonction arbitraire $v(x)$ n'a pas besoin d'être renormalisé et $P(u, v)$ sera donc nul, comme l'indique la condition (1.5). Cette même condition (1.5) nous permettra de vérifier (section 8) que le produit entre une fonction $u(x)$ qui est régulière et à support compact et une fonction arbitraire $v \in L^2$ conduit à une correction $P(u, v)$ faible, au sens que l'opérateur qui à v associe $P(u, v)$ est compact.

L'opération de renormalisation doit permettre de multiplier deux distributions tempérées arbitraires. Elle s'apparente au paraproduit, tel que J. M. Bony l'a défini. La condition (1.7) signifie qu'en se limitant à des suites bornées dans L^2, la renormalisation est compatible avec la convergence au sens des distributions.

Signalons enfin qu'il existe plusieurs solutions au problème de la renormalisation du produit. Nous avons choisi celle, présentée dans la section 3, qui conduit aux calculs les plus simples.

Revenant au théorème 1, on a automatiquement et sans tenir compte de (1.2),

$$(1.8) \qquad E_j(x)B_j(x) - P(E_j, B_j)(x) \in \mathcal{H}^1(\mathbb{R}^n), \quad 1 \le j \le n.$$

Démontrer le théorème 1 revient donc à prouver que la somme $S(x) = P(E_1, B_1) + \cdots + P(E_n, B_n)$ appartient aussi à $\mathcal{H}^1$ lorsque les conditions (1.1) et (1.2) sont satisfaites.

En fait, on a beaucoup mieux et cette somme $S(x)$ appartient à l'espace de Besov homogène $\dot{B}_1^{0,1}$ qui sera défini soigneusement dans la section suivante et qui est un sous-espace "rudimentaire" de $\mathcal{H}^1(\mathbb{R}^n)$. *L'énoncé du théorème 1 prête donc à confusion puisque le rôle joué par l'espace de Hardy $\mathcal{H}^1(\mathbb{R}^n)$ y est relativement superficiel, alors que l'espace de Besov $\dot{B}_1^{0,1}$ est au cœur du problème.*

Cette même analyse (1.3) nous fournira une démonstration particulièrement élégante du lemme "classique" du div-curl, tel que F. Murat et L. Tartar l'ont énoncé. Il s'agira de montrer que, sous les hypothèses (1.2), le produit scalaire $E \cdot B$ possède également la propriété de continuité faible décrite dans (1.7). Dans la section suivante sont rappelés quelques résultats généraux concernant les opérateurs bilinéaires et l'espace de Besov homogène $\dot{B}_1^{0,1}$. Nous pourrons alors construire toute une collection d'opérateurs P ayant les propriétés (1.4) à (1.7). A l'intérieur de cette collection, nous choisirons un exemple explicite, se reliant aux "paraproduits" de J. M. Bony et conduisant à une démonstration particulièrement simple du théorème 1 et du lemme de Murat et Tartar. Nous conclurons en indiquant (sans démonstration) les liens qui existent entre la renormalisation et les séries d'ondelettes.

2 LA DÉFINITION GÉNÉRALE
DES OPÉRATEURS BILINÉAIRES P

Pour la commodité du lecteur, nous rappelons quelques énoncés figurant dans [1] ou [5].

Les *symboles bilinéaires* servant à construire les *opérateurs bilinéaires* sont des fonctions $\tau(\xi, \eta)$ définies sur $\mathbb{R}^n \times \mathbb{R}^n$ privé de $(0, 0)$ et vérifiant les conditions usuelles

$$(2.1) \qquad |\partial_\xi^\alpha \partial_\eta^\beta \tau(\xi, \eta)| \leq C_{\alpha, \beta}(|\xi| + |\eta|)^{-|\alpha| - |\beta|}$$

où $\alpha = (\alpha_1, \ldots, \alpha_n)$ est un multi-indice de longueur $|\alpha| = \alpha_1 + \ldots + \alpha_n$ et où

$$\partial_\xi^\alpha = \left(\frac{\partial}{\partial \xi_1}\right)^{\alpha_1} \cdots \left(\frac{\partial}{\partial \xi_n}\right)^{\alpha_n}.$$

En particulier, les conditions (2.1) sont satisfaites si $\tau(\xi, \eta)$ est indéfiniment dérivable dans $\mathbb{R}^n \times \mathbb{R}^n$ privé de $(0, 0)$ et si $\tau(\lambda\xi, \lambda\eta) = \tau(\xi, \eta)$ pour tout $\lambda > 0$.

On associe à un *symbole bilinéaire* $\tau(\xi, \eta)$ l'*opérateur bilinéaire*

$$(2.2) \qquad T(f, g)(x) = (2\pi)^{-2n} \iint e^{i(\xi + \eta) \cdot x} \tau(\xi, \eta) \hat{f}(\xi) \hat{g}(\eta) \, d\xi \, d\eta$$

où $\hat{f}$, $\hat{g}$ sont les transformées de Fourier de f et g.

Les conditions (2.1) impliquent que T, qui est évidemment défini si f et g appartiennent à la classe $\mathcal{S}(\mathbb{R}^n)$ de Schwartz, se prolonge en un opérateur (encore noté T), défini sur $L^2(\mathbb{R}^n) \times L^2(\mathbb{R}^n)$, à valeurs dans $L^1(\mathbb{R}^n)$. Un cas particulier évident est le symbole $\tau = 1$ conduisant à $T(f, g) = fg$. La continuité de $T : L^2 \times L^2 \to L^1$ sous les hypothèses (2.1) est donc une généralisation de l'inégalité de Hölder. Cela amène à conjecturer que T se prolonge en un opérateur linéaire continu de $L^p \times L^q$ à valeurs dans L^r si $1 \leq r < \infty$, $1 < p \leq \infty$, $1 < q < \infty$ et $\frac{1}{p} + \frac{1}{q} = \frac{1}{r}$. Ceci est vrai (valeurs limites comprises) et la démonstration se trouve dans [5].

Une fois pour toutes, les conditions (2.1) seront supposées satisfaites dans les énoncés qui suivent (lemmes 1, 2, et 3). Dans ces conditions, la fonction τ sera appelée le symbole bilinéaire de l'opérateur bilinéaire T.

Lemme 1. *Une condition nécessaire et suffisante pour que l'opérateur bilinéaire $T : L^2 \times L^2 \to L^1$, défini par (2.2) ait la propriété suivante*

$$(2.3) \qquad T(f, g) \in \mathcal{H}^1(\mathbb{R}^n) \quad \text{lorsque } f \in L^2 \text{ et } g \in L^2$$

est que son symbole bilinéaire $\tau(\xi, \eta)$ vérifie

$$(2.4) \qquad \tau(\xi, -\xi) = 0 \qquad \text{pour tout } \quad \xi \neq 0.$$

Les conditions (2.3) et (2.4) sont aussi équivalentes à

$$(2.5) \qquad \int_{\mathbb{R}^n} T(f, g)(x) \, dx = 0 \quad \text{pour toute } f \in L^2 \text{ et toute } g \in L^2$$

ou encore à cette même condition (2.5) *lorsque* f *et* g *appartiennent à la classe de Schwartz.*

Ce lemme fournit une démonstration (fausse) du théorème 1. Pour vérifier que $E \cdot B$ appartient à $\mathcal{H}^1$, il suffirait donc de montrer que la condition (2.5) est satisfaite. Or on a $B = \nabla U(x)$ puisque curl $B(x) = 0$. Il vient $E(x) \cdot B(x) = \text{div}[U(x)E(x)]$ puisque div $E(x) = 0$. Finalement $\int E(x) \cdot B(x) \, dx = 0$. Cette démonstration est insuffisante parce que le produit scalaire $E \cdot B$ n'est pas un des opérateurs bilinéaires auxquels s'applique le lemme 1.

La preuve du lemme 1 consiste à évaluer l'intégrale $I = \int T(f, g) \cdot u(x) \, dx$ lorsque $u(x)$ appartient à VMO et que $\|u\|_{BMO} \leq 1$. Mais on a $I = \int L(f, u)g \, dx$ où l'opérateur bilinéaire L est défini par le symbole $\lambda(\xi, \eta) = \tau(\xi, -\xi - \eta)$. On a alors $\|L(f, u)\|_2 \leq C\|f\|_2\|u\|_{BMO}$ en appliquant le théorème 34 du chapitre VI de [1].

Le lemme suivant n'intervient pas dans la démonstration du théorème 1. Il nous apprendra que tous les choix de l'opérateur T que nous ferons sont équivalents.

Lemme 2. *En conservant les notations du lemme 1, une condition suffisante pour que $T(f, g)$ appartienne à l'espace de Besov homogène $\dot{B}_1^{0,1}(\mathbb{R}^n)$, pour tout couple (f, g) de deux fonctions de $L^2(\mathbb{R}^n)$, est que l'on ait*

$$(2.6) \qquad \tau(\xi, -\xi) = 0 \qquad \text{pour tout} \quad \xi \neq 0$$

et

$$(2.7) \qquad \tau(\xi, 0) = 0 = \tau(0, \eta) \qquad \text{pour} \quad \xi \neq 0, \ \eta \neq 0.$$

Avant de passer à la définition de $\dot{B}_1^{0,1}$, un dernier résultat doit être mentionné.

Lemme 3. *En conservant les notations du lemme 1, la condition* (2.4) *implique la propriété de continuité faible suivante:*

$$(2.8) \qquad \begin{aligned} &\text{si } f_j \text{ et } g_j \text{ sont deux suites bornées dans } L^2(\mathbb{R}^n) \\ &\text{et si } f_j \rightharpoonup f \text{ et } g_j \rightharpoonup g, \text{ alors } T(f_j, g_j) \rightharpoonup T(f, g). \end{aligned}$$

Le lecteur trouvera dans l'appendice les preuves très simples de ces deux lemmes.

Rappelons maintenant la définition de l'espace de Besov homogène $\dot{B}_1^{0,1}$.

On désigne par $\psi(x)$ une fonction de la classe de Schwartz $\mathcal{S}(\mathbb{R}^n)$ dont la transformée de Fourier est nulle si $|\xi| \leq 2/3$ ou si $|\xi| \geq 8/3$ et vérifie, si $\xi \neq 0$,

$$(2.9) \qquad 0 < c \leq \sum_{-\infty}^{\infty} |\hat{\psi}(2^{-j}\xi)|.$$

On désignera alors par $\psi_j(x)$ la fonction $2^{nj}\psi(2^j x)$ et par Δ_j l'opérateur de convolution avec ψ_j.

Une fonction $f \in L^1(\mathbb{R}^n)$ appartient à $\dot{B}_1^{0,1}$ si et seulement si

$$(2.10) \qquad \sum_{-\infty}^{\infty} \|\Delta_j(f)\|_1 < \infty$$

et cette condition ne dépend pas du choix des opérateurs Δ_j.

Une seconde définition équivalente s'obtient en décomposant $f(x)$ dans une base orthonormée d'ondelettes ψ_λ, $\lambda \in \Lambda$, de régularité $r \geq 1$. Alors $f \in \dot{B}_1^{0,1}$ si et seulement si

$$(2.11) \qquad f(x) = \sum_{\lambda \in \Lambda} \alpha_\lambda \psi_\lambda(x) \qquad \text{et} \qquad \sum_{\lambda \in \Lambda} |\alpha_\lambda| \, \|\psi_\lambda\|_1 < \infty.$$

Selon la terminologie de G. Weiss et de ses collaborateurs, (2.11) s'appelle une décomposition atomique en "atomes spéciaux."

Les opérateurs bilinéaires P que nous utiliserons dans la preuve du théorème 1 seront définis par des symboles bilinéaires $\pi(\xi, \eta)$ vérifiant les conditions suivantes:

$$(2.12) \qquad \pi(\xi, \eta) = \pi(\eta, \xi)$$

$$(2.13) \qquad \pi(\xi, 0) = \pi(0, \eta) = 0 \qquad \text{si } \xi \neq 0 \text{ et } \eta \neq 0$$

$$(2.14) \qquad \pi(\xi, -\xi) = 1 \qquad \text{si } \xi \neq 0.$$

Compte tenu des lemmes 1 et 3, ces propriétés entraînent les conditions (1.4) à (1.7).

En outre, si deux symboles bilinéaires π_1 et π_2 possèdent ces propriétés, alors $\pi_3 = \pi_1 - \pi_2$ vérifiera les conditions suffisantes du lemme 2.

Cela entraîne que, modulo $\dot{B}_1^{0,1}$, tous les choix de P que nous ferons seront équivalents.

3 RENORMALISATION ET PARAPRODUITS

Le but de cette section est de donner une démonstration complète du théorème 1 en utilisant un choix de l'opérateur P qui se reliera naturellement à la théorie du paraproduit de J. M. Bony.

Pour définir P on utilise la décomposition classique de l'espace de Fourier $\mathbb{R}^n \setminus \{0\}$ en couronnes dyadiques ainsi que l'analyse de Littlewood-Paley-Stein associée à cette décomposition.

On désigne pour cela par $\psi(x)$ une fonction à valeurs réelles, appartenant à la classe $\mathcal{S}(\mathbb{R}^n)$ de Schwartz et dont la transformée de Fourier vérifie les deux conditions

$$\text{(3.1)} \qquad \hat{\psi}(\xi) = 0 \qquad \text{si} \quad |\xi| \leq 2/3 \text{ ou } |\xi| \geq 8/3$$

et

$$\text{(3.2)} \qquad \sum_{-\infty}^{\infty} |\hat{\psi}(2^{-j}\xi)|^2 = 1 \qquad \text{pour tout} \quad \xi \neq 0.$$

Compte tenu de (3.1), la condition (3.2) se réduit à

$$\text{(3.3)} \qquad |\hat{\psi}(\xi)|^2 + |\hat{\psi}(2\xi)|^2 = 1 \qquad \text{si} \quad \frac{2}{3} \leq |\xi| \leq \frac{4}{3}.$$

On pose ensuite $\psi_j(x) = 2^{nj}\psi(2^j x)$, $j \in \mathbb{Z}$, et l'on désigne par Δ_j l'opérateur de convolution avec ψ_j. La condition (3.2) signifie que

$$\text{(3.4)} \qquad I = \sum_{-\infty}^{\infty} \Delta_j^* \Delta_j$$

où Δ_j^* est l'adjoint de Δ_j.

On pose finalement, si f et g appartiennent à $L^2(\mathbb{R}^n)$,

$$\text{(3.5)} \qquad P(f, g) = \sum_{-\infty}^{\infty} (\Delta_j f)(\Delta_j g).$$

Il est trivial de vérifier que $P(f, g)$ appartient à $L^1(\mathbb{R}^n)$ puisque l'on a

$$\|(\Delta_j f)(\Delta_j g)\|_1 \leq \|\Delta_j f\|_2 \|\Delta_j g\|_2 = \alpha_j \beta_j$$

et que $\sum_{-\infty}^{\infty} |\alpha_j|^2 = \|f\|_2^2$, $\sum_{-\infty}^{\infty} |\beta_j|^2 = \|g\|_2^2$, grâce à (3.2). On a donc

$$\text{(3.6)} \qquad \|P(f, g)\|_1 \leq \|f\|_2 \|g\|_2.$$

Le symbole bilinéaire de P est $\pi(\xi, \eta) = \sum_{-\infty}^{\infty} \hat{\psi}(2^{-j}\xi)\hat{\psi}(2^{-j}\eta)$. Il vérifie évidemment (2.12). La condition (2.13) provient de (3.1) tandis que (2.14) découle de (3.2) et de $\hat{\psi}(-\xi) = \hat{\psi}(\xi)$.

Pour démontrer le théorème 1, on écrit chaque produit $E_1(x)B_1(x), \ldots,$ $E_n(x)B_n(x)$ sous la forme $P(E_j, B_j) + R(E_j, B_j)$. On a $R(E_j, B_j) \in \mathcal{H}^1$ et l'amélioration du théorème 1 que nous avons en vue est le fait que

$$\text{(3.7)} \qquad S(x) = P(E_1, B_1) + \cdots + P(E_n, B_n) \in \dot{B}_1^{0,1}$$

On a

$$S(x) = \sum_{-\infty}^{\infty} (\Delta_j E) \cdot (\Delta_j B)$$

et nous allons, en désignant par $\| \cdot \|$ la norme dans l'espace de Besov $\dot{B}_1^{0,1}$, démontrer le théorème suivant:

Théorème 2. *En conservant les notations précédentes et les hypothèses du théorème 1, on a*

(3.8) $$\|(\Delta_j E) \cdot (\Delta_j B)\| \leq C(n) \|\Delta_j E\|_2 \|\Delta_j B\|_2$$

où $C(n)$ ne dépend que de la dimension n.

Il résulte évidemment de (3.8) que $S(x) \in \dot{B}_1^{0,1}$. En effet, on a $\sum_{-\infty}^{\infty} \|\Delta_j E\|_2^2 = \|E\|_2^2$ et $\sum_{-\infty}^{\infty} \|\Delta_j B\|_2^2 = \|B\|_2^2$ et cela entraîne

$$\sum_{-\infty}^{\infty} \|(\Delta_j E) \cdot (\Delta_j B)\| \leq C(n) \|E\|_2 \|B\|_2.$$

Pour établir (3.8), nous utiliserons le lemme suivant:

Lemme 4. *Soit $f(x)$ une fonction définie sur $\mathbb{R}^n$, localement intégrable, et possédant les deux propriétés suivantes:*

(3.9) $$\nabla f = \left(\frac{\partial}{\partial x_1} f, \ldots, \frac{\partial}{\partial x_n} f \right) \in L^1(\mathbb{R}^n)$$

(3.10) $$f = \operatorname{div} F \quad où \quad F = (F_1, \ldots, F_n) \in L^1(\mathbb{R}^n).$$

Alors $f(x)$ appartient à $\dot{B}_1^{0,1}(\mathbb{R}^n)$ et l'on a

(3.11) $$\|f\| \leq C(n) \big[\|\nabla f\|_1 \|F\|_1 \big]^{1/2}.$$

On a posé $\|F\|_1 = \|F_1\|_1 + \cdots + \|F_n\|_1$ et $\|\nabla f\|_1$ est définie de même. La preuve élémentaire du lemme 4 est renvoyée à la section suivante.

Pour démontrer le théorème 2, nous appliquerons le lemme 4 à la fonction $f_j(x) = (\Delta_j E) \cdot (\Delta_j B)$. La transformée de Fourier de f_j est nulle hors de la boule $|\xi| \leq \frac{16}{3} 2^j$ ce qui entraîne

(3.12) $$\|\nabla f_j\|_1 \leq \frac{16}{3} 2^j \|f_j\|_1 \leq \frac{16}{3} 2^j \|\Delta_j E\|_2 \|\Delta_j B\|_2,$$

grâce au lemme de S. Bernstein dont nous rappelons l'énoncé.

Lemme 5. *Si la transformée de Fourier $\hat{f}(\xi)$ d'une fonction $f \in L^p(\mathbb{R}^n)$, $1 \le p \le \infty$, est nulle hors de la boule $|\xi| \le R$, alors on a*

$$(3.13) \qquad \|\nabla f\|_p \le R\|f\|_p$$

et si, en revanche, $\hat{f}$ est nulle sur la boule $|\xi| \le R$, l'inégalité inverse de (3.13) est vérifiée au sens suivant:

$$(3.14) \qquad R\|f\|_p \le C(n)\|\nabla f\|_p$$

où $C(n)$ ne dépend que de la dimension n.

On observe ensuite que curl $B(x) = 0$ signifie $B(x) = \nabla u(x)$ où la fonction (scalaire) $u(x)$ est unique, modulo les fonctions constantes. Alors (3.14), appliquée à la fonction $\Delta_j u$, s'écrit

$$(3.15) \qquad \|\Delta_j(u)\|_2 \le C2^{-j}\|\Delta_j B\|_2.$$

L'hypothèse div $E(x) = 0$ entraîne

$$(3.16) \qquad f_j(x) = \text{div } F_j(x)$$

où $F_j(x) = (\Delta_j u)(\Delta_j E)$ et $\|F_j\|_1 \le C2^{-j}\|\Delta_j B\|_2\|\Delta_j E\|_2$.

Le lemme 4 entraîne donc le théorème 2.

4 LA PREUVE DU LEMME 4

On observe tout d'abord que, pour tout $a > 0$, les fonctions $f(x)$ et $a^n f(ax)$ ont la même norme dans $\dot{B}_1^{0,1}$. En choisissant convenablement $a > 0$, on peut se ramener au cas où $\|\nabla f\|_1 = \|F\|_1$.

Pour démontrer le lemme 4, il suffit alors de vérifier que, pour tout $j \le -1$, on a

$$(4.1) \qquad \|\Delta_j(f)\|_1 \le C2^j\|F\|_1$$

tandis que, pour tout $j \ge 0$, on a

$$(4.2) \qquad \|\Delta_j(f)\|_1 \le C2^{-j}\|\nabla f\|_1.$$

Ces deux estimations résultent des deux assertions du lemme de Bernstein. Pour établir (4.1), on observe que $f = \text{div } F$ et que $\Delta_j(f) = \text{div}(\Delta_j(F))$. On applique alors (3.13) à chacune des composantes du vecteur $\Delta_j(F)$ et l'on obtient (4.1).

Pour vérifier (4.2), on utilise l'inégalité inverse (3.14) que l'on applique à $\Delta_j(f)$. Ceci termine la preuve du lemme 4.

5 LE LIEN AVEC LE PARAPRODUIT DE J. M. BONY

Pour définir le paraproduit, on part d'une fonction radiale, $\varphi(x)$, appartenant à la classe de Schwartz $\mathcal{S}(\mathbb{R}^n)$, à valeurs réelles et dont la transformée de Fourier $\hat{\varphi}(\xi)$ vérifie les conditions

(5.1) $$\hat{\varphi}(\xi) = 0 \quad \text{si} \quad |\xi| \geq 4/3$$

et

(5.2) $$\hat{\varphi}(\xi) = 1 \quad \text{si} \quad |\xi| \leq 2/3.$$

On pose alors $\varphi_j(x) = 2^{nj}\varphi(2^j x)$ et l'on désigne par S_j l'opérateur de convolution avec φ_j. On définit $\psi(x)$ par $\psi(x) = 2^n\varphi(2x) - \varphi(x)$, on pose $\psi_j(x) = 2^{nj}\psi(2^j x)$ et l'opérateur de convolution avec ψ_j est $D_j = S_{j+1} - S_j$.

Soient f et g deux fonctions arbitraires, appartenant à $L^2(\mathbb{R}^n)$. On définit alors $P_1(f, g)$ par

(5.3) $$P_1(f, g) = \sum \sum_{|j-k| \leq 2} (D_j f)(D_j g).$$

Alors on a l'identité remarquable

(5.4) $$fg = P_1(f, g) + R_1'(f, g) + R_1''(f, g)$$

où

$$R_1'(f, g) = \sum_{-\infty}^{\infty} (S_{j-2}f)(D_j g)$$

et

$$R_1''(f, g) = \sum_{-\infty}^{\infty} (S_{j-2}g)(D_j f) = R_1'(g, f)$$

sont les paraproduits définis par J. M. Bony.

Le symbole bilinéaire $\pi_1(\xi, \eta)$ de P_1 vaut

$$1 - \sum_{-\infty}^{\infty} \hat{\varphi}\left(2^{-(j-2)}\xi\right)\hat{\psi}\left(2^{-j}\eta\right) - \sum_{-\infty}^{\infty} \hat{\varphi}\left(2^{-(j-2)}\eta\right)\hat{\psi}\left(2^{-j}\xi\right)$$

et vérifie les conditions (2.12), (2.13), et (2.14).

Cela signifie que l'opérateur P que nous avons utilisé dans la section 3 peut être remplacé par P_1 et que l'erreur commise appartiendra à $\dot{B}_1^{0,1}$.

6 LE LEMME DE MURAT ET TARTAR

Rappelons tout d'abord l'énoncé classique du lemme du div-curl [2].

Soient $E^{(m)}$ et $B^{(m)}$, $m \in \mathbb{N}$, deux suites de champs de vecteurs vérifiant

(6.1) $\|E^{(m)}\|_2 \leq C_0$, $\|B^{(m)}\|_2 \leq C_0$

ainsi que

(6.2) $E^{(m)} \rightharpoonup E$, $B^{(m)} \rightharpoonup B$ $(m \rightarrow +\infty)$.

Supposons, d'autre part, que div $E^{(m)} = 0$ et curl $B^{(m)} = 0$. Alors on a

Lemme 6. *Sous les hypothèses précédentes,*

(6.3) $\lim\limits_{m \to \infty} \int E^{(m)} \cdot B^{(m)}(x)u(x)\,dx = \int E(x) \cdot B(x)u(x)\,dx$

pour toute fonction de test $u(x)$, continue et nulle à l'infini.

Il suffit évidemment de démontrer (6.3) lorsque $u(x)$ est une fonction de classe C^1, à support compact.

On utilise pour cela l'opérateur P de la section 3. La propriété de continuité faible est automatiquement satisfaite pour les restes $R(E^{(m)}, B^{(m)})$, grâce à (1.7), et il suffit donc de considérer $P(E^{(m)}, B^{(m)})$.

On pose

$$I_j^{(m)} = \int (\Delta_j E^{(m)}) \cdot (\Delta_j B^{(m)})u(x)\,dx$$

et l'on doit vérifier que

(6.4) $\lim\limits_{m \to +\infty} \sum\limits_{-\infty}^{\infty} I_j^{(m)} = \sum\limits_{-\infty}^{\infty} I_j$

où

$$I_j = \int (\Delta_j E) \cdot (\Delta_j B)u(x)\,dx.$$

Pour cela, il suffit d'établir l'existence de deux constantes C_1 et C_2 telles que

(6.5) $|I_j^{(m)}| \leq C_1 2^{-j}$ si $j \geq 0$

(6.6) $|I_j^{(m)}| \leq C_2 2^{nj}$ si $j \leq 0$.

La condition (6.2) implique que, pour tout j fixé, on a

(6.7) $\lim\limits_{m \to +\infty} I_j^{(m)} = I_j$

et (6.4) résultera de (6.5), (6.6), et (6.7).

Pour établir (6.6), on écrit

(6.8) $\Delta_j E^{(m)}(x) = \int \psi_j(x - t)E^{(m)}(t)\,dt$

et l'on a

$$\|\Delta_j E^{(m)}\|_\infty \le \|\psi_j\|_2 \|E^{(m)}\|_2 \le C_0 2^{nj/2}.$$

Cela permet de majorer $|I_j^{(m)}|$ par $C_0^2 2^{nj} \|u\|_1$.

Pour démontrer (6.5), on observe que

$$(6.9) \qquad \Delta_j E^{(m)} \cdot \Delta_j B^{(m)} = \text{div}\big[(\Delta_j u^{(m)}), (\Delta_j E^{(m)})\big]$$

et l'on a donc, après intégration par parties,

$$(6.10) \qquad I_j^{(m)} = -\int (\Delta_j u^{(m)})(\Delta_j E^{(m)} \cdot \nabla u)\, dx.$$

On majore finalement cette intégrale par $\|\Delta_j u^{(m)}\|_2 \|\Delta_j E^{(m)}\|_2 \|\nabla u\|_\infty$. Or, comme nous l'avons déjà noté,

$$\|\Delta_j u^{(m)}\|_2 \le C_1 2^{-j} \qquad \text{et} \qquad \|\Delta_j E^{(m)}\|_2 \le C_0.$$

Les propriétés (6.5) et (6.6) étant démontrées, il reste à examiner (6.7).

Or la convergence faible des $E^{(m)}$ implique la convergence simple des fonctions $\Delta_j E^{(m)}(x)$. On a même convergence uniforme sur tout compact. Puisque $u(x)$ est à support compact, (6.7) en résulte.

Le lemme de Murat et Tartar est donc démontré.

7 LA RENORMALISATION PAR LES ONDELETTES

Considérons une base orthonormée, pour l'instant arbitraire, $\psi_\lambda(x)$, $\lambda \in \Lambda$, de $L^2(\mathbb{R}^n)$. Soit $u(x)$ une fonction de $L^2(\mathbb{R}^n)$. On a alors $u(x) = \sum_{\lambda \in \Lambda} \alpha_\lambda \psi_\lambda(x)$ et il en découle que

$$(7.1) \qquad |u(x)|^2 = \sum_{\lambda \in \Lambda} |\alpha_\lambda|^2 |\psi_\lambda(x)|^2 + w(x)$$

où $w(x) \in L^1(\mathbb{R}^n)$ et $\int w(x)\, dx = 0$.

On peut aussi considérer deux fonctions arbitraires $u(x)$ et $v(x)$ de $L^2(\mathbb{R}^n)$, les décomposer dans la base orthonormée ψ_λ, $\lambda \in \Lambda$, et écrire

$$(7.2) \qquad u(x)\overline{v(x)} = \sum_{\lambda \in \Lambda} \alpha_\lambda \overline{\beta_\lambda} |\psi_\lambda(x)|^2 + w(x)$$

où, là encore, $w(x) \in L^1(\mathbb{R}^n)$ et $\int w(x)\, dx = 0$.

En général la fonction $w(x)$ n'a pas d'autre propriété. Si l'on considère l'exemple du système trigonométrique $(2\pi)^{-1/2} e^{ikx}$ sur l'intervalle $[-\pi, \pi]$ au lieu de (ψ_λ) sur $\mathbb{R}^n$, alors $w(x) = u(x)\overline{v(x)} - c$ où c est une constante et $w(x)$ est donc n'importe quelle fonction de $L^1[-\pi, \pi]$ d'intégrale nulle.

Soit maintenant ψ_λ, $\lambda \in \Lambda$, une base orthonormée d'ondelettes réelles. On se reportera à [5] pour y trouver une description détaillée de ces bases. Excluons le cas particulier du système de Haar et supposons que la régularité r des fonctions ψ_λ ne soit pas nulle ($r \geq 1$ dans les notations de [5]).

Alors la fonction $w(x)$ précédente appartient à $\mathcal{H}^1$ dès que u et v appartiennent à $L^2(\mathbb{R}^n)$.

On se limite dans l'énoncé qui suit à des fonctions à valeurs réelles et, si u et v sont deux fonctions de $L^2(\mathbb{R}^n)$, on pose

$$\langle u, v \rangle = \int u(x)v(x)\,dx.$$

Alors on a [3].

Théorème 3. *Soit ψ_λ, $\lambda \in \Lambda$, une base orthonormée de $L^2(\mathbb{R}^n)$, d'ondelettes de régularité $r \geq 1$.*

Si u et v appartiennent à $L^2(\mathbb{R}^n)$, posons

$$P_2(u, v) = \sum_{\lambda \in \Lambda} \langle u, \psi_\lambda \rangle \langle v, \psi_\lambda \rangle |\psi_\lambda(x)|^2.$$

Alors cet opérateur bilinéaire P_2 possède les propriétés (1.4), (1.5), (1.6) *et* (1.7). *De plus $P_2(u, v) - P_1(u, v) \in \dot{B}_1^{0,1}$ si $u \in L^2$ et $v \in L^2$.*

L'opérateur P_2 fournit donc la renormalisation la plus simple possible du produit uv, c'est-à-dire celle où l'on soustrait le moins de termes possible. Cet opérateur peut être utilisé pour démontrer la forme précisée du théorème 1. Cependant cette démonstration, que l'on trouvera dans [3], est techniquement plus compliquée que celle que nous venons de présenter. Les difficultés techniques viennent du fait que l'opérateur P_2 ne vérifie pas la formule de Leibniz qui est satisfaite par le produit usuel et par les opérateurs P et P_1. En d'autres termes, P_2 n'appartient pas à la classe générale des opérateurs bilinéaires définis par (2.2).

8 APPENDICE (PREUVES DES LEMMES 2 ET 3)

Reprenons, pour démontrer le lemme 2, les notations de la section 5. On a donc

$$T(u, v) = \sum_{-\infty}^{\infty} \left[T(S_{j+1}u, S_{j+1}v) - T(S_j u, S_j v) \right]$$

$$= \sum_{-\infty}^{\infty} T(D_j u, S_j v) + \sum_{-\infty}^{\infty} T(S_j u, D_j v) + \sum_{-\infty}^{\infty} T(D_j u, D_j v).$$

Nous allons, en désignant par $\| \cdot \|$ la norme de l'espace de Banach $\dot{B}_1^{0,1}$, établir que

$$(8.1) \qquad \sum_{-\infty}^{\infty} \| T(D_j u, S_j v) \| \leq C \| u \|_2 \| v \|_2$$

et qu'il en est de même pour les deux autres séries.

Les trois preuves étant semblables, nous nous limiterons à (8.1). Observons que $T(D_j u, S_j v) = T_j(u, v)$ où le symbole bilinéaire $\tau_j(\xi, \eta)$ de T_j est $\hat{\psi}(2^{-j}\xi)\hat{\varphi}(2^{-j}\eta)\tau(\xi, \eta)$. Sur le support de τ_j, on a nécessairement $|\eta| \leq \frac{4}{3} 2^j$ et $\frac{2}{3} 2^j \leq |\xi| \leq \frac{8}{3} 2^j$. Cela implique

$$(8.2) \qquad \frac{2}{3} 2^j \leq |\xi| + |\eta| \leq 4.2^j$$

si bien que l'on a

$$(8.3) \qquad |\partial_\xi^\alpha \partial_\eta^\beta \tau_j(\xi, \eta)| \leq C_{\alpha, \beta} 2^{-(|\alpha|+|\beta|)j}.$$

Les symboles $\tau_j(\xi, \eta)$ s'écrivent donc $\tau_j(\xi, \eta) = \theta_j(2^{-j}\xi, 2^{-j}\eta)$ où les fonctions $\theta_j(\xi, \eta)$ sont portées par $|\xi| + |\eta| \leq 4$ et vérifient, ainsi que toutes leurs dérivées, des estimations uniformes en j.

Par hypothèse τ s'annule si $\xi = 0$ ou si $\eta = 0$ ou si $\xi + \eta = 0$. Il en est de même de τ_j et de θ. On peut donc écrire

$$\theta_j(\xi, \eta) = \sum_1^n \sum_1^n (\xi_k + \eta_k)\eta_l \theta_j^{(k,l)}(\xi, \eta)$$

où les symboles $\theta_j^{(k,l)}(\xi, \eta)$ sont portés par $|\xi| + |\eta| \leq 4$ et vérifient, ainsi que leurs dérivées, des estimations uniformes en j.

Finalement on a

$$(8.4) \qquad \tau_j(\xi, \eta) = 4^{-j} \sum_1^n \sum_1^n (\xi_k + \eta_k)\eta_l \theta_j^{(k,l)}(2^{-j}\xi, 2^{-j}\eta).$$

Pour terminer cette analyse de $\tau_j(\xi, \eta)$, on tient encore compte du support de $\tau_j(\xi, \eta)$ pour écrire $\tau_j(\xi, \eta) = \tau_j(\xi, \eta)q(2^{-j}\xi)p(2^{-j}\eta)$ où p et q sont deux fonctions de la classe $\mathcal{D}(\mathbb{R}^n)$ de Schwartz et où q est nul au voisinage de 0.

On a, en revenant à l'opérateur T_j,

$$(8.5) \qquad T_j(u, v) = 2^{-j} \operatorname{div} L_j(u_j, v_j)$$

où

$$\hat{u}_j(\xi) = \hat{u}(\xi)q(2^{-j}\xi), \qquad \hat{v}_j(\eta) = 2^{-j}\eta_l p(2^{-j}\eta)\hat{v}(\eta)$$

et où l'opérateur L_j est défini par le symbole $\theta_j^{(k,l)}(2^{-j}\xi, 2^{-j}\eta)$. On a donc, uniformément en j, $\| L_j(f, g) \|_1 \leq C \| f \|_2 \| g \|_2$.

La preuve du lemme 2 se termine en appliquant le lemme 4, comme nous l'avons fait dans la section 3. Il vient

(8.6) $\|T_j(u, v)\| \leq C\|u_j\|_2\|v_j\|_2$

et l'on conclut en observant que $\sum \|u_j\|_2^2 \leq C\|u\|_2^2$ et que $\sum \|v_j\|_2^2 \leq C\|v\|_2^2$.

Venons-en à la preuve du lemme 3. Elle repose sur l'observation suivante:

Lemme 7. *Soit R un nombre positif et $K(x, y)$, $x \in \mathbb{R}^n$, $y \in \mathbb{R}^n$, un noyau possédant les propriétés suivantes:*

(8.7) $K(x, y) = 0 \qquad si \quad |x - y| > R$

pour tout $z \in \mathbb{R}^n$, on désigne par $B(z)$ la boule de $\mathbb{R}^n \times \mathbb{R}^n$ de centre (z, z) et de rayon $2R$ et l'on demande que

(8.8) $\iint_{B(z)} |K(x, y)|^2 \, dx \, dy \quad tende\ vers\ 0\ quand\ |z| \to +\infty.$

Alors l'opérateur T défini par le noyau K est compact.

Pour démontrer ce lemme, on pose, pour tout $m \geq 1$, $K_m(x, y) = K(x, y)$ si $|x| \leq m$ et $|y| \leq m$ et $K_m(x, y) = 0$ sinon. Alors l'opérateur T_m définit par ce noyau K_m est un opérateur de Hilbert-Schmidt. A ce titre T_m est compact. Par ailleurs la norme de $T - T_m$ tend vers 0, grâce à la condition (8.8), quand m tend vers l'infini. Donc T est compact.

Revenons au lemme 3. Il s'agit de vérifier que

(8.9) $\lim_{j \to +\infty} \langle T(f_j, g_j), u \rangle = \langle T(f, g), u \rangle$

pour une classe convenable de fonctions de test u.

Nous supposerons, par exemple, que $\hat{u}$ appartient à la classe de Schwartz $\mathcal{D}(\mathbb{R}^n)$. Alors $\langle T(f, g), u \rangle = \langle L(f, u), g \rangle$ et le lemme 3 sera démontré si nous vérifions que l'opérateur défini par $f \mapsto L(f, u)$ est compact.

En passant aux transformées de Fourier, le noyau $K(\xi, \eta)$ de cet opérateur est $\hat{u}(-\xi - \eta)\tau(\xi, \eta)$. Ce noyau vérifie évidemment (8.7). Puisque $\tau(\xi, -\xi) = 0$, on a, grâce à (2.1), $K(\xi, \eta) = 0((|\xi| + |\eta|)^{-1})$ quand $|\xi| + |\eta|$ tend vers l'infini, ce qui entraîne (8.8).

Nous concluons cet appendice en reliant les algorithmes que nous avons utilisés dans la section 3 au célèbre théorème $T(1)$ de David et Journé [5]. La démonstration de ce théorème repose sur l'existence, pour toute fonction $b(x) \in BMO(\mathbb{R}^n)$, d'un opérateur de Calderón-Zygmund T_b tel que $T_b(1) = b$, ${}^tT_b(1) = 0$ et qui soit borné sur $L^2(\mathbb{R}^n)$. Cette dernière condition fait partie de la définition des opérateurs de Calderón-Zygmund que l'on trouvera dans [1] ou [5].

On a désigné par $^t L$ le transposé de l'opérateur L; le noyau-distribution de $^t L$ est $L(y, x)$ si celui de L est $L(x, y)$.

Considérons alors, en revenant aux notations de la section 3, l'opérateur S_b défini, si $b \in BMO(\mathbb{R}^n)$, par

$$(8.10) \qquad S_b(f) = \sum_{-\infty}^{\infty} (\Delta_j b)(\Delta_j f).$$

Cet opérateur est borné sur $L^2(\mathbb{R}^n)$ lorsque b appartient à $BMO(\mathbb{R}^n)$, c'est un opérateur de Calderón-Zygmund et l'on a $S_b(1) = 0$. On a $\int_{\mathbb{R}^n} S_b(f)(x) \, dx = \sum_{-\infty}^{\infty} \int (\Delta_j b)(\Delta_j f) \, dx = \int bf \, dx$ puisque $\sum_{-\infty}^{\infty} \Delta_j^* \Delta_j = I$. Cela signifie que $^t S_b(1) = b$ et que S_b est le transposé de l'opérateur T_b que l'on cherche à construire lorsqu'on démontre le théorème $T(1)$.

Lorsque $b(x)$ est une fonction de classe C^1, à support compact, on a $\|\Delta_j(b)\|_\infty \leq C2^{-j}$ si $j \geq 0$ et $\|\Delta_j(b)\|_\infty \leq C'2^{nj}$ si $j \leq 0$. En fait $C' = \|\psi\|_\infty \|u\|_1$. En outre, chaque terme $(\Delta_j b)(\Delta_j f)$ qui figure dans (8.10) est un opérateur compact de L^2 dans L^2. Il en résulte que $S_b : L^2 \to L^2$ est compact. Puisque la norme $\|S_b\|$ de $S_b : L^2 \to L^2$ est majorée par $C\|b\|_{BMO}$, l'opérateur S_b est encore compact lorsque la fonction $b(x)$ appartient à VMO.

Comme nous l'avions annoncé dans l'introduction, le produit uv entre une fonction régulière u et une fonction arbitraire $v \in L^2(\mathbb{R}^n)$ n'a pas besoin d'être renormalisé. Cette affirmation ne peut être prise au sens strict car si $P(u, v)$ était nul chaque fois que u appartient à la classe de Schwartz $\mathcal{D}(\mathbb{R}^n)$ et que v appartient à $L^2(\mathbb{R}^n)$, alors $P(u, v)$ serait, par densité, nul si u et v appartiennent à $L^2(\mathbb{R}^n)$. Mais la régularité de u influe sur l'importance de la renormalisation et le sens précis à donner à cette assertion est que l'opérateur qui à v associe $P(u, v)$ est compact si u appartient à VMO.

Yale University
University of Paris, Dauphine
University of Paris, Dauphine

REFERENCES

[1] R. Coifman et Y. Meyer. *Au delà des opérateurs pseudo-différentiels*. Astérisque 57, 2ème édition, Société Mathématique de France (1978).

[2] R. Coifman, P. L. Lions Y. Meyer, et S. Semmes. *Compacité par compensation et espaces de Hardy*. CRAS Paris, tome 309, série I (1989), 945–949.

[3] S. Dobyinsky. Thèse du CEREMADE.

[4] C. Fefferman and E. Stein. H^p *spaces of several variables*. Acta Math. 129 (1972), 137–193.

[5] Y. Meyer. *Ondelettes et opérateurs*. Tomes I, II et III. Hermann, 1990.

8

Numerical Harmonic Analysis

R. R. Coifman, Y. Meyer, and V. Wickerhauser

INTRODUCTION

The purpose of this chapter is to describe recent developments involving the numerical implementation of methods from classical harmonic analysis in signal processing and computational P.D.E.

As an example, Littlewood-Paley theory, in which a function or a Fourier multiplier is analyzed by partitioning the frequency space in dyadic blocks, has recently been translated into a powerful numerical tool through expansions in orthonormal wavelet bases. (See [1], [2].)

In this numerical setting one sees a general Calderón-Zygmund operator or ΨD.O. as given by an "almost" diagonal matrix having a simple analysis and being implementable by fast numerical algorithms (i.e., algorithms of complexity $CN \log N$, N = number of discretization points). Pseudo-differential calculus is translated into an efficient numerical calculus in which smoothing operators are represented by "small" matrices of low numerical rank (see [1]) permitting its use in explicit calculations of solutions to P.D.E. In particular, we can obtain a fast algorithm for the numerical computation of the Green's function for a variable coefficient Laplacian (with smooth coefficients).

In this exercise of translation of methods and ideas from harmonic analysis into fast computational algorithms, one soon realizes that the ability to implement efficiently an integral operator applied to a function is equivalent to a good understanding of the interaction between geometry of the underlying space and cancellation properties of the operator. In the particular case of Calderón-Zygmund operators we see an efficient computational algorithm as being a translation of the method of proof of the $T(1)$ theorem of David and Journé. For the case of fractional integrals and operators of potential theory, the need to come up with efficient computations has led V. Rokhlin to the independent discovery of various versions of Calderón-Zygmund theory as embodied in his multipole algorithms.

As it turns out, in this case, the question of fast computation is more elementary than boundedness on L^2 or other spaces. It leads directly to issues of geometry of interactions and cancellations.

This interaction between harmonic analysis and a number of concrete problems in applications, such as signal processing and computations, has opened a number of new fundamental questions in analysis.

Our goal is to describe some of these problems on a few simple examples. We start with a fundamental question of signal processing, the question of compression of a signal. Stated simply, given a function (or more precisely, a vector which is a sampled function) one would like to represent the function with as few parameters as possible (here a representation is always assumed to have a given fixed precision). Such a representation could be given in terms of expansion coefficients, Fourier, Taylor, etc., or by stating that the function solves an equation which is easy to describe (say by giving coefficients of a differential equation). The ability to represent a function simply with few parameters is not only desirable in applications for storage purposes, it is also a test of our understanding of the structure of the function and its numerical complexity. Traditionally, the first attempt to represent a signal (or a function not described analytically) would be to expand the signal in a Fourier series, or in terms of some other orthogonal (or non orthogonal expansion). This leads to a variety of problems familiar to all analysts. Assume that a smooth function is supported on a number of disjoint intervals. It is "clear" that separate Fourier expansions restricted to these intervals will be much more "efficient" than a single expansion on the union. The actual answers are not so obvious since some intervals could be close to each other and the term efficient has not been defined. We see that we are confronted with the issue of selecting an optimal expansion inside a class of possible expansions. This leads naturally to the concept of a library of orthonormal bases, as well as to precise definitions of efficiency of an expansion.

DEFINITION OF MODULATED WAVE FORM LIBRARIES

We start by observing that it is impossible to construct an orthogonal basis by localizing smoothly e^{ikx}. This is clear for the case of two adjacent windows $w_1(x)$, $w_2(x)$ since the requirement of orthogonality between $w_1(x)e^{ikx}$ and $w_2(x)e^{ijx}$ implies that

$$\int w_1(x)e^{ikx}w_2(x)e^{i(k-j)x}dx = 0$$

which implies $w_1(x)w_2(x) \equiv 0$ (if it is supported in an interval of length smaller than 2π).

Recently Daubechies, Jaffard, and Journé, as well as Malvar, observed that by taking equal windows and sines or cosines orthogonality can be maintained. It was observed in [3], [5] that the windows can be chosen to different sizes enabling adaptive constructions. (See Figures 5, 6.)

We start by defining this library of trigonometric waveforms. These are localized sine transforms associated to covering by intervals of **R** (more generally, of a manifold).

We consider a cover $\mathbf{R} = \cup_{-\infty}^{\infty} I_i, I = [\alpha_i, \alpha_{i+1}), \alpha_i < \alpha_{i+1}$, write $\ell_i = \alpha_{i+1} - \alpha_i = |I_i|$ and let $p_i(x)$ be a window function supported in $[\alpha_i - \ell_{i-1}/2, \alpha_{i+1} + \ell_{i+1}/2]$ such that

$$\sum_{-\infty}^{\infty} p_i^2(x) = 1$$

and

$$p_i^2(x) = 1 - p_i^2(2\alpha_{i+1} - x) \qquad \text{for} \quad x \text{ near } \alpha_{i+1}$$

then the functions

$$S_{i,k}(x) = \frac{2}{\sqrt{2\ell i}} p_i(x) \sin[(2k+1)\frac{\pi}{2\ell_i}(x - \alpha_i)]$$

form an orthonormal basis of $L^2(\mathbf{R})$ subordinate to the partition p_i. The collection of such bases forms a library of orthonormal bases.

It is easy to check that if H_{I_i} denotes the space of functions spanned by $S_{i,k}$, $k = 0, 1, 2, \ldots$ then $H_{I_i} + H_{I_{i+1}}$ is spanned by the functions

$$P(x) \frac{2}{\sqrt{2(\ell_i + \ell_{i+1})}} \sin[(2k+1)\frac{\pi}{2(\ell_i + \ell_{i+1})}(x - \alpha_i)]$$

where

$$P^2 = p_i^2(x) + p_{i+1}^2(x)$$

is a "window" function covering the interval $I_i \cup I_{i+1}$. This fundamental identity permits the useful implementation of the adapted window algorithm described in Figure 1. (Other possible libraries can be constructed. The space of frequencies can be decomposed into pairs of symmetric windows around the origin, on which a smooth partition of unity is constructed. Higher dimensional libraries can also be easily constructed—as well as libraries on manifolds—leading to new and direct analysis methods for linear transformations.)

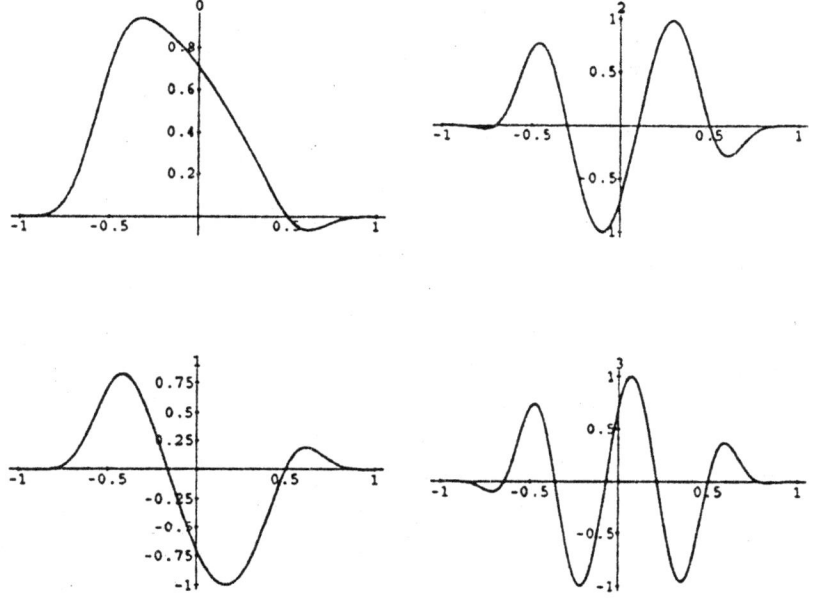

Figure 1. Local trigonometric waveforms.

RELATION TO WAVELETS—WAVELET PACKETS

We consider the frequency line $\mathbf{R}$ split as $\mathbf{R}^+ = (0, \infty)$ union $\mathbf{R}^- = (-\infty, 0)$. On $L^2(0, \infty)$ we introduce a window function $p(\xi)$ such that $\sum_{k=-\infty}^{\infty} p^2(2^{-k}\xi) = 1$ and $p(\xi)$ is supported in $(3/4, 3)$. Clearly we can view $p(2^{-k}\xi)$ as a window function above the interval $(2^k, 2^{k+1})$ and observe that

$$\sin\left[(j + \frac{1}{2})\pi \left(\frac{\xi - 2^k}{2^k} \right) \right] p(2^{-k}\xi) = s_{k,j}$$

form an orthonormal basis of $L^2(\mathbf{R}^+)$. Similarly,

$$c_{k,j} = \cos\left[(j + \frac{1}{2})\pi \left(\frac{\xi - 2^k}{2^k} \right) \right] p(2^{-k}\xi)$$

gives another basis. If we define $S_{k,j}$ as an odd extension to $\mathbf{R}$ of $s_{k,j}$ and $C_{k,j}$ as an even extension, we find $S_{k,j} \perp C_{k',j'}$ permitting us to write $C_{k,j} \pm i S_{k,j} = e^{\pm i j \pi \xi/2^k} \hat{\psi}(\xi/2^j)$ where $\hat{\psi}(\xi) = e^{i\pi/2\xi} p(\xi)$ is the Fourier transform of the base wavelet Ψ (see [4]).

We therefore see that wavelet analysis corresponds to windowing frequency space in "octave" windows $(2^k, 2^{k+1})$.

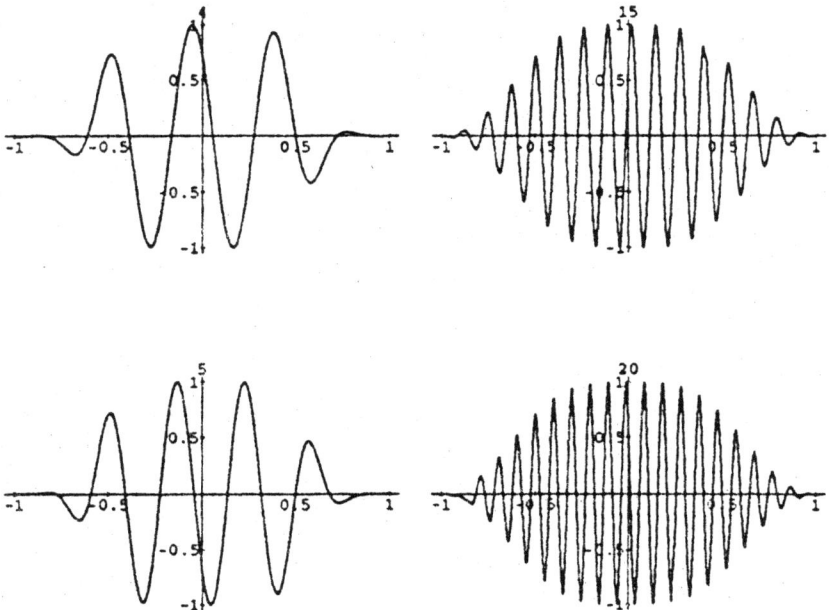

Figure 2.   Local trigonometric waveforms.

A natural extension, therefore, is provided by allowing all dyadic windows in frequency space and adapted window choice. This sort of analysis is "equivalent" to wavelet packet analysis.

The wavelet packet analysis algorithms permit us to perform an adapted Fourier windowing directly in the time domain by successive filtering of a function into different regions in frequency. The dual version of the window selection provides an adapted subband coding algorithm.

This new library of orthonormal bases constructed in the time domain is called the wavelet packet library. This library contains the wavelet basis, Walsh functions, and smooth versions of Walsh functions called wavelet packets. (See Figure 7.)

We'll use the notation and terminology of [5], whose results we shall assume.

We are given an exact quadrature mirror filter $h(n)$ satisfying the conditions of Theorem (3.6) in [5], p. 964, i.e.,

$$\sum_n h(n-2k)h(n-2\ell) = \delta_{k,\ell}, \qquad \sum_n h(n) = \sqrt{2}.$$

We let $g_k = h_{l-k}(-1)^k$ and define the operations F_i on $\ell^2(\mathbf{Z})$ into "$\ell^2(2\mathbf{Z})$"

(1.0) $$F_0\{s_k\}(i) = 2 \sum s_k h_{k-2i}$$

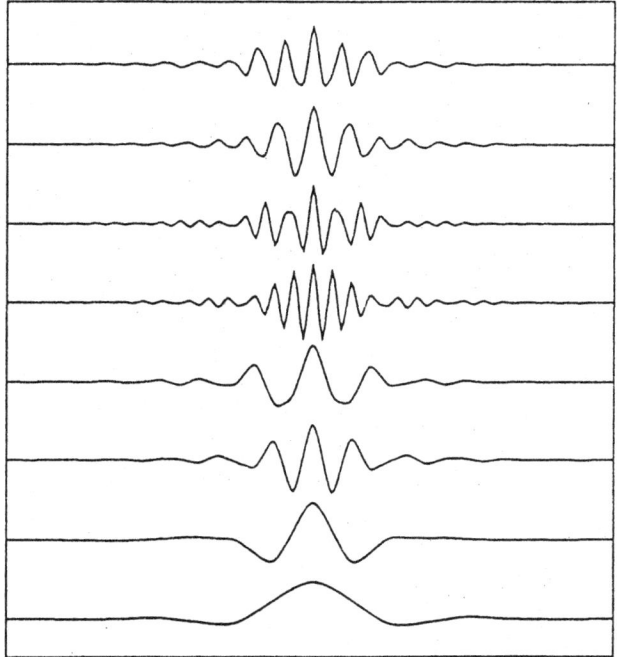

Figure 3a. Wavelet packet library.

$$F_1\{s_k\}(i) = 2 \sum s_k g_{k-2i}.$$

The map $\mathbf{F}(s_k) = F_0(s_k) \oplus F_1(s_k) \in \ell^2(2\mathbf{Z}) \oplus \ell^2(2\mathbf{Z})$ is orthogonal and

(1.1) $$F_0^* F_0 + F_1^* F_1 = I.$$

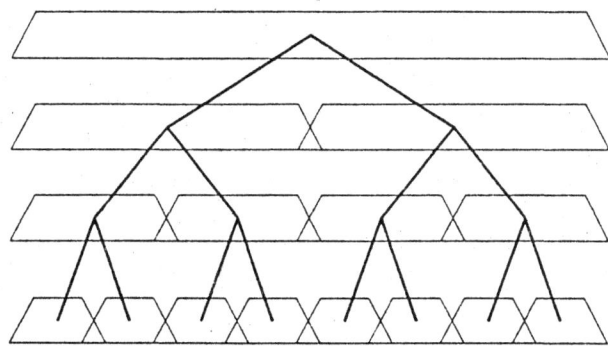

Figure 3b. Schematic Description.

We now define the following sequence of functions.

(1.2)
$$\begin{cases} W_{2n}(x) = \sqrt{2} \sum h_k W_n(2x - k) \\ W_{2n+1}(x) = \sqrt{2} \sum g_k W_n(2x - k). \end{cases}$$

Clearly the function $W_0(x)$ can be identified with the scaling function φ in **D** and W_1 with the basic wavelet ψ.

Let us define $m_0(\xi) = \frac{1}{\sqrt{2}} \sum h_k e^{-ik\xi}$ and

$$m_1(\xi) = -e^{i\xi} \bar{m}_0(\xi + \pi) = \frac{1}{\sqrt{2}} \sum g_k e^{ik\xi}.$$

Remark. The quadrature mirror condition on the operation $\mathbf{F} = (F_0, F_1)$ is equivalent to the unitarity of the matrix

$$\mathcal{M} = \begin{bmatrix} m_0(\xi) & m_1(\xi) \\ m_0(\xi + \pi) & m_1(\xi + \pi) \end{bmatrix}.$$

Taking the Fourier transform of (1.2) when $n = 0$ we get

$$\hat{W}_0(\xi) = m_0(\xi/2) \hat{W}_0(\xi/2)$$

i.e.,

$$\hat{W}_0(\xi) = \prod_{j=1}^{\infty} m_0(\xi/2^j)$$

and

$$\hat{W}_1(\xi) = m_1(\xi/2) \hat{W}_0(\xi/2) = m_1(\xi/2) m_0(\xi/4) m_0(\xi/2^3) \cdots$$

More generally, the relations (1.2) are equivalent to

(1.3)
$$\hat{W}_n(\xi) = \prod_{j=1}^{\infty} m_{\varepsilon_j}(\xi/2j)$$

and $n = \sum_{j=1}^{\infty} \varepsilon_j 2^{j-1} (\varepsilon_j = 0 \text{ or } 1)$.

The functions $W_n(x - k)$ form an orthonormal basis of $L^2(\mathbf{R}^1)$. We define a *library* of wavelet packets to be the collection of functions of the form $W_n(2^\ell x - k)$ where $\ell, k \in \mathbf{Z}, n \in N$. Here, each element of the library is determined by a scaling parameter ℓ, a localization parameter k and an oscillation parameter n. (The function $W_n(2^\ell x - k)$ is roughly centered at $2^{-\ell}k$, has support of size $\approx 2^{-\ell}$ and oscillates $\approx n$ times.)

We have the following simple characterization of subsets forming orthonormal bases.

Proposition. *Any collection of indices (ℓ, n) such that the intervals $[2^\ell n, 2^\ell n + 1)$ form a disjoint cover of $[0, \infty)$ gives rise to an orthonormal basis of L^2.*[1]

(These intervals correspond to the partition of frequency space alluded to in §1.)

Motivated by ideas from signal processing and communication theory we were led to measure the "distance" between a basis and a function in terms of the Shannon entropy of the expansion. More generally, let H be a Hilbert space. Let $v \in H$, $\|v\| = 1$ and assume

$$H = \oplus \sum H_i$$

an orthogonal direct sum. We define

$$\varepsilon^2(v, \{H_i\}) = - \sum \|v_i\|^2 \ell n \|v_i\|^2$$

as a measure of distance between v and the orthogonal decomposition.

ε^2 is characterized by the Shannon equation which is a version of Pythagoras' theorem.

Let

$$H = \oplus \left(\sum H^i\right) \oplus \left(\sum H_j\right)$$

$$= H_+ \oplus H_-.$$

H^i and H_j give orthogonal decompositions $H_+ = \sum H^i$, $H_- = \sum H_j$. Then

$$\varepsilon^2(v; \{H^i, H_j\}) = \varepsilon^2(v, \{H+, H_-\})$$

$$+ \|v_+\|^2 \varepsilon^2 \left(\frac{v_+}{\|v_+\|}, \{H^i\}\right)$$

$$+ \|v_-\|^2 \varepsilon^2 \left(\frac{v_-}{\|v_-\|}, \{H_j\}\right).$$

This is Shannon's equation for entropy (if we interpret $\|P_{H_+} v\|^2$ as the "probability" of v to be in the subspace H_+, as in quantum mechanics).

This equation enables us to search for a smallest entropy space decomposition of a given vector.

In fact, for the example of the first library restricted to covering by dyadic intervals we can start by calculating the entropy of an expansion relative to a local trigonometric basis for intervals of length one, then compare the entropy of an adjacent pair of intervals to the entropy of an expansion on their union. Pick the

---

[1] We can think of this cover as an even covering of frequency space by windows roughly localized over the corresponding intervals.

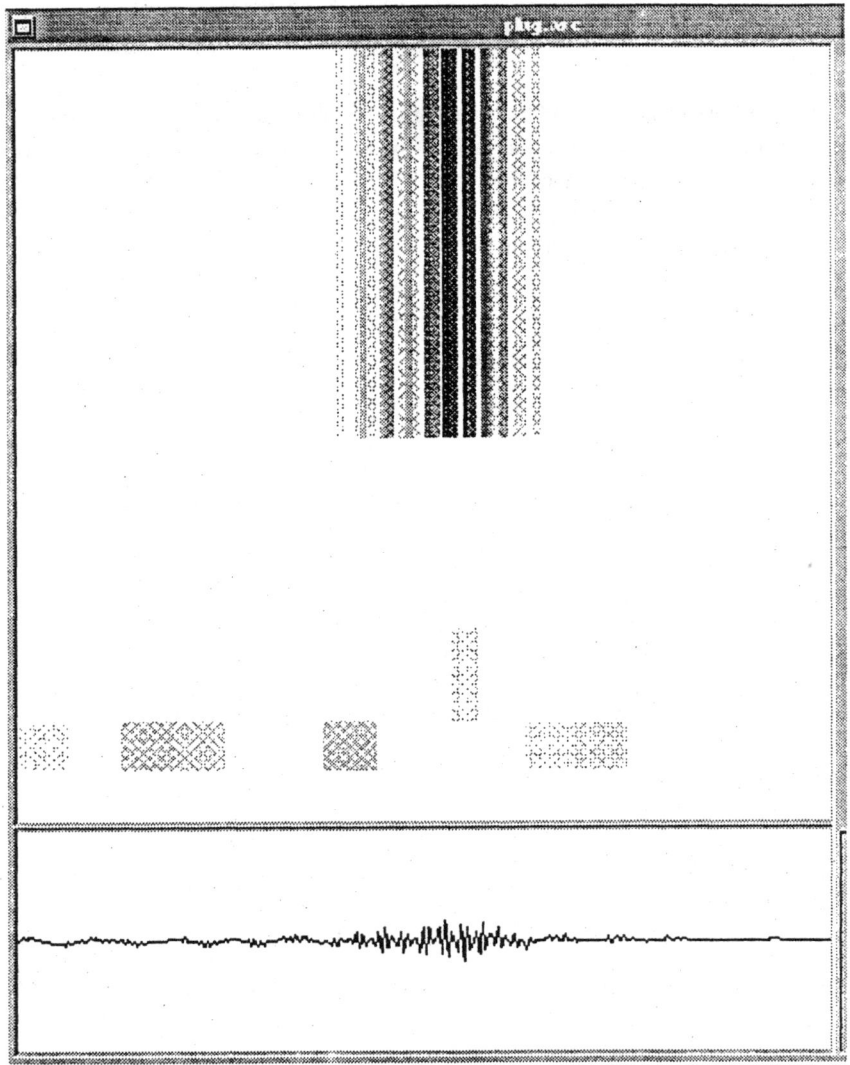

Figure 4. Wavelet coefficients of a function.

expansion of minimal entropy and continue until a minimum entropy expansion is achieved. (See Figure 4.)

Of course, while entropy is a good measure of concentration or efficiency of an expansion, various other information cost functions are possible, permitting discrimination and choice among various expansions.

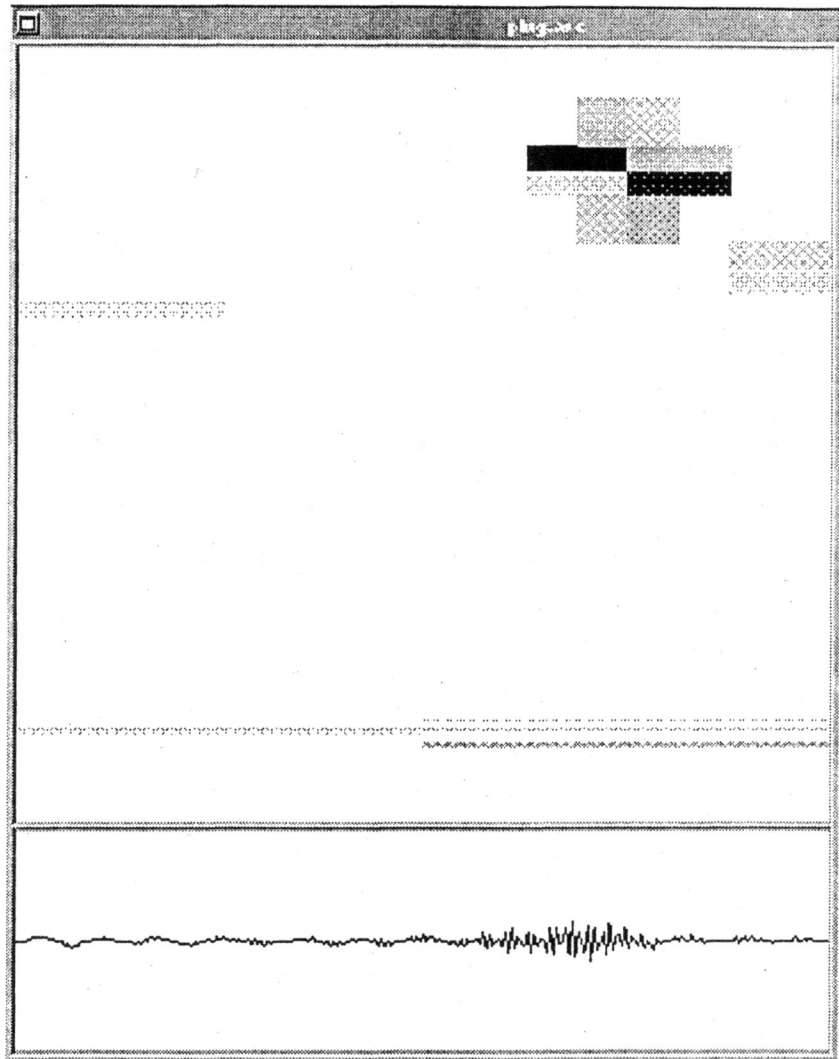

Figure 5. A best-basis wavelet packet analysis. This analysis corresponds to selection of windows in frequency space, to minimize the entropy of the expansion.

We illustrate these points, as well as the effect of various analysis methods, in figures 5–7 in which the vertical axis represents the frequency axis and the horizontal is the time (or space) axis. The signals have 512 samples (and are wrapped around). Each rectangular box in this phase space corresponds to a coefficient obtained by correlating the signal with an element of the wavelet packet

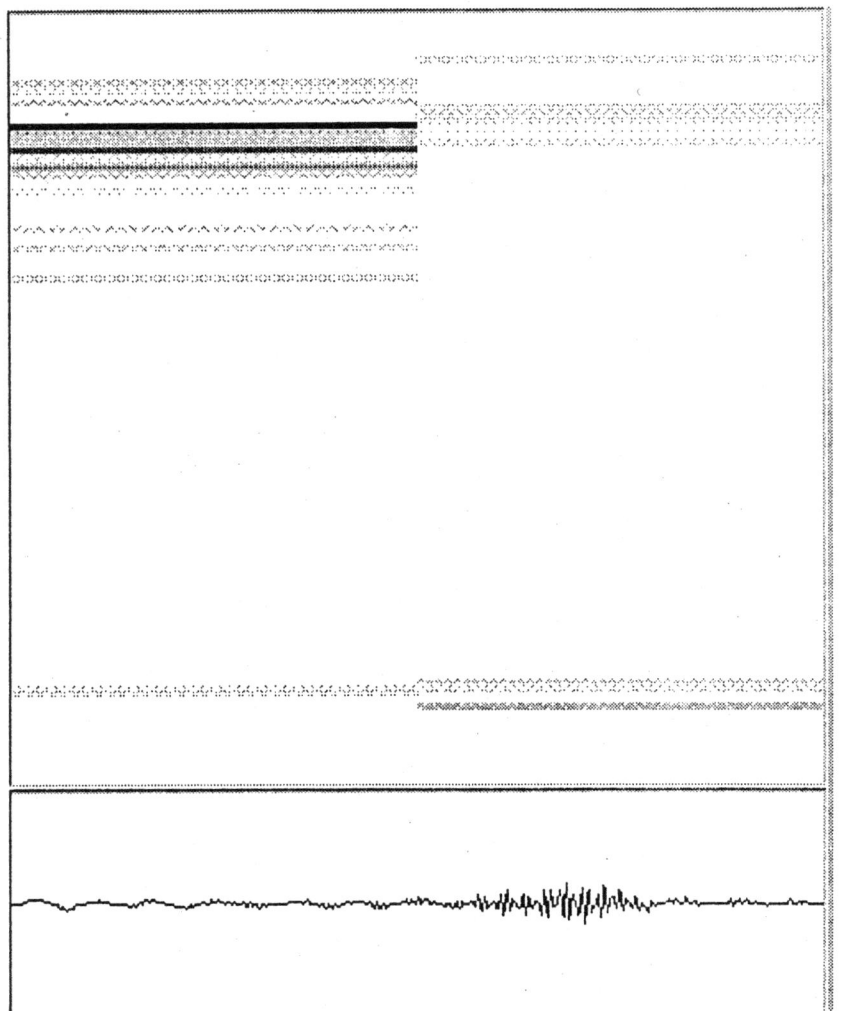

Figure 6. A two-windows expansion with no adaptation.

library whose time support lies "below" the box and whose frequency support is in the projection of the box on the vertical axis. Each box has area 512 pixels (i.e., a cover of the discrete phase plane has 512 elements).

The compression rate can be computed as the ratio of the visible gray area to the total area of the box (i.e., the relative number of visible boxes).

Of course we can try to characterize classes of functions which are well compressible, i.e., for which we can estimate the number of coefficients needed for representing the function with a prescribed accuracy. Smooth functions are ob-

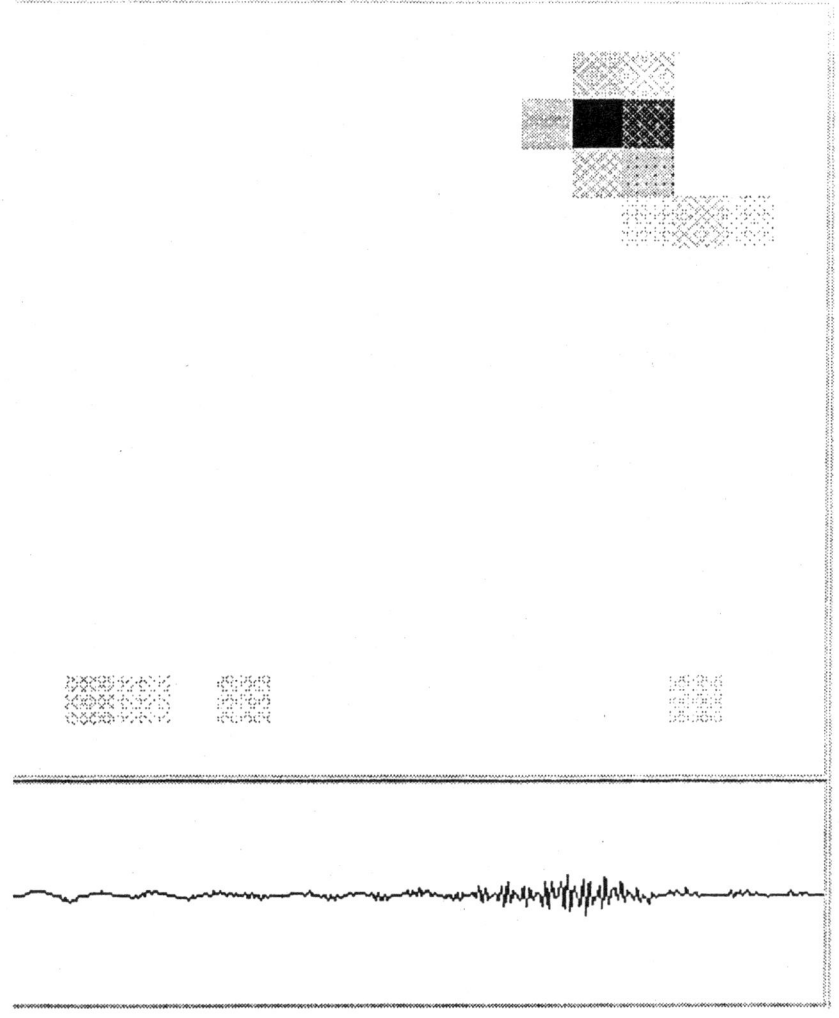

Figure 7. A best-level expansion in which a fixed window size is chosen to minimize entropy.

vious candidates as well as functions which can be well approximated locally by trigonometric polynomials of short length. Various obvious definitions come to mind. At the moment it would seem that experiment will provide a better guide.

The procedure for signal analysis described above is very similar to the usual methods of studying Fourier multipliers in which we break the multiplier by an appropriate partition of unity to simpler components whose spatial localization and structure are easier to understand. We can describe a similar procedure for integral operators

$$T(f) = \int_{\mathbf{R}} k(x, y) f(y) dy$$

which are not necessarily convolutions.

Our goal is to implement a fast discrete version of the operator.

A procedure that is equivalent to $P_t Q_t$ decompositions of Calderón-Zygmund operators (see [1]) can be obtained by trying to compress the $k(x, y)$ viewed as an image, i.e., $k(x, y)$ represents light intensity at pixel (xy). Here again the analysis consists in finding an optimal windowed expansion for $k(x, y)$ (or $\hat{k}(\xi, \eta)$) by selecting that combination of windows most efficient in capturing the kernel.

Since the kernel is represented as a sum of products of functions of x and y it is easy to convert an efficient two-dimensional representation into a corresponding efficient computation. Observe also that each box selected represents an interaction between two windows on the line.

It can be proved that for $k(x, y)$ a single or double layer potential for Helmholtz on a curve or surface (with bounded curvature), this procedure leads to an order $pN \log N$ algorithm, where p is the number of decimals desired, and N is the number of discretization points ($\approx$ number of wave lengths on the surface).

For more general curves or surfaces one has to develop specific, highly oscillatory analogues to multipoles. (Smooth bases are useless.) This has been done by Rokhlin with a resulting description of local oscillatory interactions.

Yale University
University of Paris, Dauphine
Washington University, St. Louis

REFERENCES

Software for adapted waveform analysis is available by anonymous ftp from *ceres.math.yale. edu*, Internet address 130.132.23.22.

[1] R. Coifman. *Adapted multiresolution analysis, computation, signal processing and operator theory*. ICM 90 (Kyoto).

[2] R. Coifman, Y. Meyer, and V. Wickerhauser. *Wavelet Analysis and Signal Processing*. Proceedings of the Conference on Wavelets, Lowell, Mass. 1991.

[3] R. Coifman and Y. Meyer. *Remarques sur l'analyse de Fourier à fenêtre*, série I C. R. Acad. Sci. Paris **312** (1991), 259–261.

[4] Y. Meyer. *Principes d'incertidudes, bases hilbertiennes, et algèbres d'Opérateurs*. Séminaire Bourbaki, 1985–86. *Astérisque*, 662.

[5] I. Daubechies. *Orthonormal bases of compactly supported wavelets. Comm. Pure Appl. Math.* **XLI** (1988), 909–996.

[6] E. Laeng. *Une base orthonormale de $L^2(\mathbf{R})$, dont les éléments sont bien localisés dans l'espace de phase et leurs supports adaptés à toute partition symétrique de l'espace des fréquences*. série I *C. R. Acad. Sci. Paris* **311** (1990), 677–680.

9

Some Topics from Harmonic Analysis and Partial Differential Equations

Robert A. Fefferman

It is a pleasure to celebrate, together with many colleagues, the occasion of the sixtieth birthday of Elias M. Stein. To me, Professor Stein is not only an extremely great mathematician, but he is also my teacher and friend. He gives to each of his students his time and his energy very generously. Most of all, he provides us with such an inspiring example in the way that he does mathematics. I think this is the greatest gift that a teacher can give to a student, and I shall never forget this gift.

In this article, I would like to discuss some of the mathematical issues which have interested me and have guided much of my work. They are in large part taken directly from E. M. Stein's celebrated book, *Singular Integrals and Differentiability Properties of Functions* [27] and will be motivated by the statement of three "model problems." These can be stated as follows:

Problem 1. If $\Omega \subseteq \mathbb{R}^n$ is a "nice" region, and $1 < p < \infty$, define T_Ω by

$$\widehat{T_\Omega f}(\xi) = \chi_\Omega(\xi) \widehat{f}(\xi).$$

When is it true that $\|T_\Omega f\|_{L^p(\mathbb{R}^n)} \leq C_{p,n} \|f\|_{L^p(\mathbb{R}^n)}$ where $C_{p,n}$ is a constant depending only on p and n, but not on f?

Problem 2. Understand "Marcinkiewicz multipliers." By a Marcinkiewicz multiplier $m(\xi, \eta)$, $\xi \in \mathbb{R}^n$, $\eta \in \mathbb{R}^m$, we shall mean that $m(\xi, \eta)$ satisfies all the same estimates that $m_1(\xi)m_2(\eta)$ does, where m_1 and m_2 are in some appropriate set of "classical" multipliers (such as Hörmander multipliers). We shall not assume that $m(\xi, \eta)$ takes the special form $m_1(\xi)m_2(\eta)$. This problem was first investigated by

Marcinkiewicz (see Stein [27]) who obtained L^p estimates for these multipliers, when $1 < p < \infty$. For many reasons it is natural to suspect that for the multiplier operators T corresponding to the Marcinkiewicz multipliers m, we should have estimates near L^1 of the form

$$m\{|T(f)| > \alpha\} \leq \frac{C}{\alpha} \|f\|_{L \log^+ L(\mathbb{R}^n \times \mathbb{R}^m)}$$

and these estimates are not available by the (now standard) methods presented in [1] for the L^p theory. In this article we shall be interested in the recent work in harmonic analysis which allows us to obtain finer estimates on these operators near L^1.

Problem 3. It is a very important fact from the classical theory ([27]) that if $B \subseteq \mathbb{R}^n$ is the unit ball and we are given a function $f(x)$ defined on S^{n-1} which belongs to $L^p(d\sigma)$ ($1 < p < \infty$, and $d\sigma$ is the surface area measure), then there exists a unique harmonic function u on B taking on the values $f(x)$ as non-tangential boundary values and such that $u^* \in L^p(d\sigma)$, where u^* denotes the non-tangential maximal function of u. Our third problem is to extend this classical theorem about harmonic functions to the solutions of a class of elliptic operators with bounded measurable coefficients. The solution to this problem provides at once a generalization of the theory of harmonic functions on Lipschitz domains, and also some interesting harmonic analysis involving classical weights and the Helson-Szegö theorem.

Let us begin to consider each of these problems in some detail. Starting with the first problem, we quote two results:

Theorem A (Charles Fefferman [14]). *If S is a "nice region" in $\mathbb{R}^2$ whose boundary has a point of non-zero curvature then χ_S is not an $L^p(\mathbb{R}^2)$ multiplier for any $p \neq 2$.*

Theorem B (A. Cordoba–R. Fefferman [8]). *If S is a region whose boundary consists of line segments, then the boundedness properties of the multiplier χ_S are equivalent to the boundedness properties of a certain maximal operator corresponding to S.*

In Theorem B above, the maximal operator corresponding to the multiplier χ_S is very easily described. Suppose that $\theta = \{\theta_i\}_{i=1,2,\dots}$ denotes the sequence of directions of the normal vectors to the sides of the (infinite) polygonal region S.

Let M_θ denote the maximal operator defined by

$$M_\theta f(x) = \sup_{\substack{x \in R \\ R \in \mathcal{R}_\theta}} \frac{1}{m(R)} \int_R |f(y)| \, dy$$

where $\mathcal{R}_\theta$ is the family of all rectangles in $\mathbb{R}^2$ whose side lengths are arbitrary, but whose orientations are restricted to belong to one of the directions $\theta_i \in \theta$. Then the behavior of the multiplier operator corresponding to χ_S on L^p is equivalent to the boundedness properties of M_θ on the space $L^{(p/2)'}$.

For this reason in order to answer Problem 1, it is sufficient to gain an understanding of the operators M_θ. To do this, we shall discuss a systematic approach which can often be applied to determine the action on (Orlicz) L^p classes of various maximal operators. We emphasize that just as in the case of M_θ, the maximal operators we deal with are defined by a supremum over averages with respect to a family of sets whose geometry may be considerably more complicated than in the classical case of balls or cubes.

Our method proceeds as follows. Suppose we are given a family of sets $\mathcal{R}$ in $\mathbb{R}^n$ and we wish to analyze the operator M defined by

$$M f(x) = \sup_{\substack{x \in R \\ R \in \mathcal{R}}} \frac{1}{m(R)} \int_R |f(y)| \, dy$$

and obtain (for instance) the estimate

$$m\{x \in Q \mid M(f)(x) > \alpha\} \le \frac{C}{\alpha} \|f\|_{L^\Phi(Q)}$$

(where Q is the unit cube and L^Φ denotes an Orlicz class corresponding to the convex increasing function Φ). Then we shall describe a method which often makes it possible to prove a lemma along the following lines:

Given a family of sets $\{R_i\}$, $R_i \in \mathcal{R}$, $R_i \subseteq Q$, there exists a subfamily $\widetilde{R}_i$ such that (a) $\|\sum \chi_{\widetilde{R}_i}\|_{L^\Psi(Q)} \le C$; and (b) $m(\bigcup \widetilde{R}_i) \ge cm(\bigcup R_i)$ for some pair of constants $c, C > 0$.

Here L^Ψ is the Orlicz class which is dual to the class L^Φ which occurs in the estimates on M which we are trying to prove. That this covering lemma is sufficient to prove the weak type estimate of M on L^Φ above is rather trivial. In fact, setting $E_\alpha = \{x \in Q \mid M(f)(x) > \alpha\}$, for each $x \in E_\alpha$, choose an $R_x \in \mathcal{R}$ so that $\frac{1}{m(R_x)} \int_{R_x} |f(y)| \, dy > \alpha$. Now apply the covering lemma to the collection $\{R_x\}_{x \in E_\alpha}$ to get $\{\widetilde{R}_i\}$ satisfying (a) and (b) of the lemma. It follows that

$$m(E_\alpha) \le m\left(\bigcup_{x \in E_\alpha} R_x\right) \le \frac{1}{c} m\left(\bigcup \widetilde{R}_i\right) \le \frac{1}{c} \sum_i m(\widetilde{R}_i)$$

$$\leq \frac{1}{c} \sum_i \frac{1}{\alpha} \int_{\widetilde{R}_i} |f(y)| dy = \frac{1}{c\alpha} \int_Q |f| \sum \chi_{\widetilde{R}_i} \, dy$$

$$\leq \frac{1}{c} \cdot \frac{1}{\alpha} \|f\|_{L^\Phi(Q)} \| \sum \chi_{\widetilde{R}_i} \|_{L^\Psi(Q)}$$

$$\leq \frac{C}{c} \cdot \frac{1}{\alpha} \|f\|_{L^\Phi(Q)},$$

and this is the estimate on $m(E_\alpha)$ we require.

A similar argument shows immediately that if we desire to prove that M is of weak type (p, p) for some p with $1 < p < \infty$, then it is enough to prove a covering lemma exactly as above (where M is mapping into weak L^1) except that we must replace (a) by the estimate

$$\left\| \sum \chi_{\widetilde{R}_i} \right\|_{L^{p'}(\mathbb{R}^n)} \leq C m \left(\bigcup \widetilde{R}_i \right)^{1/p'}.$$

To show how to prove covering lemmas for various types of sets, we present some concrete examples. The first of these involves proving the sharp covering lemma for rectangles in $\mathbb{R}^n$ with sides parallel to the coordinate axes. This covering lemma is exactly what is needed to prove the celebrated Jessen-Marcinkiewicz-Zygmund Strong Maximal Theorem [22]. $m\{x \in Q \mid M_n(f)(x) > \alpha\} \leq \frac{C}{\alpha} \|f\|_{L(\log^+ L)^{n-1}}(Q)$ when M_n is the strong maximal operator in $\mathbb{R}^n$ defined by

$$M_n f(x) = \sup_{x \in R} \frac{1}{|R|} \int_R |f(y)| dy,$$

the sup being taken over all rectangles in $\mathbb{R}^n$ with sides parallel to the axes. We have the following:

Covering Lemma for the Strong Maximal Function [7]. *Suppose* $\{R_i\}_{i=1,2,\dots}$ *is a given sequence of rectangles with sides parallel to the axes, with* $R_i \subseteq Q$, *where* Q *is the unit cube in* $\mathbb{R}^n$. *Then we may select a subsequence* $\{\widetilde{R}_i\}$ *from the* R_i *so that*
(a) $\| \sum \chi_{\widetilde{R}_i} \|_{\exp(L)^{1/n-1}} \leq C_n$; *and*
(b) $m(\bigcup \widetilde{R}_i) \geq c_n m(\bigcup R_i)$ *and where* c_n, C_n *are positive constants depending only on the dimension* n.

Proof. We order the rectangles R_i so that their x_n side length decreases as n increases. Select the $\widetilde{R}_i$ according to the following rule: Set $\widetilde{R}_1 = R_1$. Assuming we have chosen $\widetilde{R}_1, \widetilde{R}_2, \dots, \widetilde{R}_{k-1}$, go down the list of R_i following $\widetilde{R}_{k-1}$, until we first have a rectangle R such that

$$m \left(R \cap \left[\bigcup_{i=1}^{k-1} \widetilde{R}_i \right] \right) \bigg/ m(R) < \frac{1}{2}.$$

This rectangle R then becomes $\widetilde{R}_k$. In this way we continue to obtain the selected rectangles $\widetilde{R}_i$, $i = 1, 2, 3, \ldots$. We now claim that the $\widetilde{R}_i$ have properties (a) and (b) as stated above.

To prove estimate (a) we use duality, and fix $\varphi \geq 0$ such that $\|\varphi\|_{L(\log^+ L)^{n-1}(Q)} = 1$ and estimate $\int \sum \chi_{\widetilde{R}_i} \varphi \, dx$. To do this, set $\widetilde{E}_i = \widetilde{R}_i - \bigcup_{j<i} \widetilde{R}_j$, the disjoint part of $\widetilde{R}_i$. Then

$$\int \sum_i \chi_{\widetilde{R}_i} \varphi \, dx = \sum_i \int_{\widetilde{R}_i} \varphi \, dx \leq 2 \sum_i m(\widetilde{E}_i) \frac{1}{m(\widetilde{R}_i)} \int_{\widetilde{R}_i} \varphi \, dx.$$

(Recall that the selection process assumes that $m(\widetilde{E}_i) \geq \frac{1}{2} m(\widetilde{R}_i)$, and we have used this in our last inequality.) It is also clear that for each $x \in \widetilde{E}_i$, $\frac{1}{m(\widetilde{R}_i)} \int_{\widetilde{R}_i} \varphi \, dx \leq M_n(\varphi)(x)$ so that

$$\sum_i m(\widetilde{E}_i) \left[\frac{1}{m(\widetilde{R}_i)} \int_{\widetilde{R}_i} \varphi \, dx \right] \leq \int_Q M_n(\varphi)(x) \, dx.$$

This estimate is not useful, because, first, if $\varphi \in L(\log^+ L)^{n-1}$, then $\int_Q M_n(f)dx$ may diverge, and second, we are trying to prove a result concerning the operator M_n. Both of these difficulties can be avoided by using the geometry of the rectangles $\widetilde{R}_i$. In the argument that follows, the R_i will be assumed to be dyadic. The general case requires only minor modifications which are best left to the reader. The key point is that since the x_n side lengths of the $\widetilde{R}_i$ are decreasing, if we slice these rectangles with a hyperplane perpendicular to the x_n axis to obtain $n-1$-dimensional rectangles $\widetilde{T}_i$, then

$$m_n \left(\widetilde{R}_i \cap \left[\bigcup_{j<i} \widetilde{R}_j \right] \right) < \frac{1}{2} m_n(\widetilde{R}_i)$$

implies

$$m_{n-1} \left(\widetilde{T}_i \cap \left[\bigcup_{j<i} \widetilde{T}_j \right] \right) < \frac{1}{2} m_{n-1}(\widetilde{T}_i).$$

(The reason for this is that $m_{n-1}(\widetilde{T}_i \cap [\bigcup_{j<i} \widetilde{T}_j])$ is independent of the x_n coordinate of the hyperplane that is used to slice $\widetilde{R}_i$ as long as it passes through $\widetilde{R}_i$.) The exact same duality argument, only applied to the slices $\widetilde{T}_j$ rather than the $\widetilde{R}_j$, produces the estimate

$$\int \sum \chi_{\widetilde{T}_i} \varphi \, dx_1 \ldots dx_{n-1} \leq 2 \int M_{n-1}(\varphi) \, dx_1 \ldots dx_{n-1},$$

and integrating this in the x_n variable yields

$$\int \sum \chi_{\widetilde{R}_i} \varphi \, dx \leq 2 \int M_{n-1}(\varphi) \, dx \quad (dx \text{ is } n \text{ dimensional measure here}).$$

According to the theory of the strong maximal function M_{n-1} (which could be assumed by induction), if $\|\varphi\|_{L(\log^+ L)^{n-1}(Q)} = 1$, then $\int_Q M_{n-1}(\varphi) \, dx \leq C_n$ and so this proves part (a) of the covering lemma.

To prove part (b), suppose R is one of the R_i which have not been selected. Then by the definition of the selection criterion it follows that

$$m(R \cap [\bigcup \widetilde{R}_i]) > \frac{1}{2} m(R)$$

where, in $\bigcup \widetilde{R}_i$, the union is taken only over those $\widetilde{R}_i$ which precede R on the list of rectangles. Now, slice the rectangle R as above with a hyperplane perpendicular to the x_n direction, calling slices of R and $\widetilde{R}_i$, T and $\widetilde{T}_i$, respectively. Then just as above, since the x_n side length of each $\widetilde{R}_i$ preceding R exceeds the x_n side length of R, we have

$$m_{n-1}(T \cap [\bigcup \widetilde{T}_i]) > \frac{1}{2} m_{n-1}(T).$$

This clearly implies that on R we have

$$M_{n-1}(\chi_{\cup \widetilde{R}_i}) > \frac{1}{2},$$

or equivalently that

$$\bigcup R_i \subseteq \{M_{n-1}(\chi_{\cup \widetilde{R}_i}) > \frac{1}{2}\}.$$

By the theory of the $n-1$-dimensional strong maximal operator, M_{n-1}, this gives

$$m\left(\bigcup R_i\right) \leq Cm\left(\bigcup \widetilde{R}_i\right)$$

which is part (b) of our covering lemma. ∎

The moral of the story in the first example of our technique for proving covering lemmas is that by using the geometry of the relevant family of sets involved, we have controlled the maximal operator at hand (in this case M_n) by a simpler operator (M_{n-1}). This method will be employed in the next example of the lacunary directions maximal operator M_{lac} defined for functions in the plane.

Suppose that $\mathcal{R}_{\text{lac}}$ denotes the family of all rectangles in $\mathbb{R}^2$ whose longest side makes an angle of 2^{-k} with the positive x-direction, for some positive integer k, but whose side lengths are arbitrary. (In other words $\mathcal{R}_{\text{lac}}$ is made up of all rectangles oriented in any one of a sequence of "lacunary directions".) Define

$$M_{\text{lac}}(f)(x) = \sup_{\substack{x \in R \\ R \in \mathcal{R}_{\text{lac}}}} \frac{1}{m(R)} \int_R |f(y)| dy.$$

Then the first estimates for this operator were obtained using the method described above for the strong maximal operator. They were obtained by J. O. Stromberg [28], and involved weak type estimates on $M_{\text{lac}}(f)$, for f belonging to a certain Orlicz class near L^2. We shall present here an extension of the results in [28] where we obtain weak type estimates on $M_{\text{lac}}(f)$ for $f \in L^2$ (A. Cordoba-R. Fefferman [9]), and where the argument is considerably simpler than that in [28].

We wish to show that $m\{M_{\text{lac}}(f) > \alpha\} \leq \frac{C}{\alpha^2} \|f\|^2_{L^2(\mathbb{R}^2)}$ for $\alpha > 0$. This will be done by proving that given a sequence of rectangles $R_i \in \mathcal{R}_{\text{lac}}$, there exists a subsequence $\widetilde{R}_i$ satisfying

(1) $\| \sum \chi_{\widetilde{R}_i} \|^2_{L^2} \leq Cm(\bigcup \widetilde{R}_i)$
(2) $m(\bigcup \widetilde{R}_i) \geq cm(\bigcup R_i),$

where c and C are positive constants. In order to do this, assume that the R_i are arranged so that their longest side length is decreasing. We select the $\widetilde{R}_i$ according to a slightly different rule from that given by rectangles with sides parallel to the axes: Choose $\widetilde{R}_1 = R_1$ and assume $\widetilde{R}_1, \ldots, \widetilde{R}_{k-1}$ have already been chosen. We continue down the list of R_i until we first come upon a rectangle R so that

$$(*) \qquad \sum_{i=1}^{k-1} m(R \cap \widetilde{R}_i) \leq \frac{1}{2} m(R).$$

We then set $R = \widetilde{R}_k$. Now, we claim that the $\{\widetilde{R}_i\}$ satisfy properties (1) and (2). To prove (1), we write

$$\| \sum \chi_{\widetilde{R}_i} \|^2_{L^2} = \int_{\mathbb{R}^2} \sum_{i,j} \chi_{\widetilde{R}_i} \chi_{\widetilde{R}_j} = 2 \sum_{i<j} m(\widetilde{R}_i \cap \widetilde{R}_j) + \sum_i m(\widetilde{R}_i).$$

From the selection rule $(*)$, we see that for each fixed j,

$$2 \sum_{i=1}^{j-1} m(\widetilde{R}_i \cap \widetilde{R}_j) \leq m(\widetilde{R}_j)$$

so that

$$2 \sum_{i<j} m(\widetilde{R}_i \cap \widetilde{R}_j) \leq \sum_j m(\widetilde{R}_j) \leq 2m(\bigcup \widetilde{R}_j).$$

This gives

$$\| \sum \chi_{\tilde{R}_i} \|_{L^2}^2 \leq 4m(\bigcup_i \tilde{R}_j)$$

and establishes (1).

To prove (2) we must use the geometry of the rectangles involved, in a way quite analogous to the previous example. Let R be a rectangle which is one of the R_i that was not selected. Let S denote the smallest rectangle with sides parallel to the axes such that S contains R. Finally suppose $\widehat{S}$ denotes the concentric double of S. Then, since R was not selected,

$$\frac{1}{m(R)} \sum_{\tilde{R}_i \text{ before } R} m(R \cap \tilde{R}_i) \geq \frac{1}{2}.$$

Very simple geometry shows that

$$\frac{m(R \cap \tilde{R}_i)}{m(R)} \leq C \frac{m(\widehat{S} \cap \tilde{R}_i)}{m(\widehat{S})}$$

for all $\tilde{R}_i$ whose longest side length exceeds that of R, and whose orientation is different from that of R. Now, there are two cases:

Case 1. $\dfrac{1}{m(R)} \displaystyle\sum_{\substack{\tilde{R}_i \text{ before } R \\ \tilde{R}_i \text{ oriented differently} \\ \text{from } R}} m(R \cap \tilde{R}_i) \geq 1/4$

Case 2. $\dfrac{1}{m(R)} \displaystyle\sum_{\substack{\tilde{R}_i \text{ before } R \\ \tilde{R}_i \text{ and } R \text{ have same orientation}}} m(R \cap \tilde{R}_i) \geq 1/4.$

In Case 1, from our geometric observation, it follows that

$$\frac{1}{m(\widehat{S})} \sum_i m(\widehat{S} \cap \tilde{R}_i) \geq \frac{1}{4C},$$

so that on R,

$$M_2 \left(\sum \chi_{\tilde{R}_i} \right) \geq 1/4C.$$

In Case 2,

$$M_2^j \left(\sum_{\tilde{R}_i \in \mathcal{R}^j} \chi_{\tilde{R}_i} \right) > \frac{1}{4}$$

where M_2^j denotes the two-dimensional strong maximal operator with respect to a pair of axes oriented along the sides of R and where $\mathcal{R}^j$ denotes the collection of those $\widetilde{R}_i$ oriented in the same way as R.

It follows that

$$\bigcup R_i \subseteq \left\{ M_2 \left(\sum \chi_{\widetilde{R}_i} \right) > \frac{1}{4C} \right\} \cup \bigcup_j \left\{ M_2^j \left(\sum_{\widetilde{R}_i \in \mathcal{R}^j} \chi_{\widetilde{R}_i} \right) \geq \frac{1}{4} \right\}$$

so that

$$m(\bigcup R_i) \leq C \left\| \sum_{\text{all } i} \chi_{\widetilde{R}_i} \right\|_{L^2}^2 + \sum_j m \left\{ M_2^j \left(\sum_{\widetilde{R}_i \in \mathcal{R}^j} \chi_{\widetilde{R}_i} \right) \geq \frac{1}{4} \right\}$$

$$\leq C \left[\left\| \sum_{\text{all } i} \chi_{\widetilde{R}_i} \right\|_{L^2}^2 + \sum_j \left\| \sum_{\widetilde{R}_i \in \mathcal{R}^j} \chi_{\widetilde{R}_i} \right\|_{L^2}^2 \right].$$

Again, by the selection rule of the $\widetilde{R}_i$, and our estimates to show (1),

$$\| \sum \chi_{\widetilde{R}_i} \|_{L^2}^2 \leq Cm(\bigcup \widetilde{R}_i)$$

and for each j,

$$\left\| \sum_{\widetilde{R}_i \in \mathcal{R}^j} \chi_{\widetilde{R}_i} \right\|_{L^2}^2 \leq Cm \left(\bigcup_{\widetilde{R}_i \in \mathcal{R}^j} \widetilde{R}_i \right).$$

Summing this on j gives

$$\sum_j \| \sum_{\widetilde{R}_i \in \mathcal{R}^j} \chi_{\widetilde{R}_i} \|_{L^2}^2 \leq Cm(\bigcup_{\text{all } i} \widetilde{R}_i),$$

and this proves (2) and the desired covering lemma.

We should point out that in [26], Nagel, Stein, and Wainger were able to extend our results on L^2 for $1 < p < \infty$, and their methods are entirely different. It would be quite interesting to obtain, by a covering lemma argument, the boundedness of the M_{lac} on the full range of L^p spaces.

The reader will notice that in the preceding two examples of covering lemmas, the rule for selecting rectangles in the sparse subcollection changed. In the case of the strong maximal operator, M_n, we chose a rectangle R if and only if

$$m(R \cap [\bigcup_{i<k} \widetilde{R}_i]) \leq \frac{1}{2} m(R).$$

In the second example of M_{lac}, a rectangle was selected if and only if

$$\sum_{i<k} m(R \cap \widetilde{R}_i) \leq \frac{1}{2} m(R),$$

which is a slightly stronger control of the overlap of R with the preceding selected rectangles $\widetilde{R}_i$ than in the case of M_n above. In our last example, the reader will notice a further strengthening of the selection rule.

Consider, in $\mathbb{R}^3$, the collection of all rectangles $\mathcal{R}_Z$ with sides parallel to the coordinate axes, contained in the unit cube, whose side lengths are of the form s, t, and $\psi(s, t)$ (in the x, y, and z directions, respectively) where ψ is increasing in each of the s and t variables separately. Then, A. Zygmund conjectured many years ago that if we set

$$M_Z f(p) = \sup_{p \in R \in \mathcal{R}_Z} \frac{1}{m(R)} \int_R |f|$$

then we get the estimate

$$m\{p \in Q \,|\, M_Z f(p) > \alpha\} \leq \frac{C}{\alpha} \|f\|_{L(\log^+ L)(Q)}.$$

In 1978, A. Cordoba [6], using the methods we have described, proved this conjecture using a beautiful argument which we reproduce here.

In fact, assuming that $\{R_i\}$ is a sequence of rectangles in $\mathcal{R}_Z$ we prove that there exists a subcollection $\{\widetilde{R}_i\}$ so that

(1) $\|\sum \chi_{\widetilde{R}_i}\|_{\exp(L)} \leq C$, and
(2) $m(\bigcup \widetilde{R}_i) \geq cm(\bigcup R_i)$.

To do this we order the R_i so that the z side lengths of the R_i are decreasing. (We shall also assume in what follows that the R_i are dyadic.) Then set $\widetilde{R}_1 = R_1$, and, assuming that $\widetilde{R}_i$, $i < k$ have been chosen, choose the first rectangle R on the list such that

$$\frac{1}{m(R)} \int_R \left[e^{\sum_{i=1}^{k-1} \chi_{\widetilde{R}_i}} - 1 \right] dx \leq \frac{1}{2}.$$

It is a simple matter to verify that $e^{\sum_i \chi_{\widetilde{R}_i}} - 1 \in L^1(Q)$ and $\|e^{\sum_{\text{all } i} \chi_{\widetilde{R}_i}} - 1\|_{L^1(Q)} \leq Cm(\bigcup \widetilde{R}_i)$ for some constant C. We need to verify now that $m(\bigcup R_i) \leq \mathcal{C}m(\bigcup \widetilde{R}_i)$. To do this, suppose R is an unselected rectangle so that

$$\frac{1}{m(R)} \int_R [e^{\sum \chi_{\widetilde{R}_i}} - 1] > \frac{1}{2},$$

where the sum only extends over those $\widetilde{R}_i$ before R on the list. Slice R and $\widetilde{R}_i$ with a plane perpendicular to the z axis, and call the two-dimensional slice of R

and $\widetilde{R}_i$, S and $\widetilde{S}_i$, respectively. Then, because the z side lengths of the $\widetilde{R}_i$ before R exceed the z side length of R, we have

$$\frac{1}{m_2(S)} \int_S [e^{\Sigma \chi_{\widetilde{S}_i}} - 1] > \frac{1}{2}$$

where again the sum only extends over those i for which $\widetilde{R}_i$ comes before R on our list of R_i. It is at this point that we make use of the fact that all of the rectangles in question (in $\mathbb{R}^3$) have side lengths of the form s, t, and $\psi(s, t)$. It follows from this that each of the $\widetilde{S}_i$ in the sum above has either its x side length greater than the x side length of S (call such an $\widetilde{S}_i$ of type I) or $\widetilde{S}_i$ has its y side length greater than the y side of S (type II).

Then

$$\frac{1}{m_2(S)} \int_S \left[e^{\Sigma \chi_{\widetilde{S}_i}} - 1 \right] = \frac{1}{m_2(S)} \int_S \left[e^{(\Sigma_I \chi_{\widetilde{S}_i} + \Sigma_{II} \chi_{\widetilde{S}_i})} - 1 \right]$$

(here $\sum_I$ indicates that we sum over only the $\widetilde{S}_i$ of type I, and similarly for $\sum_{II}$)

$$= \frac{1}{m_2(S)} \int_S \left[e^{\Sigma_I \chi_{\widetilde{S}_i}} - 1 \right] \left[e^{\Sigma_{II} \chi_{\widetilde{S}_i}} - 1 \right]$$

$$+ \frac{1}{m_2(S)} \int_S (e^{\Sigma_I \chi_{\widetilde{S}_i}} - 1) + \frac{1}{m_2(S)} \int_S (e^{\Sigma_{II} \chi_{\widetilde{S}_i}} - 1).$$

Since the sum above exceeds $1/2$, at least one of the terms must exceed $1/6$.

In case

$$\frac{1}{m_2(S)} \int_S \left[e^{\Sigma_I \chi_{\widetilde{S}_i}} - 1 \right] \left[e^{\Sigma_{II} \chi_{\widetilde{S}_i}} - 1 \right] dx dy > 1/6,$$

since $e^{\Sigma_I \chi_{\widetilde{S}_i}} - 1$ is a function of the y variable only considered as a function on S, and $e^{\Sigma_{II} \chi_{\widetilde{S}_i}} - 1$ is a function of x only, we have (writing $S = J \times K$)

$$\frac{1}{m_2(S)} \iint_S \left[e^{\Sigma_I \chi_{\widetilde{S}_i}} - 1 \right] \left[e^{\Sigma_{II} \chi_{\widetilde{S}_i}} - 1 \right] dx dy$$

$$= \frac{1}{m_1(K)} \int_K \left[e^{\Sigma_I \chi_{\widetilde{S}_i}} - 1 \right] dy \frac{1}{m_1(J)} \int \left[e^{\Sigma_{II} \chi_{\widetilde{S}_i}} - 1 \right] dx > \frac{1}{6}.$$

Thus in this case at least one of the inequalities

$$\frac{1}{m_1(K)} \int_K \left[e^{\Sigma_I \chi_{\widetilde{S}_i}} - 1 \right] dy > \frac{1}{\sqrt{6}}$$

or

$$\frac{1}{m_1(J)} \int_J \left[e^{\Sigma_{II} \chi_{\widetilde{S}_i}} - 1 \right] dx > \frac{1}{\sqrt{6}}$$

is valid.

In case

$$\frac{1}{m_2(S)} \int_S \left[e^{\Sigma_I \chi_{\widetilde{S_i}}} - 1 \right] dx\, dy > \frac{1}{6},$$

it follows that

$$\frac{1}{m_1(K)} \int_K \left[e^{\Sigma_I \chi_{\widetilde{S_i}}} - 1 \right] dy > \frac{1}{6},$$

and a similar analysis holds in the remaining case where

$$\frac{1}{m_2(S)} \int_S \left[e^{\Sigma_{II} \chi_{\widetilde{S_i}}} - 1 \right] dx\, dy > \frac{1}{6}.$$

Putting all of this together yields

$$\bigcup R_i \subseteq \left\{ M_x \left(e^{\Sigma \chi_{\widetilde{R_i}}} - 1 \right) > 1/6 \right\} \cup \left\{ M_y \left(e^{\Sigma \chi_{\widetilde{R_i}}} - 1 \right) > 1/6 \right\}.$$

(Here M_x and M_y denote one-dimensional maximal functions in the x and y directions, respectively.) Using the estimate $\| e^{\Sigma \chi_{\widetilde{R_i}}} - 1 \|_{L^1(Q)} \leq Cm(\bigcup \widetilde{R_i})$ and the weak type L^1 estimates for M_x and M_y, we see that $m(\bigcup R_i) \leq C(\bigcup \widetilde{R_i})$ finishing the proof of the covering lemma.

As the reader can see, it is far from routine to apply the covering lemma method to get information about a particular maximal operator. Nevertheless, this method has had a number of deep applications, and can be expected to have more in the future. Before moving on to consider the second "model problem" it should also be pointed out that in order to do the operator theory of operators which are invariant with respect to certain actions of $\mathbb{R}^n$ (certain groups of dilations, rotations, etc.) it is essential to understand the geometry of the relevant sets associated to these operators and for this, covering lemmas must be proved. It is not enough just to know the sharp estimates for the associated maximal operator. This will be illustrated quite clearly in the following part of this article dealing with the analysis of singular integrals on product spaces.

Now, let us consider the second problem, namely obtaining sharp estimates for Marcinkiewicz-type operators. We shall begin by explaining the precise meaning of this class of operators, but the reader should be aware that the details of the definition could be altered considerably without making essential changes in their basic theory.

Recall that a function $m(\xi), \xi \in \mathbb{R}^n$ is called a "Hörmander multiplier" provided all the dilates $m(\delta\xi), \delta > 0$ satisfy, uniformly in δ, estimates of the form

$$|m(\xi)| \leq C, \qquad \xi \in \mathbb{R}^n$$

and

$$\| D^\alpha [m(\xi)\varphi(|\xi|)] \|_{L^2(\mathbb{R}^n)} \leq C$$

for all multiindices α with $|\alpha| \leq [\frac{n}{2}] + 1$; here $\varphi(x)$ is any function $C^\infty(\mathbb{R}^1)$ which satisfies $\varphi(x) = 1$ for all $x \in [\frac{1}{2}, 2]$ and $\varphi(x) = 0$ for all $x \notin [\frac{1}{4}, 4]$.

It is a fundamental fact from the theory of singular integrals (see Stein [27]) that the multiplier operators T_m corresponding to the Hörmander class are bounded on $L^p(\mathbb{R}^n)$, $1 < p < \infty$, and are of weak type (1,1). Now, to motivate the definition of the Marcinkiewicz multipliers, the reader should consider the estimates satisfied, on $\mathbb{R}^n \times \mathbb{R}^m$, by multipliers of the form $m_1(\xi)m_2(\eta)$, $\xi \in \mathbb{R}^n$, $\eta \in \mathbb{R}^m$, where m_1 and m_2 are Hörmander multipliers on $\mathbb{R}^n$ and $\mathbb{R}^m$, respectively. These estimates on a function m defined on $\mathbb{R}^n \times \mathbb{R}^m$ are as follows:

The functions $m(\delta_1\xi, \delta_2\eta)$ satisfy (uniformly in δ_1 and δ_2) the estimates

$$|m(\xi, \eta)| \leq C, \qquad \text{for} \quad \xi \in \mathbb{R}^n, \ \eta \in \mathbb{R}^m$$

$$\|D_\xi^\alpha[m(\xi, \eta)\varphi(|\xi|)]\|_{L^2(\mathbb{R}^n)} \leq C$$

$$\text{for each } \eta \in \mathbb{R}^m \text{ whenever } |\alpha| \leq [\frac{n}{2}] + 1$$

$$\|D_\eta^\beta[m(\xi, \eta)\varphi(|\eta|)]\|_{L^2(\mathbb{R}^m)} \leq C$$

$$\text{for each fixed } \xi \in \mathbb{R}^n \text{ whenever } |\beta| \leq [\frac{m}{2}] + 1$$

and

$$\|D_\xi^\alpha D_\eta^\beta[m(\xi, \eta)\varphi(|\xi|)\varphi(|\eta|)]\|_{L^2(\mathbb{R}^n \times \mathbb{R}^m)}$$

$$\text{for all } \alpha, \beta \text{ such that } |\alpha| \leq [\frac{n}{2}] + 1 \text{ and } |\beta| \leq [\frac{m}{2}] + 1.$$

Functions $m(\xi, \eta)$ satisfying these estimates are called here "Marcinkiewicz-type multipliers," and their corresponding multiplier operators, the "Marcinkiewicz operators."

The problem we investigate now is that of obtaining the weak type inequality

$$m\{(x, y) \in Q \,|\, |T_m(f)(x, y)| > \alpha\} \leq \frac{C}{\alpha} \|f\|_{L \log^+ L(Q)}$$

for all $\alpha > 0$, where Q denotes the unit cube of $\mathbb{R}^n \times \mathbb{R}^m$, and T_m is the Marcinkiewicz operator.

This estimate is easily derived by iteration in case $m(\xi, \eta)$ takes the special form $m_1(\xi)m_2(\eta)$. However the general case is much more difficult, and we shall explain why this is the case.

One of the best ways to obtain the L^p theory ($1 < p < \infty$) of Hörmander operators is via the pointwise majorization of Stein:

(#) $$g(T_m f)(x) \leq C g_\lambda^*(f)(x)$$

where g and g_λ^* are the classical Littlewood-Paley functions (Stein [27]). The quantity $\lambda > 1$ in the inequality depends on how much smoothness the multiplier has. If the multiplier is minimally smooth ($[\frac{n}{2}] + 1$ derivatives in L^2 away from 0 and ∞) then λ is not much bigger than 1. Unfortunately g_λ^* is not bounded on $L^p(\mathbb{R}^n)$ unless $\lambda p > 2$, i.e., in this case g_λ^* is not of weak type (1,1). Thus when we are interested in the action of our multipliers near L^1 we must resort to something other than (#). The other way to handle T on $L^p(\mathbb{R}^n)$ is to realize that T is essentially a classical Calderón-Zygmund operator. One then uses the decomposition into "good" and "bad" parts, or the atomic decomposition of the classical Hardy spaces $H^p(\mathbb{R}^n)$ in order to obtain L^p boundedness and the weak type estimates on $L^1(\mathbb{R}^n)$. Our point here is that the classical Calderón-Zygmund decomposition and atomic decomposition cannot be used to analyze the more complicated Marcinkiewicz operators.

The approach we take will be to study different spaces $H^p(\mathbb{R}^n \times \mathbb{R}^m)$ (we write $H^p(\mathbb{R}^{n+m})$ to denote the Charles Fefferman-E. M. Stein classical H^p space, and $H^p(\mathbb{R}^n \times \mathbb{R}^m)$ to denote our non-classical Hardy spaces) adapted to the study of the operators at hand. We construct the appropriate theory of BMO($\mathbb{R}^n \times \mathbb{R}^m$) and the atomic decomposition of $H^p(\mathbb{R}^n \times \mathbb{R}^m)$, and study the interpolation properties of these spaces. Once all this is well understood, we will be in a good position to solve our second problem.

Let us begin with the space $H^p(\mathbb{R}^n \times \mathbb{R}^m)$, for $p \leq 1$. Suppose that f is a tempered distribution on $\mathbb{R}^n \times \mathbb{R}^m$. Let $\varphi_1(x) \in C_0^\infty(\mathbb{R}^n)$, $\varphi_2(y) \in C_0^\infty(\mathbb{R}^m)$ with $\int \varphi_i = 1$, $i = 1, 2$. Set $\varphi_{\delta_1,\delta_2}(x, y) = \varphi_1(x/\delta_1)\varphi(y/\delta_2)\delta_1^{-n}\delta_2^{-m}$, for $\delta_1, \delta_2 > 0$. We define the product non-tangential maximal function $f^*(x, y)$ by

$$f^*(x, y) = \sup_{\substack{\delta_1,\delta_2>0 \\ |\bar{x}-x|<\delta_1 \\ |\bar{y}-y|<\delta_2}} |f * \varphi_{\delta_1,\delta_2}(\bar{x}, \bar{y})|$$

and the product Lusin area integral $S(f)(x, y)$ by

$$S^2(f)(x, y) = \iint_{\Gamma(x) \times \Gamma(y)} |f * \psi_{\delta_1,\delta_2}(\bar{x}, \bar{y})|^2 \frac{d\bar{x}d\delta_1 \, d\bar{y}d\delta_2}{\delta_1^{n+1}\delta_2^{m+1}}$$

where $\psi(x, y) = \psi_1(x)\psi_2(y)$ and $\psi_1 \in C_0^\infty(\mathbb{R}^n)$, $\psi_2 \in C_0^\infty(\mathbb{R}^m)$, $\int \psi_i = 0$ and ψ_i are suitably non-trivial. Then we have the following fundamental theorem.

Theorem of Gundy-Stein [20]. *For all $p > 0$, $S(f) \in L^p(\mathbb{R}^n \times \mathbb{R}^m)$ if and only if $f^* \in L^p(\mathbb{R}^n \times \mathbb{R}^m)$. In this case the ratio $\|S(f)\|_{L^p} \big/ \|f^*\|_{L^p}$ is bounded above and below by strictly positive finite constants depending only on p, n, and m.*

This theorem forms the basis for the definition of product Hardy spaces: $f \in H^p(\mathbb{R}^n \times \mathbb{R}^m)$ if and only if $f^* \in L^p(\mathbb{R}^n \times \mathbb{R}^m)$ or equivalently $S(f) \in L^p(\mathbb{R}^n \times \mathbb{R}^m)$ for $p > 0$. We define $\|f\|_{H^p(\mathbb{R}^n \times \mathbb{R}^m)} = \|f^*\|_{L^p(\mathbb{R}^n \times \mathbb{R}^m)}$. There are certain obvious examples of functions in product H^p, namely tensor products of H^p functions of the x and y variables, respectively:

$$f(x, y) = f_1(x) f_2(y) \qquad \text{where} \quad f_1 \in H^p(\mathbb{R}^n), \; f_2 \in H^p(\mathbb{R}^m).$$

In particular, when $p \leq 1$ if $f_1 = a_1$ and $f_2 = a_2$, where the a_i are classical H^p atoms, then $a(x, y) = a_1(x) a_2(y)$ is an $H^p(\mathbb{R}^n \times \mathbb{R}^m)$ function which serves as a model for what we shall call an $H^p(\mathbb{R}^n \times \mathbb{R}^m)$ "rectangle atom." (The necessity of inserting the word "rectangle" will be clarified shortly.) Clearly these $a(x, y)$ have the following properties.

(1) $a(x, y)$ is supported on the rectangle $R = I \times J$ where $I \subseteq \mathbb{R}^n$ and $J \subset \mathbb{R}^m$ are cubes;

(2) $\int_I a(x, y) x^\alpha \, dx = 0$ for all multiindices α with $|\alpha| \leq N_p(n)$ and for each $y \in J$, where $N_p(n) = [n(\frac{1}{p} - 1)]$;

(3) $\int_J a(x, y) y^\beta \, dy = 0$ for all multiindices β with $|\beta| \leq N_p(m)$ and for each $x \in I$, where $N_p(m) = [m(\frac{1}{p} - 1)]$; and finally,

(4) $\|a\|_{L^2(R)} \leq m(R)^{\frac{1}{2} - \frac{1}{p}}$.

Such a function $a(x, y)$ on $\mathbb{R}^n \times \mathbb{R}^m$ is called an $H^p(\mathbb{R}^n \times \mathbb{R}^m)$ rectangle atom.

For a while, it was thought that these rectangle atoms would span $H^p(\mathbb{R}^n \times \mathbb{R}^m)$ in a way that would immediately reduce the harmonic analysis of product spaces to the consideration of the action of the relevant operators on these very simple atoms. The conjecture was that every $f \in H^p(\mathbb{R}^n \times \mathbb{R}^m)$ could be written as

$$f = \sum_k \lambda_k a_k$$

where a_k are rectangle atoms of $H^p(\mathbb{R}^n \times \mathbb{R}^m)$ and λ_k are scalars satisfying $\sum_k |\lambda_k|^p \leq C \|f\|_{H^p(\mathbb{R}^n \times \mathbb{R}^m)}^p$.

In 1974, in [31], Lennart Carleson disproved this conjecture, suggesting that the harmonic analysis of product spaces had to proceed along lines very different from the ones of the classical theory. The fact that rectangle atoms did not properly span $H^p(\mathbb{R}^n \times \mathbb{R}^m)$ also suggested that the product theory could not be nearly as simple and satisfactory in its final form as the classical one. We now know that both of these suggestions are wrong, and to see this, the reader need only consider the following:

Theorem [15]. *Let T be an $L^2(\mathbb{R}^n \times \mathbb{R}^m)$ bounded linear operator. Suppose, further, that whenever a is a rectangle atom of $H^p(\mathbb{R}^n \times \mathbb{R}^m)$ $(0 < p \leq 1)$*

supported on the rectangle R, we have

$$\int_{{}^{c}\widetilde{R}_{\gamma}} |T(a)|^{p}\, dxdy \leq C\gamma^{-\delta} \qquad \textit{for all} \quad \gamma \geq 2 \textit{ and some } \delta > 0.$$

(Here $\widetilde{R}_{\gamma}$ denotes the concentric dilate of R by the factor γ.) Then T is a bounded operator from $H^{p}(\mathbb{R}^{n} \times \mathbb{R}^{m})$ to $L^{p}(\mathbb{R}^{n} \times \mathbb{R}^{m})$.

The meaning of this is that, if T is to map H^{p} boundedly to L^{p}, then it is obviously necessary that

(‡) $$\int_{\mathbb{R}^{n} \times \mathbb{R}^{m}} |T(a)|^{p}\, dxdy \leq C \qquad \text{for all rectangle atoms} \quad a.$$

Carleson's counterexample shows that (‡) is not sufficient for T to be bounded between H^{p} and L^{p}. Nevertheless, if we trivially strengthen the assumption above to say that not only is the integral (‡) bounded, but the contribution to the integral from integrating over that part of space which is away from the support of the atom vanishes as we look farther and farther from this support, then this stronger condition is sufficient. Thus, it is not important that rectangle atoms fail to span $H^{p}(\mathbb{R}^{n} \times \mathbb{R}^{m})$. By examining the action of an operator on these atoms, we may still conclude its boundedness. Below, we shall give the main elements in the proof of this result.

The first thing we must discuss is the correct notion of an $H^{p}(\mathbb{R}^{n} \times \mathbb{R}^{m})$ atom. This atomic decomposition can be found in S. Y. Chang-R. Fefferman [4]. We set the following notation: If $\Omega \subseteq \mathbb{R}^{n} \times \mathbb{R}^{m}$ is an open set of finite measure, then the maximal dyadic subrectangles of Ω will be denoted by $\mathcal{M}(\Omega)$. If R is a rectangle, then $\widetilde{R}$ denotes its concentric double, and if Ω is an open set, then Ω^{*} denotes $\{M_{s}(\chi_{\Omega}) > \frac{1}{2^{n+m}}\}$ where M_{s} is the strong maximal function in $\mathbb{R}^{n} \times \mathbb{R}^{m}$ defined by a supremum of averages over rectangles (products of cubes in $\mathbb{R}^{n}$ and $\mathbb{R}^{m}$, respectively).

Let $0 < p \leq 1$ and suppose that $\Omega \subseteq \mathbb{R}^{n} \times \mathbb{R}^{m}$ is an open set of finite measure. A function $a(x, y)$ on $\mathbb{R}^{n} \times \mathbb{R}^{m}$ is called an $H^{p}(\mathbb{R}^{n} \times \mathbb{R}^{m})$ atom (with associated open set Ω) if and only if:

(1) a is supported in Ω^{*};
(2) $\|a\|_{L^{2}(\mathbb{R}^{n} \times \mathbb{R}^{m})} \leq m(\Omega)^{\frac{1}{2} - \frac{1}{p}}$;
(3) a can be decomposed as $a = \sum_{R \in \mathcal{M}(\Omega)} \alpha_{R}$, where each α_{R} has the following properties:

    (i) α_{R} is supported in $\widetilde{R}$;

    (ii) $\int \alpha_{R}(x, y)x^{\nu}\, dx = 0$ for each $y \in \mathbb{R}^{m}$ and for all multi-indices ν such that $|\nu| \leq [n(\frac{1}{p} - 1)] = N_{p}(n)$ and $\int \alpha_{R}(x, y)y^{\nu}\, dy = 0$ for each $x \in \mathbb{R}^{n}$ and for all ν such that $|\nu| \leq N_{p}(m) = [m(\frac{1}{p} - 1)]$;

(iii) $\left[\sum_{R\in\mathcal{M}(R)} \|\alpha_R\|_2^2\right]^{1/2} \le m(\Omega)^{\frac{1}{2}-\frac{1}{p}}$.

Observe that (i) and (ii) describe α_R as a rectangle atom except without any normalizing condition on $\|\alpha_R\|_2$. This is, of course, compensated for.

The atomic decomposition in [4] then states that every $f \in H^p(\mathbb{R}^n \times \mathbb{R}^m)$ (here $0 < p \le 1$) can be decomposed as

$$f = \sum_h \lambda_k a_k$$

where λ_k are scalars, a_k are $H^p(\mathbb{R}^n \times \mathbb{R}^m)$ atoms, and

$$\sum_k |\lambda_k|^p \le C\|f\|_{H^p(\mathbb{R}^n\times\mathbb{R}^m)}^p.$$

The second ingredient in the proof of our theorem is a geometric lemma due to J. L. Journé [23], which he used in order to obtain estimates on operators such as product commutators. We set the following notation in what follows: Let $\Omega \subseteq \mathbb{R}^n \times \mathbb{R}^m$ be an open set of finite measure. $\mathcal{M}_1(\Omega)$ will denote the collection of all dyadic subrectangles of Ω which are maximal in the x coordinate. A similar definition applies to $\mathcal{M}_2(\Omega)$. Let Q be a cube. We denote by $\widetilde{Q}_\gamma$ the γ fold concentric dilate of Q. Then, given R, a dyadic subrectangle of Ω, $R = I \times J$ ($I \subseteq \mathbb{R}^n$, $J \subseteq \mathbb{R}^m$ are cubes), we define $\gamma_1(R) = \sup\{t \ge 1|\widetilde{I}_t \times J \subseteq \Omega^*\}$. Then we have:

Journé's Lemma. *If $\Omega \subseteq \mathbb{R}^n \times \mathbb{R}^m$ is an open set of finite measure, and $\delta > 0$, then*

$$\sum_{R\in\mathcal{M}_2(\Omega)} m(R)\gamma_1(R)^{-\delta} \le C_\delta m(\Omega).$$

Of course, if Ω is an open set in $\mathbb{R}^1$, then the sum of the lengths of the maximal dyadic subintervals of Ω equals the measure of Ω. Although for Ω open in $\mathbb{R}^2$ it is not true that the sum of the areas of the maximal dyadic subrectangles of Ω is finite (or less than $Cm(\Omega)$), it is true if in the sum, we multiply the area of each maximal dyadic subrectangle by a suitable small constant depending (roughly speaking) on how much we may expand the rectangle without leaving Ω (more precisely Ω^*). This is the meaning of Journé's result.

Now let us prove the theorem. Let T be $L^2(\mathbb{R}^n \times \mathbb{R}^m)$ bounded and linear, and suppose that

$$\int_{^c\widetilde{R}_\gamma} |T(a)|^p \, dxdy \le C\gamma^{-\delta}$$

for all $\gamma \geq 2$ whenever a is an $H^p(\mathbb{R}^n \times \mathbb{R}^m)$ rectangle atom supported in R. We claim that for $f \in H^p(\mathbb{R}^n \times \mathbb{R}^m)$, $p \leq 1$, we have

$$\|Tf\|_{L^p(\mathbb{R}^n \times \mathbb{R}^m)} \leq C_p \|f\|_{H^p(\mathbb{R}^n \times \mathbb{R}^m)}.$$

By the atomic decomposition this will follow once we prove that

$$\|T(a)\|_{L^p(\mathbb{R}^n \times \mathbb{R}^m)} \leq C_p$$

whenever a is an $H^p(\mathbb{R}^n \times \mathbb{R}^m)$ atom. So, suppose a is such an atom with associated open set Ω. Then

$$\int_{\mathbb{R}^n \times \mathbb{R}^m} |T(a)|^p \, dxdy = \int_{\Omega^{**}} |T(a)|^p \, dxdy + \int_{c\Omega^{**}} |T(a)|^p \, dxdy.$$

Clearly

$$\int_{\Omega^{**}} |T(a)|^p \, dxdy \leq \left(\int_{\mathbb{R}^n \times \mathbb{R}^m} |T(a)|^2 \, dxdy \right)^{p/2} m(\Omega^{**})^{1/(\frac{2}{p})'}$$

$$\leq m(\Omega)^{\frac{p}{2}-1} m(\Omega^{**})^{1-p/2} \leq C_p.$$

As for $\int_{c\Omega^{**}} |T(a)|^p \, dxdy$, we write $a = \sum_{R \in \mathcal{M}(\Omega)} \alpha_R$ as in the definition of an atom. Then we proceed as follows: Suppose $R \in \mathcal{M}(\Omega)$, $R = I \times J$. Let $\widehat{I}$ denote the largest dyadic cube containing I so that $\widehat{I} \times J \subseteq \Omega^*$. Then, let $\widehat{J}$ denote the largest dyadic cube containing J so that $\widehat{I} \times \widehat{J} \subseteq \Omega^{**}$. We then estimate

$$\int_{c\Omega^{**}} |T(a)|^p \, dxdy = \int_{c\Omega^{**}} \left| T\left(\sum_{R \in \mathcal{M}(\Omega)} \alpha_R \right) \right|^p \, dxdy$$

$$\leq \int_{c\Omega^{**}} \sum_{R \in \mathcal{M}(\Omega)} |T(\alpha_R)|^p \, dxdy$$

$$\leq \sum_{R \in \mathcal{M}(\Omega)} \int_{c(\widetilde{\widehat{I}} \times \widetilde{\widehat{J}})} |T(\alpha_R)|^p \, dxdy$$

$$\leq \sum_{R \in \mathcal{M}(\Omega)} \int_{c(\widetilde{\widehat{I}}) \times \mathbb{R}^m} |T(\alpha_R)|^p \, dxdy$$

$$+ \int_{\mathbb{R}^n \times c(\widetilde{\widehat{J}})} |T(\alpha_R)|^p \, dxdy.$$

We now use the fact that $\alpha_R / \left(\|\alpha_R\|_2 m(R)^{(1/p)-(1/2)} \right)$ is an $H^p(\mathbb{R}^n \times \mathbb{R}^m)$ rectangle atom, so that by the hypothesis of our theorem, the right-hand side above is dominated by

$$\sum_{R\in\mathcal{M}(\Omega)} m(R)^{1-p/2}\|\alpha_R\|_2^p \gamma_1^{-\delta}(R) + \sum_{R\in\mathcal{M}(\Omega)} m(R)^{1-p/2}\|\alpha_R\|_2^p \gamma_2^{-\delta}(\widehat{R})$$

$$= (A) + (B).$$

We estimate (A) by Hölder's inequality:

$$(A) = \left(\sum_{R\in\mathcal{M}(\Omega)}\|\alpha_R\|_2^2\right)^{p/2}\left(\sum_{R\in\mathcal{M}(\Omega)} m(R)^{(1-p/2)(2/p)'}\gamma_1^{-\delta(2/p)'}(R)\right)^{1/(2/p)'}$$

$$\le m(\Omega)^{\frac{p}{2}-1}\left(\sum_{R\in\mathcal{M}(\Omega)}\|\alpha_R\|_2^2\right)^{p/2}$$

$$\le C_p \quad \text{by Journé's lemma.}$$

In order to estimate (B) we proceed similarly.

$$(B) \le \left(\sum_{R\in\mathcal{M}(R)}\|\alpha_R\|_2^2\right)^{p/2}\left(\sum_{R\in\mathcal{M}(\Omega)} m(R)^{(1-p/2)(2/p)'}\gamma_2^{-\delta''}(\widehat{R})\right)^{1/(2/p)'}$$

$$\le m(\Omega)^{p/2-1}\left[\sum_{R\in\mathcal{M}(\Omega)} m(R)\gamma_2^{-\delta''}(\widehat{R})\right]^{1/(2/p)'}.$$

Then

$$\sum_{R\in\mathcal{M}(\Omega)} m(R)\gamma_2^{-\delta''}(\widehat{R}) = \sum_{S\in\mathcal{M}_1(\Omega^*)} \gamma_2^{-\delta''}(S) \sum_{\substack{R\in\mathcal{M}(\Omega)\\ \widehat{R}=S}} m(R).$$

Clearly any two dyadic $R \in \mathcal{M}(\Omega)$ such that $\widehat{R} = S$ must be disjoint, so

$$\sum_{\substack{R\in\mathcal{M}(\Omega)\\ \widehat{R}=S}} m(R) \le m(S),$$

and so

$$\sum_{S\in\mathcal{M}_1(\Omega^*)} \gamma_2^{-\delta''}(S) \sum_{\substack{R\in\mathcal{M}(\Omega)\\ \widehat{R}=S}} m(R) \le \sum_{S\in\mathcal{M}_1(\Omega^*)} \gamma_2^{-\delta''}(S)m(S) \le Cm(\Omega^*)$$

by Journé's lemma. This proves that

$$(B) \leq Cm(\Omega)^{p/2-1} m(\Omega^*)^{1/(2/p)'} \leq C'.$$

This completes the proof of the boundedness of T from $H^p(\mathbb{R}^n \times \mathbb{R}^m)$ to $L^p(\mathbb{R}^n \times \mathbb{R}^m)$.

Once we understand that in order to check the boundedness of an operator on product H^p, it is enough to check its action on rectangle atoms, it is not difficult to settle the question of the boundedness properties of Marcinkiewicz operators near L^1. This proof of the weak type estimate

$$m\{(x, y) \in Q \mid |Tf(x, y)| > \alpha\} \leq \frac{C}{\alpha} \|f\|_{L(\log^+ L)(Q)}$$

(Q denotes the unit cube of $\mathbb{R}^n \times \mathbb{R}^m$) where T is a Marcinkiewicz operator proceeds much in the same spirit as the argument given in detail above for H^p. In fact, it involves an atomic decomposition of $L(\log^+ L)(Q)$ much like that of $H^p(\mathbb{R}^n \times \mathbb{R}^m)$. For the details, the reader should consult [19].

Again, we wish to stress that it is not the technical details that are important here, but rather the philosophy that although the function spaces in the product theory are inherently more complicated than in the classical case, the operator theory is not. This philosophy has some very concrete applications besides the one we have been discussing. As another example, we should mention the weighted inequalities for product singular integrals, which can be found in [15] and [16]. It is quite non-trivial to prove weighted norm inequalities for these singular integrals, because there are no known analogues of the classical Burkholder-Gundy good λ inequalities in this setting. One studies this problem by determining rather precisely the way that the strong maximal operator M_s controls the singular integral. In what follows below we shall be a little more precise.

To begin with, let us recall that a weight $w(x, y)$ defined on $\mathbb{R}^n \times \mathbb{R}^m$ is said to belong to product A^p (written $A^p(\mathbb{R}^n \times \mathbb{R}^m)$) if and only if for all rectangles R (products of cubes in $\mathbb{R}^n$ and in $\mathbb{R}^m$, respectively) we have

$$\left(\frac{1}{m(R)} \int_R w(x, y) \, dx \, dy \right) \left(\frac{1}{m(R)} \int_R w(x, y)^{-\frac{1}{p-1}} \, dx \, dy \right)^{p-1} \leq C.$$

This is equivalent to w satisfying a uniform classical A^p condition in the x and y variables separately fixing the other variable. Then the way we prove weighted inequalities for general product singular integrals (not necessarily convolutions) in [15] and [16] is by studying a variant of the C. Fefferman-Stein sharp function, $f^\#$ in the product setting. Thus, if we wish to prove, for very general classes of singular integrals T, that

$$\int_{\mathbb{R}^n \times \mathbb{R}^m} |T(f)(x, y)|^p w(x, y)dxdy \leq C \int_{\mathbb{R}^n \times \mathbb{R}^m} |f(x, y)|^p w(x, y)dxdy$$

for $1 < p < \infty$ when $w \in A^p(\mathbb{R}^n \times \mathbb{R}^m)$, then we need to have an analogue of the fact, from classical singular integrals, that

(##) $$(Tf)^{\#}(x) \leq C[M(f^{1+\varepsilon})(x)]^{1/1+\varepsilon}.$$

Here T is a Calderón-Zygmund singular integral, M is the Hardy-Littlewood maximal operator, and

$$f^{\#}(x) = \sup_{x \in Q} \frac{1}{m(Q)} \int_Q |f(y) - f_Q|dy,$$

(where $f_Q = \frac{1}{m(Q)} \int_Q f$ and the sup is taken over all cubes $Q \subseteq \mathbb{R}^n$ containing x) is the C. Fefferman-Stein sharp function. In the classical theory, the value of the inequality (##) lies in the fact that $f^{\#} \in L^p(\mathbb{R}^n)$ implies $f \in L^p(\mathbb{R}^n)$. Unfortunately, as we would expect, this is not true in product domains. To be more specific, suppose we define the mean oscillation of a function $f(x, y)$ on $\mathbb{R}^n \times \mathbb{R}^m$ over the rectangle R by

$$\text{osc}_R(f) = \inf_{f_1, f_2} \left(\frac{1}{m(R)} \int_R |f(x, y) - f_1(x) - f_2(y)|^2 dxdy \right)^{1/2}$$

where the inf is taken over all pairs of functions f_1 and f_2 depending only on the x and y variables respectively. If we then define the (product) sharp function of f, $f^{\#}(x, y)$ by

$$f^{\#}(x, y) = \sup_{(x,y) \in R} \text{osc}_R(f)$$

where the sup is taken over all rectangles containing (x, y), then it is simply not true that $f^{\#} \in L^p(\mathbb{R}^n \times \mathbb{R}^m)$ implies that $f \in L^p(\mathbb{R}^n \times \mathbb{R}^m)$. In order to use the product sharp function as a useful tool to prove weighted inequalities, we must transfer the definition to the context of operators, rather than functions.

Intuitively, we would like to define the sharp operator S of an operator T by

$$Sf(x, y) = \sup_{(x,y) \in R} \text{osc}_R T(f).$$

As we stated above, the boundedness of S on $L^p(\mathbb{R}^n \times \mathbb{R}^m)$ does not imply the boundedness of T on this space. However, suppose, in addition, we require that whenever the support of f is far away from the rectangle R, i.e., if $\text{supp}(f) \subseteq {}^c\widetilde{R}_\gamma$ then for some $\delta > 0$, and $\gamma \geq 1$

$$Sf(x, y) \geq \gamma^\delta \text{osc}_R(Tf) \qquad \text{when} \quad (x, y) \in R.$$

Then we say that S is a sharp operator for T. It is shown in [15] that if $p \geq 2$ then if S is bounded on $L^p(\mathbb{R}^n \times \mathbb{R}^m)$, then so is T. Also, if T is a product singular integral, then $M_s(f^2)^{1/2}$ is a sharp operator for T. This precise expression of the control of product singular integrals by the strong maximal operator can then be used in order to prove the product version of the Hunt-Muckenhoupt-Wheeden weighted inequalities for singular integrals, above. It is another graphic illustration that in product domains, one can use the philosophy mentioned above in order to bring things into a very simple and satisfactory form quite like that of the classical theory.

Our last problem concerns elliptic differential equations, but the mathematical issues involved have quite a lot in common with model problems 1 and 2. In fact, we will see that just as in the first two problems, the theory of maximal functions, singular integrals, and especially the topic of weighted inequalities will play a central role.

We shall begin to discuss the third problem by formulating it precisely. We consider elliptic operators in the unit ball B of $\mathbb{R}^n$ in divergence form. This means that $Lu = \mathrm{div}(A\nabla u)$ where $A(x)$, $x \in B$, is a symmetric $n \times n$ matrix valued function satisfying

$$\lambda|\xi|^2 \leq a_{ij}(x)\xi_i\xi_j \leq \lambda^{-1}|\xi|^2, \quad A(x) = (a_{ij}(x))$$

for all $\xi \in \mathbb{R}^n$ and some fixed $\lambda > 0$. We will assume, in general, no regularity of the coefficients $a_{ij}(x)$ except for measurability. The question we investigate is the solvability of the L^p Dirichlet problem. To define this properly, we require several definitions. First, we say that a function $u \in H^1_{\mathrm{loc}}(B)$ ($H^1_{\mathrm{loc}}(B)$ denotes those functions in B which are, together with their distributional first derivatives, locally in L^2) is a solution to $Lu = 0$ provided $a(u, \varphi) = 0$ for all $\varphi \in C^\infty_0(B)$, and where $a(u, v) = \int_B a_{ij}(x)\partial_i u(x)\partial_j v(x)\,dx$. We denote by $\Gamma(x)$ the right circular non-tangential cone with fixed aperture having vertex at $x \in \partial B$. Then, if u is a function in B, we denote its non-tangential maximal function by $u^*(x)$ where

$$u^*(x) = \sup_{z \in \Gamma(x)} |u(z)|, \quad \text{for} \quad x \in \partial B.$$

Then, by the L^p Dirichlet problem for L in B we mean the following. Let there be given a function $f \in L^p(d\sigma)$ ($d\sigma$ is the surface area measure on ∂B). Find a solution $u(z)$, $z \in B$ to $Lu \equiv 0$ in B, satisfying

$$\lim_{\substack{z \to x \\ z \in \Gamma(x)}} u(z) = f(x) \quad \text{for} \quad x \in \partial B,$$

and also the estimate

$$\|u^*\|_{L^p(d\sigma)} \leq C\|f\|_{L^p(d\sigma)}.$$

The question we ask is: Given an elliptic operator L, when does there exist a solution u to the problem above given any $f \in L^p(d\sigma)$? Here p is a fixed exponent satisfying $1 < p < \infty$. The estimate on the non-tangential maximal function is included to insure uniqueness of solutions to the problem.

Before giving a summary of the known results concerning this question, we shall discuss some general background information on the solutions to the elliptic equations we are considering. It is a celebrated fact from the classical theory that positive solutions of these equations satisfy the Harnack principle (Moser [25]). This means that if $K \subseteq B$ is a compact set, and $u(z) > 0$ is a solution of $Lu = 0$ in B, then there is a constant C depending only on K and λ such that

$$\sup_{z \in K} u(z) \le C \inf_{z \in K} u(z).$$

Another beautiful result from the classical theory of these equations is that the classical Dirichlet problem is always solvable for any elliptic L (Littman-Stampacchia-Weinberger [24]):

Given any continuous function $f(x)$ on ∂B, there exists a unique function $u(z)$ continuous on $\overline{B}$ so that

$$Lu \equiv 0 \text{ in } B \text{ and } u(x) = f(x) \qquad \text{for all} \quad x \in \partial B.$$

There is also the maximum principle for solutions to $Lu = 0$ in B: If u is continuous on $\overline{B}$ and satisfies $Lu = 0$ in B then

$$\max_{z \in \overline{B}} u(z) = \max_{x \in \partial B} u(x).$$

By the previous results it is possible to define the harmonic measure ω_L^z associated to the operator L with respect to the point z in a manner exactly analogous to the familiar case of the Laplace operator: Suppose u_f, $f \in C(\partial B)$ denotes the unique solution to $Lu = 0$ in B with $u_f(x) = f(x)$, $x \in \partial B$. Then by the maximum principle, the map $f \to u_f(z)$ is a positive linear functional on $C(\partial B)$. Hence, according to the Riesz Representation Theorem, there exists a unique probability measure ω_L^z so that

$$u_f(z) = \int_{\partial B} f \, d\omega_L^z, \quad f \in C(\partial B).$$

This measure $d\omega_L^z$ is the harmonic measure. It is a trivial consequence of the Harnack principle that if $z_1, z_2 \in B$, then the ratio $\omega_L^{z_1}(E)/\omega_L^{z_2}(E)$ is bounded above and below by a pair of finite, strictly positive constants for all Borel sets $E \subseteq \partial B$. Thus, the exact choice of $z \in B$ makes virtually no difference when considering the properties of harmonic measure. In fact, if $z = 0$ we will write $d\omega_L^0$ often simply as $d\omega_L$ (or even $d\omega$ if L is understood). Now the nature of the harmonic measure $d\omega_L$ is of critical importance. To be more exact about this, we recall the classical weights from the theory of singular integrals and maximal functions.

We say that a measure μ on ∂B is in the class $B^p(d\sigma)$ if and only if $\mu = k\,d\sigma$ where

$$\left(\frac{1}{\sigma(\Delta)} \int_\Delta k^p\,d\sigma \right)^{1/p} \leq C \left(\frac{1}{\sigma(\Delta)} \int_\Delta k\,d\sigma \right)$$

for all surface balls $\Delta \subseteq \partial B$.

A measure $\mu = k\,d\sigma$ is said to belong to the Muckenhoupt A^p class if and only if

$$\left(\frac{1}{\sigma(\Delta)} \int_\Delta k\,d\sigma \right) \left(\frac{1}{\sigma(\Delta)} \int_\Delta k^{-\frac{1}{p-1}}\,d\sigma \right)^{p-1} \leq C$$

for all surface balls $\Delta \subseteq \partial B$.

Then the key point for our purposes here is the following:

> The $L^p(d\sigma)$ Dirichlet problem is solvable for L if and only if the harmonic measure $d\omega_L$ belongs to $B^{p'}(d\sigma)$ where $\frac{1}{p} + \frac{1}{p'} = 1$.

Unfortunately, it is not true that $d\omega_L$ is always in some class $B^p(d\sigma)$ for an arbitrary elliptic operator L. In fact in Caffarelli, Fabes, and Kenig [1] an example of an elliptic operator L is given whose harmonic measure $d\omega_L$ is singular with respect to $d\sigma$. Thus, the nature of the solvability of the L^p Dirichlet problem depends on the operator we are dealing with. Considerable progress has been made on this issue. In what follows, we shall give a quick overview of some of the highlights of the theory.

One of the motivations for considering the class of elliptic operators with bounded measurable coefficients is to achieve a better understanding of the harmonic functions on Lipschitz domains.

One strategy for solving the Dirichlet problem for the Laplacian on a Lipschitz domain is to change variables in such a way that a harmonic function on the Lipschitz domain is pulled back via the change of variables to a solution, in the upper half space, of an elliptic operator with bounded measurable coefficients. (The coefficients of this operator come by taking the derivative of the defining function of the Lipschitz domain; hence, they are merely in L^∞.) The coefficients of the operator gotten in this way are independent of the x_n coordinate in $\mathbb{R}^n_+$. It is a well-known result of B. Dahlberg [10] that the L^p Dirichlet problem is solvable for the Laplacian in a Lipschitz domain when $p \geq 2$. Corresponding to this (stated for the unit ball rather than the upper half space) is a basic positive result for the theory we are now considering:

Theorem of Jerison and Kenig [21]. *Suppose* $Lu = \text{div}(A\nabla u)$, *is an elliptic operator whose coefficients* $A(x)$ *are bounded measurable functions in* $B \subseteq \mathbb{R}^n$. *Suppose further that* $A(x)$ *is independent of the radial variable, i.e.,* $A(x) = A(y)$ *whenever* $x/|x| = y/|y|$. *Then the harmonic measure* $d\omega_L \in B^2(d\sigma)$.

Thus, if the coefficients of the operator are "radially independent" then the L^p Dirichlet problem is solvable when $p \geq 2$, and this can be thought of as an extension of Dahlberg's result on Lipschitz domains.

A bit later, E. Fabes, D. Jerison, and C. Kenig [13] proved the following result: Suppose $Lu = \text{div}(A\nabla u)$ is elliptic, and that $A(x)$ is continuous on $\overline{B}$. Let $\omega(\delta) = \sup_{\substack{|x-y| \leq \delta \\ x,y \in \overline{B}}} |A(x) - A(y)|$ denote the modulus of continuity of A on $\overline{B}$. Assume that

$$\int_0^1 \frac{\omega(\delta)^2}{\delta} \, d\delta < \infty.$$

Then $\omega_L \in B^p(d\sigma)$ for every $p < \infty$.

This result holds only for continuous coefficients, and, in particular, does not cover the operators which arise via a change of variables from the Laplacian on Lipschitz domains. In order to remedy this, B. Dahlberg proved a perturbation theorem which applied equally well to the case of discontinuous coefficients. In order to state his result, we recall a definition. Suppose that μ is a positive measure on B. Then we say that μ is a Carleson measure provided, for any $x \in \partial B$ and $r > 0$,

$$\frac{\mu(B(x;r) \cap B)}{r^{n-1}} \leq C.$$

We say that μ is a Carleson measure of vanishing trace provided

$$\frac{\mu(B(x;r) \cap B)}{r^{n-1}} \to 0 \text{ uniformly in } x \text{ as } r \to 0.$$

Theorem of Dahlberg [11]. *Suppose* L_0 *and* L_1 *are elliptic operators with coefficients* $A_i(x)$ *bounded and measurable in* B. *For each* $z \in B$, *set* $a(z) = \sup_{y \in B(z; \frac{1-|z|}{2})} |A_0(y) - A_1(y)|$. *Suppose that the measure* $a^2(z) \frac{dz}{\delta(z)}$ *is a Carleson measure of vanishing trace (here* $\delta(z) = 1 - |z|$). *Then* $\omega_{L_0} \in B^p(d\sigma)$ *implies that* $\omega_{L_1} \in B^p(d\sigma)$ *for any* $1 < p < \infty$.

In Dahlberg's theorem, we have an easily-checked criterion that tells us that the coefficients of two operators are close enough near the boundary so that the solvability of the L^p Dirichlet problem for one of the operators guarantees its solvability for the other operator. This result depends in a crucial way on the smallness of the Carleson measure on small balls. If the vanishing trace assumption is removed, then simple counterexamples show that the conclusion of the theorem is no longer valid. Nevertheless, in the absence of the vanishing trace condition, Dahlberg made the following conjecture:

Dahlberg's Conjecture. Assume that L_0 and L_1 are elliptic in B with bounded measurable coefficients A_0 and A_1, respectively. As above, set

$$a(z) = \sup_{y \in B(z; \frac{\delta(z)}{2})} |A_0(y) - A_1(y)|$$

and assume only that $a^2 \frac{dz}{\delta}$ is a Carleson measure. Then if $\omega_{L_0} \in B^{p_0}(d\sigma)$ (for some $p_0 > 1$) then it follows that $\omega_{L_1} \in B^{p_1}(d\sigma)$ for some $p_1 > 1$.

This says that when we no longer assume smallness of the relevant Carleson measure, then what is preserved is the solvability of the L^p Dirichlet problem in some range of p not necessarily the same for both operators.

In what follows we wish to sketch the proof of this conjecture, which is given in [18] (R. Fefferman, C. Kenig, and J. Pipher).

Let us begin by mentioning the method of proof of Dahlberg in his vanishing trace theorem.

Assume that L_0 and L_1 are elliptic operators, and that $a^2 \frac{dz}{\delta}$ is a Carleson measure of vanishing trace. Dahlberg introduces the one-parameter family $L_t = (1-t)L_0 + tL_1$, and sets $Q(t)$ equal to the $B^p(d\sigma)$ norm of the harmonic measure ω_t of L_t. Then the argument proceeds by showing that for some large N we have $|Q'(t)| \leq CQ(t)^N$ where $C \to 0$ as the Carleson norm of the measure $a^2 \frac{dz}{\delta} \to 0$. Since we are assuming that $a^2 \frac{dz}{\delta}$ has vanishing trace, we may essentially assume that C in the differential inequality above is as small as we like; hence, $Q(1)$ is controlled by $Q(0)$, and the proof will be finished. What remains is the proof of the differential inequality, or more precisely, something essentially equivalent to it. Fix $t \in [0, 1]$, and consider the $B^p(d\sigma)$ constant of $\omega_t = k_t d\sigma$, namely $Q(t)$. By definition, there exists a surface ball on ∂B, Δ, so that

$$Q(t) \leq 2 \left(\frac{1}{\sigma(\Delta)} \int_\Delta k_t^p \, d\sigma \right)^{1/p} \Big/ \frac{1}{\sigma(\Delta)} \int_\Delta k_t d\sigma.$$

Suppose that Δ has radius r, and has center $x_0 \in \partial B$. Take $A = B \cap B(x_0; 2r)$. Then it is a simple consequence of the so-called "comparison theorem" (see Caffarelli, Fabes, Mortola and Salsa [2]) that if we restrict all our operators to the region A, then the resulting harmonic measure is equivalent on Δ to that of the operator considered on all of B. (Equivalent in this context means that the ratio of the two measures applied to any Borel set in Δ is bounded above and below by strictly positive constants.) We shall abuse notation by henceforth denoting all the harmonic measures of L_t on A by ω_t. Then this rescales everything so that

$$c \le \omega_t(\Delta) \le 1,$$

where c depends only on the ellipticity constant of L_t (see [2]). This means that to estimate $Q(t)$ it suffices to estimate (using duality)

$$\int_\Delta f k_t \, d\sigma$$

under the assumption that $\|f\|_{L^{p'}(\frac{d\sigma}{\sigma(\Delta)})} = 1$ and f is a non-negative function on ∂A supported on Δ.

The next step in the proof is that if $\varphi \ge 0$ is a smooth bump function with compact support in A, which is suitably normalized, then by Harnack's principle,

$$\int_\Delta f k_t \, d\sigma \le C_0 \int_A u_t \varphi \, dz$$

where $L_t u_t = 0$ and $u_t|_{\partial A} = f$. It is therefore enough to estimate $\int_A u_t \varphi \, dz$, and it is actually this quantity whose t derivative we shall bound.

In fact, letting a dot stand for differentiation in t, we see that

$$\int_A L_t(u_t) h_t \, dz \equiv 0 \quad \forall t \text{ where } L_t h_t = \varphi, \ h_t|_{\partial A} = 0,$$

so that

$$0 = \int_A L_t(u_t) h_t dz = \int_A \dot{L}_t \varepsilon(u_t) h_t dz + \int_A L_t(\dot{u}_t) h_t dz + \int_A L_t(u_t) \dot{h}_t \, dz.$$

Now the last of these three terms is 0, since $L_t u_t = 0$, and integration by parts in the second term transforms it into

$$\int_A u_t \dot{\varphi} \, dz.$$

Integrating by parts in the first term gives the identity

$$\int_A u_t \dot{\varphi} \, dz = \int_A \varepsilon_{ij}(z) \partial_i u_t(z) \partial_j h_t(z) \, dz$$

where $(\varepsilon_{ij}) = \varepsilon$ and $\varepsilon = A_1 - A_0$ We must then prove that if $\varepsilon(z) = \max_{i,j} |\varepsilon_{ij}(z)|$, then

$$\int_A \varepsilon |\nabla u_t| \, |\nabla h_t| dz \le C Q(t)^N.$$

It is quite easy to show that

$$\int_A \varepsilon |\nabla u_t| \, |\nabla h_t| \, dz \le C \int_{x \in \Delta} S_t(f)(x) H(k_t) \, d\sigma$$

where S_t denotes the Lusin area integral, i.e.,

$$S_t^2(f)(x) = \int_{\Gamma(x)} |\nabla u_t|^2 \delta^{2-n} dz$$

and where H is a classical Calderón-Zygmund singular integral. Now to finish the proof, Dahlberg uses Hölder's inequality to arrive at

$$\left| \int_A u_t \dot{\varphi}\, dz \right| \leq C \|S_t f\|_{L^{p'}(d\sigma)} \|H(k_t)\|_{L^p(d\sigma, \Delta)}.$$

By the boundedness of singular integrals on $L^p(d\sigma)$, we have

$$\|H(k_t)\|_{L^p(d\sigma, \Delta)} \leq C \|k_t\|_{L^p(d\sigma, \Delta)} \leq \sigma(\Delta)^{1/p}.$$

As for $\|S_t f\|_{L^{p'}(d\sigma)}$, according to a theorem of Dahlberg, Jerison, and Kenig [12] the area integral S_t is bounded on all $L^p(d\omega_t)$ for $1 < p < \infty$ for all elliptic operators. We need to estimate here the quantity $\|S_t f\|_{L^{p'}(d\sigma)}$, not $\|S_t f\|_{L^p(d\omega_t)}$. However, by a good λ argument one can show [12] that the operator S_t is also bounded on all $L^p(d\mu)$, where μ is any measure satisfying an A^∞ condition with respect to ω_t. Thus $\|S_t f\|_{L^{p'}(d\sigma)} \leq C \|f\|_{L^{p'}(d\sigma)}$ where the constant C is dominated by a constant depending only on p times the norm $\|d\sigma\|_{A^\infty(d\omega_t)}$ and this last quantity turns out to be less than $C Q(t)^N$ for a large power N. This competes the proof of the differential inequality and Dahlberg's theorem.

Now, let us pass to the proof of the conjecture, where we only assume that the measure $a^2 \frac{dz}{\delta}$ is a Carleson measure in B, with no smallness assumption. We assume that the L^p Dirichlet problem is solvable for the operator L_0 when $p = p_0$, and then seek to show that the L^p Dirichlet problem can be solved for L_1 when $p = p_1$, and where p_1 might be much larger than p_0. In terms of the harmonic measure ω_i, $i = 0, 1$ of the operators L_i, this means $\omega_0 \in B^{p_0}(d\sigma)$ implies $\omega_1 \in B^{p_1}(d\sigma)$ for some $1 < p_0, p_1 < \infty$. It will be important to think of this in terms of the so-called A^∞ class. This class of weights can be defined in a great many different ways (see Coifman-C. Fefferman [5]). Here we recall that

$$A^\infty(d\sigma) = \bigcup_{p>1} B^p(d\sigma),$$

so our conjecture assumes that $\omega_0 \in A^\infty(d\sigma)$, and we now show that $\omega_1 \in A^\infty(d\sigma)$ which proves it.

To do this we must first single out an important special case: Assume rather than $a^2 \frac{dz}{\delta}$ a Carleson measure, the stronger condition

(A) $$\int_0^1 a^2((1-t)x) \frac{dt}{t} \leq C \qquad \text{for all} \quad x \in \partial B.$$

Then we show under assumption (A) that $\omega_1 \in A^\infty(d\sigma)$ (see R. Fefferman [17]). To do this we use a novel characterization of $A^\infty(d\sigma)$: Observe that $\omega = k\,d\sigma \in A^\infty(d\sigma)$ if and only if it satisfies

$$\|k\|_{L\log^+ L(\Delta;\,\frac{d\sigma}{\sigma(\Delta)})} \le C\|k\|_{L^1(\Delta;\,\frac{d\sigma}{\sigma(\Delta)})}$$

and we let $L_t = (1-t)L_0 + tL_1$ as before, but here we set

$$Q(t) = \sup_\Delta \frac{\|k_t\|_{L\log^+ L(\Delta;\,\frac{d\sigma}{\sigma(\Delta)})}}{\|k_t\|_{L^1(\Delta;\,\frac{d\sigma}{\sigma(\Delta)})}} \qquad \text{for} \quad t \in [0,1].$$

The idea is then to show that because $Q(t)$ is now defined in terms of an Orlicz class near L^1, we actually have

$$|\dot{Q}(t)| \le CQ(t)$$

and this controls $Q(1)$ in terms of $Q(0)$ without any smallness of C. We do this as follows: We show that the quantity $Q(t)$ can be bounded by $\int_\Delta fk_t\,d\sigma$ for some choice of Δ and a non-negative f supported in Δ satisfying $\|f\|_{\mathrm{BMO}} + \|f\|_{L^1(\frac{d\sigma}{\sigma(\Delta)})} \le 1$. Then an analysis similar to Dahlberg's shows that

$$|\dot{Q}(t)| \le C\int_A \varepsilon(z)|\nabla u_t(z)|\,|\nabla h_t(z)|\,dz$$

where

$$L_t u_t = 0 \text{ in } A, \qquad u_t\big|_{\partial A} = f \text{ and}$$

$$L_t h_t = \varphi \text{ in } A, \qquad u_t\big|_{\partial A} = 0.$$

We estimate $\int_A \varepsilon|\nabla u_t|\,|\nabla h_t|dz$ by breaking this up into two parts $A = A_1 \cup A_2$ where we write $\Delta = B(x_0;r) \cap \partial B$ for some $x_0 \in \partial B$ and $r > 0$, and then define A_1 by $A_1 = B(x_0; 3/2r) \cap B$ and $A_2 = A - A_1$. The main estimate is the term $\int_{A_1} \varepsilon|\nabla u_t|\,|\nabla h_t|dz$ which we carry out below.

$$\int_{A_1} \varepsilon|\nabla u_t|\,|\nabla h_t|\,dz$$

$$\le C_\alpha \int_{x\in\widetilde{\Delta}} \left(\int_{\Gamma_\alpha(x)} \delta^{1-n}(z)\varepsilon(z)|\nabla u_t(z)|\,|\nabla h_t(z)|\,dz \right) d\sigma(x).$$

Now we estimate

$$\int_{\Gamma_\alpha(x)} \delta^{1-n}(z)\varepsilon(z)|\nabla u_t(z)|\,|\nabla h_t(z)|\,dz$$

$$\le \int_0^1 \int_{B((1-\rho)x;\alpha\rho)} \delta^{1-n}(z)\varepsilon(z)|\nabla u_t(z)|\,|\nabla h_t(z)|\,dz\,\frac{d\rho}{\rho}$$

$$\le C \int_0^1 \rho^{1-n} a_\alpha((1 - \rho)x) \left(\int_{B((1-\rho)x;\alpha\rho)} |\nabla u_t(z)|^2 \, dz \right)^{1/2}$$

$$\left(\int_{B((1-\rho)x;\alpha\rho)} |\nabla h_t(z)|^2 \, dz \right)^{1/2} \frac{d\rho}{\rho}$$

(here $a_\alpha(z) = \sup_{y \in B(z;\alpha\delta(z))} |\varepsilon(z)|$)

$$\le C \int_0^1 \rho^{1-n} a_\alpha((1 - \rho)x) \left(\int_{B((1-\rho)x;\alpha\rho)} |\nabla u_t(z)|^2 \, dz \right)^{1/2}$$

(∗)
$$\rho^{-1} \rho^{n/2} \rho^{2-n} \omega_t(\Delta_{x,\rho}) \frac{d\rho}{\rho}$$

where $\Delta_{x,\rho}$ is the surface ball centered at x of radius ρ, and we have used Cacciopoli's inequality as well as the estimate

$$h_t((1 - \rho)x; \alpha\rho) \le C_\alpha \rho^{2-n} \omega_t(\Delta_{x,\rho}).$$

Then

$$(\ast) \le \int_0^1 a_\alpha((1 - \rho)x) \left(\int_{B((1-\rho)x;\alpha\rho)} |\nabla u_t(z)|^2 \delta^{2-n} \, dz \right)^{1/2} \rho^{1-n} \omega_t(\Delta_{x,\rho}) \frac{d\rho}{\rho}.$$

Thus

$$\int_{A_1} \varepsilon |\nabla u_t(z)| \, |\nabla h_t(z)| \, dz \le \iint_{A_1} a_\alpha(z) S_z \omega_t(\Delta_z) \frac{dz}{\delta^n}$$

where $S_z = \left(\iint_{B(z;\alpha\delta(z))} |\nabla u_t(y)|^2 \delta^{2-n}(y) \, dy \right)^{1/2}$ and $\Delta_z = \Delta(\frac{z}{|z|}; \delta(z))$. The last integral above is easily seen, by Fubini's theorem and the Cauchy-Schwarz inequality, to be bounded by

$$C \int_{\partial A} S(f)(x) d\omega_t(x).$$

Therefore our claim is reduced to showing that

$$\int_{\partial A} S(f)(x) d\omega_t(z) \le C Q(t).$$

We proceed as follows:

$$\int_{\partial A} S(f)(x) d\omega_t(x) \le \left[\int_{\partial A} S^2(f) d\omega_t(z) \right]^{1/2} \le C \left(\int_{\partial A} f^2 \, d\omega_t \right)^{1/2}.$$

Let $f_{\Delta,\omega_t} = \frac{1}{\omega_t(\Delta)} \int_\Delta f \, d\omega_t$. Then

$$\left(\int_{\partial A} f^2 \, d\omega_t \right)^{1/2} \leq \left(\int_\Delta |f - f_{\Delta,\omega_t}|^2 \, d\omega_t \right)^{1/2} + \left(\int_\Delta f_{\Delta,\omega}^2 \, d\omega_t \right)^{1/2}$$

$$\leq C \left[\left(\frac{1}{\omega_t(\Delta)} \int_\Delta |f - f_{\Delta,\omega_t}|^2 \, d\omega_t \right)^{1/2} + |f_{\Delta,\omega_t}| \right].$$

We observe that

$$f_{\Delta,\omega_t} = \frac{1}{\omega_t(\Delta)} \int_\Delta f k_t \, d\sigma \leq \frac{1}{\omega_t(\Delta)} \left(\int_\Delta |f - f_\Delta| k_t \, d\sigma + f_\Delta \right)$$

where $f_\Delta = \frac{1}{\sigma(\Delta)} \int_\Delta f \, d\sigma \leq 1$. Also

$$\frac{1}{\omega_t(\Delta)} \int_\Delta |f - f_\Delta| k_t \, d\sigma = \frac{1}{\left[\frac{\omega_t(\Delta)}{\sigma(\Delta)} \right]} \frac{1}{\sigma(\Delta)} \int_\Delta |f - f_\Delta| k_t \, d\sigma$$

$$\leq C \|f\|_{\text{BMO}(d\sigma)} \frac{\|k_t\|_{L \log^+ L(d\sigma/\sigma(\Delta))}}{\int_\Delta k_t \, d\sigma/\sigma(\Delta)}$$

$$\leq C Q(t).$$

The term

$$\left(\frac{1}{\omega_t(\Delta)} \int_\Delta |f - f_{\Delta,\omega_t}|^2 \, d\omega_t \right)^{1/2}$$

is dominated by a constant depending only on the doubling constant of ω_t (which, by [2] depends only on the ellipticity constant of L_t) times the quantity $\|f\|_{\text{BMO}(d\omega_t)}$. Now, we claim $\|f\|_{\text{BMO}(d\omega_t)}$ is bounded by $C Q(t)$. To see this we estimate: For $\Delta_0 \subseteq \Delta$

$$\frac{1}{\omega_t(\Delta_0)} \int_{\Delta_0} |f - f_{\Delta_0}| d\omega_t = \frac{1}{\omega_t(\Delta_0)} \sigma(\Delta_0) \frac{1}{\sigma(\Delta_0)} \int_{\Delta_0} |f - f_{\Delta_0}| k_t \, d\sigma$$

$$\leq C \|f\|_{\text{BMO}(d\sigma)} \frac{\|k_t\|_{L \log^+ L(d\sigma/\sigma(\Delta_0))}}{\left[\frac{\omega_t(\Delta_0)}{\sigma(\Delta_0)} \right]}$$

$$\leq C Q(t).$$

This completes the proof of the conjecture under the more stringent hypothesis that $\int_0^1 a^2((1 - t)x) \frac{dt}{t} \in L^\infty(d\sigma(x))$. In order to reduce the general case where $a^2 \frac{dz}{\delta}$ is a Carleson measure to this one we proceed as follows:

Choose a surface ball $\Delta = \Delta(x_0; r) \subseteq \partial B$ and a subset $F \subseteq \Delta$ with $\frac{\sigma(F)}{\sigma(\Delta)} > 1 - \frac{1}{10^{10}}$. We will show that $\omega_1 \in A^\infty(d\sigma)$ by proving that $\frac{\omega_1(F)}{\omega_1(\Delta)} > \varepsilon$ for some $\varepsilon > 0$.

Observe that

$$\frac{1}{\sigma(\Delta)} \int_{x \in \Delta} \left(\int_0^r a^2((1-t)x) \frac{dt}{t} \right) d\sigma(x) \leq C$$

by the Carleson condition we are assuming. This means that on a set E such that $\sigma(E) \geq \frac{1}{2}\sigma(\Delta)$ we have

$$\int_0^r a^2((1-t)x) \frac{dt}{t} \leq C' = 2C \qquad \text{for each} \quad x \in E.$$

Define $\Gamma_\alpha^r(x)$ to be the right circular cone with vertex at $x \in \partial B$, aperture α (suitably small) and truncated at height r. Set $\mathcal{R} = \bigcup_{x \in E} \Gamma_\alpha^r(x)$. Then define the intermediate operator $Mu = \text{div}(A\nabla u)$ by

$$A(z) = \begin{cases} A_0(z) & \text{if } z \notin \mathcal{R} \\ A_1(z) & \text{if } z \in \mathcal{R} \end{cases}.$$

A simple geometric argument shows that if

$$\widetilde{\varepsilon}(z) = |A(z) - A_0(z)|, \quad z \in B \text{ and if}$$

$$\widetilde{a}(z) = \sup_{y \in B(z;\alpha\delta(z))} \widetilde{\varepsilon}(y)$$

then for α sufficiently small, we have

$$\int_0^1 \widetilde{a}^2((1-t)x) \frac{dt}{t} \leq C' \qquad \text{for all} \quad x \in \partial B.$$

Thus, by the special case treated above, the harmonic measure of the operator M, $\omega_M \in A^\infty(d\sigma)$. Now the main question is "What is the relationship between ω_M and ω_{L_1}?" This is answered by a lemma from [12].

Lemma. *Suppose that $\Delta \subseteq \partial B$ is a surface ball of radius r. Let $\mathcal{R} = \bigcup_{x \in S} \Gamma_\alpha^r(x)$ for some set $S \subseteq \Delta$. Then there exists a constant $C > 0$ and $\theta > 0$ so that for all subsets $E \subseteq S$ and operators M and L_1 whose coefficients agree on $\mathcal{R}$ we have*

$$\frac{\omega_{L_1}(E)}{\omega_{L_1}(\Delta)} \geq C \left[\frac{\omega_M(E)}{\omega_M(\Delta)} \right]^\theta.$$

Applying this lemma in our context gives

$$\frac{\omega_1(F)}{\omega_1(\Delta)} \geq \frac{\omega_1(F \cap E)}{\omega_1(\Delta)} \geq C \left[\frac{\omega_M(F \cap E)}{\omega_M(\Delta)} \right]^\theta \geq C\eta^\theta$$

if $\eta > 0$ is so small that by the A^∞ property of ω_M,

$$\frac{\sigma(H)}{\sigma(\Delta)} > \frac{1}{4} \text{ implies } \frac{\omega_M(H)}{\omega_M(\Delta)} > \eta \qquad \text{for all} \quad H \subseteq \Delta.$$

Thus, under the assumption that $\frac{\sigma(F)}{\sigma(\Delta)} > 1 - \frac{1}{10^{10}}$, we have proven that $\frac{\omega_1(F)}{\omega_1(\Delta)} > C\eta^\theta$ and this shows that $\omega_1 \in A^\infty(d\sigma)$ completing the proof of Dahlberg's conjecture.

Finally, we would like to discuss a beautiful connection between the Dahlberg Conjecture and the Helson-Szegö Theorem from harmonic analysis.

It is a simple matter to construct, given any doubling measure $\mu = f\,dx$ on $\mathbb{R}^1$, an elliptic operator L on $\mathbb{R}^2_+$, whose harmonic measure ω_L is μ. The difference (using the notation discussed above) between the coefficients of this L and the identity matrix $a(x, t)$ for $(x, t) \in \mathbb{R}^2_+$ is essentially given by

$$a(x, t) = \frac{|f * \psi_t(x)|}{f * \varphi_t(x)}.$$

Here $\psi, \varphi \in C_0^\infty(\mathbb{R}^1)$, with $\int \psi\,dx = 0$ and $\varphi \geq 0$ with $\int \varphi = 1$. It is a special case of Dahlberg's Conjecture that if

$$\frac{|f * \psi_t(x)|^2}{(f * \varphi_t(x))^2} \frac{dxdt}{t}$$

is a Carleson measure in $\mathbb{R}^2_+$, then $f \in A^\infty(dx)$. Actually this condition gives a complete characterization of $A^\infty(\mathbb{R}^n)$ for any dimension n.

Theorem [18]. *Let $\mu = f\,dx$ be a positive measure on $\mathbb{R}^n$ which satisfies the doubling condition $\mu(B(x; 2r)) \leq C\mu(B(x; r))$ for all $x \in \mathbb{R}^n$, and $r > 0$. Then $f \in A^\infty(\mathbb{R}^n)$ if and only if*

$$\left| \frac{f * \psi_t(x)}{f * \varphi_t(x)} \right|^2 \frac{dxdt}{t} \text{ is a Carleson measure on } \mathbb{R}^{n+1}_+.$$

It is interesting to note that the assumption that the measure $f\,dx$ is doubling is crucial, and the theorem is incorrect if this assumption is deleted.

Now how is this related to the Helson-Szegö Theorem? The latter states:

Helson-Szegö Theorem. *A measure $\mu = f\,dx$ on $\mathbb{R}^1$ is such that the Hilbert transform is bounded on $L^2(d\mu)$ if and only if*

$$\log f = b_1 + H(b_2) \qquad where \quad b_i \in L^\infty(\mathbb{R}^1), i = 1, 2, \text{ and } \|b_2\|_\infty < \pi/2.$$

Taking into account the celebrated Hunt-Muckenhoupt-Wheeden Theorem on weighted norm inequalities for the Hilbert transform, as well as the C. Fefferman-Stein Decomposition of BMO, we see that except for the restriction $\|b_2\|_\infty < \pi/2$, the Helson-Szegö Theorem would say that $f \in A^2(\mathbb{R}^1)$ if and only if $\log f \in$ BMO($\mathbb{R}^1$). In a sense this says that $\log f \in$ BMO is almost enough to guarantee that f is a weight in A^2, and it tells us exactly the difference between these two conditions. What we have in our theorem above is an analogue of this for A^∞ weights in all dimensions. To see this, let us write that $\log f \in$ BMO using C. Fefferman's characterization of BMO in terms of Carleson measures:

$$\log f \in \text{BMO}(\mathbb{R}^1) \qquad \text{if and only if} \qquad |(\log f) * \psi_t|^2 \frac{dxdt}{t}$$

is a Carleson measure in $\mathbb{R}^2_+$. Suppose $\psi = \varphi'$. Then this last condition can be written

$$\left| t(\log f) * \frac{\partial \varphi_t}{\partial x} \right|^2 \frac{dxdt}{t} = \left| t \frac{\partial}{\partial x}[(\log f) * \varphi_t] \right|^2 \frac{dxdt}{t} \text{ is a Carleson measure.}$$

Now suppose we commute the taking of the logarithm with the process of averaging (convolving with φ_t). We get (the inequivalent) condition

$$\left| t \frac{\partial}{\partial x}[\log(f * \varphi_t)] \right|^2 \frac{dxdt}{t} = \left| \frac{f * \psi_t}{f * \varphi_t} \right|^2 \frac{dxdt}{t} \text{ is a Carleson measure.}$$

But this is exactly the condition which we proved in [18] is equivalent to $f \in A^\infty$. This tells us that our Carleson condition, shown to characterize in all dimensions the space $A^\infty(\mathbb{R}^n)$ is nothing more than a commutation away from $\log f \in$ BMO($\mathbb{R}^n$). In this way we have a version of the Helson-Szegö Theorem for the space A^∞.

Not only does the Dahlberg Conjecture imply a special case of our characterization of A^∞, but the characterization tells us that the Dahlberg conjecture is completely sharp. Thus, for instance, if we start with the Laplace operator and run through all elliptic operators whose coefficients satisfy

$$a^2 \frac{dz}{\delta} \text{ is a Carleson measure,}$$

then the class of harmonic measures which results is precisely the class A^∞. The conclusion of Dahlberg, given his assumption, is therefore absolutely sharp. Similarly, the equivalence of the A^∞ condition with the Carleson condition shows that under any assumption weaker than that made by Dahlberg, we cannot conclude that the resulting operator will have its harmonic measure in $A^\infty(d\sigma)$.

University of Chicago

REFERENCES

[1] L. Caffarelli, E. Fabes, and C. Kenig. "Completely Singular Elliptic Harmonic Measures." *Indiana Univ. Math. J.* **30** (1981).

[2] L. Caffarelli, E. Fabes, S. Mortola, and S. Salsa. "Boundary Behavior of Non-Negative Solutions of Elliptic Operators in Divergence Form." *Indiana Univ. Math. J.* **30** (1981).

[3] L. Carleson. "A Counterexample for Measures Bounded on H^p for the Bidisc." *Mittag-Leffler Report* No. 7 (1974).

[4] S. Y. Chang and R. Fefferman. "A Continuous Version of the Duality of H^1 and BMO on the Bi-Disc." *Ann. of Math.* **112** No. 2 (1980).

[5] R. Coifman and C. Fefferman. "Weighted Norm Inequalities for Maximal Functions and Singular Integrals." *Studia Math.* **51** (1974).

[6] A. Cordoba. "Maximal Functions, Covering Lemmas, and Fourier Multipliers." *Proc. Symp. Pure Math.* **35**, Part I. Amer. Math. Soc., 1978.

[7] A. Cordoba and R. Fefferman. "A Geometric Proof of the Strong Maximal Theorem." *Ann. of Math.* **102** (1975).

[8] A. Cordoba and R. Fefferman. "On the Equivalence Between the Boundedness of Certain Classes of Maximal and Multiplier Operators in Fourier Analysis." *Proc. Nat. Acad. Sci.* **74** No. 2 (1977).

[9] A. Cordoba and R. Fefferman. "On Differentiation of Integrals." *Proc. Nat. Acad. Sci.* **74** No. 6 (1977).

[10] B. Dahlberg. "On Estimates of Harmonic Measure." *Arch. Rat. Mech. Anal.* **65** (1977).

[11] B. Dahlberg. "On the Absolute Continuity of Elliptic Measures." *Amer. J. Math.* **108** (1986).

[12] B. Dahlberg, D. Jerison, and C. Kenig. "Area Integral Estimates for Elliptic Differential Operators with Non-Smooth Coefficients." *Arkiv. Math.* **22** (1984).

[13] E. Fabes, D. Jerison, and C. Kenig. "Necessary and Sufficient Conditions for Absolute Continuity of Elliptic Harmonic Measure." *Ann. of Math.* **119** (1984).

[14] C. Fefferman. "The Multiplier Problem for the Ball." *Ann. of Math.* **94** (1971).

[15] R. Fefferman. "Calderón-Zygmund Theory for Product Domains: H^p Spaces." *Proc. Nat. Acad. Sci.* **83** (1986).

[16] R. Fefferman. "A^p Weights and Singular Integrals." *Amer. J. Math.* **110** (1988).

[17] R. Fefferman. "A Criterion for the Absolute Continuity of the Harmonic Measure Associated with an Elliptic Operator." *J. Amer. Math. Soc.* **2** (1989).

[18] R. Fefferman, C. Kenig, and J. Pipher. "The Theory of Weights and the Dirichlet Problem for Elliptic Equations." *Ann. of Math.* **134** (1991).

[19] R. Fefferman and K. C. Lin. "A Sharp Marcinkiewicz Multiplier Theorem." To appear in *Ann. Fourier Inst. Grenoble*.

[20] R. Gundy and E. M. Stein. "H^p-Theory for the Polydisk." *Proc. Nat. Acad. Sci.* **76** (1979).

[21] D. Jerison and C. Kenig. "The Dirichlet Problem in Nonsmooth Domains." *Ann. of Math.* **113** (1981).

[22] B. Jessen, J. Marcinkiewicz, and A. Zygmund. "A Note on Differentiability of Multiple Integrals." *Fund. Math.* **25** (1935).

[23] J. L. Journé. "Calderón-Zygmund Operators on Product Spaces." *Rev. Mat. Iber.* **1** (1985).

[24] W. Littman, G. Stampacchia, and H. Weinberger. "Regular Points for Elliptic Equations With Discontinuous Coefficients." *Ann. Scuola Norm. Sup. Pisa* **17** (1963).

[25] J. Moser. "On Harnack's Theorem for Elliptic Differential Equations." *Comm. Pure Appl. Math.* **14** (1961).

[26] A. Nagel, E. M. Stein, and S. Wainger. "Differentiation in Lacunary Directions." *Proc. Nat. Acad. Sci.* **75** (1978).

[27] E. M. Stein. *Singular Integrals and Differentiability Properties of Functions*. Princeton University Press, 1970.

[28] J. Stromberg. "Weak Estimates on Maximal Functions with Rectangles in Certain Directions." *Arkiv. Math.* **15** (1977).

10

Function Spaces on Spaces
of Homogeneous Type

Yongsheng Han and Guido Weiss

1 INTRODUCTION

There are many topological vector spaces that arise naturally in Analysis. In the
theory of partial differential equations and in harmonic analysis these spaces con-
sist of functions, or, more generally, distributions defined on the n-dimensional
real Euclidean space $\mathbb{R}^n$. Some of the best known examples are the Lebesgue
spaces $L^p(\mathbb{R}^n)$, $1 \leq p \leq \infty$, the Lipschitz spaces Λ_α, $0 < \alpha < 1$, the
Hardy spaces H^p, and the Sobolev spaces L^p_α, $\alpha > 0$, $1 \leq p < \infty$. These
are examples of a scale of spaces that are naturally defined by a method known
as *Littlewood-Paley-Stein Theory*. Our purpose is to describe how this theory
can be extended so that this scale of spaces can be introduced in the general setting of *spaces of homogeneous type*. There are, in fact, two scales:
the *Besov spaces* and the *Triebel-Lizorkin spaces*. The former includes the
Lipschitz spaces and the latter includes the Hardy, Lebesgue, and Sobolev
spaces.

We shall not present here a detailed description of the Littlewood-Paley theory
in its various classical settings. We refer the reader to [FJW] for a presentation of
this subject that is appropriate for our purposes. Another account of this theory
that, in addition, explains E. M. Stein's considerable contribution to this subject
can be found in [CW1]. In order to explain our program, however, we do have to
present those aspects of the classical Littlewood-Paley theory associated with $\mathbb{R}^n$
that are needed for the extension of the spaces in question in the setting of spaces
of homogeneous type.

We begin with a version of the *Calderón reproducing formula*. It is clear that we can construct a function $\varphi \in S$ satisfying

(1.1)
$$\text{Supp }\hat{\varphi} \subset \{\xi: 1/2 \leq |\xi| \leq 2\},$$

$$|\hat{\varphi}(\xi)| \geq c > 0 \qquad \text{if} \quad 3/5 \leq |\xi| \leq 5/3.$$

The result that we call the Calderón reproducing formula then asserts that we can find two such functions φ and ψ in S such that

(1.2)
$$f = \sum_{k \in \mathbb{Z}} \psi_k * \tilde{\varphi}_k, *f,$$

where $\psi_k(x) = 2^{kn}\psi(2^k x)$, $\varphi_k(x) = 2^{kn}\varphi(2^k x)$ and $\tilde{\varphi}(x) = \overline{\varphi(-x)}$ (see [FJ]). The convergence of the series can be taken in several senses: for example, in L^2 and, much more generally, $S'/\mathcal{P}$ (that is, the space of tempered distributions modulo the polynomials). For each dyadic cube Q of side length $\ell(Q) = 2^{-k}$,

$$Q = Q_{kj} = \left\{ x \in \mathbb{R}^n : 2^{-k}j_i \leq x_i \leq 2^{-k}(j_i + 1), \quad i = 1, 2, \ldots, n \right\},$$

where $k \in \mathbb{Z}$ and $j = (j_1, j_2, \ldots, j_n) \in \mathbb{Z}^n$, let

$$\varphi_Q = 2^{kn/2}\varphi(2^k x - j)$$

and, similarly,

$$\psi_Q = 2^{kn/2}\psi(2^k x - j).$$

Then it is not hard to show, once (1.2) is established, that

(1.3)
$$f = \sum_{Q \in \mathcal{Q}} \langle f, \varphi_Q \rangle \psi_Q,$$

where $\langle g, h \rangle = g(\bar{h})$ when $g \in S'$ and $h \in S$ (this is the ordinary L^2 inner product when $g \in L^2$), and $\mathcal{Q}$ denotes the collection of all dyadic cubes. The relatively simple argument establishing (1.3) is an extension of one that gives us the Shannon Sampling Theorem and can be found in [FJW].

The function φ and the coefficients $\langle f, \varphi_Q \rangle$ can be used to characterize the spaces that are of interest to us. For the moment we shall restrict ourselves to the *homogeneous* versions of the Besov and Triebel-Lizorkin spaces. For $\alpha \in \mathbb{R}$, $p \neq \infty, 0 < p, q \leq \infty$, the Besov space $\dot{B}_p^{\alpha,q}$ consists of all elements f in $S'/\mathcal{P}$ such that

(1.4)
$$\|f\|_{\dot{B}_p^{\alpha,q}} = \left[\sum_{k \in \mathbb{Z}} \left(2^{k\alpha} \|\varphi_k * f\|_p \right)^q \right]^{1/q} < \infty.$$

The *Triebel-Lizorkin* space $\dot{F}_p^{\alpha,q}$ consists of all elements f in $\mathcal{S}'/\mathcal{P}$ such that

$$(1.5) \qquad \|f\|_{\dot{F}_p^{\alpha,q}} = \left\| \left[\sum_{k \in \mathbb{Z}} \left(2^{k\alpha} |\varphi_k * f| \right)^q \right]^{1/q} \right\|_p < \infty.$$

The "dot" denotes the fact that we are considering the homogeneous version of these spaces.

It is most important to show that these definitions are independent of the choice of the function $\varphi \in \mathcal{S}$ satisfying (1.1). This independence follows from the Calderón reproducing formula. To see that (1.4) defines the Besov space $\dot{B}_p^{\alpha,q}$ independently of the choice of φ, it suffices to show that there exists a constant C such that

$$(1.6) \qquad \left[\sum_{k \in \mathbb{Z}} \left(2^{k\alpha} \|\eta_k * f\|_p \right)^q \right]^{1/q} \leq C \left[\sum_{k \in \mathbb{Z}} \left(2^{k\alpha} \|\varphi_k * f\|_p \right)^q \right]^{1/q}$$

whenever η is another function satisfying (1.1). For simplicity we show this when $-1 < \alpha < 1$ and $1 < p, q < \infty$. By interchanging the roles of φ and $\tilde{\varphi}$ in (1.2), however, we can find ψ satisfying (1.1) such that

$$(1.7) \qquad \eta_k * f = \eta_k * \sum_{\ell \in \mathbb{Z}} \psi_\ell * \varphi_\ell * f.$$

Thus,

$$\|\eta_k * f\|_p \leq \sum_{\ell \in \mathbb{Z}} \|\eta_k * \psi_\ell * \varphi_\ell * f\|_p \leq \sum_{\ell \in \mathbb{Z}} \|\eta_k * \psi_\ell\|_1 \|\varphi_\ell * f\|_p.$$

An easy calculation shows that

$$(1.8) \qquad |(\eta_k * \psi_\ell)(x)| \leq C 2^{-|k-\ell|} \frac{2^{-k} \vee 2^{-\ell}}{(2^{-k} \vee 2^{-\ell} + |x|)^{n+1}},$$

where $a \vee b = \max\{a, b\}$. This estimate clearly implies

$$\|\eta_k * \psi_\ell\|_1 \leq C 2^{-|k-\ell|}$$

and, consequently, the left side of (1.6) does not exceed

$$C \left\{ \sum_{k \in \mathbb{Z}} \left(\sum_{\ell \in \mathbb{Z}} 2^{k\alpha} 2^{-|k-\ell|} \|\varphi_\ell * f\|_p \right)^q \right\}^{1/q}$$

$$= C \left\{ \sum_{k \in \mathbb{Z}} \left(\sum_{\ell \in \mathbb{Z}} \left\{ 2^{\ell\alpha} \|\varphi_\ell * f\|_p \left[2^{(k-\ell)\alpha - |k-\ell|} \right]^{1/q} \right\} \right. \right.$$

$$\left. \left. \left\{ 2^{(k-\ell)\alpha - |k-\ell|} \right\}^{1/q'} \right)^q \right\}^{1/q}$$

$$\leq C \left\{ \sum_{k \in \mathbb{Z}} \left\{ \sum_{\ell \in \mathbb{Z}} 2^{\ell \alpha q} \| \varphi_\ell * f \|_p^q 2^{(k-\ell)\alpha - |k-\ell|} \right\} \right.$$

$$\left. \left\{ \sum_{\ell \in \mathbb{Z}} 2^{(k-\ell)\alpha - |k-\ell|} \right\}^{q/q'} \right\}^{1/q}$$

$$\leq C \left\{ \sum_{\ell \in \mathbb{Z}} 2^{\ell \alpha q} \| \varphi_\ell * f \|_p^q \sum_{k \in \mathbb{Z}} 2^{(k-\ell)\alpha - |k-\ell|} \right\}^{1/q}$$

$$= C \left\{ \sum_{\ell \in \mathbb{Z}} 2^{\ell \alpha q} \| \tilde{\varphi}_\ell * f \|_p^q \right\}^{1/q},$$

where $(1/q) + (1/q') = 1$. Hölder's inequality was used to establish the first inequality, and C denotes a constant (depending on α and q) that varies appropriately.

Similarly, the finiteness of the expression (1.5) that defines the norms for the Triebel-Lizorkin spaces is independent of the choice of the test function φ satisfying (1.1). Again, this can be shown with the aid of the Calderón reproducing formula: By (1.7) and (1.8) we have the pointwise estimate

$$|\eta_k * f(x)| \leq C \sum_{\ell \in \mathbb{Z}} 2^{-|k-\ell|} M(\varphi_\ell * f)(x),$$

where M is the Hardy-Littlewood maximal function operator. Thus,

$$\left\| \left\{ \sum_{k \in \mathbb{Z}} (2^{k\alpha} |\eta_k * f|)^q \right\}^{1/q} \right\|_p \leq C \left\| \left\{ \sum_{k \in \mathbb{Z}} \left(\sum_{\ell \in \mathbb{Z}} 2^{k\alpha} 2^{-|k-\ell|} M(\varphi_\ell * f) \right)^q \right\}^{1/q} \right\|_p.$$

But, applying first Hölder's inequality and the assumption $-1 < \alpha < 1$ and then the Fefferman-Stein vector valued maximal function inequality [FS] (valid here since $1 < p, q < \infty$), we see that the last expression is dominated by

$$C \left\| \left\{ \sum_{\ell \in \mathbb{Z}} (2^{\ell \alpha} M(\varphi_\ell * f))^q \right\}^{1/q} \right\|_p \leq C \left\| \left\{ \sum_{\ell \in \mathbb{Z}} (2^{\ell \alpha} |\varphi_\ell * f|)^q \right\}^{1/q} \right\|_p.$$

The equivalence of the norms defined in terms of η and φ clearly follows from these inequalities.

Let us now return to equality (1.3). We have stated that the coefficients $s_Q = \langle f, \varphi_Q \rangle$ can be used to characterize the spaces we have introduced. Toward this end we introduce the spaces $\dot{b}_p^{\alpha,q}$ and $\dot{f}_p^{\alpha,q}$ of sequences $s = \{s_Q\}$ indexed by the dyadic cubes Q and satisfying

$$\| s \|_{\dot{b}_p^{\alpha,q}} \equiv \left\{ \sum_{k \in \mathbb{Z}} \left(\sum_{\ell(Q)=2^{-k}} \left[|Q|^{\frac{1}{p} - \frac{\alpha}{n} - \frac{1}{2}} |s_Q| \right]^p \right)^{q/p} \right\}^{1/q} < \infty$$

and

$$\|s\|_{\dot{f}_p^{\alpha,q}} \equiv \left\| \left(\sum_Q \left[|Q|^{-\frac{\alpha}{n}-\frac{1}{2}} |s_Q| \chi_Q \right]^q \right)^{1/q} \right\|_p < \infty.$$

The following theorem gives us this characterization (see [FJW]):

Theorem (1.9). *Suppose* $\alpha \in \mathbb{R}$, $0 < p, q \leq \infty$, *and the functions* φ *and* ψ *are as in* (1.1) *and* (1.2). *Given* $f \in S'/\mathcal{P}$ *let* $s_Q = \langle f, \varphi_Q \rangle$, *for each dyadic cube Q (so that equality* (1.3) *holds). Then* $f \in \dot{B}_p^{\alpha,q}$ *if and only if the sequence* $s = \{s_Q\} \in \dot{b}_p^{\alpha,q}$; *and* $f \in \dot{F}_p^{\alpha,q}$ *if and only if the sequence* $s = \{s_Q\} \in \dot{f}_p^{\alpha,q}$, *provided* $p < \infty$. *When this is the case we have the norm equivalences*

$$\|s\|_{\dot{f}_p^{\alpha,q}} \approx \|f\|_{\dot{F}_p^{\alpha,q}} \qquad and \qquad \|s\|_{\dot{b}_p^{\alpha,q}} \approx \|f\|_{\dot{B}_p^{\alpha,q}}.$$

There are several reasons why these spaces are "natural" and why Theorem (1.9) is useful. The space L^2 is rather special in many ways; for example, it is endowed with an inner product with respect to which the Fourier transform is a unitary operator, and the bounded convolution operators on it are characterized by the fact that their kernels have bounded Fourier transforms. These properties can be used to give rather simple proofs of the boundedness of many operators on L^2. While these properties are no longer true for L^p, $p \neq 2$, several of these boundedness results still hold when $1 < p < \infty$, but they fail to be true for either L^1 or L^∞. There are, however, appropriate replacements for these two spaces on which such boundedness properties are valid. In the one dimensional case, the Hilbert transform provides an example of this situation, where the Hardy space H^1 replaces L^1 and BMO replaces L^∞. The scale of spaces $\dot{F}_p^{\alpha,q}$ includes the spaces L^p for $1 < p < \infty$ ($\dot{F}_p^{0,2}$ is L^p in this case) and it also includes H^1 ($= \dot{F}_1^{0,2}$) and BMO ($= \dot{F}_\infty^{0,2}$). In addition to this the Hilbert space techniques that simplify these problems in the case $p = 2$ have somewhat successful extensions to the scale of spaces we are describing. This is a consequence of Theorem (1.9) and the "almost orthogonal" expansion (1.3). In a well defined sense the operators of interest correspond to matricial operators on the sequence spaces $\dot{b}_p^{\alpha,q}$ and $\dot{f}_p^{\alpha,q}$ that are "almost diagonal" (see [FJW] and [T] where many of these features of the Triebel-Lizorkin and Besov spaces are discussed).

As stated at the beginning, our purpose is to describe how one can define Triebel-Lizorkin and Besov spaces in the setting of spaces of homogeneous type. We are now in a position to be more specific about our goal. In this general setting we do not have an underlying group that gives us a convolution which can be used in the definitions of the norms given in (1.4) and (1.5). The lack of a convolution structure is also an obstacle for obtaining the reproducing formula (1.2). Nevertheless, the original motivation for the introduction of spaces of homogeneous type was to introduce a most general setting in which the ideas of harmonic analysis can be extended. Moreover, many important convolution operators, such as the Calderón-Zygmund Singular Integral Operators, have been extended to operators defined by more general kernels, such as the Calderón-Zygmund Operators (see [CM] and [J]). Thus, it is natural to try to extend more general operators and their properties when the underlying space is of homogeneous type. Such extensions can be found in the literature (see [DJS], for example); however, these results do not include the scale of spaces that we are considering. Very recently considerable progress has been made in this direction. In this paper we describe this effort. We shall not prove all the results we announce; however, we will give complete references of the work now in progress. This paper should be considered to be a companion to [HS], where the details that are not contained here are given. On the other hand, much of the motivation for the material developed in [HS] is presented here.

2 MOTIVATION FOR A CALDERÓN REPRODUCING FORMULA ON SPACES OF HOMOGENEOUS TYPE

A *quasi-metric* on a set X is a function $\rho\colon X \times X \to \mathbb{R}_+ = \{r \in \mathbb{R}\colon r \geq 0\}$ satisfying

(a) $\rho(x, y) = 0$ if and only if $x = y$,

(b) $\rho(x, y) = \rho(y, x)$ for all $x, y \in X$,

(c) $\rho(x, y) \leq K[\rho(x, z) + \rho(z, y)]$ for some $K < \infty$ and all x, y, and z in X.

That is, a quasi-metric satisfies the properties of a metric except that the triangle inequality is replaced by the more general condition (c) (clearly K must be at least 1 if X contains more than one point). A space of homogeneous type (which we will denote as an *HT space*) consists of a measure space X that, in addition to a measure μ, is endowed with a quasi-metric ρ; moreover, μ and ρ are related in the following way:

(i) $\mu(B(x, r)) < \infty$ for all $x \in X$ and $r > 0$, where $B(x, r) = \{y \in X\colon \rho(x, y) < r\}$,

(ii) there exists a positive constant c such that for all $x \in X$ and $r > 0$

$$\mu(B(x, 2r)) \leq c\mu(B(x, r)).$$

If we want to introduce norms that are analogous to the ones in (1.4) and (1.5), we must find appropriate substitutes for the convolution operators $f \to \varphi_k * f$. Moreover, in order to develop the theory of the spaces defined by these new norms along the lines we indicated above, we also need a version of the Calderón reproducing formula that has the same basic features as the one involved in equality (1.2), that does not depend on a convolution. We claim that this can be done in $\mathbb{R}^n$ by using certain families of Calderón-Zygmund operators. We begin by describing a version of this program that can be found in [HJTW].

In this context the operators we consider are linear and map the space of test functions $\mathcal{D}(\mathbb{R}^n)$ continuously into the space of distributions $\mathcal{D}'(\mathbb{R}^n)$. By the Schwartz kernel theorem each such operator T is associated with a distribution $K \in \mathcal{D}'(\mathbb{R}^n \times \mathbb{R}^n)$ such that

$$\langle T\theta, \eta \rangle = \langle K, \eta \otimes \theta \rangle,$$

for all $\theta, \eta \in \mathcal{D}$. K is called a *Calderón-Zygmund kernel* if its restriction to the set of all $(x, y) \in \mathbb{R}^n \times \mathbb{R}^n$ such that $x \neq y$ is a continuous function $K(x, y)$ that satisfies:

$$(2.1) \qquad |K(x, y)| \leq c \frac{1}{|x - y|^n} \text{ for all } x \neq y,$$

$$(2.2) \quad |K(x, y) - K(x, y')| \leq c \frac{|y - y'|^\epsilon}{|x - y|^{n+\epsilon}} \text{ whenever } 2|y - y'| \leq |x - y|,$$

$$(2.3) \quad |K(x, y) - K(x', y)| \leq c \frac{|x - x'|^\epsilon}{|x - y|^{n+\epsilon}} \text{ whenever } 2|x - x'| \leq |x - y|,$$

for some constant $c > 0$ and $\epsilon \in (0, 1]$. In this case T is called a *Calderón-Zygmund operator* and we write $T \in CZO$ or, if we wish to emphasize the "Lipschitz condition of order ϵ," we write $T \in CZO(\epsilon)$. If $T \in CZO, \theta \in \mathcal{D}$, and $x \notin \text{Supp} \, \theta$, then

$$T\theta(x) = \int_{\mathbb{R}^n} K(x, y)\theta(y)dy.$$

The family of operators that can be used to give a more general Calderón reproducing formula consists of such integral operators, where the above equality holds without the support restriction. We call such a family, $\{S_k\}_{k\in\mathbb{Z}}$, an *approximation to the identity*. It consists of a sequence of integral operators with kernels $S_k(x, y)$, $k \in \mathbb{Z}$, satisfying the following conditions:

$$(2.4) \qquad |S_k(x, y)| \leq c \frac{2^{-k\epsilon}}{(2^{-k} + |x - y|)^{n+\epsilon}}, \quad \text{for all } x, y \in \mathbb{R}^n;$$

$$(2.5) \quad |S_k(x, y) - S_k(x', y)| \leq c \frac{2^{-k\epsilon}}{(2^{-k} + |x - y|)^{n+\epsilon}} \left[\frac{|x - x'|}{2^{-k} + |x - y|} \right]^\epsilon$$

for $|x - x'| \leq \frac{1}{2}(2^{-k} + |x - y|)$;

$$(2.6) \quad |S_k(x, y) - S_k(x, y')| \leq c \frac{2^{-k\epsilon}}{(2^{-k} + |x - y|)^{n+\epsilon}} \left[\frac{|y - y'|}{2^{-k} + |x - y|} \right]^\epsilon$$

for $|y - y'| \leq \frac{1}{2}(2^{-k} + |x - y|)$;

(2.7) $\displaystyle\int_{\mathbb{R}^n} S_k(x, y)dy = 1$ for all $x \in \mathbb{R}^n$ and $k \in \mathbb{Z}$;

(2.8) $\displaystyle\int_{\mathbb{R}^n} S_k(x, y)dx = 1$ for all $y \in \mathbb{R}^n$ and $k \in \mathbb{Z}$.

Christ and Journé [CJ] considered families such as these that satisfied conditions (2.4) and (2.5). When the last three properties are added one can use $\{S_k\}$ to obtain norms that are equivalent to the ones given in (1.5) and (1.6). More precisely, the following result is shown in [HJTW]:

Theorem (2.9). *Suppose that $\{S_k\}, k \in \mathbb{Z}$, is an approximation to the identity and $D_k = S_k - S_{k-1}$. Then for $-\epsilon < \alpha < \epsilon$ and $1 \leq p, q \leq \infty$,*

(2.10) $$\|f\|_{\dot{B}_p^{\alpha,q}} \approx \left[\sum_{k\in\mathbb{Z}}(2^{k\alpha}\|D_k f\|_p)^q\right]^{1/q}$$

and

(2.11) $$\|f\|_{\dot{F}_p^{\alpha,q}} \approx \left\|\left[\sum_{k\in\mathbb{Z}}(2^{k\alpha}|D_k f|)^q\right]^{1/q}\right\|_p.$$

The role played by a more general Calderón reproducing formula in establishing these equivalences is the following. Suppose we want to show that

(2.12) $$\|f\|_{\dot{B}_p^{\alpha,q}} \leq c\left\{\sum_{k\in\mathbb{Z}}(2^{k\alpha}\|D_k f\|_p)^q\right\}^{1/q}.$$

We can do this by making use of an idea due to R. R. Coifman. He observed that it follows from the properties (2.4)–(2.7) that

(2.13) $$I = \sum_{k\in\mathbb{Z}} D_k.$$

From this equality (which is easily established when, say, the operators involved are applied to test functions), we obtain

(2.14) $I = \left\{\displaystyle\sum_{k\in\mathbb{Z}} D_k\right\}\left\{\displaystyle\sum_{\ell\in\mathbb{Z}} D_\ell\right\} = \displaystyle\sum_{|k-\ell|>N} D_k D_\ell + \displaystyle\sum_{|k-\ell|\leq N} D_k D_\ell \equiv R_N + T_N,$

where N is a fixed positive integer. In [DJS] it is shown that the operators T_N are bounded on $L^2(\mathbb{R}^n)$ and converge to I, as $N \to \infty$, in the strong operator topology. Moreover, it is shown in [HJTW] that

(2.15) $$\|T_N f\|_{\dot{B}_p^{\alpha,q}} \leq c\left\{\sum_{k\in\mathbb{Z}}(2^{k\alpha}\|D_k f\|_p)^q\right\}^{1/q}.$$

The operators R_N, on the other hand, belong to $CZO(\epsilon')$ for $0 < \epsilon' < \epsilon$ and their kernels $R_N(x, y)$ satisfy the following conditions: There exist $c, \delta > 0$ such that

(i) $$|R_N(x, y)| \leq c2^{-N\delta}|x - y|^{-n};$$

$$|R_N(x, y) - R_N(x', y)| \leq c2^{-N\delta}|x - x'|^{\epsilon'}|x - y|^{-(n+\epsilon')}$$

(ii) $$\text{for} \quad |x - x'| \leq \frac{1}{2}|x - y|;$$

$$|R_N(x, y) - R_N(x, y')| \leq c2^{-N\delta}|y - y'|^{\epsilon'}|x - y|^{-(n+\epsilon')}$$

(iii) $$\text{for} \quad |y - y'| \leq \frac{1}{2}|x - y|;$$

moreover, these operators satisfy the following version of the weak boundedness property

(iv) $$|\langle R_N f, g\rangle| \leq c2^{-N\delta}t^n(\|f\|_\infty + t\|\nabla f\|_\infty)(\|g\|_\infty + t\|\nabla g\|_\infty)$$

for all $f, g \in \mathcal{D}$ having support in a cube Q whose diameter is $t > 0$, as well as the David-Journé condition

(v) $$R_N 1 = R_N^* 1 = 0 \quad \text{modulo constants.}$$

We can then apply the "T1 theorem" for Besov and Triebel-Lizorkin spaces that is proved in [FTW], [T], or [HJTW] to conclude that

(2.16) $$\|R_N f\|_{\dot{B}_p^{\alpha,q}} \leq c2^{-N\delta}\|f\|_{\dot{B}_p^{\alpha,q}},$$

(2.17) $$\|R_N f\|_{\dot{F}_p^{\alpha,q}} \leq c2^{-N\delta}\|f\|_{\dot{F}_p^{\alpha,q}},$$

when $-\alpha < \epsilon < \alpha$ and $1 \leq p, q \leq \infty$. These estimates applied to $T_N^{-1} = (I - R_N)^{-1} = \sum_{m=0}^\infty R_M^m$ show that this operator is bounded on $\dot{B}_p^{\alpha,q}$ and on $\dot{F}_p^{\alpha,q}$ for these parameters α, p, and q. We thus obtain another Calderón-type reproducing formula:

(2.18) $$f = T_N^{-1}T_N f = T_N^{-1}\sum_{k\in\mathbb{Z}} D_k^N D_k f,$$

where $D_k^N = \sum_{|j|\leq N} D_{k+j}$ (so that $T_N = \sum_{|k-\ell|\leq N} D_\ell D_k = \sum_{k\in\mathbb{Z}} D_k^N D_k$). We can now give a simple proof of (2.12): Since T_N^{-1} is bounded on $\dot{B}_p^{\alpha,q}$,

$$\|f\|_{\dot{B}_p^{\alpha,q}} = \|T_N^{-1}T_N f\|_{\dot{B}_p^{\alpha,q}} \leq c\|T_N f\|_{\dot{B}_p^{\alpha,q}} \leq c\left[\sum_{k\in\mathbb{Z}}(2^{k\alpha}\|D_k f\|_p)^q\right]^{1/q},$$

where the last inequality follows from (2.15).

The inequality in the opposite direction to the one in (2.12),

$$(2.19) \qquad \left\{ \sum_{k \in \mathbb{Z}} (2^{k\alpha} \| D_k f \|_p)^q \right\}^{1/q} \le c \| f \|_{\dot{B}_p^{\alpha,q}},$$

is easier to establish. In fact, it can be done by exactly the same argument used for proving (1.6) by replacing the operator $g \to \eta_k * g$ with the operator $g \to D_k g$. All that is needed is to observe that inequality (1.8) remains valid if $\eta_k * \psi_\ell$ is replaced by $D_k \psi_\ell$. This, then, gives us (2.10). Completely analogous estimates give us (2.11). We are now ready to extend these ideas and results to HT spaces.

3 THE CALDERÓN REPRODUCING FORMULA ON SPACES OF HOMOGENEOUS TYPE

When working on HT spaces it is often quite useful to introduce the *measure distance* $m(x, y)$, equal to the measure of the smallest ball containing x and y. Without loss of great generality we can assume that $m(x, y)$ is the quasi-metric metric $\rho(x, y)$ and that it satisfies a Lipschitz smoothness condition

$$|\rho(x, y) - \rho(x', y)| \le c\rho(x, x')^\beta [\rho(x, y) + \rho(x', y)]^{1-\beta}$$

(see [CW2] and [MS]). For example, replacing the usual Euclidean distance in $\mathbb{R}^n$ by its nth power gives us such a measure distance which does not change the topology of the underlying space. The estimates we shall present require a smoothness property of this type (though much of what we present is also true when the HT space is discrete) and their form and derivations are simpler if we are dealing with a measure distance.

There are many ways of introducing approximations to the identity in HT spaces. A construction by Coifman gives us a "continuous approximate identity"

$$P_r f(x) = \int_X p_r(x, y) f(y) d\mu(y),$$

where $p_r(x, y) = 0$ for $\rho(x, y) > cr$, $p_r(x, y) = p_r(y, x)$, $|p_r(x, y) - p_r(x, y')| \le \rho(y, y')^\beta / r^{1+\beta}$, $0 \le p_r(x, y) \le \frac{c}{r}$, and $\int p_r(x, y) d\mu(y) = 1$. It can be easily shown that $\lim_{r \to 0} P_r f = f$ in $L^2(X)$ (but we shall be interested in other types of convergence, which will be discussed in a moment). Let the "variation" of P_r, D_r, be defined by either $D_r f = (P_r - P_{r/2}) f$ or by $D_r f = r \frac{\partial}{\partial r} P_r f$. A discrete version is then obtained by letting $r = 2^{-k}$, $k \in \mathbb{Z}$. Then, as in (2.13), we have

$$(3.1) \qquad f = \sum_{k \in \mathbb{Z}} D_k f.$$

Our task, then, is to carry out the program described in the previous two sections. We want to establish an analog of the Calderón reproducing formulas described in §1 and in §2. Moreover, we have to show how this analog can be used to introduce a class of Besov and Triebel-Lizorkin spaces associated with HT spaces. Guided by the material in §2 the definition the spaces $\dot{B}_p^{\alpha,q}$ and $\dot{F}_p^{\alpha,q}$ should be, simply, that these are the spaces of all f such that

$$(3.2) \qquad \|f\|_{\dot{B}_p^{\alpha,q}} \equiv \left[\sum_{k \in \mathbb{Z}} (2^{k\alpha} \|D_k f\|_p)^q\right]^{1/q} < \infty$$

and

$$(3.3) \qquad \|f\|_{\dot{F}_p^{\alpha,q}} \equiv \left\|\left[\sum_{k \in \mathbb{Z}} (2^{k\alpha} |D_k f|)^q\right]^{1/q}\right\|_p < \infty.$$

We must, however, give meaning to these inequalities. In a general HT space we do not have a theory of distributions; thus, it is not clear what we mean by f. Moreover, we do not have a class of test functions available. Finally, and most importantly, we must show that the definition of these spaces $\dot{B}_p^{\alpha,q}$ and $\dot{F}_p^{\alpha,q}$ we have just given in terms of the norms (3.2) and (3.3) is independent of the choice of the family $\{D_k\}$. In this section we give the properties of the class of operator families $\{D_k\}$ that can be used in these definitions and describe in detail the program that we are following. As stated before, the technical details are contained in the companion paper [HS].

We begin by introducing the classes of test functions that are needed for our purposes. All this is in strong analogy with the theory of Hardy spaces associated with HT spaces that we developed in [CW2]. There the test functions were elements of Lipschitz spaces that turned out to be the duals of the Hardy spaces. In the present situation the test functions are "localized smooth molecules." We say that a function f on X is a *strong smooth molecule of type* (β, γ), *centered at* $x_0 \in X$, *and having width* $d > 0$ if and only if

(i)
$$|f(x)| \leq c \frac{d^\gamma}{(d + \rho(x, x_0))^{1+\gamma}},$$

(ii)
$$|f(x) - f(x')| \leq c \left\{\frac{\rho(x, x')}{d + \rho(x, x_0)}\right\}^\beta \frac{d^\gamma}{[d + \rho(x, x_0)]^{1+\gamma}},$$

$$\text{for} \quad \rho(x, x') \leq \frac{d + \rho(x, x_0)}{2K},$$

(iii)
$$\int_X f(x) d\mu(x) = 0.$$

Denote by $\mathfrak{M}^{(\beta,\gamma)}(x_0, d)$ the collection of all such molecules and let the "norm" of such an f be the infimum of all constants c for which (i) and (ii) hold. If we choose a fixed $x_0 \in X$, then the space $\mathfrak{M}^{(\beta,\gamma)}(x_0, d)$ is our space of test functions. The theory of Calderón-Zygmund operators has been extended to HT spaces by several authors. (For a good account of this see the monograph by Christ [Ch].) In [HS] it is shown that they are bounded on the spaces $\mathfrak{M}^{(\beta,\gamma)}$, where the indices β and γ are adapted to the degree of smoothness of the kernels of these operators.

The analog of (2.18) can be derived from the following result in [HS]:

Theorem (3.4). *Suppose* $\{S_k\}, k \in \mathbb{Z}$, *is an approximation to the identity on an HT space and* $D_k = S_k - S_{k-1}$. *Set* $T_N = \sum_{k \in \mathbb{Z}} D_k^N D_k$, *where* $D_k^N = \sum_{|j| \leq N} D_{k+j}$ *and* N *is a positive integer. Then for* N *large enough,* T_N^{-1} *is a Calderón-Zygmund operator,* $T_N^{-1}(1) = (T_N^{-1})^*(1) = 0$ *and* T_N^{-1} *has the weak boundedness property.*

In the setting of $\mathbb{R}^n$ this theorem follows from results of David and Journé [DJ] and techniques described to us by Tchamitchian that involve aspects of the theory of wavelets that are not available to us for the HT space case. The method used in [HS] parallels the approach we described in §2 (see the estimates (i), (ii), (iii), (iv), and (v) that follow inequality (2.15)). More precisely, Theorem (3.4) is obtained from the following estimates on the operators $R_N = \sum_{|\ell| > N} \sum_{k \in \mathbb{Z}} D_{\ell+k} D_k$: if $0 < \epsilon' < \epsilon$ then there exists a constant c such that for all $m = 1, 2, 3, \ldots$,

(i) $$|R_N^m(x, y)| \leq c^m 2^{-N\delta m} \rho(x, y)^{-1};$$

(ii) $$|R_N^m(x, y) - R_N^m(x', y)| \leq c^m 2^{-N\delta m} \rho(x, x')^{\epsilon'} \rho(x, y)^{-(1+\epsilon')}$$

$$\text{for} \quad \rho(x, x') \leq \frac{\rho(x, y)}{2K};$$

(iii) $$|R_N^m(x, y) - R_N^m(x, y')| \leq c^m 2^{-N\delta m} \rho(y, y')^{\epsilon'} \rho(x, y)^{-(1+\epsilon')}$$

$$\text{for} \quad \rho(y, y') \leq \frac{\rho(x, y)}{2K};$$

(iv) $$|\langle R_N^m f, g \rangle| \leq c^m 2^{-N\delta m} \mu(B(x_0, r))$$

for all f and g having support within $B(x_0, r)$, with Lipschitz norms not exceeding $r^{-\eta}$ and L^∞ norms bounded by 1;

v $$R_N^m(1) = (R_N^m)^*(1) = 0,$$

where $R_N^m(x, y)$ is the kernel of R_M^m, $\delta > 0$ and $0 < \eta < 1$.

These estimates and the fact that $T_N^{-1} = \sum_{m=0}^{\infty} R_N^m$ are the basic tools needed to prove Theorem (3.4). We can now announce a version of the Calderón reproducing formula for HT spaces:

Theorem (3.5). *Suppose* $\{S_k\}$, $k \in \mathbb{Z}$, *is an approximation to the identity on a space of homogeneous type and* $D_k = S_k - S_{k-1}$. *Then there exist families of operators* $\{\tilde{D}_k\}$ *and* $\{\tilde{\tilde{D}}_k\}$, $k \in \mathbb{Z}$, *such that*

$$f = \sum_{k \in \mathbb{Z}} \tilde{D}_k D_k f$$

and

$$f = \sum_{k \in \mathbb{Z}} D_k \tilde{\tilde{D}}_k f,$$

where these series converge in the topologies of $\mathfrak{M}^{(\beta, \gamma)}$. *Moreover, the kernels* $\tilde{D}_k(x, y)$ *of these operators satisfy*

(i) $$|\tilde{D}_k(x, y)| \le c \frac{2^{-k\epsilon'}}{(2^{-k} + \rho(x, y))^{1+\epsilon'}};$$

(ii) $$|\tilde{D}_k(x, y) - \tilde{D}_k(x', y)| \le c \left[\frac{\rho(x, x')}{2^{-k} + \rho(x, y)} \right]^{\epsilon'} \frac{2^{-k\epsilon'}}{(2^{-k} + \rho(x, y))^{1+\epsilon'}}$$

$$for \quad \rho(x, x') \le \frac{2^{-k} + \rho(x, y)}{2K};$$

(iii) $$\int_X \tilde{D}_k(x, y) d\mu(x) = 0 \quad for\ all \quad y \in X\ and\ k \in \mathbb{Z} \quad .$$

The kernels $\tilde{\tilde{D}}_k(x, y)$ *satisfy the same conditions except that in* (ii) $\tilde{D}_k(x', y)$ *is replaced by* $\tilde{D}_k(x, y')$ *and* $\rho(x, x')$ *is replaced by* $\rho(y, y')$.

These are the basic definitions and results that are needed to carry out the known program on $\mathbb{R}^n$ to the new setting of HT spaces. As mentioned above, the details are carried out in [HS]. There one can also find the atomic and molecular decompositions of these general Besov and Triebel-Lizorkin spaces. The theory of interpolation of operators acting on these spaces is also extended to this general setting. We hope that this exposition furnishes sufficient motivation to make it easier to go through the rather complicated technical details that are needed for the thorough presentation contained in [HS].

University of Windsor, Windsor, Ontario
Washington University, St. Louis

REFERENCES

[Ch] M. Christ. *Lectures on singular integral operators*. CBMS Regional Conference Series, Number 77 (1991), 1–132.

[ChJ] M. Christ and J.-L. Journé. "A boundedness criterion for generalized Calderón-Zygmund operators." *Acta Math.* 159 (1987), 51–80.

[C] R. R. Coifman. "Multiresolution analysis in non-homogeneous media." In *Wavelets, Time-Frequency Methods and Phase Space.* Combes, Grossman, and Tchamitchian, eds. Springer Verlag, 1989.

[CM] R. R. Coifman and Y. Meyer. "Au délà des opérateurs pseudo-différentiels." *Astérisque* 57 (1978), 1–185.

[CW1] R. R. Coifman and Guido Weiss. "Littlewood-Paley and Multiplier Theory, by Edwards and Gaudry." *Bull. Amer. Math. Soc.* 84, Number 2 (1978), 242–50.

[CW2] R. R. Coifman and Guido Weiss. "Extensions of Hardy spaces and their use in analysis." *Bull. Amer. Math. Soc.* 83 (1977), 569–645.

[DJS] G. David, J.-L. Journé, and S. Semmes. "Calderón-Zygmund operators, para-accretive functions and interpolation." *Rev. Mat. Iber.* 1 (1985), 1–56.

[FJ] M. Frazier and B. Jawerth. "A discrete transform and decomposition of distribution spaces." *J. Funct. Anal.* 93 (1990), 34–170.

[FJW] M. Frazier, B. Jawerth, and Guido Weiss. *Littlewood-Paley theory and the study of function spaces.* CBMS Regional Conference Series, Number 79 (1991), 1–132.

[FS] C. Fefferman and E. M. Stein. "H^p spaces of several variables." *Acta Math.* 129 (1972), 137–193.

[HJTW] Y.-S. Han, B. Jawerth, M. Taibleson, and Guido Weiss. "Littlewood-Paley theory and ϵ-families of operators." *Colloq. Math.* LX/LXI (1990), 321–59.

[HS] Y.-S. Han and E. T. Sawyer. "Littlewood-Paley theory on spaces of homogeneous type and the classical function spaces." *Memoirs of the AMS* **110** (July 1994), 1–126.

[J] J.-L. Journé. *Calderón-Zygmund operators, pseudo-differential operators and the Cauchy integral of Calderón.* Lecture Notes in Mathematics, no. 994. Springer Verlag, 1983.

[MS] R. Macías and C. Segovia. "Lipschitz functions on spaces of homogeneous type." *Adv. Math.* 33 (1979), 257–70.

[T] R. Torres. "Boundedness results for operators with singular kernels on distribution spaces." *Mem. Amer. Math. Soc.* 442 (1991), 1–172.

11

The First Nodal Set
of a Convex Domain

David Jerison*

INTRODUCTION

Our purpose in this paper is to prove that the first nodal set of a long, thin convex domain touches the boundary. Let

$$B(z, r) = \{y \in \mathbf{R}^n : |y - z| < r\}$$

$$\text{inradius}(\Omega) = \max\{r : B(z, r) \subset \Omega \text{ for some } z\}.$$

Theorem. *Let Ω be a convex, open subset of $\mathbf{R}^n$. There is a dimensional constant C such that if* $\text{diameter}(\Omega) / \text{inradius}(\Omega) \geq C$*, then the nodal set for the second Dirichlet eigenfunction for Ω touches the boundary. In other words, if $\Delta u = -\lambda u$ in Ω, $u = 0$ on $\partial\Omega$, and λ is the second eigenvalue for the Dirichlet problem in Ω, then $\Lambda = \{z \in \Omega : u(z) = 0\}$ satisfies $\overline{\Lambda} \cap \partial\Omega \neq \emptyset$. In the case $n = 2$, $\overline{\Lambda} \cap \partial\Omega$ consists of exactly two points.*

This theorem was announced in [J] in the case of planar domains. A few months later, A. Melas [M] proved the result for all smooth convex planar domains, not just ones with large eccentricity. (He also showed by a limiting argument that even if the domain is not smooth, the nodal line touches the boundary in at least one point.) Our method of proof is much more crude, but it has some advantages. First of all, it applies in higher dimensions, as we show here. Secondly, it applies

*This research was supported in part by a Presidential Young Investigator Award and National Science Foundation grants DMS-9106507 at the Massachusetts Institute of Technology and DMS-9100383 at the Institute for Advanced Study.

directly to all convex domains, with no smoothness hypothesis. Finally, it is more quantitative than the one given by Melas, in that it gives some indication as to where the nodal set is. We hope to return to this in the future.

L. E. Payne made the conjecture that the first nodal line touches the boundary for any planar domain [P]. S.-T. Yau made the conjecture for all convex planar domains [Y]. The best previous results on the first nodal line are as follows. L. E. Payne [P1] proved that the nodal line touches the boundary of a convex domain in the plane which is symmetric under a reflection. This result was extended by C.–S. Lin [L] to the case of convex domains which are symmetric with respect to a rational rotation instead of a reflection.

A basic notion underlying our approach is that eigenfunctions behave like harmonic functions. This was motivated in part by results of [KP]. Another result that motivated this work is the well-known lower bound for the first eigenvalue in terms of the inradius due to W. K. Hayman [H]. We would like to thank Carlos Kenig for pointing out the generalized maximum principle (Proposition 1). The use of this maximum principle simplified our original proof which was based on comparisons with harmonic functions using a Neumann series. We would also like to thank Charles Fefferman for pointing out an error in Remark 2 of the original manuscript.

OUTLINE OF THE PROOF

F. John's theorem [dG, p. 139] says that any convex domain is comparable to a rectangle. More precisely, there is a dimensional constant C_1 such that Ω can be rotated and translated so that

$$(1) \qquad \{z \in \mathbf{R}^n : |z_k| \leq L_k\} \subset \Omega \subset \{z \in \mathbf{R}^n : |z_k| \leq C_1 L_k\}.$$

We may assume without loss of generality that the sidelengths L_k are increasing with k. After a dilation we may assume that $L_1 = 1$. Let $n = p + q$. We will consistently split the variables into $z = (x, y) \in \mathbf{R}^p \times \mathbf{R}^q$. The theorem follows from

Claim. For any M there exists N (depending on M and n) such that if $L_p \leq M$ and $L_{p+1} \geq N$, then the nodal set for Ω touches the boundary.

We prove by induction that the claim implies the theorem. Let N_1 be the value given in the claim for $M = 1$. If $L_2 \geq N_1$, then the nodal set touches. If not, then since $L_2 < N_1$ we can choose N_2 according to the proposition corresponding to $M = N_1$. If $L_3 \geq N_2$, then the nodal set touches. If not, then we may continue in this fashion until finally the nodal set touches the boundary provided $L_n \geq N_{n-1}$.

Because N_{n-1} is a dimensional constant and because $L_n = L_n/L_1$ is comparable to the ratio of the diameter to the inradius of Ω, the theorem is proved.

We will prove the claim by contradiction. Assume that the nodal set does not touch the boundary. Denote $\Omega_- = \{z \in \Omega : u(z) < 0\}$ and $\Omega_+ = \{z \in \Omega : u(z) > 0\}$. Without loss of generality we may assume that Ω_- is the component that is relatively compact in Ω. Denote $u_-(z) = -u(z)$ for $z \in \Omega_-$ and $u_-(z) = 0$ elsewhere. Denote $u_+(z) = u(z)$ for $z \in \Omega_+$ and $u_+(z) = 0$ elsewhere. Denote $\Omega(y) = \{x \in \mathbf{R}^p : (x, y) \in \Omega\}$, $\Omega_-(y) = \{x \in \mathbf{R}^p : (x, y) \in \Omega_-\}$, and similarly for $\Omega_+(y)$. For any domain D let $\lambda_1(D)$ denote the first eigenvalue of D, and let $\lambda_2(D)$ denote the second eigenvalue. Let

$$\lambda_0 = \inf_{y \in \mathbf{R}^q} \lambda_1(\Omega(y)).$$

The proof goes as follows:

1. When N is large, $\lambda_2(\Omega)$ is very close to λ_0 (Lemma 1).
2. It follows from Step 1 that for "most" points y in a large cube Q', $\lambda_1(\Omega_-(y))$ is very close to $\lambda_1(\Omega(y))$ and λ_0 (Lemma 4).
3. It follows from Step 2 that $\Omega_-(y)$ is "most" of $\Omega(y)$, that is, the Lebesgue measure, $m(\Omega_+(y))$, of $\Omega_+(y)$ is very small (Lemma 3). In other words, Ω_+ is very thin in some places.
4. By the Hopf lemma (Lemma 5) and Carleson lemma (Proposition 3) max u_- dominates max u_+ on the thin portion of Ω_+. (See (12).)
5. $u_+(z)$ decays exponentially as z approaches the center of the thin portion of Ω_+. On the other hand, because we have assumed that Ω_- is relatively compact in Ω, $\Omega_-(y)$ is relatively compact in $\Omega(y)$. In particular, $\Omega_+(y)$ is nonempty, and, using the Hopf lemma again, max u_+ on the central part of the thin portion of Ω_+ dominates max u_- nearby. Thus we have come around in a circle, and max u_- on a certain subset is bounded by a factor times itself. The exponential decay (the factor 2^{-N^c} for some $c > 0$) dominates all of the earlier bounds (which grow at most as fast as powers of N) so that for large N the factor is less than 1, and we obtain a contradiction.

The reader may also wish to consult the paper [J] for the very similar argument in the case $n = 2$. The main differences between the two-dimensional argument and the one in higher dimensions are these. In $\mathbf{R}^2$, Step 3 is trivial because the cross-sections $\Omega(y)$ are one–dimensional. In $\mathbf{R}^n$, $n \geq 3$, Step 4 requires an extra technical lemma (Lemma 6) to show that very thin filaments of Ω_- do not prevent the control of max u_+ by max u_- via the Hopf lemma and Carleson lemma.

The rest of the paper consists of one long section giving the details of the argument for $n \geq 2$. Then in the final section we indicate the modifications needed to prove that when $n = 2$ the nodal line touches the boundary in two points.

Throughout the proof we will use the convention that the constants $c > 0$, and $C < \infty$ are either dimensional constants or depend on dimension and M and may vary from line to line. Lower case c will be used for lower bound constants and upper case C for upper bounds. The numbered constants c_1, c_2, C_1, etc., do not change from line to line, which should help the reader to follow their interdependence.

THE PROOF

Lemma 1. *There is a constant C_2 depending only on n such that the second eigenvalue $\lambda = \lambda_2(\Omega)$ satisfies*

$$\lambda_0 \le \lambda \le \lambda_0 + C_2 N^{-2/3}.$$

Proof. It is easy to see that $\lambda_1(\Omega) \ge \lambda_0$. In fact, if v is the first eigenfunction for Ω then

$$\int_\Omega |\nabla v|^2 \ge \int_{\mathbf{R}^q} \int_{\Omega(y)} |\nabla_y v(x, y)|^2 dx dy$$

$$\ge \lambda_0 \int_{\mathbf{R}^q} \int_{\Omega(y)} |v(x, y)|^2 dx dy$$

$$= \lambda_0 \int_\Omega v^2.$$

Moreover, $\lambda \ge \lambda_1(\Omega)$, so the lower bound follows.

Choose $D = \Omega(y)$ so that $\lambda_1(D) \le \lambda_0$. By (1), $(0, y') \in \Omega$ provided $|y'_k| \le N$, $k = 1, \ldots, q$. Therefore, by convexity, $((1 - N^{-2/3})x, (1 - N^{-2/3})y + N^{-2/3}y') \in \Omega$ for all $x \in D$. In other words, $D' \times Q \subset \Omega$ where $D' = (1 - N^{-2/3})D$ and $Q = (1 - N^{-2/3})y + \{y' : |y'_k| \le N^{1/3}\}$, a cube of sidelength $2N^{1/3}$. Separation of variables shows that $\lambda_2(D' \times Q) = \lambda_0(1 - N^{-2/3})^{-2} + \pi^2(q + 3)(2N^{1/3})^{-2}$. (1) also implies that at least one cross section $\Omega(y)$ contains a cube of sidelength 2 so that $\lambda_0 \le \pi^2 p/4$, a dimensional constant. Thus $\lambda_2(D' \times Q) \le \lambda_0 + C_2 N^{-2/3}$ for some dimensional constant C_2. The min-max principle implies that $\lambda \equiv \lambda_2(\Omega) \le \lambda_2(D' \times Q)$, so the lemma is proved. ∎

Let us recall the generalized maximum principle for solutions to eigenfunction equations. (See [PW] for a more detailed discussion.)

Proposition 1 (Generalized Maximum Principle). *Let h be a bounded function. Let u and w satisfy $\Delta u + hu = 0$ and $\Delta w + hw \le 0$ in a bounded domain D.*

Suppose that u and w are continuous in $\overline{D}$. Finally, suppose that u > 0 in D and w > 0 in $\overline{D}$. Then

$$\max_{\overline{D}} u/w \leq \max_{\partial D} u/w.$$

To prove this, observe that $v = u/w$ satisfies the inequality $\Delta v + 2\nabla \log w \cdot \nabla v \geq 0$. The inequality then follows from the usual maximum principle. (See [PW].)

Remark 1. Let $B = B(x, r)$, a ball of radius $r \leq 1$. The boundary problem $\Delta u + u = 0$ in B and $u = f$ on ∂B can be solved for every continuous function f and

$$u(x) = c_n(r) \frac{1}{\sigma(\partial B)} \int_{\partial B} f \, d\sigma,$$

and $c_n(r) \leq 2$. (Here, and elsewhere s will denote surface measure.)

Proof. The existence of a solution follows from the generalized maximum principle and the existence of a solution to the equation that is positive in the closed ball $\overline{B}$. Such a solution exists for $r \leq 1$, and even for r less than the first zero of an appropriate Bessel function. We will need a few facts about Bessel functions later on anyway, so let us write down the solution in detail. Denote

$$F_k(r) = \int_0^1 \cos(sr)(1 - s^2)^{k-1/2} ds.$$

The relationship of F_k with the Bessel functions is $J_k(r) = c_k r^k F_k(r)$ for some constant c_k ([SW] pp. 153–154). If $k = (n - 2)/2$, then Bessel's equation implies $(\Delta + 1)F_k(|x|) = 0$. Because $1/2 < \cos(t) \leq 1$ for $0 \leq t \leq 1$, we have $F_k(0) \leq 2F_k(r)$ for all r, $0 \leq r \leq 1$. In particular, $F_k(r) > 0$ for $0 \leq r \leq 1$. Now, we can confirm the formula for $u(x)$. By symmetry we see that $u(x)$ is some constant multiple $c_n(r)$ of the average of f over the boundary. The constant can be evaluated using the radial solution $F_k(|z - x|)$ for $k = (n - 2)/2$:

$$c_n(r) = F_k(0)/F_k(r) \leq 2. \qquad \blacksquare$$

Proposition 2 (Harnack Inequality). *There is a dimensional constant C_3 such that if $0 < r < 1/(\max h)^{1/2}$, $\Delta u + hu = 0$ and $u > 0$ in $B(x, 2r)$, then*

$$\max_{\overline{B}(x,r)} u \leq C_3 \min_{\overline{B}(x,r)} u.$$

(If $h \leq 0$, then the inequality holds for all $r > 0$.)

Proposition 2 can be proved by using the Bessel function $w(z) = F_{(n-2)/2}(|z - x|)$ and the Harnack inequality due to Serrin [Se] for the function $v = u/w$. (See [PW].)

In Lemma 3 below we will use Propositions 1 and 2 for a function h which is a constant plus the characteristic function of a ball. However in all other cases we will have $h = \lambda$, a constant. In that case there is another way to prove Propositions 1 and 2, namely, to deduce them from their counterparts for parabolic equations using the observation that if $(\Delta + \lambda)u = 0$, then the function $e^{-\lambda t}u$ satisfies the heat equation $(\partial_t - \Delta)(e^{-\lambda t}u) = 0$. (For other applications of this idea, see [KP].) The parabolic analogues of Propositions 1 and 2 are well known. (See [PW].)

Proposition 3 (Carleson lemma). *Let $\phi : \mathbf{R}^{n-1} \to \mathbf{R}$ satisfy the Lipschitz condition $|\nabla \phi| \leq M$ and $\phi(0) = 0$. Denote $D = \{x \in \mathbf{R}^n : x_n > \phi(x_1, \ldots, x_{n-1})\}$. Let $0 < r \leq 1/\sqrt{\lambda}$ and u is a positive solution to $(\Delta + \lambda)u = 0$ in $B(0, 2r) \cap D$ such that u vanishes on $B(0, 2r) \cap \partial D$. Let $z = (0, r)$. There is a constant C_4 depending only on M and n such that*

$$\max_{\overline{B}(0,r) \cap D} u \leq C_4 u(z).$$

The parabolic version of the Carleson lemma was proved by Salsa [S], and Proposition 3 is an immediate consequence, as in the case of constant h for Propositions 1 and 2.

Proposition 3 was proved by Carleson [C] and Hunt and Wheeden [HW] in the case of harmonic functions. (In that case, $\lambda = 0$, and the upper bound on r is superfluous.) There are many subsequent versions. An examination of the proofs in [C] and [CFMS] shows that the result only depends on the scale-invariant Harnack inequality (Proposition 2) and a uniform Hölder continuity for solutions at the boundary, which is obtained by a majorization of solutions by an explicit supersolution (barrier) on a cone, that can be constructed using Bessel functions. We will not carry out the proof, since another one is available. But we will need the barriers in annuli given by the next lemma.

Lemma 2. *For any $\epsilon > 0$ and $0 < \rho < 1/4$ there exists a function $w_\epsilon(r)$ defined for $\rho \leq r \leq 2\rho$ satisfying*
(a) $w_\epsilon(r) > 0$ *for $\rho \leq r \leq 2\rho$,*
(b) $(\Delta + 1)w_\epsilon(|x|) \leq 0$ *for $\rho \leq |x| \leq 2\rho$ $(x \in \mathbf{R}^n)$,*
(c) $w_\epsilon(r) = 1 + \epsilon$ *for $r = 2\rho$,*
(d) *As ϵ tends to zero, w_ϵ tends uniformly to a function w_0 such that*

$$w_0(\rho) = 0, \quad w_0(2\rho) = 1 \quad and \quad \frac{\partial w_0}{\partial r}(\rho) \leq C/\rho$$

for some dimensional constant C.

Proof. For $n \geq 3$, define

$$w_\epsilon(r) = (r/2\rho)^{1-\frac{n}{2}} \left[\frac{\log r/\rho}{\log 2} + \epsilon \right].$$

Then $\Delta w_\epsilon(|x|) = -\left(\frac{n}{2} - 1\right)^2 |x|^{-2} w_\epsilon(|x|)$ and the properties listed are easy to check. In particular, $w_0'(\rho) = 2^{n/2-1}/\rho \log 2$.

For $n = 2$, define

$$w_\epsilon(r) = \frac{F_0(r)}{F_0(2\rho)} \left[\frac{\log r/\rho}{\log 2} + \epsilon \right].$$

Then $(\Delta + 1)w_\epsilon(|x|) = F_0'(|x|)/|x|F_0(2\rho) \log 2$. This last expression is negative for $|x| < 1$ because

$$F_0'(r) = - \int_0^1 s(\sin rs)(1 - s^2)^{-1/2} ds \leq 0$$

for $0 \leq r \leq \pi$ since the integrand is positive. Once again, the other properties are easy to check. For example, because F_0 is decreasing on the interval under consideration,

$$w_0'(\rho) = \frac{F_0(\rho)}{F_0(2\rho)\rho \log 2} \leq \frac{F_0(0)}{\rho F_0(1/2) \log 2} \leq \frac{2}{\rho}. \qquad \blacksquare$$

Lemma 3. *Let D be an open convex subset of $\mathbf{R}^p$ such that $B(0, 1) \subset D \subset B(0, M)$. There exist constants c_1 and K depending only on M and p such that for any open subset V of D,*

$$\lambda_1(D \backslash \overline{V}) \geq \lambda_1(D) + c_1 m(V)^K.$$

Proof. Suppose that w is the first eigenfunction for D, with $\mu = \lambda_1(D)$. Normalize w so that $\max w = 1$. Let $B = \overline{B}(0, 1/2)$. Consider a dilation of the function w_ϵ of Lemma 2 in dimension p so that it satisfies $(\Delta + \lambda)w_\epsilon \leq 0$. This changes the bound on the derivative in part (d) by only a bounded amount because λ is bounded above by a dimensional constant. Now translate w_ϵ so that the inner circle is tangent to ∂D and contained in the complement of D. This is possible because D is convex. The generalized maximum principle for the intersection of D with the annulus implies that $w \leq \overline{w}_\epsilon$, where $\overline{w}_\epsilon$ is the translated and dilated w_ϵ. Taking the limit, we find that $w \leq \overline{w}_0$. Thus, there is a dimensional constant C such that $w(x) \leq C \, \text{dist}(x, \partial D)$. It follows that w attains its maximum, 1, at a point whose distance to ∂D is greater than $1/C$. It then follows from Harnack's inequality that there is a constant $c_2 > 0$ depending only on M such that

$$(2) \qquad \min_B w \geq c_2.$$

Now consider the first eigenfunction ψ for the operator $Lu = -\Delta u + \chi_B u$. Thus $L\psi = \mu_1 \psi$ in D and $\psi = 0$ on ∂D. Normalize ψ so that $\max \psi = 1$. Since

w_ϵ in the proof above is only required to be a supersolution, the proof applies to ψ, so that (2) holds with ψ in place of w. In particular, there is a constant $c_3 > 0$ depending only on M for which

$$\int_B \psi^2 \geq c_3 \int_D \psi^2.$$

Therefore,

$$\int_D |\nabla \psi|^2 + \int_B \psi^2 \geq (\mu + c_3) \int_D \psi^2.$$

Hence, $\mu_1 \geq (\mu + c_3)$ and we have

$$(3) \qquad \int_D |\nabla v|^2 + \int_B v^2 \geq (\mu + c_3) \int_D v^2$$

for any function v that vanishes on ∂D.

The function w is superharmonic. We can bound from below the rate of vanishing of w at the boundary of D by comparing it with the rate of vanishing of a positive harmonic function that vanishes on the boundary of a cone. The harmonic function vanishes at most as fast as a power of the distance to the vertex, and the power is bounded above in terms of a lower bound on the aperture of the cone, that is, in terms of M. So we deduce from (2) that there is K_1 depending only on M for which $w(x) \geq c \operatorname{dist}(x, \partial D)^{K_1}$. It follows that

$$m(D \cap \{x \in D : w(x) < t\}) \leq Ct^\delta,$$

for some constants C and $\delta > 0$ depending on M. Choose t by the equation $m(V) = 3Ct^\delta$. Then

$$m(V \cap \{x \in D : w(x) > t\}) \geq 2Ct^\delta.$$

Choose an open subset U of $V \cap \{x \in D : w(x) > t\}$ with a smooth boundary and such that $m(U) \geq Ct^\delta$. Since $\lambda_1(D \backslash \overline{V}) \geq \lambda_1(D \backslash \overline{U})$ we will confine our attention to $D_1 = D \backslash \overline{U}$. The boundary of D_1 consists of a smooth part, ∂U, and a convex part, ∂D. In particular, it is a Lipschitz domain (without any control on the Lipschitz constant). Thus we can solve

$$\Delta \phi = 0 \text{ in } D_1; \quad \phi = t \text{ on } \partial U, \ \phi = 0 \text{ on } \partial D.$$

Denote by v the first eigenfunction in D_1, with $v > 0$ and $\Delta v = -\nu v$ in D_1. Because D_1 is a Lipschitz domain the following calculation based on Green's formula is valid. (Here, $\partial / \partial n$ denotes the outer normal derivative of ∂D_1. Note that $\partial v / \partial n \leq 0$.)

$$(\nu - \mu) \int_{D_1} vw = \int_{D_1} v(\Delta w) - (\Delta v)w = -\int_{\partial D_1} w \frac{\partial v}{\partial n} \, d\sigma$$

$$\geq -t \int_{\partial U} \frac{\partial v}{\partial n} \, d\sigma = \int_{\partial D_1} \left(\frac{\partial \phi}{\partial n} v - \phi \frac{\partial v}{\partial n} \right) d\sigma$$

$$= \int_{D_1} (\Delta \phi) v - (\Delta v) \phi = v \int_{D_1} v\phi.$$

In order to obtain a lower bound for the last integral, consider any point $x \in B$. By Fubini's theorem there is a dilate $D^r \equiv r(D - x) + x$ of D, with respect to x as the origin, for which

$$\frac{\sigma(\partial D^r \cap U)}{\sigma(\partial D^r)} \geq \frac{m(U)}{m(D)}$$

for some $r, 0 < r < 1$. On the other hand, the A_∞ estimate for harmonic measure in Lipschitz domains due to Dahlberg [D,CF] implies that the harmonic measure ω for D^r at x satisfies

$$\omega(S) \geq c \left(\frac{\sigma(S)}{\sigma(\partial D^r)} \right)^{K_2}$$

for some constants $c > 0$ and K_2 depending on M, and for every Borel subset S of ∂D^r. But if $S = \partial D^r \cap U$, then by the maximum principle, $\phi(x) \geq t\omega(S)$. Therefore,

$$\min_B \phi \geq c_4 t^{K_3}$$

with $K_3 = 1 + K_2 \delta$. Extend v by zero outside D_1. It follows that

(4)
$$\int_{D_1} v\phi \geq c_4 t^{K_3} \int_B v.$$

Define θ by the equation

$$\int_B v = \theta \int_{D_1} v.$$

We consider two cases.

Case 1. $\theta < c$ for a constant c, depending on M, to be specified later.

Let us first verify that there are dimensional constants $A > 0$ and C_5 such that

(5)
$$\int_B f^2 \leq C_5 \epsilon \int_B |\nabla f|^2 + C_5 \epsilon^{-A} \left(\int_B |f| \right)^2$$

for all $\epsilon > 0$. Let $p > 2$. Hölder's inequality and the fact that $A_1 A_2^\alpha \leq A_1^{1+\alpha} + A_2^{1+\alpha}$ for positive numbers A_1, A_2, and $\alpha = (p - 2)/p > 0$ yields

$$\int_B g^2 \leq \left(\int_B g^p \right)^{1/(p-1)} \left(\int_B g \right)^{(p-2)/(p-1)}$$

$$\leq \epsilon \left(\int_B g^p \right)^{2/p} + \epsilon^{-p/(p-2)} \left(\int_B g \right)^2$$

for any non-negative function g. Recall the Sobolev estimate

$$\left(\int_B |f - a|^p \right)^{2/p} \leq C \int_B |\nabla f|^2$$

where $a = \frac{1}{m(B)} \int_B f$, the average value of f, and C is a dimensional constant. Applying this inequality and the previous one for $g = |f - a|$, we have

$$\int_B |f - a|^2 \leq C\epsilon \int_B |\nabla f|^2 + \epsilon^{-p/(p-2)} \left(\int_B |f - a| \right)^2$$

$$\leq C\epsilon \int_B |\nabla f|^2 + 4\epsilon^{-p/(p-2)} \left(\int_B |f| \right)^2 .$$

Therefore,

$$\int_B f^2 \leq 2 \int_B |f - a|^2 + a^2$$

$$\leq 2C\epsilon \int_B |\nabla f|^2 + 4\epsilon^{-p/(p-2)} \left(\int_B |f| \right)^2 + 2m(B) \left(\int_B |f| \right)^2 ,$$

and (5) is proved.

Now, we apply (5) and then (3) to the function v extended by 0 outside D_1 to get

$$\int_{D_1} |\nabla v|^2 \geq (1 - C_5\epsilon) \int_{D_1} |\nabla v|^2 - C_5\epsilon^{-A} \left(\int_B v \right)^2 + \int_B v^2$$

$$\geq (1 - C_5\epsilon)(\mu + c_3) \int_{D_1} v^2 - C_5\epsilon^{-A} \left(\int_B v \right)^2$$

$$\geq (1 - C_5\epsilon)(\mu + c_3) \int_{D_1} v^2 - C_5\epsilon^{-A}\theta \left(\int_{D_1} v \right)^2$$

We can choose ϵ depending only on M so that $(1 - C_5\epsilon)(\mu + c_3) > \mu + c_3/2$. Then there is a constant c depending only on M such that if $\theta < c$ then $\mu + c_3/2 - C\epsilon^{-A}\theta > \mu + c_3/4$. It follows that $\lambda_1(D\backslash\overline{V}) \geq \nu > \mu + c_3/4$ and the lemma is proved in this case.

Case 2. $\theta \geq c$. Then $\int_B v \geq c \int_{D_1} vw$ and consequently

$$\nu - \mu \geq cc_4 \nu t^{K_3} \geq c_1 m(V)^{K_3/\delta},$$

for some constant c_1 depending only on M. This concludes the proof of Lemma 3.∎

Remark 2. Let Q be a cube in $\mathbf{R}^q$ of sidelength s. Let $E \subset Q$. There are dimensional constants C_6 and $K(q)$ such that

$$\int_Q f(y)^2 dy \le C_6 \left(\frac{m(Q)}{m(E)} \right)^{K(q)} \left(s^2 \int_Q |\nabla f(y)|^2 dy + \int_E f(y)^2 dy \right).$$

Proof. Without loss of generality we may assume that $s = 1$. We will prove by induction on q that for $0 < r \le 1$,

$$\int_Q |\nabla f(y)|^2 dy + r \int_E f(y)^2 dy \ge c(q)(rm(E))^{K(q)} \int_Q f(y)^2 dy.$$

The remark is the special case $r = 1$. For $q = 1$ we let $I = [0, 1]$. It is easy to show that

$$\int_I f^2 \le 5 \int_I (f')^2 + 5 \min_I f^2.$$

Hence,

$$\int_I (f')^2 + r \int_E f^2 \ge rm(E) \int_I (f')^2 + rm(E) \min_I f^2 \ge \frac{rm(E)}{5} \int_I f^2.$$

For the induction step, let $Q = Q' \times I$ where Q' is the unit cube in $\mathbf{R}^{q-1}$. Let $E(x') = \{t \in I : (x', t) \in E\}$ and $F = \{x' \in Q' : m(E(x')) > m(E)/10\}$. Then

$$m(E) = \int_{Q' \setminus F} m(E(x'))dx' + \int_F m(E(x'))dx' \le \frac{1}{10} m(E) + m(F).$$

Thus $m(F) > m(E)/2$. If ∇' denotes the gradient in the first $q - 1$ variables, then using the inequality for $q = 1$ and the induction hypothesis we have

$$\int_Q |\nabla f|^2 + r\chi_E f^2 = \int_Q |\nabla' f|^2$$

$$+ \int_{Q'} \left(\int_I |\partial_t f(x', t)|^2 dt + r \int_{E(x')} f(x', t)^2 dt \right) dx'$$

$$\ge \int_Q |\nabla' f|^2 + \int_{Q'} \frac{r}{5} m(E(x')) \int_I f(x', t)^2 dt dx'$$

$$\ge \int_Q |\nabla' f|^2 + \int_{Q'} \frac{r}{5} \frac{m(E)}{10} \chi_F(x') \int_I f(x', t)^2 dt dx'$$

$$= \int_I \left(\int_{Q'} |\nabla' f(x', t)|^2 + \frac{rm(E)}{50} \chi_F(x') f(x', t)^2 dx' \right) dt \ge$$

$$\geq \int_I c(q-1) \int_{Q'} \left[r\frac{m(E)}{50} m(F) \right]^{K(q-1)} f(x', t)^2 dx' dt$$

$$\geq c(q-1) \left[r\frac{m(E)^2}{100} \right]^{K(q-1)} \cdot \int_Q f^2,$$

which proves the remark.   ■

We can now show that Ω_- is very wide (in the x variable) for many values of y.

Lemma 4. *Let* $\alpha = 1/10K(q)$. *There is a constant C depending only on M and n such that for any $N \geq C$, there exists a cube Q in $\mathbf{R}^q$ satisfying*
(a) $m(Q) = N^\alpha$;
(b) $m(E \cap Q) \leq N^{-\alpha}$, *where $E = \{y \in \mathbf{R}^q : \lambda_1(\Omega_-(y)) \geq \lambda_0 + N^{-1/3}\}$;*
(c) *If B denotes the unit cube with the same center as Q, then*

$$N \int_{y \in B} \int_{x \in \Omega(y)} u_-(x, y)^2 dx dy \geq \int_{y \in Q} \int_{x \in \Omega(y)} u_-(x, y)^2 dx dy$$

(d) *If Q' denotes the cube concentric with Q of half the diameter, then for $y \in Q'$ the eccentricity of $\Omega(y)$ is controlled: the inradius is bounded below by a dimensional constant and the diameter is bounded by a constant depending on M.*

Proof. Choose a tiling of $\mathbf{R}^q$ by cubes Q of measure N^α, that is, a family $\mathcal{F}$ of cubes that overlap only on their boundaries and whose union is all of $\mathbf{R}^q$. Let s denote the sidelength of Q. Thus $s = N^{\alpha/q}$. Define $\epsilon(Q)$ by

$$\int_Q \int_{\Omega(y)} |\nabla_y u_-(x, y)|^2 dx dy = \epsilon(Q) \int_Q \int_{\Omega(y)} u_-(x, y)^2 dx dy.$$

We consider three subfamilies of $\mathcal{F}$.

$$\mathcal{F}_1 = \{Q \in \mathcal{F} : \epsilon(Q) \leq 5C_2 N^{-2/3} \text{ and (b) fails for } Q\},$$

$$\mathcal{F}_2 = \{Q \in \mathcal{F} : \epsilon(Q) > 5C_2 N^{-2/3}\},$$

$$\mathcal{F}_3 = \{Q \in \mathcal{F} : \text{(c) fails for } Q\}.$$

For Q in $\mathcal{F}_1$ we have $m(Q)/m(E \cap Q) \leq N^{2\alpha}$. Fix x, and define $f(y) = u_-(x, y)$ if (x, y) belongs to Ω_- and zero otherwise. If we apply Remark 2 with E replaced by $E \cap Q$ and integrate in x we find

$$\int_Q \int_{\Omega(y)} u_-(x, y)^2 dx dy \leq C_6 s^2 N^{1/5} \int_Q \int_{\Omega(y)} |\nabla_y u_-(x, y)|^2 +$$

$$+ C_6 N^{1/5} \int_{E \cap Q} \int_{\Omega(y)} u_-(x, y)^2 dx dy$$

$$\leq C_6 s^2 \epsilon(Q) N^{1/5} \int_Q \int_{\Omega(y)} u_-(x, y)^2 dx dy$$

$$+ C_6 N^{1/5} \int_{E \cap Q} \int_{\Omega(y)} u_-(x, y)^2 dx dy$$

For N sufficiently large, $C_6 s^2 N^{1/5} \epsilon(Q) < \frac{1}{2}$, and, hence,

(6)
$$\int_Q \int_{\Omega(y)} u_-(x, y)^2 dx dy \leq 2 C_6 N^{1/5} \int_{E \cap Q} \int_{\Omega(y)} u_-(x, y)^2 dx dy.$$

On the other hand, making use of (6) and bounds in terms of ∇_x, we have

$$\int_Q \int_{\Omega(y)} u_-^2 = (1 - t) \int_{Q \backslash E} \int_{\Omega(y)} u_-^2 + t \int_{Q \backslash E} \int_{\Omega(y)} u_-^2 + \int_{E \cap Q} \int_{\Omega(y)} u_-^2$$

$$\leq \frac{(1 - t)}{\lambda_0} \int_{Q \backslash E} \int_{\Omega(y)} |\nabla_x u_-|^2 + (1 + 2 t C_6 N^{1/5}) \int_{E \cap Q} \int_{\Omega(y)} u_-^2$$

$$\leq \frac{(1 - t)}{\lambda_0} \int_{Q \backslash E} \int_{\Omega(y)} |\nabla_x u_-|^2 + \frac{(1 + 2 t C_6 N^{1/5})}{\lambda_0 + N^{-1/3}}$$

$$\int_{E \cap Q} \int_{\Omega(y)} |\nabla_x u_-|^2$$

$$\leq \max \left(\frac{(1 - t)}{\lambda_0}, \frac{(1 + 2 t C_6 N^{1/5})}{\lambda_0 + N^{-1/3}} \right) \int_Q \int_{\Omega(y)} |\nabla_x u_-|^2.$$

Let $t = 10 C_2 \lambda_0^{-1} N^{-2/3}$. Then, noting that $\lambda_0 \geq 1/M^2$, we find that for sufficiently large N,

(7)
$$(\lambda_0 + 5 C_2 N^{-2/3}) \int_Q \int_{\Omega(y)} u_-^2 \leq \int_Q \int_{\Omega(y)} |\nabla_x u_-|^2$$

for all Q in $\mathcal{F}_1$.

Next, we clearly have

(8)
$$(\lambda_0 + 5 C_2 N^{-2/3}) \int_Q \int_{\Omega(y)} u_-^2 \leq \int_Q \int_{\Omega(y)} |\nabla_{(x, y)} u_-|^2$$

for all Q in $\mathcal{F}_2$.

For Q in $\mathcal{F}_3$ we use Remark 2 again but with E replaced by B:

$$\int_Q \int_{\Omega(y)} u_-^2 = (1 - t) \int_Q \int_{\Omega(y)} u_-^2 + t \int_Q \int_{\Omega(y)} u_-^2$$

$$\leq \frac{(1-t)}{\lambda_0} \int_Q \int_{\Omega(y)} |\nabla_x u_-|^2 + t C_6 s^2 N^{1/10} \int_Q \int_{\Omega(y)} |\nabla_y u_-|^2$$

$$+ t C_6 N^{1/10} \int_B \int_{\Omega(y)} u_-^2$$

$$\leq \max \left(\frac{1-t}{\lambda_0}, t C_6 s^2 N^{1/10} \right) \int_Q \int_{\Omega(y)} |\nabla_{(x,y)} u_-|^2$$

$$+ t C_6 N^{-9/10} \int_Q \int_{\Omega(y)} u_-^2.$$

Therefore, if $t = N^{-1/2}$ and N is sufficiently large,

$$(1 - t C_6 N^{-9/10}) \int_Q \int_{\Omega(y)} u_-^2 \leq \frac{1-t}{\lambda_0} \int_Q \int_{\Omega(y)} |\nabla_{(x,y)} u_-|^2.$$

For large N, $(1 - t)/(1 - C_6 t N^{-9/10}) < 1 - 10 C_2 \lambda_0^{-1} N^{-2/3}$, and, consequently, (8) holds for all Q in $\mathcal{F}_3$ as well.

Now if every cube in $\mathcal{F}$ is in one of the three families, then we can sum over all cubes and obtain

$$(\lambda_0 + 5 C_2 N^{-2/3}) \int_\Omega u_-^2 \leq \int_\Omega |\nabla u_-|^2 = \lambda \int_\Omega u_-^2$$

which contradicts Lemma 1. So there must be a cube that does not belong to any of the three families, in other words, a cube for which (a), (b), and (c) of Lemma 4 hold.

Finally, let us deduce (d) from (b). The theorem of F. John (1) implies that the diameter of $\Omega(y)$ is bounded by a constant times M for any y. λ_0 is bounded above by a dimensional constant. Therefore there is a dimensional constant c_p such that every domain contained in a rectangle with side of length less than c_p has first eigenvalue greater than $\lambda_0 + 1$. By the John theorem applied to the convex set $\Omega(y)$, there is a dimensional constant c' such that if the inradius of $\Omega(y)$ is less than c', then $\Omega(y)$ is contained in a rectangle with a side smaller than c_p, so that $\lambda_1(\Omega(y)) > \lambda_0 + 1$. Now suppose that there is a point y^0 in Q' such that the inradius is smaller than c'. Then $\lambda_1(\Omega_-(y^0)) \geq \lambda_1(\Omega(y^0)) > \lambda_0 + 1$. In particular, y^0 belongs to E. The set of all y for which the inradius is larger than c' is convex. Therefore, there is an entire halfspace containing y^0 for which the inradius is smaller than c'. But this half-space intersects Q in a set of large measure, and all of its points belong to E. This contradicts (b). (The proof of (d) can be simplified using a theorem of E. Lieb that $\lambda_1(\Omega(y))$ is a convex function of y.)

Let B and Q be given by Lemma 4. Denote

$$E^2 = \int_B \int_{\Omega_-(y)} u^2.$$

Let Q'' be the cube with the same center as Q' but with half the diameter. Define $R' = \{(x, y) \in \Omega : y \in Q'\}$ and $R'' = \{(x, y) \in \Omega : y \in Q''\}$. We claim that

(9) $$\max_{R'} u_- \leq C_8 N^{1/2} E.$$

To prove (9), let $\zeta = (x, y) \in R'$. By Lemma 4(c) and Fubini's theorem there is a dimensional constant C such that for some r, $1/2\sqrt{\lambda} \leq r \leq 1/\sqrt{\lambda}$,

$$\frac{1}{\sigma(\partial B(\zeta, r))} \int_{\partial B(\zeta, r)} u_- d\sigma \leq C N^{1/2} E.$$

Consider the function ψ that solves

$$(\Delta + \lambda)\psi = 0 \text{ in } B(\zeta, r); \quad \psi(z) = u_-(z) + E \text{ for } z \in \partial B(\zeta, r).$$

The maximum principle implies that $u_- \leq \psi$ in $B(\zeta, r) \cap \Omega_-$, and by Remark 1,

$$u_-(\zeta) \leq \psi(\zeta) \leq \frac{2}{\sigma(\partial B(\zeta, r))} \int_{\partial B(\zeta, r)} (u_- + E) d\sigma \leq C_8 N^{1/2} E. \qquad \blacksquare$$

Lemma 5. *Suppose that* $0 < r < 1/4\sqrt{\lambda}$, $B(\zeta, r) \subset \Omega_+$ *and* $\partial B(\zeta, r) \cap \Lambda \neq \emptyset$. *Then*

$$u(\zeta) \leq C_9 \max_{\overline{B}(\zeta, 2r)} u_-.$$

Proof. Let $S = \max_{\overline{B}(\zeta, 2r)} u_-$. Let $\overline{w}_\epsilon(z) = w_\epsilon(\sqrt{\lambda}(z - \zeta))$ where w_ϵ is the function given in Lemma 2 with $\rho = r\sqrt{\lambda} < 1/4$. Then $(\Delta + \lambda)\overline{w}_\epsilon = 0$ in $B(\zeta, 2r) \backslash B(\zeta, r)$, and w_ϵ is positive in the closed annulus. Therefore, by the maximum principle for the region $\Omega_- \cap B(\zeta, 2r) \backslash B(\zeta, r)$, $u_-(z) \leq S\overline{w}_\epsilon(z)$. Taking the limit as ϵ tends to 0, we have $u_-(z) \leq S\overline{w}_0(z)$. Let η belong to $\partial B(\zeta, r) \cap \Lambda$. Harnack's inequality implies that there is a dimensional constant C such that

$$u(\zeta) \leq C \min_{\overline{B}(\zeta, r/2)} u.$$

The function u is superharmonic in the ball $B(\zeta, r)$. Therefore, u is larger than a dimensional constant times the harmonic function $G(z) = u(\zeta)(|x/r|^{2-n} - 1)$ on $B(\zeta, r) \backslash B(\zeta, r/2)$. (When $n = 2$, we put $G(z) = \log|z/r|$.) Since both G and u vanish at η, it follows that $|\nabla G(\eta)| \leq C|\nabla u(\eta)|$, and hence

$$u(\zeta) \leq Cr|\nabla u(\eta)|.$$

In particular, $\nabla u(\eta) \neq 0$, so that, near η, Λ is a smooth hypersurface tangent to $\partial B(\zeta, r)$. Because both u and $\overline{w}_0$ vanish at η, it follows that $|\nabla u(\eta)| \leq |S\nabla \overline{w}_0(\eta)|$. Thus, the upper bound from Lemma 2(d) implies $u(\zeta) \leq C_9 S$, as desired. ∎

Remark 3. Suppose that $0 < v \leq 1$ and $(\Delta + \lambda)v = 0$ in D. Let $0 < r \leq 1/\sqrt{\lambda}$. If $\zeta \in D$, $m(B(\zeta, r) \cap D) \leq m(B(\zeta, r))/4$, and $v = 0$ on $B(\zeta, r) \cap \partial D$, then $v(\zeta) \leq 1/2$.

Proof. By Fubini's theorem, there exists s, $0 < s \leq r$, such that $\sigma(\partial B(\zeta, s) \cap D) \leq \sigma(\partial B(\zeta, s))/4$. For any $\epsilon \geq 0$, let v_ϵ satisfy $(\Delta + \lambda)v_\epsilon = 0$ in $B(\zeta, s)$, $v_\epsilon = v + \epsilon$ on $\partial B(\zeta, s) \cap D$ and $v_\epsilon = \epsilon$ on $\partial B(\zeta, s) \setminus D$. By the generalized maximum principle in $D \cap B(\zeta, s)$, $v \leq v_\epsilon$ provided $\epsilon \geq 0$. Passing to the limit as ϵ tends to 0, we find $v \leq v_0$. Remark 1 implies

$$v(\zeta) \leq v_0(\zeta) \leq \frac{2}{\sigma(\partial B(\zeta, s))} \int_{\partial B(\zeta, s)} v_0 d\sigma \leq \frac{1}{2}. \qquad ∎$$

Lemma 6. *Denote* $Y = \max_{R'} u_-$. *Choose* $z_0 \in \partial\Omega$ *such that* $|z_0 - \zeta| = r = \mathrm{dist}(\zeta, \partial\Omega)$. *There is a constant* C_{10} *depending on* M *such that if* ζ *belongs to* R', $B(\zeta, 10r) \cap \Omega \subset R'$, *and*

(10) $u(\zeta) > C_{10}Y$

then $B(z_0, 2r) \cap \Lambda = \emptyset$.

Proof. If $\overline{B}(\zeta, r)$ is not contained in Ω_+, then it intersects Λ. Choose $s < r$ such that $B(\zeta, s)$ is contained in Ω_+ and $\partial B(\zeta, s)$ intersects Λ. Then by Lemma 5 and Harnack's inequality the opposite inequality to (10) holds. So (10) implies $B(\zeta, r) \subset \Omega_+$.

Denote $B_0 = B(z_0, 4r)$ and

$$B_k = B(z_0, r_k) \qquad \text{with} \quad r_k = 2(1 + 2^{-k})r.$$

Thus the intersection of B_k for all k is $\overline{B}(z_0, 2r)$, and $r_{k+1} = r_k - 2^{-k}r$. Denote the convex hull of z with $B(\zeta, r)$ by $\Gamma(z)$. We claim that

(i) $\sup_{B_k} u_- \leq C_{11} 2^{-2^{k-1}} Y$

and

(ii) $\Lambda \cap \Gamma(z) \subset B(z, \theta 2^{-2k}r) \qquad \text{for all} \quad z \in B_k \cap \partial\Omega,$

where $\theta > 0$ will be specified later. We prove the claim by induction. For $k = 0, 1, \ldots 5$ part (i) is true by the definition of Y, with a suitable choice of C_{11}. Let B be the ball of largest radius in $\Gamma(z)$ centered on the axis of $\Gamma(z)$ such that

$B \subset \Omega_+$, but ∂B intersects Λ. It follows from Lemma 5 that $u(z_2) \le C_9 Y$, where z_2 is the center of B. If we assume that part (ii) fails for some $k \le 5$, then the radius of B is greater than $2^{-20}\theta r$. On the other hand, by following the cone as it expands, this ball can be connected by a chain of balls which overlap in a large fraction of their volume to the ball $B(\zeta, r)$. The number of such balls is at most a dimensional constant times $\log 1/\theta$. Therefore, the Harnack inequality applied $\log 1/\theta$ times shows that $u(\zeta) \le A\theta^{-A} u(z_2)$ for some dimensional constant A. To assure that (ii) is true for $k = 0, 1, \ldots 5$ we assume $C_{10} \ge A\theta^{-A}$.

Next, for the induction step, suppose that (i) and (ii) hold for some $k \ge 5$. Suppose that $z_1 \in \Omega_- \cap B(z_0, r_k - 10 \cdot 2^{-2k}r)$. We now show that

(11) $m(B(z_1, 2^{-2k}r) \cap \Omega_-) < C_{12}\theta m(B(z_1, 2^{-2k}r))$,

by showing that every $z \in B(z_1, 2^{-2k}r) \cap \Omega_-$ is within $\theta 2^{-2k}r$ of $\partial\Omega$. The ray from ζ to z meets $\partial\Omega$ at a point z_3 and because $|z_3 - \zeta| \le 5r$ and the convex set Ω contains the ball $B(\zeta, r)$, the angle of the ray with a plane tangent to $\partial\Omega$ at z_2 is at least $1/5$. It follows that $z_3 \in B_k \cap \partial\Omega$ and by part (ii) of the induction hypothesis, $z \in B(z_2, \theta 2^{-2k}r)$. In particular, z is within $\theta 2^{-2k}r$ of $\partial\Omega$. Furthermore, the fact that $B(\zeta, r)$ is contained in Ω implies that the Lipschitz constant of $\partial\Omega \cap B_k$ is less than 5, so that (11) holds for some dimensional constant C_{12}.

If we choose θ so that $C_{12}\theta < 1/4$, we can apply Remark 3 repeatedly to obtain for $j = 1, 2, \ldots$

$$u_-(z) \le 2^{-j}Y \qquad \text{for all} \quad z \in B(z_0, r_k - (10 + j)2^{-2k}r).$$

In particular, if $10 + j = 2^k$ we have

$$u_-(z) \le 2^{10}2^{-2^k}Y \qquad \text{for all} \quad z \in B_{k+1},$$

which is the induction step for part (i). Now consider any $z \in B_{k+1} \cap \partial\Omega$. If there is a point of $\Gamma(z) \cap \Lambda$ at a distance greater than $\theta 2^{-2k-2}r$ from z, then there is a ball B in $\Gamma(z)$ of radius at least a dimensional constant times $\theta 2^{-2k-2}r$ which is tangent to Λ and such that the convex hull of B with $B(\zeta, r)$ is contained in Ω_+. By Lemma 5, the value of u at the center of B is less than $C_9 2^{10}2^{-2^k}Y$. Moreover, B can be connected by a chain of overlapping balls to the ball $B(\zeta, r)$ as above. The number of balls in the chain is less than a dimensional constant times $\log(2^{2k+2}/\theta)$. Therefore, by Harnack's inequality,

$$u(\zeta) \le A(2^{-2k-2}\theta)^{-A}C_9 2^{10}2^{-2^k}Y.$$

Choosing C_{10} sufficiently large, we have a contradiction. This proves the induction step for part (ii).

Now, take the intersection over all k to obtain $B(z_0, 2r) \cap \Lambda = \emptyset$, which proves Lemma 6.   ■

We can now deduce that if R'' denotes the subset of Ω that projects onto a cube Q'' concentric with Q' with half the diameter, then

(12) $$\max_{R''} u_+ \le C_{13} \max_{R'} u_-.$$

For any ζ in $\Omega_+ \cap R''$, either $u(\zeta) \le C_{10}Y$ and we are done, or else the opposite inequality holds and the hypothesis of Lemma 6 is satisfied. With z_0 as in Lemma 6, consider the ball $B(z_0, s)$ disjoint from Λ but such that $\partial B(z_0, s)$ intersects Λ. Thus $s \ge 2r$, but because of the estimates of Lemma 4, s must be smaller than some negative power of N. In particular, $s < 1/\sqrt{\lambda}$. Consider a point $z_4 \in B(z_0, s)$ at a distance, say s/M from $\partial(\Omega \cap B(z_0, s))$. By Lemma 6 and Harnack's inequality, $u(z_4) \le CC_{10}Y$, for some constant C depending on M. Now by Proposition 3, $u(\zeta) \le CC_4C_{10}Y$, which proves (12).

Note that $\Omega_+ \cap R' \subset \{(x, y): y \in E \cap Q'\} \cup \{(x, y): x \in \Omega_+(y) \text{ and } y \in Q'\backslash E\}$. Lemma 4 implies $m(\{(x, y): y \in E \cap Q'\}) \le (C_1M)^pN^{-1/10}$. Lemma 4(d) and Lemma 3 imply that for $y \in Q'\backslash E$,

$$\lambda_1(\Omega(y)) + c_1m(\Omega_+(y))^K \le \lambda_1(\Omega_-(y)) < \lambda_0 + N^{-1/3}.$$

It follows that $m(\Omega_+(y)) \le c_1^{-1/K}N^{-1/3K}$. Let $z_5 \in \Omega$ project onto the center of the unit cube B given by Lemma 4. All the points of $B(z_5, T)$, where $T = c_5N^{\alpha/q}$, project onto Q''. Choose a dimensional constant $c_7 < \min\{1/\sqrt{\lambda}, 1/10\}$. The estimates above for the measure of E and for $\Omega_+(y)$ when y is in the complement of E show that if N is sufficiently large, then for any $z \in B(z_5, T)$,

$$m(B(z, c_7) \cap \Omega_+) \le \frac{1}{4}m(B(z, c_7)).$$

Therefore, we can apply Remark 3 repeatedly to conclude that for $j = 1, 2, \ldots$

$$u_+(z) \le 2^{-j} \max_{R''} u_+ \qquad \text{for all} \quad z \in B(z_5, T - jc_7).$$

In particular,

$$u_+(z) \le 2^{-T} \max_{R''} u_+ \qquad \text{for all} \quad z \in B(z_5, T/2).$$

Now take any point $z \in \Omega_-$ that projects onto B of Lemma 4. Because we assumed that Ω_- is relatively compact in Ω, there is an open ball centered at z contained in Ω_- whose boundary intersects Λ. By the same reasoning as in Lemma 5 (with the roles of u_+ and u_- reversed), we find that

$$u_-(z) \le C_9 2^{-T} \max_{R'} u_+.$$

Integrating and using (9) and (12), we have

$$E \le C_{14}2^{-T} \max_{R'} u_+ \le C_{14}C_{13}2^{-T}C_8N^{1/2}E.$$

If N is sufficiently large, the 2^{-T} term dominates, and this inequality cannot hold unless $E = 0$. But if E vanishes, then u vanishes on an open subset of Ω, a contradiction. This concludes the proof that the nodal set touches the boundary.

THE CASE $n = 2$

We would now like to show that in $\mathbf{R}^2$ the nodal line must touch the boundary in exactly two points. The Courant nodal domain theorem says that the sets Ω_+ and Ω_- are connected. Therefore, Λ cannot intersect $\partial\Omega$ in more than two points. In the very last stage of the proof above we used the fact that any ball in Ω_- can be expanded until it touches Λ. In the case $n = 2$, we prove a variant of Lemma 4 which will imply that Ω_+ must touch *both* sides of $\partial\Omega$ somewhere in the region that projects onto Q.

We will make a slightly different normalization of the domain Ω for notational convenience. We claim that we can assume that

$$\Omega \subset \{(x, y) \in \mathbf{R}^2 : 0 < x < 1\},$$

the diameter of Ω is N and the inradius is at least $1 - N^{-1}$. In fact, suppose first that the inradius is $1/2$. Choose a disk D of radius $1/2$ in Ω. If ∂D touches $\partial\Omega$ in two points on the same diameter of D, then Ω is contained between the two tangent lines perpendicular to the diameter, which are a distance 1 apart. If no such pair of points exists, then $\partial D \cap \partial\Omega$ contains at least three points and the tangent lines at those points form a triangle containing Ω. The longest side of the triangle has length at least N and it is easy to check that the altitude perpendicular to that side has length at most $1 + N^{-1}$. Thus, Ω is contained in a rectangle of length N parallel to the longest side and width $1 + N^{-1}$. If we dilate by the factor $(1 + N^{-1})^{-1}$ and rotate so that the rectangle is horizontal, we have our normalization.

As in the case of Lemma 1, an easy consequence of this normalization is

Lemma 1′. $\pi^2 \le \lambda \le \pi^2 + C_2 N^{-2/3}$.

Lemma 4′. *Let Ω be a convex subset of $\mathbf{R}^2$ with the normalization above. There is a constant C_{15} such that if $N \ge C_{15}$, then there exists an interval J of length comparable to $N^{1/4}$ such that*

(a) $\qquad\qquad$ *the rectangle $I \times J \subset \Omega_-$,*

where $I = \{x : N^{-1/8} < x < 1 - N^{-1/8}\}$. Moreover, if y_0 denotes the center of J, the square

$$Q_1 = \{(x, y) : |x - 1/2| \le 1/8, |y - y_0| \le 1/8\}$$

satisfies

(b)
$$\int_{\Omega_-} u^2 \le N^2 \int_{Q_1} u^2.$$

The differences between Lemma 4′ and Lemma 4 are that we show here that an entire rectangle is contained in Ω_-, instead of showing that all but a subset of small measure of a rectangle is contained in Ω_- and also that we can control the integral of u^2 on a large set by the integral on a "unit–sized" set Q_1 whose double is contained in Ω_-.

Suppose that Λ intersects $\partial\Omega$ in at most one point. Then, since $\partial\Omega_-$ is a simple closed curve that encloses R_1, either for every y in J there exists x such that $0 < x < N^{-1/8}$ and $(x, y) \in \Lambda$ or else for every y in J there exists x such that $1 - N^{-1/8} < x < 1$ and $(x, y) \in \Lambda$. In either case, one can use the same argument as above to control u_+ by u_-. Then one finds exponential decay of u_+ along a thin strip of Ω_+ either in $0 < x < N^{-1/8}$ or $1 - N^{-1/8} < x < 1$. Then one can control u_- on Q_1 in terms of u_+ in either one of the strips by translating a disk of radius $1/8$ around a point of Q_1 to the left (decreasing x) or to the right (increasing x) until its boundary touches Λ. This yields the same type of contradiction as above, and concludes the proof that the nodal line touches in two points.

It remains to prove Lemma 4′.

Lemma 7. *Suppose that $0 < \epsilon < 1/2$ and $(\epsilon, 0) \notin \Omega_-$. There exists η, $|\eta| < \epsilon$ such that if $E = \{y \in \mathbf{R} : \eta < y < \eta + \epsilon/8\}$, then*

$$\pi^2 \int_E \int_{\Omega(y)} u_-^2 \, dx dy \le (1 - \epsilon) \int_E \int_{\Omega(y)} |\nabla u_-|^2 dx dy.$$

The proof makes use of the inequality

(13)
$$\pi^2 \int_0^L f(s)^2 ds \le L^2 \int_0^L f'(s)^2 ds,$$

for all functions f such that $f(0) = f(L) = 0$. (The minimizer is $f(s) = \sin(\pi s/L)$.) Also,

(14)
$$\pi^2 \int_0^L f(s)^2 ds \le 4L^2 \int_0^L f'(s)^2 ds,$$

for all functions f such that $f(0) = 0$. (The minimizer is $f(s) = \sin(\pi s/2L)$.)

Because Ω_- is simply–connected, there is a continuous curve in the complement of Ω_- that joins $(\epsilon, 0)$ to the real axis. Denote $V = \{(x, y) : |x - \epsilon| < \epsilon/4, |y| < \epsilon/8\}$. Let γ denote the component of the curve from $(\epsilon, 0)$ to the first time that the curve exits V. Let

$$\eta = \min_{(x,y)\in\gamma} y \quad \text{and} \quad \eta_1 = \max_{(x,y)\in\gamma} y.$$

If $\eta \le y \le \eta_1$ then there exists x such that $|x - \epsilon| < \epsilon/4$ and $u_-(x, y) = 0$. Therefore,

$$\pi^2 \int_0^1 u_-(s, y)^2 ds = \pi^2 \int_0^x u_-(a, y)^2 ds + \pi^2 \int_x^1 u_-(a, y)^2 ds$$

$$\le x^2 \int_0^x (\partial_1 u_-(s, y))^2 ds + (1 - x)^2 \int_x^1 (\partial_1 u_-(s, y))^2 ds$$

$$\le (1 - 3\epsilon/4)^2 \int_0^1 (\partial_1 u_-(s, y))^2 ds,$$

where $\partial_1 u_-(x, y)$ is the partial derivative of u_- with respect to x. If $\eta_1 - \eta \ge \epsilon/8$, then we are done. If not, then γ exits V on one of its vertical sides. Suppose that the side is the one for which $x = 3\epsilon/4$. (The case in which γ exits V on the side $x = 5\epsilon/4$ is similar.) We divide the rectangle $R = \{(x, y) : \eta < y < \eta + \epsilon/8, 0 < x < 1\}$ into four parts. Denote $h = \eta - \eta_1 + \epsilon/8$, $P = (\epsilon - h, \eta + \epsilon/8)$, $P_1 = (\epsilon, \eta_1)$, and $P_2 = (\epsilon - 2h, \eta + \epsilon/8)$. Let S_1 be the segment from P to P_1, and let S_2 be the segment from P to P_2. The region R_1 is defined as the union of horizontal segments with left endpoint on S_1 and right endpoint on the line $x = 1$ with the shortest vertical segments with upper endpoint on S_1 and lower endpoint on γ. R_2 is union of horizontal segments with right endpoint on the line $x = 0$ and left endpoint on S_2 with the shortest vertical segments with upper endpoint on S_2 and lower endpoint on γ. R_3 is the union of the shortest horizontal line segments with right endpoints $(1, y)$ for all $\eta < y < \eta_1$ and left endpoint on γ. Finally, R_4 is the rest of the rectangle R. It is easy to see that R_4 is a disjoint union of vertical line segments at least one of whose endpoints belongs to γ.

The region R_1 is foliated by curves that start at $x = 1$, move horizontally until they reach S_1, and then drop vertically to γ. These curves have length at most $1 - 5\epsilon/8$. We claim that this implies that

(15) $$\pi^2 \int_{R_1} u_-^2 \le (1 - 5\epsilon/8)^2 \int_{R_1} |\nabla u_-|^2.$$

In fact, we can make an area preserving change of variable of the form

$$F(s, t) = \begin{cases} (s, t) & s + t \ge 0 \\ (-t, s + 2t) & s + t \le 0 \end{cases}.$$

Note that

$$\frac{\partial}{\partial s} v(F(s, t)) = \begin{cases} \partial_1 v(F(s, t)) & s + t > 0 \\ \partial_2 v(F(s, t)) & s + t < 0 \end{cases}$$

if v is defined on a set U and v vanishes on ∂U. For any numbers $a < b$ let $W = U \cap \{F(s, t): a < t < b \text{ and } s \in \mathbf{R}\}$. Let

$$L = \max_{a \leq t \leq b} \text{length}(W \cap \{F(s, t): s \in \mathbf{R}\}),$$

the length of the longest "right angle" curve in W. Using (13) and the change of variable F we find

$$\pi^2 \int_W v^2 \leq L^2 \int_{F^{-1}(W)} \frac{\partial}{\partial s} v(F(s, t))^2 ds dt \leq L^2 \int_W |\nabla v|^2.$$

The segment S_1 has slope -1, so a translate of this change of variables yields (15). The region R_2 is also foliated by right angles with corner on S_2. These curves have length at most $9\epsilon/8$, so that by the same argument as for (15),

$$\pi^2 \int_{R_2} u_-^2 \leq (9\epsilon/8)^2 \int_{R_2} |\nabla u_-|^2.$$

The region R_3 is foliated by horizontal line segments of length at most $1 - 3\epsilon/4$, so that by (13),

$$\pi^2 \int_{R_3} u_-^2 \leq (1 - 3\epsilon/4)^2 \int_{R_3} |\nabla u_-|^2.$$

Finally, R_4 is foliated by vertical line segments of length at most $\epsilon/4$, so that by (14),

$$\pi^2 \int_{R_4} u_-^2 \leq 4(\epsilon/4)^2 \int_{R_4} |\nabla u_-|^2.$$

Now, since all of the coefficients on the right–hand integrals over R_1, R_2, R_3, and R_4 are less than $1 - \epsilon$, Lemma 7 is proved.

We also need the inequality

(16) $$\pi^2 \int_0^1 f(s)^2 ds \leq \frac{7}{8} \int_0^1 f'(s)^2 ds + \int_{3/8}^{5/8} f(s)^2 ds,$$

for all functions f for which $f(0) = f(1) = 0$. (The essential feature here is that $7/8 < 1$, so that there is an improvement on the eigenvalue for the Dirichlet problem on the interval if one adds the extra integral on the right–hand side.) Inequality (16) can be proved by direct computation of the lowest eigenfunction; the calculation is left to the reader.

We can now proceed with the proof of Lemma 4' in much the same way as we proved Lemma 4. Let J be an interval of $\mathbf{R}$ of length $N^{1/4}$. Define $\epsilon(J)$ by

$$\epsilon(J) \int_J \int_0^1 u_-(x, y)^2 dxdy = \int_J \int_0^1 (\partial_2 u_-(x, y))^2 dxdy.$$

If $\epsilon(J) > 1/2C_6 N^{1/2}$, then since

$$\pi^2 \int_J \int_0^1 u_-^2 dxdy \leq \int_J \int_0^1 (\partial_1 u_-)^2 dxdy,$$

we have

(17)
$$\left(\pi^2 + \frac{1}{2C_6 N^{1/2}}\right) \int_J \int_0^1 u_-^2 dxdy \leq \int_J \int_0^1 |\nabla u_-|^2 dxdy.$$

If $\epsilon(J) \leq 1/2C_6 N^{1/2}$ and (a) fails for an interval J' with the same center as J, but of length $N^{1/4} - 1$, then by Lemma 7 with $\epsilon = N^{-1/8}$, there is an interval $E \subset J$ of length $N^{-1/8}/8$ for which

$$\pi^2 \int_E \int_0^1 u_-^2 dxdy \leq (1 - N^{-1/8}) \int_E \int_0^1 |\nabla u_-|^2 dxdy.$$

Remark 2 implies

$$\int_J f(y)^2 dy \leq C_6 \left\{ N^{1/2} \int_J f'(y)^2 dy + 8N^{3/8} \int_E f(y)^2 \right\},$$

and, hence,

$$\int_J \int_0^1 u_-^2 dxdy \leq C_6 \left\{ N^{1/2} \int_J \int_0^1 (\partial_2 u_-)^2 dxdy + 8N^{3/8} \int_E \int_0^1 u_-^2 dxdy \right\}$$

$$= C_6 \left\{ \epsilon(J) N^{1/2} \int_J \int_0^1 (u_-)^2 dxdy + 8N^{3/8} \int_E \int_0^1 u_-^2 dxdy \right\}.$$

It follows that

$$\int_E \int_0^1 u_-^2 dxdy \leq 16C_6 N^{3/8} \int_E \int_0^1 u_-^2 dxdy.$$

Now

$$\int_E \int_0^1 u_-^2 dxdy = (1 - t) \int_{J\backslash E} \int_0^1 u_-^2 + t \int_{J\backslash E} \int_0^1 u_-^2 + \int_E \int_0^1 u_-^2$$

$$\leq \frac{1-t}{\pi^2} \int_{J\backslash E} \int_0^1 (\partial_1 u_-)^2 + (1 + 16C_6 N^{3/8} t) \int_E \int_0^1 u_-^2$$

$$\leq \frac{1-t}{\pi^2} \int_{J\backslash E} \int_0^1 (\partial_1 u_-)^2 + \pi^{-2}(1 + 16C_6 N^{3/8} t).$$

$$\cdot (1 - N^{-1/8}) \int_E \int_0^1 |\nabla u_-|^2.$$

Choose $t = N^{-5/8}$. Then for large N,

$$(1 + 16C_6 N^{3/8} t)(1 - N^{-1/8}) \leq 1 - N^{-1/8}/2 \leq 1 - N^{-5/8},$$

and, hence,

(18) $$\pi^2 \int_J \int_0^1 u_-^2 \leq (1 - N^{-5/8}) \int_J \int_0^1 |\nabla u_-|^2.$$

Finally, consider the case in which $\epsilon(J) \leq 1/2C_6 N^{1/2}$ and (b) fails for J. Then the interval $E = [y_0 - 1/8, y_0 + 1/8] \subset J$, and, by (16),

$$\pi^2 \int_E \int_0^1 u_-^2 dx dy \leq (1 - 1/8) \int_E \int_0^1 (\partial_1 u_-)^2 dx dy + \int_E \int_{3/8}^{5/8} u_-^2.$$

The same argument as the one leading to (18) yields

$$\pi^2 \int_J \int_0^1 u_-^2 \leq (1 - N^{-1/2}) \int_J \int_0^1 |\nabla u_-|^2 + \int_E \int_{3/8}^{5/8} u_-^2.$$

And since (b) fails,

(19) $$\pi^2 \int_J \int_0^1 u_-^2 \leq (1 - N^{-1/2}) \int_J \int_0^1 |\nabla u_-|^2 + N^{-2} \int_\Omega u_-^2.$$

Now if Lemma 4' were false there would be a disjoint covering of the projection onto the y-axis of Ω by $N^{3/4}$ intervals J satisfying one of (17), (18), or (19). Taking the sum, we obtain

$$\pi^2 \int_\Omega u_-^2 \leq (1 - N^{-5/8}) \int_\Omega |\nabla u_-|^2 + \frac{N^{3/4}}{N^2} \int_\Omega u_-^2.$$

Therefore,

$$(\pi^2 - N^{-5/4})(1 - N^{-5/8}) \int_\Omega u_-^2 \leq \int_\Omega |\nabla u_-|^2 = \lambda \int_\Omega u_-^2,$$

which contradicts Lemma 1' when N is sufficiently large.

Institute for Advanced Study
Massachusetts Institute of Technology

REFERENCES

[C] L. Carleson. "On existence of boundary values for harmonic functions in several variables." *Arkiv Math.* 4 (1961), 393–399.

[CF] R. R. Coifman and C. Fefferman. "Weighted norm inequalities for maximal functions and singular integrals." *Studia Math.* **51** (1974), 214–250.

[CFMS] L. A. Caffarelli, E. B. Fabes, S. Mortola, S. Salsa. "Boundary behavior of nonnegative solutions of elliptic operators in divergence form." *Indiana Univ. J. Math.* **30** (1981), 621–640.

[D] B. E. J. Dahlberg. "Estimates for harmonic measure." *Arch. Rat. Mech. Anal.* **65** (1977), 275–283.

[dG] M. deGuzman. *Differentiation of Integrals in $\mathbf{R}^n$.* Lecture Notes in Mathematics, no. 481. Springer Verlag, 1975.

[H] W. K. Hayman. "Some bounds for principal frequency." *Appl. Anal.* **17** (1978), 247–254.

[HW] R. Hunt and R. Wheeden. "On the boundary values of harmonic functions." *Trans. Amer. Math. Soc.* **132** (1968), 307–322.

[J] D. Jerison. "The first nodal line of a convex planar domain." *Int. Math. Res. Not.* **1** (1991), 1–5.

[KP] C. E. Kenig and J. Pipher. "The h-path distribution of the lifetime of conditioned Brownian motion for non-smooth domains." *Prob. Theo. Rel. Fields* **82** (1989), 615–624.

[L] C.-S. Lin. "On the second eigenfunction of the Laplacian in $\mathbf{R}^2$." *Comm. Math. Phys.* **111** (1987), 161–166.

[M] A. Melas. "On the nodal line of the second eigenfunction of the Laplacian in $\mathbf{R}^2$." *J. Diff. Geom.* To appear.

[P] L. E. Payne. "Isoperimetric inequalities and their applications." *S.I.A.M. Rev.* **9** (1967), 453–488 (Conjecture 5).

[P1] L. E. Payne. "On two conjectures in the fixed membrane eigenvalue problem." *J. Appl. Math. and Phys.* (JAMP) **24** (1973), 721–729.

[PW] M. H. Protter and H. F. Weinberger. *Maximum Principles in Differential Equations.* Springer Verlag, 1984.

[S] S. Salsa. "Some properties of nonnegative solutions of parabolic differential operators." *Ann. Mat. Pura Appl.* **128** (1981), 193–206.

[Se] J. B. Serrin. "On the Harnack inequality for linear elliptic equations." *J. d'Anal. Math.* **4** (1954–56), 292–308.

[SW] E. M. Stein and G. Weiss. *Introduction to Fourier Analysis on Euclidean Spaces.* Princeton University Press, 1971.

[Y] S.-T. Yau. "Problem Section." *Seminar on Differential Geometry*, S.-T. Yau, ed. Annals of Mathematics Studies 102. Princeton University Press, 1982.

12

On Removable Sets for
Sobolev Spaces in the Plane

*Peter W. Jones**

1 INTRODUCTION

Let K be a compact subset of $\bar{\mathbf{C}} = \mathbf{R}^2$ and let K^c denote its complement. We say $K \in HR$, K is holomorphically removable, if whenever $F : \bar{\mathbf{C}} \to \bar{\mathbf{C}}$ is a homeomorphism and F is holomorphic off K, then F is a Möbius transformation. By composing with a Möbius transform, we may assume $F(\infty) = \infty$. The contribution of this paper is to show that a large class of sets are HR. Our motivation for these results is that these sets occur naturally (e.g., as certain Julia sets) in dynamical systems, and the property of being HR plays an important role in the Douady-Hubbard description of their structure. (See [4].)

To prove that the sets in question are HR we establish what may be a stronger result. A compact set K is said to be removable for $W^{1,2}$ if every f which is continuous on $\mathbf{R}^2$ and in the Sobolev space $W^{1,2}(K^c)$ (one derivative in L^2 on K^c) is also in $W^{1,2}(\mathbf{R}^2)$. It is a fact that if K is removable for $W^{1,2}$, K is HR. We do not know the answer to the following question:

If K is HR, is K removable for $W^{1,2}$?

To prove the fact we first show that the two dimensional Lebesgue measure of K, $|K|$, is zero. If not, let $F_n = \frac{1}{\pi z} * (e^{in(x+y)} \chi_K(z))$. Then $\lim_{n \to \infty} \|F_n\|_{L^\infty(\mathbf{R}^2)} = 0$ and F_n is continuous. Since $|\partial F_n| = \chi_n$, $\|\bar{\partial} F_n\|_{L^2(\mathbf{R}^2)} = |K|^{1/2}$. On the other hand, $F_n(z) \to 0$ for $z \notin K$, so L^4 bounds on convolution with $\frac{1}{\pi z^2}$ when combined with Hölder's inequality show $\|F_n'\|_{L^2(K^c)} \to 0$. (See [11], Chapter 1.) Taking a sum of functions like the F_n, we obtain a globally continuous $F \in$

*This research was supported by the National Science Foundation under grant DMS-8916968.

$W^{1,2}(K^c)$, $F \notin W^{1,2}(\mathbf{R}^2)$. Now using the fact that $|K| = 0$, we deduce $K \in HR$. Take a homeomorphism F with $F'(\infty) = 1$. Then $f(z) = F(z) - z \in W^{1,2}(K^c)$ because integrating $|F'|^2$ gives the area of the image. Now $f \in W^{1,2}(\mathbf{R}^2)$ and $\bar\partial f = 0$ except on a set of measure zero implies (Weyl's lemma) f is holomorphic. Therefore, $F(z) = z + a$.

We recall some elementary facts concerning HR. If $|K| > 0$, it follows from the "measurable Riemann mapping theorem" (see [1]) that there is a nontrivial quasiconformal mapping F which is holomorphic off K. (Thus $K \notin HR$.) If K has Hausdorff dimension less than 1, $\mathrm{Dim}(K) < 1$, the fact that $K \in HR$ follows from the Cauchy integral formula (Painlevé's theorem). Similarly, if K is a rectifiable curve, Morera's theorem implies $K \in HR$. Kaufman [7] has produced examples of curves where $\mathrm{Dim}(K) = 1$ but $K \notin HR$. The "difficult" case is the one that occurs in conformal dynamics: K is connected and has some "fractal" properties. (The case of "pure" Cantor-type sets is easy; they are HR. By a pure Cantor set, we mean, e.g., one arising from a Cantor construction with a constant ratio of disection, or the Julia set for $z^2 + c$ where c is not in the Mandelbrot set.) We also point out that the case where K is a quasicircle seems to be folklore—again, $K \in HR$. That the property of being HR is related to quasiconformal mappings is seen from the following:

Remark. K is HR if and only if whenever F is a homeomorphism of $\bar{\mathbf{C}}$ which is M quasiconformal on K^c, F is globally quasiconformal (and hence M quasiconformal). (See [8], page 200.)

To prove the remark, first assume that K is HR. By the measurable Riemann mapping theorem there is a globally quasiconformal mapping G such that $G \circ F$ is holomorphic off K. Since $G \circ F$ is a Möbius transformation and $|K| = 0$, F is globally (M) quasiconformal. For the other direction, standard L^p estimates (see [1]) show that necessarily $|K| = 0$. If F is a homeomorphism which is analytic off K, F is globally quasiconformal and hence ($|K| = 0$) a Möbius transformation.

Let Ω be a domain on the Riemann sphere and let $z_0 \in \Omega$. Then Ω is a John domain (with center z_0) if there is $\varepsilon > 0$ such that for all $z_1 \in \Omega$ there is an arc $\gamma \subset \Omega$ which connects z_0 to z_1 and has the property that

$$d(z) \geq \varepsilon d(z, z_1), \quad z \in \gamma.$$

Here $d(z, z_1)$ is the chordal distance from z to z_1 and $d(z)$ is the chordal distance of z to $\partial\Omega$. We call such an arc γ a John arc. In this paper we will choose coordinates so that $z_0 = \infty$, and this allows us to replace $d(z)$, $d(z, z_1)$ by the corresponding Euclidean distances. The property of being a John domain is preserved under globally quasiconformal mappings. If Ω is a simply connected John domain, it is easy to show that the arc γ may be taken to be the hyperbolic geodesic from z_0 to ∞. (See [9] for an exposition of properties of John domains.) The main result of this paper is

Theorem 1. *If Ω is a John domain and $K = \partial\Omega$, then K is removable for $W^{1,2}$.*

Notice that the hypothesis demands that $K = \partial\Omega$, but says *nothing* about the other components of $C\backslash K$. This is because the hypothesis will be seen to force some geometry on those other components. (For example, the interior of a cardioid is a John domain while the exterior is not. The parabolic basin for $z^2 + \frac{1}{4}$ is also a John domain, while the basin for ∞—the exterior domain—is not.) It is of some philosophical interest to note the similarities between Theorem 1 and the results of [6] on extension problems for Sobolev spaces.

Since the John condition is quasiconformally invariant, we obtain directly (see also "Remark")

Corollary 1. *If Ω is a John domain and $K = \partial\Omega$, any global homeomorphism which is quasiconformal off K is globally quasiconformal (with the same constant of quasiconformality).*

We say that a polynomial $P(z)$ is subhyperbolic on its Julia set J if there is a metric $\lambda(z)|dz|$ such that $\lambda(z) - \sum_j |z - z_j|^{-\alpha_j}$ is C^∞ for some numbers $\alpha_j < 1$, and $P(z)$ is hyperbolic on J in the metric λ. In other words there are $c, \varepsilon > 0$ such that for all $n \geq 1$,

$$\lambda(z)^{-1}\lambda(P_n(z))|\frac{d}{dz} P_n(z)| \geq c(1 + \varepsilon)^n.$$

Here $P_n(z) = P \circ \cdots \circ P(z)$ is the nth iterate of P. (This definition may be a bit restrictive, but it is all we will need for this paper.) The following question is open:

> If J is the Julia set for a polynomial, is $J \in HR$?

It is proven in [3] that whenever a polynomial $P(z)$ is subhyperbolic on its Julia set J, then A_∞, the basin of attraction at ∞ for P, is a John domain. Since $J = \partial A_\infty$, we obtain

Corollary 2. *If $P(z)$ is subhyperbolic on its Julia set J, then $J \in HR$.*

The corollary answers a question of A. Douady and J. Hubbard and was the starting point of this investigation. Douady posed the question to the author for the particular (subhyperbolic) case where $P(z) = z^2 + c$ has the (Misiurewicz)

property that the origin is preperiodic but not periodic (e.g., $z^2 + i$). This case is not fundamentally different for the general case of subhyperoblic polynomials. An amusing feature of our proof is that the Julia set for a Misiurewicz point (from the family $z^2 + c$) is actually easier to deal with than those arising from the hyperbolic case. (When $K^c = \Omega$, our argument is a bit simpler. The arguments of sections 5 and 6 are not needed.)

The proof of Theorem 1 starts by proving it in the case where Ω is simply connected on $\bar{\mathbf{C}}$, i.e., K is connected. The general case then follows from

Theorem 2. *If Ω is an (ε) John domain, there is a $(c(\varepsilon))$ John domain Ω' with Ω' simply connected and*

$$\partial\Omega \subset \partial\Omega'.$$

While the proof of Theorem 2 is perhaps not immediately obvious, it turns out to follow from a simple construction with planar graphs.

Section 2 contains background material, and sections 3–7 are devoted to the proof of Theorem 1. The idea is to redefine F near K so that it is C^∞ near K and so that the Sobolev norm does not change much. Theorem 2 is proven in section 8.

2 BACKGROUND MATERIAL

Let $F \in W^{1,2}(K^c)$ be continuous on $\mathbf{C}$. An easy argument with the Dirichlet principle shows that to prove $F \in W^{1,2}(\mathbf{R}^2)$ it is sufficient to treat the case we now assume, where F is harmonic near K. We also assume the reader is familiar with elementary properties of logarithmic capacity, which we denote by $\mathrm{Cap}(\circ)$. See, e.g., [10] for the first two of the next three lemmata. Let $f : \mathbf{D} \to \mathbf{C}$ be univalent, $f(0) = 0$, $f'(0) = 1$. Then f has a Fatou extension to $\mathbf{T} = \partial\mathbf{D}$ and this extension is always defined except on a set of capacity (and hence Lebesgue measure) zero. In our applications, all image domains will have locally connected boundaries, and, hence, f will be continuous on $\bar{\mathbf{D}}$. The following results are due to Beurling. The values of c below are various universal constants.

Lemma 2.1. *If $E \subset \mathbf{T}$,*

$$\mathrm{Cap}(f(E)) \geq c\,\mathrm{Cap}(E)^2.$$

Lemma 2.2. *Let $g_\theta = f(\{re^{i\theta} : 0 \leq r < 1\})$. Then if $\ell(\cdot)$ denotes arclength,*

$$\mathrm{Cap}(\{e^{i\theta} : \ell(g_\theta) > \lambda\}) \leq c\lambda^{-1/2}$$

Lemma 2.3. *Suppose H is harmonic and continuous in* $\mathbf{D}$*, and* $(|\nabla H|^2 = |H_x|^2 + |H_y|^2)$*,*

$$\iint_{\mathbf{D}} |\nabla H|^2 dx dy = 1.$$

Then

$$\mathrm{Cap}(\{e^{i\theta} : |H(e^{i\theta}) - H(0)| \geq \lambda\}) \leq c e^{-\pi\lambda^2}.$$

This last lemma can be found on page 30 of [2]. We next require some elementary geometric facts about simply connected John domains. For the next result see [5].

Lemma 2.4. *If g is a Poincaré geodesic from ∞ to $z_0 \in \partial\Omega$ where Ω is an (ε) John domain, then g is an arc of a $K(\varepsilon)$ quasicircle.*

Suppose now Ω is a bounded (ε) John domain and suppose the John center z_0 satisfies $d(z_0) = 1$, where

$$d(z) = \text{distance}(z, \partial\Omega).$$

Then diameter$(\Omega) \sim 1$. Let $f : \mathbf{D} \to \Omega$, $f(0) = z_0$ be any choice of Riemann mapping, and define for $E \subset \partial\Omega$,

$$\mathrm{Cap}(E, z_0, \Omega) \equiv \mathrm{Cap}(\{e^{i\theta} : f(e^{i\theta}) \in E\}).$$

Lemma 2.5. *For any Borel set $E \subset \partial\Omega$,*

$$\mathrm{Cap}(E, z_0, \Omega) \sim \mathrm{Cap}(E).$$

In the last line we mean that $A \sim B$ if there is a constant $M = M(\varepsilon)$ such that

$$M^{-1}A^M \leq B \leq MA^{1/M}.$$

Proof. Let $G(z) = G(z, z_0)$ be Green's function for Ω with pole at z_0. Then it follows from the John condition and the Koebe $\frac{1}{4}$ theorem that

$$G(z) \geq cd(z)^\alpha, \quad \alpha = \alpha(\varepsilon),$$

whenever $|z - z_0| \geq \frac{1}{2}$. Suppose now that $z_j \in E$, $z_j = f(\zeta_j)$, $j = 1, 2$. Fix a point $\zeta_3 \in \mathbf{D}$ such that

$$(1 - |\zeta_3|) \sim |\zeta_1 - \zeta_3| \sim |\zeta_2 - \zeta_3| \sim |\zeta_1 - \zeta_2|$$

and let $z_3 = f(\zeta_3)$. Then by the John condition

$$|z_1 - z_2| \leq Cd(z_3),$$

while by our last estimate,

$$d(z_3) \leq CG(z_3)^{1/\alpha} \sim C|\varsigma_1 - \varsigma_2|^{1/\alpha}.$$

In other words,

$$|\varsigma_1 - \varsigma_2| \geq c|z_1 - z_2|^\alpha,$$

and it follows from the definition of logarithmic capacity that

$$\mathrm{Cap}(E, z_0, \Omega) \geq M^{-1} \mathrm{Cap}(E)^M.$$

The other direction of the lemma follows from Lemma 2.1.   ∎

Lemma 2.6. *Suppose* Ω_j *are* (ε) *John domains with centers* z_j, $j = 1, 2$, *and suppose* $d(z_1), d(z_2) \sim 1$. *Suppose also that* F *is harmonic on* $\Omega_1 \cup \Omega_2$ *and continuous on* $\bar{\Omega}_1 \cup \bar{\Omega}_2$. *Then if* $E \subset \partial\Omega_1 \cap \partial\Omega_2$ *satisfies*

$$\mathrm{Cap}(E) \geq \delta > 0,$$

there are geodesics $g_j \subset \Omega_j$ *from* z_j *to* $\partial\Omega_j$ *such that* g_1 *and* g_2 *terminate at the same point* $\zeta \in \partial\Omega_1 \cap \partial\Omega_2$, *and*

$$|F(\zeta) - F(z_j)| \leq A(\varepsilon, \delta)(\iint_{\Omega_j} |\nabla F|^2 dxdy)^{1/2}, \quad j = 1, 2.$$

Proof. Let

$$E_j = \{z \in \partial\Omega_j : |F(z) - F(z_j)| \geq \lambda(\iint_{\Omega_j} |\nabla F|^2 dxdy)^{1/2}\}.$$

If λ is large enough, Lemmata 2.3 and 2.5 show $\mathrm{Cap}(E_1 \cup E_2) < \delta$. Then $E \backslash (E_1 \cup E_2) \neq \phi$, so we may select ζ from that set.   ∎

3 QUASICIRCLES

We now give a quick outline of our proof for the case where K is a quasicircle. This represents the only idea of the paper. The rest of the sections contain only technical arguments which make the same philosophy work for the general case.

Let Ω_+ and Ω_- denote, respectively, the unbounded and bounded components of $\bar{\mathbf{C}} \backslash K$. Fix two points $z_\pm \in \Omega_\pm$ satisfying

$$\delta(z_+) \sim \delta(z_-) \sim |z_+ - z_-| \sim \delta,$$

and build domains $\mathcal{D}_\pm \subset \Omega_\pm$ which are bounded by quasicircles and such that $\partial\mathcal{D}_+ \cap \partial\mathcal{D}_-$ is a subarc of K with diameter $\sim \delta$. The points $z_\pm$ are made to be the "centers" of $\mathcal{D}_\pm$. Then by Lemma 2.6 there is A such that

$$|F(z_+) - F(z_-)| \leq A(\iint_{\mathcal{D}_+ \cup \mathcal{D}_-} |\nabla F|^2 dxdy)^{1/2}.$$

Standard smoothing techniques now show there is $\tilde{F} \in W^{1,2}(\mathbf{R}^2)$ such that $\tilde{F} = F$ outside of $K_\delta = \{z : d(z) \leq \delta\}$, $\tilde{F}$ is C^∞ near K, and

$$\iint_{K_\delta} |\nabla \tilde{F}|^2 dxdy \leq c \iint_{K_{c\delta}} |\nabla F|^2 dxdy.$$

Sending δ to zero we see that $F \in W^{1,2}(\mathbf{R}^2)$ and

$$\iint_{\mathbf{R}^2} |\nabla F|^2 dxdy = \iint_{K^c} |\nabla F|^2 dxdy.$$

If F is M quasiconformal on K^c, Lemma 2.2 and an argument similar to the one above show that F is globally quasiconformal. The point of this vague remark is that, *whatever argument we use*, it should show that F being M quasiconformal on K^c implies F is globally quasiconformal. (See the "Remark" in Section 1.)

4 SOME GEOMETRY

In this section we construct certain domains related to a point $x_0 \in K$ and a scale r. Since the John condition is scale invariant, we may assume $x_0 = 0$ and $r = 1$. We will add to K certain curves to obtain a new set $\hat{K}$ so that, in a certain sense, $\mathbf{C} \setminus \hat{K}$ looks like a union of quasidisks of diameter about 1 (near K).

Let $f : \mathbf{D}^* = \{|z| > 1\} \to \Omega$ be univalent with $f(\infty) = \infty$. Since $K = \partial\Omega$ is locally connected, f is continuous up to $\mathbf{T}$. Select angles $0 = \theta_0 < \theta_1 < \theta_2 < \cdots < \theta_N = 2\pi$ so that

$$|f(e^{i\theta}) - f(e^{i\theta_j})| \leq 1, \quad \theta_j \leq \theta \leq \theta_{j+1},$$

and

$$|f(e^{i\theta_j}) - f(e^{i\theta_{j+1}})| \geq \frac{1}{2}.$$

Now fix $M \geq 1$ and let $r_j < 1$ be the largest value of r so that

$$|f(re^{i\theta_j}) - f(e^{i\theta_j})| = M.$$

Setting $L_j = \{re^{i\theta_j}, r_j \leq r < 1\}$ we see that

(4.1) $$\text{distance}(L_j, L_k) \geq c(1 - r_j), \quad j \neq k,$$

for otherwise the John condition would be violated for the corresponding geodesics in Ω.

Lemma 4.1. $|1 - r_j| \sim |1 - r_{j+1}| \sim \text{distance}(L_j, L_{j+1})$.

Proof. We show that $|\theta_{j+1} - \theta_j| \leq C(1 - r_j)$. The proof that $|\theta_{j+1} - \theta_j| \leq C(1 - r_{j+1})$ is the same. The lemma will then follow from (4.1). Let $I = \{e^{i\theta} : \theta_j \leq \theta \leq \theta_j + \pi\}$. By symmetry

$$\omega(\zeta_j, I, .\mathbf{D}^*) = \frac{1}{2},$$

where $\zeta_j \equiv r_j e^{i\theta_j}$. Here $\omega(z, E, \mathcal{D})$ denotes the harmonic measure at z of $E \subset \partial\mathcal{D}$ in $\mathcal{D}$. By Beurling's so-called $\frac{1}{2}$ theorem [10], if we set $I_j = \{e^{i\theta} : \theta_j \leq \theta \leq \theta_{j+1}\}$,

$$\omega(\zeta_j, I_j, \mathbf{D}) = \omega(f(\zeta_j), f(I_j), \Omega) \leq CM^{-\frac{1}{2}},$$

because diameter $(f(I_j)) \leq 1$ and distance $(f(\zeta_j), f(I_j)) \geq M - 1$. Thus

$$\omega(\zeta_j, I \backslash I_j, \mathbf{D}^*) \geq \frac{1}{4}$$

if M is large enough, and the lemma follows from simple estimates on the Poisson kernel. ∎

Let $\mathcal{D}_j$ be the domain bounded by $\mathbf{T}$, L_j, L_{j+1}, and the line segment $[\zeta_j, \zeta_{j+1}]$ and let $\tilde{\zeta}_j = R_j e^{i\varphi_j}$ where $R_j - 1 = \frac{1}{2} \min(r_j - 1, r_{j+1} - 1)$, and $\varphi_j = \frac{1}{2}(\theta_j + \theta_{j+1})$. Then since we are assuming diameter $(K) >> 1$, each $\mathcal{D}_j$ looks like a quadrilateral (in $\mathbf{D}^*$ with one side on $\mathbf{T}$) with bounded geometry.

Lemma 4.2. $\Omega_j = f(\mathcal{D}_j)$ is an (ε') John domain with John center $z_j = f(\tilde{\zeta}_j)$.

Proof. Let $\zeta \in \mathcal{D}_j$ and let $L = [\zeta, \tilde{\zeta}_j]$ be the line segment from ζ to $\tilde{\zeta}_j$. Then $L \subset \mathcal{D}_j$ and if $\zeta' \in L$, distance$(\zeta', \partial\mathcal{D}_j) \geq c|\zeta' - \zeta|$. (This follows from the elementary geometry of $\mathcal{D}_j$.) Now if L' is the geodesic from ζ to ∞ in $\mathbf{D}^*$, $L' = \{R\zeta : R \geq 1\}$, $\rho(\zeta', L') \leq C$ for all $\zeta' \in L$, where ρ is the hyperbolic metric on $\mathbf{D}^*$. The lemma now follows from the John property on the arc $f(L')$ and the distortion theorem for f. The details are left to the reader.

Lemma 4.2 is actually a special case of the following fact:

If $f : \mathbf{D} \to \Omega$, $f(0) = z_0$, and Ω is an (ε) John domain with John center z_0, and if $\mathcal{D} \subset \mathbf{D}$ is a (δ) John domain with John center the origin, then $f(\mathcal{D})$ is an $\eta(\varepsilon, \delta)$ John domain with John center z_0.

We leave a proof of this statement as an exercise for the reader. ∎

At this point we remark that $\hat{\Omega}_j = \text{interior of } \bar{\Omega}_j$ is a $\delta(\varepsilon)$ quasicircle if $\bar{\mathbf{C}} \backslash K = \Omega$. (This is, e.g., the case for the Julia set corresponding to $z^2 + i$.) A most unfortunate complication is that this statement is easily seen to be false if $\bar{\mathbf{C}} \backslash K$ is allowed to have bounded components. This necessitates the technical construction of our next section. The reader interested only in the case where $\Omega = \bar{\mathbf{C}} \backslash K$ may skip to Section 7, noticing that Proposition 6.1 has already been proven for quasicircles.

5 SOME ADDITIONAL CURVES

We now add some additional curves to K. Let $\mathcal{O}_j$ be a bounded component of $\bar{\mathbf{C}} \backslash K$. Then by the definition of the domains Ω_k, each $\partial \Omega_k$ intersects $\partial \mathcal{O}_j$ in either a connected set or the empty set. Let us for the moment reorder the Ω_k so that $\partial \Omega_1$, $\partial \Omega_N$ are exactly those domains such that $\partial \Omega_n \cap \partial \mathcal{O}_j$ consists of more than one point (and hence an arc). Let $\delta > 0$ be a small constant to be fixed later and fix a Riemann mapping $f_j : \mathbf{D} \to \mathcal{O}_j$ so that $I_1, \ldots, I_N$ are intervals with $f_j(I_n) = \partial \Omega_n \cap \partial \mathcal{O}_j$. By selecting $f_j(0)$ to lie very close to $\partial \Omega_1 \cap \partial \mathcal{O}_j$ we may assume $\ell(I_1) \approx 2\pi$. Let T_1 be the tent-shaped region bounded by $\mathbf{T} \backslash I_1$ and two straight lines in $\mathbf{D}$ which intersect $\mathbf{T} \backslash I_1$ at angle δ. The T_1 is a "thin sliver." Define $\mathcal{U}_1 = \hat{\mathcal{U}}_1 = \mathbf{D} \backslash T_1$ so that $\partial \mathcal{U}_1 \cap \mathbf{T} = I_1$. For $n \geq 2$ let L_n^1, L_n^2 be the two lines which start at the endpoints of I_n, go into $\mathbf{D}$, and make angle $= \delta$ with $\mathbf{T} \backslash I_n$. Let $J_n = \{(1 - \delta^{-1}\ell(I_n))e^{i\theta} : 0 \leq \theta \leq 2\pi\}$, and let $\hat{\mathcal{U}}_n$ be the domain bounded by the four arcs I_n, L_n^1, L_n^2, J_n. Then $\hat{\mathcal{U}}_n$ almost fills up a rectangle with length (along $\mathbf{T}$) $= \delta^{-2}\ell(I_n)$ and width (in the direction orthogonal to $\mathbf{T}$) $= \delta^{-1}\ell(I_n)$. Then by elementary estimates on the Poisson kernel,

$$(5.1) \qquad \{\omega(z, I_n, \mathbf{D}) \geq c_1\delta\} \subset \hat{\mathcal{U}}_n \subset \{\omega(z, I_n, \mathbf{D}) > c_2\delta\}.$$

By reordering we may now assume that

$$\ell(I_2) \geq \ell(I_3) \geq \cdots \geq \ell(I_N).$$

Define $\mathcal{U}_n = \hat{\mathcal{U}}_n \backslash \bigcup_{k=1}^n \mathcal{U}_k$ so that $\bigcup_{n=1}^N \mathcal{U}_n = \bigcup_{n=1}^N \hat{\mathcal{U}}_n$. Recall that a domain $\mathcal{U}$ is called an M Lipschitz domain if there is $z_0 \in \mathcal{U}$ and $R > 0$ such that

$$\partial \mathcal{U} = \{z_0 + Rr(\theta)e^{i\theta} : 0 \leq \theta \leq 2\pi\}$$

where

$$(1 + M)^{-1} \leq r(\theta) \leq 1 \qquad \text{for all} \quad \theta$$

and

$$|r(\theta) - r(\theta')| \leq M|\theta - \theta'|.$$

Lemma 5.1. *$\mathcal{U}_n$ is a $M(\delta)$ Lipschitz domain, $1 \leq n \leq N$. Furthermore, if $\zeta \in \mathbf{D} \cap \partial \mathcal{U}_n$, there is $\varphi \in [0, 2\pi]$ such that the line segment $\bar{\mathbf{D}} \cap \{\zeta + re^{i\theta} : r \geq 0\}$*

lies in $\bar{\mathcal{U}}_n$ and has endpoint on I_n whenever

$$|\theta - \varphi| \leq c\delta^2.$$

The proof of the lemma is an exercise in elementary geometry. Now let $\mathcal{O}_j^n = f_j(\mathcal{U}_n)$. If we consider any Ω_k, we have for each $\mathcal{O}_j$, such that $\partial\mathcal{O}_j \cap \partial\Omega_k$ is an arc, obtained a domain $\mathcal{O}_j^k \subset \mathcal{O}_j$ (sometimes $\mathcal{O}_j^k = \mathcal{O}_j$) with the property that $\partial\mathcal{O}_j^k \cap \partial\mathcal{O}_j \subset \partial\Omega_k$. Let $\mathcal{F}_k = \{\mathcal{O}_j^k : \partial\mathcal{O}_j \cap \partial\Omega_k \text{ is an arc}\}$ and let $\tilde{\Omega}_k = $ interior of closure of $\Omega_k \cup \bigcup_{\mathcal{F}_k} \mathcal{O}_j^k$.

Lemma 5.2. *$\partial\tilde{\Omega}_k$ is an $\eta(\varepsilon, \delta)$ quasicircle.*

Proof. Let $\gamma_j^k = \mathcal{O}_j \cap \partial\mathcal{O}_j^k$. Then $\partial\tilde{\Omega}_k = \partial\Omega_k \cup \bigcup_{\mathcal{F}_k} \gamma_j^k$. We first claim $\tilde{\Omega}_k$ is an $\eta(\varepsilon, \delta)$ John domain. It is only necessary to find for every $z_0 \in \partial\tilde{\Omega}_k$ an arc $\gamma \subset \tilde{\Omega}_k$ which has endpoints z_0 and z_k such that

$$\text{distance}(z, \partial\tilde{\Omega}_k) \geq \eta|z - z_0|, \quad z \in \gamma.$$

If $z_0 \in \partial\Omega_k$ this is clear by Lemma 3.3. We therefore assume $z_0 \in \gamma_j^k$ for some j. By Lemmata 2.2, 2.3, and 5.1 there are angles $\varphi_{-1} < \varphi_0 < \varphi_1$ such that $|\varphi_\ell - \varphi_m| \sim \delta^2, \ell \neq m$, such that

$$\mathbf{D} \cap \{f_j^{-1}(z_0) + re^{i\theta}, r > 0\} \subset \mathcal{U}_j,$$

whenever $\varphi_{-1} \leq \theta \leq \varphi_1$, and such that

$$\ell(\Gamma_m) \equiv \ell(f_j(\{f_j^{-1}(z_0) + re^{i\varphi_m} : r > 0\})) \leq Cd(z_0).$$

Furthermore, Lemma 5.1 allows us to assume that Γ_m is a Jordan arc and if $z \in \Gamma_\ell$ and $|z - z_0| \geq \frac{1}{2}d(z_0)$,

(5.2) $$\text{distance}(z, \Gamma_m) \geq cd(z_0), \quad \ell \neq m.$$

Now let δ_m be the endpoint of Γ_m on $\partial\Omega_k$ and let γ_{-1} (resp. γ_1) be the John geodesic from ζ_{-1} (resp. ζ_1) in Ω_k to z_k (the John center of Ω_k). Then the curve $\gamma = \Gamma_{-1} \cup \Gamma_1 \cup \gamma_{-1} \cup \gamma_1$ surrounds ζ_0 and by the John condition on Ω_k,

(5.3) $$\text{distance}(\zeta_0, \gamma) \geq cd(z_0).$$

(Notice here that we are implicitly using the fact that $z \in \mathcal{O}_j^k$ implies $d(z) \leq C$. This, in turn, follows from (5.1) and either Lemma 2.2 or 2.3.) Notice also that the interior of γ must lie entirely in $\tilde{\Omega}_k$.

Let γ_0 be the John geodesic in Ω_k from ζ_0 to z_k. We claim that the John condition for $\tilde{\Omega}_k$ holds on $\Gamma_0 \cup \gamma_0$. First suppose that $z \in \Gamma_0$ and $|z - z_0| \leq \frac{1}{2}d(z_0)$. Then by the distortion theorem for f_j,

$$\text{distance}(z, \partial\tilde{\Omega}_k) \geq \text{distance}(z, \gamma) \geq c|z - z_0|.$$

Now by inequality (5.2),

$$d(z, \partial\tilde{\Omega}_k) \geq d(z, \gamma) \geq c|z - z_0|$$

whenever $z \in \Gamma_0$ and $|z - z_0| \geq \frac{1}{2}d(z_0)$. (Here we have used the John property on Ω_k to obtain distance $(z, \gamma_{-1} \cup \gamma_1) \geq c|z - z_0|$.)

We must finally check the John condition on γ_0. If $z \in \gamma_0$ and $|z - \zeta_0| \leq cd(z_0)$, the inequality on distance $(z, \partial\tilde{\Omega}_k)$ follows from (5.3) and the fact that $|z_0 - \zeta_0| \leq Cd(z_0)$. If $z \in \gamma_0$ and $|z - \zeta_0| > cd(z_0)$, the inequality on distance$(z, \partial\tilde{\Omega}_k)$ follows from the John condition distance$(z, \partial\Omega_k) \geq c|z - \zeta_0|$. We have thus established that $\tilde{\Omega}_k$ is a John domain.

We now claim that $G_k = \bar{\mathbf{C}} \backslash (\tilde{\Omega}_k)$ is a John domain. We note that by the definition of $\tilde{\Omega}_k$, $\partial\tilde{\Omega}_k = \partial G_k$. Now fix a point $z_0 \in G_k$.

Case A. $z_0 \in \overline{(\tilde{\Omega}_j)}$ for some j. (Then $j \neq k$.) First draw the John geodesic in Ω_j from z_0 to z_j. We then draw the geodesic (in the Poincaré metric of Ω) from z_j to ∞. By the construction of the domains Ω_j (Lemma 3.2) this is a John geodesic. The union of these two geodesics provides the arc joining z_0 to ∞.

Case B. $z_0 \in \Omega \backslash \bigcup_j \tilde{\Omega}_j$. Let γ be the Poincaré geodesic in Ω from z_0 to ∞. Then by Lemma 3.2, γ is a John geodesic in G_k.

Case C. $z_0 \notin \Omega \cup \bigcup_j \tilde{\Omega}_j$. Then $z_0 \in \mathcal{O}_{j_0}$ for some j_0. Let

$$A = \sup_j \text{diameter } \Omega_j,$$

so that $A \sim 1$. If $d(z_0) \geq 2A$ there is a half line γ (to ∞ from z_0) which is a John geodesic in G_k. If $d(z_0) < 2A$ there is a hyperbolic geodesic (which is also a John arc in G_k) γ_1 from z_0 to $z_1 \in \mathcal{O}_{j_0,\ell}$ where $\ell \neq k$, $d(z) \geq 1$ on γ_1, and $\ell(\gamma) \leq C$. (This follows from the definition of the domains $\mathcal{O}_{j,k}$.) By Case A there is a John geodesic γ_2 from z_1 to ∞ in G_k. The curve $\gamma = \gamma_1 \cup \gamma_2$ is the required John arc.

The proof is now completed by first observing that a simply connected domain $\mathcal{D}$ with $\partial\mathcal{D}$ locally connected, $\mathcal{D} = \text{Interior}(\bar{\mathcal{D}})$, and $\mathbf{C}\backslash\bar{\mathcal{D}}$ connected is bounded by a Jordan curve, and then invoking the following fact (see [9]):

A Jordan domain $\mathcal{D}$ is bounded by a quasicircle if and only if $\mathcal{D}$ and $\bar{\mathbf{C}}\backslash\bar{\mathcal{D}}$ are John domains.   ∎

6 AN ESTIMATE ON CAPACITY

We now seek to imitate the proof given in Section 3. What is required is an estimate implying that $|F(z) - F(z_j)|$ is not too large on $\partial\tilde{\Omega}_j$, except for a set of small capacity. While $\tilde{\Omega}_j$ is a quasidisk, $\tilde{\Omega}_j \cap K^c$ is not necessarily connected. This means we cannot simply apply Lemma 2.3. We state our result as a proposition; its proof will be broken into several steps. The result we state is far from optimal, but it is all we need. Let $\tilde{\tilde{\Omega}}_k$ be the domain obtained by adding to $\tilde{\Omega}_k$ the set

$$\bigcup_j \{z \in \mathcal{O}_j : \rho(z, \partial\mathcal{O}_{j,k}) < 1\},$$

where ρ is the hyperbolic metric on $\mathcal{O}_j$. The domains $\tilde{\tilde{\Omega}}_k$ then satisfy

$$\sum \chi_{\tilde{\tilde{\Omega}}_k} \leq C.$$

Proposition 6.1. *Suppose H is continuous on the closure of $\tilde{\tilde{\Omega}}_k$ and harmonic on $\tilde{\tilde{\Omega}}_k \backslash K$. Then if*

$$\iint_{\tilde{\tilde{\Omega}}_k \backslash K} |\nabla H|^2 dx dy = 1,$$

we have the estimate

$$\text{Cap}(\{z \in \partial\tilde{\Omega}_k : |H(z) - H(z_k)| > \lambda\}) = o(1)$$

as $\lambda \to \infty$.

Proof. Let $E_1 = \{z \in \partial\tilde{\Omega}_k \cap \partial\Omega_k : |H(z) - H(z_k)| > \lambda\}$ and let $E_2 = \{z \in \partial\tilde{\Omega}_k \backslash \partial\Omega_k : |H(z) - H(z_k)| > \lambda\}$. Then by Lemmata 2.3 and 2.5, $\text{Cap}(E_1) = o(1)$ as $\lambda \to \infty$, so it is sufficient to show $\text{Cap}(E_2) = o(1)$ as $\lambda \to \infty$.

Step 1. Construction of Some Special Points. Let $\{z_n\}$ be a collection of points in $\partial\tilde{\Omega}_k \backslash \partial\Omega_k$ satisfying

$$|z_n - z_m| \geq \frac{1}{4} d(z_n), \quad \forall n, m$$

and

$$\inf_n |z - z_n| \leq \frac{1}{2} d(z), \quad \forall z \in \partial\tilde{\Omega}_k \backslash \partial\Omega_k.$$

We will now form for each z_n a point $z_n^* \in \Omega_k$. For an arbitrary point $z \in \partial\mathcal{O}_{j,k} \backslash \partial\Omega_k$ we let $K_z = \{\zeta \in \partial\tilde{\Omega}_k : |z - \zeta| \leq 2d(z)\}$ so that $\text{Cap}(K_z) \geq c d(z)$. By the John condition and Lemma 2.5, there is a point $z^* \in \Omega_k$ such that $d(z) \sim d(z^*) \sim |z - z^*|$ and there is a set $\tilde{K}_z \subset K_z$ such that

(6.1) $$\text{Cap}(\tilde{K}_z, z^*, \Omega), \text{Cap}(\tilde{K}_z, z, \mathcal{O}_j) \geq c.$$

Denote by f a Riemann mapping from Ω_k to **D** with $f(z_k) = 0$, where z_k is the "center" of Ω_k. We can move z^* slightly so that $f(z^*)$ has the form

(6.2) $$f(z^*) = (1 - 2^{-\ell}) \exp\{im2^{-\ell}\pi\}$$

for some $\ell, m \in \mathbf{N}$. By this method we produce from our collection $\{z_n\}$ a new collection $\{z_n^*\}$. Notice that it is possible that $z_n^* = z_m^*$ even if $n \neq m$, but then $d(z_n) \sim d(z_m) \sim |z_n - z_m|$.

Step 2. Another Geometric Construction. Let $\{z_n\}$ be the collection of points in Step 1. Let $\{I_n\}$ be a collection of subarcs of $\partial\tilde{\Omega}_k \backslash \partial\Omega_k$ such that $\bigcup_n I_n = \partial\tilde{\Omega}_k \backslash \partial\Omega_k$, $I_n \cap I_m = \phi$ when $n \neq m$, diameter$(I_n) \sim d(z_n)$, and $|z - z_n| \leq \frac{1}{2} d(z_n)$ for $z \in I_n$. We also define I_n^* to be the arc

$$f^{-1}(\{(1 - 2^{-\ell}) \exp\{i(m + t)2^{-\ell}\pi\} : 0 \leq t \leq 1\}).$$

See (6.2) for notation. Then I_n^* has diameter $\sim d(z_n)$ and if $z \in I_n^*$ the hyperbolic distance from z to z_n^* (in Ω_k) is bounded by C.

With the notation of (6.1) we also denote by J_n the subarc of $\mathbf{T}$

$$J_n = \{e^{i\theta} : m2^{-\ell}\pi < \theta \leq (m + 1)2^{-\ell}\pi\}$$

and we denote by Q_n the "square"

$$Q_n = \{re^{i\theta} : (1 - 2^{-\ell}) \leq r \leq 1, e^{i\theta} \in J_n\}.$$

We now use the standard terminology that an arc J_m is maximal in a subcollection $\mathcal{F}$ of $\{J_n\}$ if $J_m \in \mathcal{F}$ and $J_\ell \in \mathcal{F}$, $\ell \neq m$, imply either $J_\ell \cap J_m = \phi$ or $J_\ell \subset J_m$. Notice (by the John condition) that if $J_\ell \subset J_m$,

$$(6.3) \qquad\qquad |z_\ell - z_m^*| \leq Cd(z_m).$$

Finally, if $\hat{\mathcal{F}}$ is a subcollection of $\{I_n\}$ and $E = \bigcup_{I_n \in \hat{\mathcal{F}}} I_n$ we denote by E^* the set

$$E^* = \bigcup_{J_n \in \mathcal{F}} I_n^*$$

where

$$\mathcal{F} = \{J_n : I_n \in \hat{\mathcal{F}} \text{ and } J_n \text{ is maximal}\}.$$

Step 3. A Capacitary Estimate. Let E and E^* be sets as in the previous paragraph.

Lemma 6.2. $\mathrm{Cap}(E^*) \geq c\,\mathrm{Cap}(E)$.

Proof. Let μ be a probability measure on E satisfying

$$\int \log \frac{1}{|z - \zeta|} d\mu(\zeta) \leq \gamma, \quad z \in \mathbf{C}.$$

We relabel the intervals $J_n \in \mathcal{F}$ so that $\mathcal{F} = \{J_1, J_2, \ldots\}$ and $d(z_1) \geq d(z_2) \geq \cdots \geq d(z_n) \geq d(z_{n+1}) \geq \cdots$. Define

$$E_n = \{z \in E : |z - z_n| \leq Cd(z_n) \text{ and } |z - z_m| > Cd(z_m), m < n\},$$

so that by (6.3),

$$E = \bigcup_n E_n.$$

Notice that the sets E_n are pairwise disjoint.

Now define a probability measure μ^* by setting μ^* to have uniform distribution on the center half $\frac{1}{2} I_n^*$ of I_n^*, $\mu^*(I_n^* \setminus \frac{1}{2} I_n^*) = 0$, and

$$\mu^*(I_n^*) = \mu(E_n).$$

By the construction of I_n^* and $\frac{1}{2} I_n^*$,

$$\text{dist}(\frac{1}{2} I_n^*, \frac{1}{2} I_m^*) \geq c d(z_n), \quad \forall \, n \neq m.$$

Let $z \in E$ and $z' \in \frac{1}{2} I_n$ where n satisfies $z^* \in I_n$. Then

$$\int \log \frac{1}{|z' - \zeta|} \, d\mu^*(\zeta) = \int_{\{|z'-\zeta| \leq Ad(z)\}} + \int_{\{|z'-\zeta| > Ad(z)\}}$$

$$\leq c + \int_{\{|z'-\zeta| \leq Ad(z)\}} \log \frac{1}{|z - \zeta|} \, d\mu(\zeta)$$

$$+ c + \int_{\{|z'-\zeta| > Ad(z)\}} \log \frac{1}{|z' - \zeta|} \, d\mu(\zeta)$$

$$\leq 2c + \gamma,$$

and Lemma 6.2 is established. ∎

Step 4. Proof of the Proposition. By Lemmata 2.3 and 2.5 and by estimate (6.1),

$$|H(z) - H(z^*)| \leq 1.$$

Now let $\mathcal{D} = \Omega_k \setminus \bigcup_{\mathcal{F}} \hat{Q}_n$, where the Q_n are as defined in Step 2 and $\hat{Q}_n = f^{-1}(Q_n)$. Then by Lemma 4.1, $\mathcal{D}$ is an (ε') John domain. We define our collection $\mathcal{F}$ to be $\{I_n : \exists z \in I_n, |H(z) - H(z_k)| \geq \lambda\}$. Then if $I_n \in \mathcal{F}, |H(\zeta) - H(z_k)| \geq \lambda - c$ for all $\zeta \in I_n$. (This is why we slightly enlarge $\tilde{\Omega}_k$ to $\tilde{\tilde{\Omega}}_k$.) By our previous estimate,

$$|H(z) - H(z_k)| \geq \lambda - 2c \quad \text{on} \quad I_n^*,$$

for any $I_n \in \mathcal{F}$. Setting as before

$$E = \bigcup_{\mathcal{F}} I_n$$

$$\text{and} \quad E^* = \bigcup I_m^*,$$

we have $\text{Cap}(E^*) \geq c \, \text{Cap}(E)$. Now since

$$\iint_D |\nabla H|^2 dx dy \leq 1,$$

it follows from Lemmata 2.3 and 2.5 that

$$\text{Cap}(E^*) = o(1) \qquad \text{as} \quad \lambda \to \infty.$$

This completes the proof of Proposition 6.1. ∎

7 PROOF OF THEOREM 1

Let $\tilde{\Omega}_j$ and $\tilde{\Omega}_k$ be two domains satisfying

$$\text{distance}(\tilde{\Omega}_j, \tilde{\Omega}_k) \leq 1.$$

It is an exercise to find domains $\tilde{\Omega}_{j_1}, \ldots, \tilde{\Omega}_{j_N}$ where $j_1 = j$, $j_N = k$, and $\partial \tilde{\Omega}_{j_m} \cap \partial \tilde{\Omega}_{j_{m+1}}$ is an arc of diameter $\geq c$, and $N \leq C$. (Use the fact that Ω is a John domain and each $\tilde{\Omega}_j$ is an η quasicircle, i.e., Lemma 5.2.) Then by Proposition 6.1,

$$|F(z_j) - F(z_k)| \leq \sum_{m=1}^{N-1} |F(z_{j_m}) - F(z_{j_{m+1}})|$$

$$\leq C \sum_{m=1}^{N} \left(\iint_{\tilde{\Omega}_{j_m} \setminus K} |\nabla F|^2 dx dy \right)^{1/2}$$

$$\leq C' \left(\iint_{\{z \in K^c : |z - z_j| \leq C\}} |\nabla F|^2 dx dy \right)^{1/2}.$$

We also notice by (6.1) that if $z \in \tilde{\Omega}_k$ and $d(z) \geq 1$,

$$|F(z) - F(z_k)| \leq C \left(\iint_{\tilde{\Omega}_k \setminus K} |\nabla F|^2 dx dy \right)^{1/2}.$$

Putting our last two estimates together we see there is $\tilde{F} \in C^\infty(\mathbf{R}^2)$ such that $\tilde{F}(z) = F(z)$ when $d(z) \geq 1$ and

$$\iint_{\{z \in K^c : |d(z)| \leq 1\}} |\nabla \tilde{F}|^2 dx dy \leq C \iint_{\{z \in K^c : |d(z)| \leq C\}} |\nabla F|^2 dx dy.$$

Here we are using the fact that, by the construction of the $\tilde{\Omega}_k$, $\{z \in K^c : d(z) \leq 1\} \subset \bigcup_k \tilde{\Omega}_k$. Since the John condition is dilation invariant, we may now build a sequence $\tilde{F}_n \in C^\infty(\mathbf{R}^2)$ with $\tilde{F}_n(z) = F(z)$ when $d(z) \geq \frac{1}{n}$ and

$$\iint_{\{z \in K^c \,:\, d(z) \leq \frac{1}{n}\}} |\nabla \tilde{F}_n|^2 dxdy \leq C \iint_{\{z \in K^c \,:\, d(z) \leq \frac{c}{n}\}} |\nabla F|^2 dxdy.$$

Since $|K| = 0$, it follows that $F \in W^{1,2}(\mathbf{R}^2)$ and

$$\iint_{\mathbf{R}^2} |\nabla F|^2 dxdy = \iint_{K^c} |\nabla F|^2 dxdy.$$

8 PROOF OF THEOREM 2

Let Ω be an (ε) John domain with compact boundary K of diameter one, let $\{Q_j\}$ denote the Whitney decomposition of Ω into dyadic squares [11], and let z_j be the center of Q_j. Let $A = A(\varepsilon)$ be a large constant and define $\mathcal{F}_n = \{z_j : A^{-n} \leq d(z_j) \leq A\}$. It is an exercise with the John condition to construct a connected graph G_0 such that every edge in G_0 is of the form $[z_j, z_k]$ where $\partial Q_j \cap \partial Q_k \neq \phi$, and where the vertices V_0 of G_0 satisfy

$$\mathcal{F}_0 \subset V_0 \subset \{z_j : 1 \leq d(z_j) \leq A^2\}.$$

We also build G_0 so that

(8.1) If $d(z_j) \geq 1$ and $z_j \notin V_0$ then Q_j is in the unbounded component of $\mathbf{C} \backslash \bigcup_{z_k \in \mathcal{F}_0} Q_k$.

It is now an exercise (with induction) to construct connected graphs G_n with the following properties:

(8.2) Every edge in G_n is of the form $[z_j, z_k]$ for some $z_j, z_k \in \mathcal{F}_{n+1} \cup V_0$ where $\partial Q_j \cap \partial Q_k \neq \phi$.

(8.3) Every $z_j \in \mathcal{F}_n$ is in V_n, the vertices of G_n.

(8.4) If $\rho(z_j, z_k)$ is the graph distance on G_n,

$$\inf_{z_k \in \mathcal{F}_n} \rho(z_j, z_k) \leq C, \quad z_k \in G_n.$$

(8.5) $G_n \subset V_{n+1}$.

Notice that we have chosen G_0 to be connected. Let $z_0 \in V_0$ be an extreme point of the (planar set) convex hull (G_0). We may assume by induction that each G_n is actually a *directed* graph in the following sense. Each edge $[z_j, z_k]$ is directed in the sense that (perhaps switching j and k)

(8.6) $\rho(z_j, z_0) = \rho(z_k, z_0) + 1$.

Such an edge is an outgoing edge from z_j. It is not hard to see that we may choose the G_n so that

(8.7) Each $z_j \neq z_0$ has exactly one outgoing edge.

Lemma 8.1. *The graph G_n is simply connected, i.e., it contains no loops.*

Proof. Suppose, to the contrary, that there is a loop in G_n. Let z_j be a vertex in the loop maximizing $\rho(z_j, z_0)$. Then z_j has two outgoing edges (by (8.6)) and this contradicts (8.7).

Let $G = \lim_n G_n$ be the limiting graph, so that G is simply connected. It is clear that $K \cup G$ is connected. Notice by (8.3) that

(8.8) For every $z_j \in G$ there is an arc $\gamma \subset G$ from z_j to z_0 which satisfies the ε' John condition in Ω.

In other words, G is a John graph.

For a Whitney square Q_j with $z_j \in G$ let $\{\mathcal{L}_k^j\}$ denote all the edges of G with one endpoint being z_j. Define

$$I_k^j = \{z \in \partial Q_j : \text{distance}(z, \mathcal{L}_k^j) < \delta \, \text{diam}(Q_j)\},$$

where δ is a small constant, and put

$$S_j = \partial Q_j \backslash \bigcup_k I_k^j, \quad j \neq 0.$$

For the special point $z_0 \in G$ we select a Whitney square Q_ℓ such that $z_\ell \notin G$, $\partial Q_0 \cap \partial Q_\ell \neq \phi$, and we put

$$S_0 = \partial Q_0 \backslash (I_\ell^0 \cup \bigcup_k I_k^0),$$

$$\hat{\Omega} = \Omega \backslash \bigcup_{z_j \in G} S_j. \qquad \blacksquare$$

Lemma 8.2. $\hat{\Omega}$ *is simply connected.*

Proof. Let $\hat{\Omega}_+ = \cup\{\hat{\Omega} \cap Q_j : z_j \notin G\}$, $\hat{\Omega}_- = \cup\{\hat{\Omega} \cap Q_j : z_j \in G\}$ so that

$$\hat{\Omega} = \hat{\Omega}_+ \cup \hat{\Omega}_- \cup I_\ell^0.$$

By condition (8.1), $\hat{\Omega}_+$ is simply connected (in $\bar{\mathbb{C}}$), so it is only necessary to check that $\hat{\Omega}_-$ is simply connected.

We first verify that $\hat{\Omega}_-$ is connected. Let $z \in Q_j \cap \hat{\Omega}_-$ and let γ be an arc in G connecting z_j to z_0. Then $\gamma' = [z, z_j] \cup \gamma$ is an arc in $\hat{\Omega}_-$ which connects z to z_0.

Now suppose that γ is a loop in $\hat{\Omega}_-$ that is not homologous to zero. It is then an elementary exercise to homotopy γ to γ', a loop in G that is not homologous to zero. This contradicts Lemma 8.1.

It is clear from the construction of $\hat{\Omega}$ that $\partial\Omega \subset \partial\hat{\Omega}$. To verify that $\hat{\Omega}$ is a John domain we must look at two cases.

Case 1. $z \in \hat{\Omega}_+ \cap Q_j$. There is arc γ from z_j to some $z_k \notin G$ such that length$(\gamma) \leq C$, $d(z) \geq 1$ on γ, and z_k is not in the convex hull of $\partial\hat{\Omega}$. By selecting a suitable ray R from z_k to ∞ we then see that

$$[z, z_j] \cup \gamma \cup R$$

is the required John arc.

Case 2. $z \in \hat{\Omega}_- \cup I_\ell^0$. Let γ be a John arc from z_ℓ (the center of the special Whitney square Q_ℓ adjacent to z_0) to ∞. Then if $z \in Q_j$ and $\gamma' \subset G$ is the John arc from z_j to z_0 guaranteed by condition (8.7), we see that

$$[z, z_j] \cup \gamma' \cup [z_0, z_\ell] \cup \gamma$$

is the required John arc. ∎

Yale University

REFERENCES

[1] L. V. Ahlfors. *Lectures on Quasiconformal Mappings.* Wadsworth, 1966; 1987.

[2] ———. *Conformal Invariants.* McGraw-Hill, 1973.

[3] L. Carleson and P. W. Jones. "On coefficient problems for univalent functions and conformal dimension." *Duke Math J.* **66** (1992), 169–206.

[4] A. Douady and J. H. Hubbard. "Étude dynamique des polynomes complexes, I, II." *Publ. Math. d'Orsay*, 84–102.

[5] F. W. Gehring and K. Hag. "Quasi-hyperbolic geodesics in John domains." *Math. Scand.* **65** (1989), 75–92.

[6] P. W. Jones. "Quasiconformal mappings and extendibility of functions in Sobolev spaces." *Acta Math.* **147** (1981), 71–88.

[7] R. Kaufman. "Fourier-Stieltjes coefficients and continuation of functions." *Ann. Acad. Sci. Fenn.* **9**, 27–31.

[8] O. Lehto and K. I. Virtanen. *Quasiconformal Mappings in the Plane.* Springer Verlag, 1973.

[9] R. Näkki and J. Väisälä. "John Disks." *Exp. Math.* **9** (1991), 3–43.

[10] Ch. Pommerenke. *Univalent Functions.* Göttingen, 1975.

[11] E. M. Stein. *Singular Integrals and Differentiability Properties of Functions.* Princeton University Press, 1970.

13

Oscillatory Integrals and Non-Linear Dispersive Equations

*Carlos E. Kenig**

In this chapter I will describe a collection of results, obtained jointly with G. Ponce and L. Vega, on well posedness and non-linear scattering, with data in Sobolev spaces, for solutions to a variety of non-linear dispersive equations.

I hope that I will be able to convey, in describing these results, and the methods used in their proofs, how many of the ideas pioneered and developed by E. M. Stein, such as complex interpolation, the application of oscillatory integrals to the study of restriction theorems for the Fourier transform, and the application of the Kolmogorov-Seliverstov-Plessner method to study maximal operators in which curvature is present, can be used in a natural way in the study of non-linear dispersive partial differential equations. The results thus obtained are sharp in many instances, and do not seem to be attainable by any other approach.

When considering an initial value problem (I.V.P.) of the form

$$
\begin{cases}
\dfrac{\partial u}{\partial t} + A(u) = 0 \\[2mm]
u|_{t=0} = u_0
\end{cases}
$$

where A is a linear or non-linear operator and $u_0 \in X$, a Banach space, we say that the problem is globally well posed in X if given $u_0 \in X$, there exists a unique $u \in C((-\infty, +\infty); X) \cap L^\infty((-\infty, +\infty); X)$ satisfying $\frac{\partial u}{\partial t} + A(u) = 0$, with $u(0) = u_0$, and such that the mapping

$$
u_0 \mapsto u \in C((-\infty, +\infty); X) \cap L^\infty((-\infty, +\infty); X)
$$

*This research was supported, in part, by the National Science Foundation.

is continuous. If this is true when the time interval $(-\infty, +\infty)$ is replaced by $(-T, T)$, $T = T(u_0) > 0$, we say that the problem is locally well-posed.

From now on, $X = H^s(\mathbb{R}^n) = \{f \in L^2(\mathbb{R}^n) : (-\Delta)^{s/2} f \in L^2(\mathbb{R}^n)\}$, or $X = \dot{H}^s(\mathbb{R}^n) = \{f \in L^2_{\text{loc}}(\mathbb{R}^n) : (-\Delta)^{s/2} f \in L^2(\mathbb{R}^n)\}$. To illustrate these concepts, we present two families of well studied examples.

Example 1 (Generalized Burger's equation).

$$(1) \qquad \begin{cases} \dfrac{\partial u}{\partial t} + u^k \dfrac{\partial u}{\partial x} = 0 & k \geq 1 \\[2mm] u|_{t=0} = u_0. \end{cases}$$

Then (see [58]), (1) is locally well posed in $H^s(\mathbb{R})$, $s > 3/2$, and is not locally well posed in $H^s(\mathbb{R})$, $s \leq 3/2$, for any $k \geq 1$.

Example 2 (Semilinear Schrödinger equation in $\mathbb{R} \times \mathbb{R}$).

$$(2) \qquad \begin{cases} \dfrac{\partial u}{\partial t} + i \dfrac{\partial^2 u}{\partial x^2} + i|u|^k u = 0, & k > 0 \\[2mm] u|_{t=0} = u_0. \end{cases}$$

Then (see [58], [21] and [62]), for $0 < k < 4$, (2) is globally well posed in $L^2(\mathbb{R})$, while for $4 \leq k$, (2) is locally well posed in $H^s(\mathbb{R})$, $s \geq s_k = \frac{k-4}{2k}$, and globally well posed for small data in $\dot{H}^{s_k}(\mathbb{R})$. Analogous results hold in $\mathbb{R}^n$; there 4 is replaced by $4/n$, and $\frac{k-4}{2k}$ by $\frac{kn-4}{2k}$.

Example 1 is treated by using energy estimates. Example 2 is treated using the extensions due to R. Strichartz ([57]) of the Stein-Tomas ([61]) restriction theorem for the Fourier transform.

In our work, the general strategy when studying a non-linear dispersive problem is to use oscillatory integral estimates to obtain sharp inequalities for the homogeneous and inhomogeneous associated linear problems. Many times, these are estimates that involve fractional derivatives and are measured in mixed norms, where the time variable norms are taken first. In order to apply those estimates to our non-linear problems, we develop mixed norm (and weighted norm) Leibniz and chain rules for fractional derivatives. Using our linear estimates, we then construct suitable space-time Banach spaces, in which we solve our non-linear problem by using the contraction mapping principle.

In the remainder of this chapter, I will describe the most salient results thus obtained. I will then illustrate the proofs by sketching the arguments used in one particular case. This case will clearly show the impact of the ideas of E. M. Stein on the subject at hand.

We start out with our results on the generalized Korteweg-de Vries equation:

$$(3) \qquad \begin{cases} \dfrac{\partial u}{\partial t} + \dfrac{\partial^3 u}{\partial x^3} + u^k \dfrac{\partial u}{\partial x} = 0, \qquad k \geq 1 \\ u|_{t=0} = u_0 \end{cases}.$$

For $k = 1$, this equation was derived by Korteweg-de Vries as a model for long waves propagating in a channel. Subsequently, the cases $k = 1, 2$ have been found to be relevant in a number of different physical systems. Moreover, these equations have been studied because of their relation to inverse scattering theory and to algebraic geometry.

Theorem 4.

(i)   For $k = 1$, (3) is locally well posed in $H^s(\mathbb{R})$, $s > 3/4$, with time interval of length $T = T(\|u_0\|_{H^s})$.

(ii) For $k = 2$, (3) is locally well posed in $H^s(\mathbb{R})$, $s \geq 1/4$, $T = T(\|u_0\|_{H^{1/4}})$.

(iii) For $k = 3$, (3) is locally well posed in $H^s(\mathbb{R})$, $s \geq 1/12$, $T = T(\|u_0\|_{H^{1/12}})$.

(iv) For $k = 1, 2, 3$, if $s \geq 1$, then we have well posedness on any time interval $[-T, T]$, and the solution belongs to $L^\infty(\mathbb{R}; H^{[s]}(\mathbb{R}))$.

(v)   For $k \geq 4$, (3) is locally well posed in $H^s(\mathbb{R})$, $s \geq s_k = (k - 4)/2k$.

(vi) For $k \geq 4$, (3) is globally well posed for small data in $\dot{H}^{s_k}(\mathbb{R})$. It is also globally well posed in $H^s(\mathbb{R})$, $s \geq s_k$, for data small in $\dot{H}^{s_k}(\mathbb{R})$.

(i) was first obtained in [33], (iv) for $k = 2, 3$ was obtained in [31], while for $k = 1$, it was obtained in [33]. The remaining results are in [36].

We have also obtained a sharp non-linear scattering result.

Theorem 5. Let $k \geq 4$, and assume that $\|u_0\|_{\dot{H}^{s_k}(\mathbb{R})} \leq \varepsilon_k$, ε_k the "smallness" prescribed by Theorem 4(vi). Then, there exist unique $w_0^\pm \in \dot{H}^{s_k}(\mathbb{R})$ such that

$$(6) \qquad \lim_{t \to \pm\infty} \|u(t) - W(t)w_0^\pm\|_{\dot{H}^{s_k}} = 0,$$

where $u(t)$ is the solution to (3) given in Theorem 4(vi), and $W(t)w_0 = w(t)$ is the solution to the associated linear problem

$$(7) \qquad \begin{cases} \dfrac{\partial w}{\partial t} + \dfrac{\partial^3 w}{\partial x^3} = 0 \\ w|_{t=0} = w_0 \end{cases}.$$

Remarks on Theorems 4 and 5. (iv) in Theorem 4 follows from (i), (ii), (iii) because of the conservation laws

$$I_2(u) = \int_{-\infty}^{+\infty} u^2(x, t)\, dx; \qquad I_3(u) = \int_{-\infty}^{+\infty} \left((\frac{\partial u}{\partial x})^2 - c_k u^{k+2} \right)(x, t)\, dx.$$

Previously, ([4], [5], [50], and [27]) the energy method showed local well posedness in $H^s(\mathbb{R})$, $s > 3/2$, just as in example 2. Moreover, using this result and energy estimates for the second derivative, it was shown in [5] and [50] that there is global well posedness in $H^s(\mathbb{R})$, $s \geq 2$ (with small H^1 norm for $k \geq 4$).

In [59] and [28], using the conservation laws I_2 and I_3, global weak solutions with H^1 data were constructed (for small H^1 data if $k \geq 4$). Moreover, in [28], taking advantage of the term $\frac{\partial^3 u}{\partial x^3}$ in (3), a "local smoothing effect" was discovered (which is not true for solutions of (1)). Thus, it was proven that there exists a weak solution u, with data $u_0 \in H^1(\mathbb{R})$ (small data if $k \geq 4$) such that

$$(8) \qquad \int_{|x|<R} \int_{-R}^{R} |\frac{\partial^2 u}{\partial x^2} u(x, t)|^2\, dxdt \leq C(R, \|u_0\|_{H^1})$$

and that, (for $1 \leq k \leq 3$) if $u_0 \in L^2(\mathbb{R})$, there exists a weak solution such that

$$(9) \qquad \int_{|x|<R} \int_{-R}^{R} \left| \frac{\partial u}{\partial x}(x, t) \right|^2\, dxdt \leq C(R, \|u_0\|_{L^2}).$$

In [23], it was shown that, for $k \geq 2$, the weak solution with H^1 data is unique. Finally, in [22], and using the methods of [32], [33], and [35], results for $k \geq 4$, weaker than those in Theorem 4(v), are announced.

Note that the results in Theorem 4 are very similar to the ones in example 2. The novelty here is that the non-linear term includes derivatives, complicating the analysis.

The results in Theorem 4 are sharp (in a suitable sense) for $k \geq 4$. To explain this, and to clarify Theorem 5, let us fix our attention on the case $k = 4$. A perfect balance between the dispersive effect and the non-linearity in (3) is represented by the existence of solitary wave solutions

$$u_{c,k}(x, t) = \phi_{c,k}(x - ct), \qquad c > 0$$

where

$$(10) \qquad \phi_{c,k}(x) = \left\{ \frac{(k + 2)c}{2} \operatorname{sech}^2 \left(\frac{k}{2} \sqrt{c}x \right) \right\}^{1/4}.$$

A simple computation shows that, for $k \geq 4$, and $s_k = \frac{k-4}{2k}$,

$$\|\phi_{c,k}\|_{\dot{H}^{s_k}} = a_k,$$

where $a_k > 0$ is independent of c, and if $s \neq s_k$, then

$$\|\phi_{c,k}\|_{\dot{H}^s} \to 0 \qquad \text{as} \quad c \to \text{ either } 0 \text{ or } +\infty.$$

Returning to the case $k = 4$, the solution that we construct to (3) for $\|u_0\|_{L^2} \leq \varepsilon_0$ has the property that

$$\int_{-\infty}^{+\infty} \int_{-\infty}^{+\infty} |D^{1/6}u(x,t)|^2 \, dxdt < \infty.$$

This result fails for powers $k \neq 4$ as can be seen from $u_{c,k}$, and also for large $L^2(\mathbb{R})$ norm as can be seen from $u_{1,4}$. Similarly, the scattering result of Theorem 5 cannot hold in the L^2 norm for $k \neq 4$, or for the L^2 norm $k = 4$ and large data.

Next, we turn our attention to the results that we have obtained for the hierarchy of the generalized KdV equations. Thus, we consider

$$\textbf{(11)} \qquad \begin{cases} \dfrac{\partial u}{\partial t} + \dfrac{\partial^{2j+1} u}{\partial x^{2j+1}} + Q\left(u, \dfrac{\partial u}{\partial x}, \dots, \dfrac{\partial^{2j} u}{\partial x^{2j}}\right) = 0 \\[2mm] u\big|_{t=0} = u_0 \end{cases}$$

where j is a positive integer, $x, t \in \mathbb{R}$ and

$$Q : \mathbb{R}^{2j+1} \longrightarrow \mathbb{R}$$

is a polynomial having no constant or linear terms, i.e.,

$$\textbf{(12)} \qquad Q(z) = Q(z_1, z_2, \dots, z_{2j+1}) = \sum_{|\alpha| \geq k}^{\rho} a_\alpha z^\alpha, \qquad \text{with} \quad k \geq 2.$$

Theorem 13 ([37]).
(i) *There exist $m, s_0 \in \mathbb{N}$, and $\delta > 0$ such that, for any $s \geq s_0$, and any $u_0 \in H^s(\mathbb{R}) \cap L^2(|x|^m dx)$, of norm less than δ in that space, (11) is well posed on a time interval $(-T, T)$, $T = T(Q; \|u_0\|_{H^{s_0} \cap L^2(|x|^m \, dx)})$.*
(ii) *If k in (12) ≥ 3, we can take $m = 0$ in (i).*
(iii) *If Q does not depend on z_{2j+1}, the results in (i) and (ii) hold without any smallness assumption on u_0.*
(iv) *The results in (ii) are global in time if one of the following two hypotheses is verified:*
   (a) $k \geq 4j + 3$ *in (12), or*
   (b) $k \geq 5$ *and $Q(z) = Q(z_\ell, \dots, z_{2j+1})$ with $\ell \geq (2j - 1)/4$.*

Remarks on Theorem 13. (11) generalizes the KdV hierarchy introduced in [41]. In [18], it was shown that the eigenvalues of the time independent Schrödinger operator

$$L(g) = \frac{d^2}{dx^2} - q(x)$$

remains unchanged when the potential $q(\cdot) = u(\cdot, t)$ evolves according to the KdV equation $\frac{\partial u}{\partial t} + \frac{\partial^3 u}{\partial x^3} + u \frac{\partial u}{\partial x} = 0$. This discovery was the starting point of the inverse scattering method. In [41], it was shown that the same principle holds for the sequence (KdV hierarchy)

$$(14) \qquad \frac{\partial u}{\partial t} + [B_j; L(u)] = 0,$$

(with $[A; M] = AM - MA$), and B_j denotes the skew-symmetric operator

$$B_j = \alpha_j \frac{d^{2j+1}}{dx^{2j+1}} + \sum_{\ell=0}^{j-1} \left(b_{j\ell} \frac{d^{2\ell+1}}{dx^{2\ell+1}} + \frac{d^{2\ell+1}}{dx^{2\ell+1}} b_{j\ell} \right),$$

and the coefficients $b_{j\ell} = b_{j\ell}(u)$ are chosen so that the differential operator $[B_j; L(u)]$ has order zero. Equations of the type considered in (13) also appear as higher order models in water waves, in elastic media with microstructure ([42]), and in recent developments in physics ([67]).

It is interesting to note that the classical approaches—such as energy estimates, abstract semi-group theory, or space-time estimates—used to study other evolution equations cannot be applied to (13) except for very particular forms of the polynomial Q. Previously, existence results for the KdV hierarchy (14) were obtained in [48], and uniqueness for (14) was shown in [47]. Both sets of results depended heavily on the Hamiltonian form of (14). In [45], for some fifth order equations of the type appearing in (11) both in Hamiltonian and non-Hamiltonian form, well posedness results were obtained.

Next, we turn to higher dimensional results. We have been able to establish local well posedness results for non-linear Schrödinger equations, with first order derivatives in the non-linearity. Thus, we consider

$$(15) \qquad \begin{cases} \dfrac{\partial u}{\partial t} = i \Delta u + P(u, \nabla_x u, \overline{u}, \nabla_x \overline{u}) & t \in \mathbb{R}, \ x \in \mathbb{R}^n \\ \\ u\big|_{t=0} = u_0 \end{cases}$$

where $u = u(x, t)$ is a complex valued function, and

$$P : \mathbb{C}^{2n+2} \longrightarrow \mathbb{C}$$

is a polynomial, $P(\vec{z}) = P(z_1, \ldots, z_{2n+2}) = \sum_{d \leq |\alpha| \leq \rho} a_\alpha z^\alpha$, and $d \geq 2$. (We always assume that there exists a_{α_0}, $|a_{\alpha_0}| \neq 0$, $|\alpha_0| = d$.)

Theorem 16 ([35]).

(i)   *Assume that $d = 2$. Then, there exists $\delta = \delta(P) > 0$ such that, for any $u_0 \in H^s(\mathbb{R}^n) \cap H^{2n+3}(\mathbb{R}^n : |x|^{2n+2} \, dx)$, with $s \geq s_0 = 3n + 4 + 1/2$, and $\|u_0\|_{H^{s_0} \cap H^{2n+3}(|x|^{2n+2} dx)} = \delta_0 \leq \delta$, then (15) is locally well posed on $(-T, T)$, $T = T(\delta_0)$.*

(ii) *Assume that $d \geq 3$. Then there exists $\delta = \delta(P) > 0$ such that, for any $u_0 \in H^s(\mathbb{R}^n)$, with $s \geq s_0 = n + 2 + 1/2$ and $\|u_0\|_{H^{s_0}} \leq \delta$, then (15) is locally well posed on $(-T, T)$, $T = T(\|u_0\|_{H^{s_0}})$.*

Remarks on Theorem 16. As mentioned in the discussion of Example 2, the semilinear case $P = f(|u|)u$ (with f a real valued function) has been studied extensively. The general initial value problem (15) had been treated before mainly when one could either use energy estimates or deal with analytic data and analytic solutions. Energy estimates for (15) can be established when, using integration by parts, one can show that

$$\left| \sum_{|\alpha| \leq s} \int_{\mathbb{R}^n} \left(\frac{\partial}{\partial x} \right)^\alpha P(u, \nabla_x u, \overline{u}, \nabla_x \overline{u}) \left(\frac{\partial}{\partial x} \right)^\alpha u \, dx \right| \leq C_s (1 + \|u\|_{H^s}^\rho) \|u\|_{H^s}^2,$$

for any $u \in H^s(\mathbb{R}^n)$, with $s > \frac{n}{2} + 1$, $\rho = \rho(P) \in \mathbb{Z}^+$. The above estimate can be guaranteed only if P exhibits an appropriate symmetry. For example, $n = 1$, $P = \frac{\partial}{\partial x}(|u|^k u)$, $k \in \mathbb{Z}^+$ (see [63], [64]), $n \geq 1$ and $\partial_{\frac{\partial u}{\partial x_j}} P, \partial_{\frac{\partial \overline{u}}{\partial x_j}} P$, $j = 1, \ldots, n$ are real valued functions. In this case the local well posedness in $H^s(\mathbb{R}^n)$, $s > \frac{n}{2} + 1$ follows from the argument used for quasilinear symmetric hyperbolic systems ([26]). Indeed, for those P's the same proof works if one removes the Laplacian term from the equation in (15). Also, when $n = 1$, $P = \frac{\partial}{\partial x}(|u|^2 u)$, in [30] the inverse scattering method is applied.

The approach that deals with analytic data uses analytic function techniques to overcome the loss of derivatives introduced by the non-linearity ([25]).

Our method works without any special assumptions on P. It works for real or complex valued functions, or for systems of equations. The same method applies to non-linearities given by smooth functions $F(u, \nabla_x u, \overline{u}, \nabla_x \overline{u})$ with Taylor expansion at the origin having no constant or linear terms. We have not attempted to obtain the best results provided by our method, since, in any case, it is not clear that they would be optimal.

Nevertheless, in cases where both the energy estimate and our method apply, our method seems to give better results. An instance of this is our study of the initial value problem for the generalized Benjamin-Ono equations

(17)
$$\begin{cases} \dfrac{\partial u}{\partial t} + u^k \dfrac{\partial u}{\partial x} - \dfrac{\partial}{\partial x} D_x u = 0, & t, x \in \mathbb{R}, \ k \in \mathbb{Z}^+ \\ u\big|_{t=0} = u_0 \end{cases}$$

where $D_x = (-\frac{\partial^2}{\partial x^2})^{1/2}$. In the case $k = 1$ this equation was deduced in [3] and [43] as a model in internal wave theory. The generalized Benjamin-Ono equation presents the interesting fact that the dispersive effect is described by a non-local operator and is weaker than the one exhibited by the Korteweg-de Vries equations. The energy method proves local well posedness in H^s, $s > 3/2$. Our result here is

Theorem 18 ([34]).
(i)   *Let $k \geq 2$ and s be such that*

$$\begin{cases} s > 1 & k = 2 \\ s > 5/6 & k = 3 \\ s \geq 3/4 & k \geq 4. \end{cases}$$

There exists $\delta = \delta(k) > 0$ such that, for any $u_0 \in H^s(\mathbb{R})$, $\|u_0\|_s \leq \delta$, (17) is locally well posed on $(-T, T)$, $T = T(\|u_0\|_s, k)$.
(ii) *For $k \geq 4$, $s \geq 1$ there is global well posedness, for small initial data.*
(iii) *For $k \geq 4$, $u_0 \in H^1(\mathbb{R})$, $\|u_0\|_1 \leq \delta$, if u is the solution of (17) there exist unique $\omega_{0\pm} \in H^1(\mathbb{R})$ such that*

$$\lim_{t \to \pm\infty} \|u(t) - V(t)\omega_{0\pm}\|_{H^1} = 0,$$

where $V(t)\omega_0$ is the solution of

$$\begin{cases} \dfrac{\partial v}{\partial t} - \dfrac{\partial}{\partial x} D_x v = 0 \\ v\big|_{t=0} = \omega_0 \end{cases}.$$

Remarks on Theorem 18. The previously known results for (17) were: For $k = 1$ or 2, $u_0 \in H^s(\mathbb{R})$, $s = 0$ or $1/2$, there exists a weak solution u with $u \in L^2_{loc}(\mathbb{R}; H^{s+1/2}_{loc})$ (local smoothing effect) ([20]). For $k = 1$, the same is true for $u_0 \in H^1(\mathbb{R})$ ([20], [60]). For $k = 1$, (17) is globally well posed in $H^{3/2}(\mathbb{R})$

([44]). For $k > 1$, (17) is locally well posed in $H^s(\mathbb{R})$, $s > 3/2$ ([1], [24]). Note that none of these results requires a restriction on the size of the initial data.

The results of Theorem 18 extend also to the initial value problem

(19)
$$\begin{cases} \dfrac{\partial u}{\partial t} + i\dfrac{\partial^2 u}{\partial x^2} + P_1(u, \overline{u})\dfrac{\partial u}{\partial x} + P_2(u, \overline{u})\dfrac{\partial \overline{u}}{\partial x} = 0 \\[2mm] u\big|_{t=0} = u_0 \end{cases}$$

where $P_j : \mathbb{C}^2 \to \mathbb{C}$, $j = 1, 2$ are polynomials, with $P_j(z_1, z_2) = \sum_{\alpha+\beta \geq 2} a_{j\alpha\beta} z_1^\alpha z_2^\beta$.

Next, in order to illustrate our method, we will sketch a proof of Theorem 4(v), case $k = 8$. Thus, we will prove

Theorem 20. *There exists $\delta > 0$ such that for any $u_0 \in H^{1/4}(\mathbb{R})$ with $\|D^{1/4}u_0\|_{L^2(\mathbb{R})} < \delta$, there exists a unique strong solution of*

$$\begin{cases} \dfrac{\partial u}{\partial t} + \dfrac{\partial^3 u}{\partial x^3} + u^8\dfrac{\partial u}{\partial x} = 0 \\[2mm] u\big|_{t=0} = u_0 \end{cases}$$

satisfying

$$u \in C(R; \dot{H}^{1/4}(\mathbb{R})) \cap L^\infty(\mathbb{R}; \dot{H}^{1/4}(\mathbb{R})); \qquad \|D^{1/4}\tfrac{\partial u}{\partial x}\|_{L_x^\infty L_t^2} < +\infty;$$

$$\|u\|_{L_t^{12}L_x^\infty} < +\infty; \qquad \sup_{\gamma \in \mathbb{R}} \|D^{i\gamma}u\|_{L_x^4 L_t^\infty} < +\infty.$$

Moreover, the map $u_0 \mapsto u(t)$ from $\{u_0 \in \dot{H}^{1/4}(\mathbb{R}) : \|u_0\|_{\dot{H}^{1/4}} < \delta\}$ into

$$B = \Bigg\{ f \in C((-\infty, +\infty); H^{1/4}(\mathbb{R})) : \|f\|_B =$$

$$\max\Bigg[\|f\|_{L^\infty(\mathbb{R}; \dot{H}^{1/4}(\mathbb{R}))}; \|D^{1/4}\tfrac{\partial f}{\partial x}\|_{L_x^\infty L_t^2}; \|f\|_{L_t^{12}L_x^\infty}; \sup_\gamma \|D^{i\gamma}f\|_{L_x^4 L_t^\infty} \Bigg]$$

$$< +\infty \Bigg\}$$

is Lipschitz.

In order to prove Theorem 20, we introduce the quantities

$$\beta_1(f) = \sup_{-\infty < t < +\infty} \|D^{1/4}f\|_{L^2}$$

$$\beta_2(f) = \|D^{1/4}\frac{\partial f}{\partial x}\|_{L_x^\infty L_t^2}$$

$$\beta_3(f) = \|f\|_{L_t^{12}L_x^\infty}$$

$$\beta_4(f) = \sup_\gamma \|D^{i\gamma}f\|_{L_x^4 L_t^\infty}$$

and for $a > 0$, we let $B_a = \{f \in B : \|f\|_B \le a\}$. We have the following two claims:

Claim 21. If $u_0 \in \dot{H}^{1/4}(\mathbb{R})$, and we let $f_0 = W(t)u_0$ (i.e., $f_0 = \int_{-\infty}^{+\infty} e^{i(t\xi^3+x\xi)}\widehat{u_0}(\xi)\,d\xi$ is the solution of (7)), then

$$(22) \qquad\qquad \|f_0\|_B \le C\|u_0\|_{\dot{H}^{1/4}}.$$

Claim 23. If for $u_0 \in \dot{H}^{1/4}(\mathbb{R})$, $v \in B_a$ we define

$$u(t) = \Phi_{u_0}(v)(t) = W(t)u_0 - \int_0^t W(t-t')(v^8\frac{\partial v}{\partial x})(t')\,dt'$$

then there exists $\delta > 0$, $a = a(\delta)$ such that, if $\|u_0\|_{\dot{H}^{1/4}} < \delta$, then $\Phi(B) \subset B$, and Φ is a contraction.

Notice that once 23 is established, Theorem 20 follows from the contraction mapping principle and Duhamel's formula.

We start out with a verification of (22). These are inequalities for oscillatory integrals, which are versions of the "local smoothing effect" of Kato ([28]), the maximal function estimates of L. Carleson and others ([7], [38]), and the L^2 restriction theorem of Tomas-Stein ([61]) and Strichartz ([57]). Note first that the inequality

$$\beta_1(f_0) \le \|u_0\|_{\dot{H}^{1/4}}$$

is an immediate consequence of the fact that $W(t)$ is a unitary group on $L^2(\mathbb{R})$.

Lemma 24. *Let $W(t)w_0$ be defined as in* (7). *Then*

$$\left\|\frac{\partial W}{\partial x}(t)w_0\right\|_{L_x^\infty L_t^2} \le C\|w_0\|_{L^2}.$$

Proof. $\frac{\partial}{\partial x}W(t)w_0 = \int e^{i(t\xi^3+x\xi)}(i\xi)\widehat{w_0}(\xi)\,d\xi$. Now, fix x, and perform the change of variables $\eta = \xi^3$. Plancherel's theorem in t and a new change of variables shows that, for each fixed x,

$$\int_{-\infty}^{+\infty} \left|\frac{\partial}{\partial x}W(t)w_0\right|^2 dt = c\|w_0\|_{L^2}^2.$$

The lemma follows from this.   ∎

Note that this is, for the solution of (7), a refinement of the estimate (9) obtained by Kato for solutions of the Korteweg-de Vries equation. Local smoothing effects for linear dispersive equations have attracted a lot of attention recently ([13], [14], [66], [51]). This sharp form of it was first observed in [65].

Corollary 25. $\beta_2(W(t)u_0) \leq c\|u_0\|_{\dot{H}^{1/4}}$.

Lemma 26. *For $u_0 \in \dot{H}^{1/4}$ we have the estimate*

$$\|W(t)u_0\|_{L_t^{12}L_x^\infty} \leq C\|u_0\|_{\dot{H}^{1/4}}.$$

This is a version of the Strichartz type inequalities ([57]) for the Airy equation (7). Its connection with the restriction theorem for the curve $\Gamma = (\xi, \xi^3)$ is that, since $W(t)u_0 = \int e^{i[t\xi^3 + x\xi]}\widehat{u}_0(\xi)d\xi$, the solution operator to (7) is the dual of the operator restriction of the Fourier transform to Γ (i.e., the extension operator) applied to the density $\widehat{u}_0(\xi)d\xi$ viewed as a measure on Γ. Lemma 26 says that if $\int |\xi|^{1/2}|\widehat{u}_0(\xi)|^2 d\xi < +\infty$, the extension operator yields a function in $L_t^{12}L_x^\infty$. By duality this is equivalent to a restriction estimate. The main tool in the proof of Lemma 26 is an estimate for oscillatory integrals.

Lemma 27. *Let $I_\delta^t(x) = \int_{-\infty}^{+\infty} e^{i(t\xi^3 + x\xi)} \frac{d\xi}{|\xi|^\delta}$. Then,*

(28) $$|I_\delta^t(x)| \leq \frac{C_\delta}{|x|^{1-\delta}} \quad for \quad \frac{1}{2} \leq \delta < 1$$

(29) $$|I_\delta^t(x)| \leq \frac{C_\delta}{|t|^{(1-\delta)/3}} \quad for \quad -\frac{1}{2} \leq \delta < 1.$$

The proof of Lemma 27 is an application of the version of Van der Corput's lemma found in E. M. Stein's article ([54]). See [31] and [36] for the details.

Once we have Lemma 27, the proof of Lemma 26 follows the lines of the proof of the Tomas-Stein ([61]) L^2 restriction theorem (see [36]). In fact, 26 is equivalent to the estimate

(30) $$\|D^{-1/4}W(t)u_0\|_{L_t^{12}L_x^\infty} \leq C\|u_0\|_{L^2}.$$

We claim that (30) is in turn equivalent to

(31) $$\left\|\int_{-\infty}^{+\infty} D^{-1/2}W(t-t')g(-, t')dt'\right\|_{L_t^{12}L_x^\infty} \leq C\|g\|_{L_t^{12/11}L_x^1}.$$

This is because, by duality, (30) is equivalent to

(32)
$$\left\| \int_{-\infty}^{+\infty} D^{-1/4} W(t) g(-,t) dt \right\|_{L_x^2} \le C \|g\|_{L_t^{12/11} L_x^1}.$$

In turn, the left-hand side of (32) squared equals

$$\int \left(\int D^{-1/4} W(t) g(-,t) \, dt \right) \overline{\left(\int D^{-1/4} W(t') g(-,t') \, dt' \right)} dx$$

$$= \iint g(x,t) \left(\int D^{-1/2} W(t-t') \overline{g(-,t')} dt' \right) dx \, dt.$$

Hence, (32) and (31) are equivalent. In order to prove (31), we apply Lemma 27, estimate (29), $\delta = \frac{1}{2}$, to obtain

$$\left\| \int_{-\infty}^{+\infty} D^{-1/2} W(t-t') g(-,t') dt' \right\|_{L_x^\infty} \le C \int_{-\infty}^{+\infty} \frac{1}{|t-t'|^{1/6}} \|g(-,t')\|_{L_x^1} dt'.$$

Fractional integration in t now finishes the proof.

Corollary 33. $\beta_3(W(t)u_0) \le C \|u_0\|_{\dot{H}^{1/4}}.$

Lemma 34. *For $u_0 \in \dot{H}^{1/4}$ we have the estimate*

$$\|W(t)u_0\|_{L_x^4 L_t^\infty} \le C \|u_0\|_{\dot{H}^{1/4}}.$$

Lemma 34 is a maximal function inequality for the solution of (7). It, of course, implies that $\lim_{t \downarrow 0} W(t)u_0 = u_0$ a.e., for $u_0 \in \dot{H}^{1/4}$. This kind of a.e. convergence problem was first studied by L. Carleson ([7]), who showed that, for the linear Schrödinger equation on $\mathbb{R}$, $u_0 \in \dot{H}^{1/4}$ suffices for a.e. convergence. Later on, in [15] it was observed that the exponent $1/4$ is sharp, and in [38] the maximal inequality in Lemma 34, for the case of the Schrödinger operator is proved. It was E. M. Stein who first noticed that inequalities as those in 34 are simple consequences of the Kolmogorov-Seliverstov-Plessner method. Lemma 34 was first proven in [65]. To prove 34, note that, arguing as in the proof of Lemma 26, it suffices to show that

(35)
$$\left\| D^{-1/2} \int_{-\infty}^{+\infty} W(t-t') g(-,t') dt' \right\|_{L_x^4 L_t^\infty} \le C \|g\|_{L_x^{4/3} L_t^1}.$$

To show (35), we use Lemma 27, estimate (28), with $\delta = 1/2$, to obtain

$$\left\| D^{-1/2} \int_{-\infty}^{+\infty} W(t - t')g(-, t')dt' \right\|_{L_t^\infty} \leq C \frac{1}{|x|^{1/2}} * \|g(-, t)\|_{L_t^1},$$

and (35) follows by fractional integration in x.

Corollary 36. $\beta_4(W(t)u_0) \leq C\|u_0\|_{\dot{H}^{1/4}}.$

This finishes the proof of (22). In order to prove (23) we need a non-linear estimate, which is contained in the following:

Lemma 37. *If* $v \in B_a$, *then*

$$(38) \qquad\qquad \int_{-\infty}^{+\infty} \left\| D^{1/4}(v^8 \frac{\partial v}{\partial x}) \right\|_{L_x^2} dt \leq Ca^9.$$

Taking (38) for granted, in order to establish 23, we see that, by Minkowski's inequality

$$\|u(t)\|_B \leq \|f_0\|_B + \int_{-\infty}^{+\infty} \|W(t - t')(v^8 \frac{\partial v}{\partial x})\|_B dt'$$

$$\leq C\delta + C \int_{-\infty}^{+\infty} \|D^{1/4}(v^8 \frac{\partial v}{\partial x})\|_{L^2} dt'$$

by (22). Using (38) we see that

$$\|u\|_B \leq C\delta + Ca^9.$$

Using the following variant of (38)

$$(39) \qquad \int_{-\infty}^{+\infty} \left\| D^{1/4} \left(v_1^8 \frac{\partial v_1}{\partial x} - v_2^8 \frac{\partial v_2}{\partial x} \right) \right\|_{L_x^2} dt \leq C\|v_1 - v_2\|_B a^8,$$

where $v_i \in B_a$, we see that

$$\|u_1 - u_2\|_B \leq C\|v_1 - v_2\|_B a^8,$$

and hence 23 follows.

We will now indicate the proof of (38). What is needed here are vector valued versions of the Leibniz and chain rules for fractional derivatives. Previous results in this direction are due to [56], [29], [9], and [58]. Our results are a consequence of the ideas of Coifman-Meyer ([10]) and Bony ([6]), and the vector valued inequalities of C. Fefferman-Stein ([16]), Benedek-Calderón-Panzone ([2]), and Rubio de Francia, Ruiz, and Torrea ([46]). To deal with the endpoint results, we also need the H^p theory as developed by C. Fefferman and E. Stein ([17]), as well as the "tent spaces" of Coifman-Meyer-Stein ([12]). The results needed here are:

Lemma 40. *Let F be of class C^1, $F(0) = 0$. Let $\alpha \in (0, 1)$, $p, q, p_1, p_2, q_2 \in (1, \infty)$, $q_1 \in (1, \infty]$ be such that $\frac{1}{p} = \frac{1}{p_1} + \frac{1}{p_2}$, $\frac{1}{q} = \frac{1}{q_1} + \frac{1}{q_2}$. Then*

$$\text{(41)} \qquad \|D^\alpha F(f)\|_{L_x^p L_t^q} \leq C\|F'(f)\|_{L_x^{p_1} L_t^{q_1}} \|D^\alpha f\|_{L_x^{p_2} L_t^{q_2}}.$$

Also, if $r > 1$, $h \in L_{loc}^{rp}(\mathbb{R})$, then

$$\text{(42)} \qquad \|D^\alpha F(f)h\|_p \leq C\|F'(f)\|_\infty \|D^\alpha(f)M(h^{rp})^{1/rp}\|_p,$$

where M denotes the Hardy-Littlewood maximal operator.

Lemma 43. *Let $\alpha \in (0, 1)$, $\alpha_1, \alpha_2 \in [0, \alpha]$ with $\alpha = \alpha_1 + \alpha_2$. Let $p, p_1, p_2, q, q_1, q_2 \in (1, \infty)$ be such that $\frac{1}{p} = \frac{1}{p_1} + \frac{1}{p_2}$, $\frac{1}{q} = \frac{1}{q_1} + \frac{1}{q_2}$. Then*
(i) $\|D^\alpha(fg) - fD^\alpha g - gD^\alpha f\|_{L_x^p L_t^q} \leq C\|D^{\alpha_1} f\|_{L_x^{p_1} L_t^{q_1}} \|D^{\alpha_2} g\|_{L_x^{p_2} L_t^{q_2}}.$
   Moreover, for $\alpha_1 = 0$, the value $q_1 = \infty$ is allowed.
(ii) $\|D^\alpha(fg) - fD^\alpha g - gD^\alpha f\|_p \leq C\|g\|_\infty \|D^\alpha f\|_p.$

Let us now explain why (38) should hold. By the Leibniz rule,

$$D^{1/4}\left(v^8 \frac{\partial v}{\partial x}\right) \, `=' \, v^8 D^{1/4} \frac{\partial v}{\partial x} + D^{1/4}(v^8) \frac{\partial v}{\partial x}.$$

For the first term, note that

$$\|v^8 D^{1/4} \frac{\partial v}{\partial x}\|_{L_x^2} \leq \|v\|_{L_x^\infty}^6 \|v^2 D^{1/4} \frac{\partial v}{\partial x}\|_{L_x^2}$$

and so

$$\leq \int_{-\infty}^{+\infty} \|v^8 D^{1/4} \frac{\partial v}{\partial x}\|_{L_x^2} \leq \left(\int_{-\infty}^{+\infty} \|v\|_{L_x^\infty}^{12}\right)^{1/2} \left(\iint |v^2 D^{1/4} \frac{\partial v}{\partial x}|^2 \, dx dt\right)^{1/2}$$

$$\leq \beta_3(v)^6 \beta_4(v)^2 \beta_2(v).$$

For the second term, by the chain rule, it "equals" $v^7 D^{1/4}(v) \frac{\partial v}{\partial x}$. Stein's complex interpolation theorem ([52]) now shows that

$$\|D^{1/4}(v)\|_{L_x^\alpha L_t^\beta} \leq C\beta_2(v)^\theta \beta_4(v)^{1-\theta}$$

$$\|\frac{\partial v}{\partial x}\|_{L_x^\mu L_t^\nu} \leq C\beta_2(v)^\eta \beta_4(v)^{1-\eta},$$

where

$$\frac{1}{\alpha} = \frac{\theta}{\infty} + \frac{(1-\theta)}{4} \quad \frac{1}{\beta} = \frac{\theta}{2} + \frac{(1-\theta)}{\infty}, \quad \frac{1}{4} = \theta(1 + \frac{1}{4}) + (1-\theta)0$$

$$\frac{1}{\mu} = \frac{\eta}{\infty} + \frac{(1-\eta)}{4} \quad \frac{1}{\nu} = \frac{\eta}{2} + \frac{(1-\eta)}{\infty}, \quad 1 = \eta(1 + \frac{1}{4}) + (1-\eta)0.$$

(It is in this step that $D^{i\gamma}f$ in the definition of β_4 is needed.) Then,

$$\|v^7 D^{1/4}(v) \frac{\partial v}{\partial x}\|_{L_x^2} \le C \|v\|_{L_x^\infty}^6 \|v D^{1/4}(v) \frac{\partial v}{\partial x}\|_{L_x^2}.$$

One then proceeds as before, and then has to estimate

$$\left(\iint |v D^{1/4}(v) \frac{\partial v}{\partial x}|^2 \, dx dt \right)^{1/2}.$$

This is accomplished by means of Hölder's inequality and the above estimates. Finally, the error terms are handled by using (40) and (43) appropriately. The details are given in [36]. (39) is proven in a similar manner, and this concludes our proof.

University of Chicago

REFERENCES

[1] L. Abdelouhab, J. L. Bona, M. Felland, and J. C. Saut. "Nonlocal models for nonlinear dispersive waves." *Physica D* **40** (1989), 360–392.

[2] A. Benedek, A. P. Calderón, and R. Panzone. "Convolution operators on Banach space valued functions." *Proc. Nat. Acad. Sci.* **48** (1962), 356–365.

[3] T. B. Benjamin. "Internal waves of permanent form in fluids of great depth." *J. Fluid Mech.* **29** (1967), 559–592.

[4] J. L. Bona and R. Scott. "Solutions of the Korteweg-de Vries equation in fractional order Sobolev spaces." *Duke Math. J.* **43** (1976), 87–99.

[5] J. L. Bona and R. Smith. "The initial value problem for the Korteweg-de Vries equation." *Roy. Soc. London, Ser A* **278** (1978), 555–601.

[6] J. M. Bony. "Calcul symbolique et propagation des singularités pour les équations aux dérivées partielles non linéaires." *Ann. Sci. E. N. S.* **114** (1981), 209–246.

[7] L. Carleson. "Some analytical problems related to statistical mechanics, Euclidean harmonic analysis." Lecture Notes in Mathematics, no. 779. Springer Verlag, 1979.

[8] T. Cazenave and F. B. Weissler. "The Cauchy problem for the critical nonlinear Schrödinger equation in H^s." *Nonlinear Anal. TMA* **14** (1990), 807–836.

[9] F. M. Christ and M. I. Weinstein. "Dispersion of small amplitude solutions of the generalized Korteweg-de Vries equation." To appear in *J. Funct. Anal.*

[10] R. R. Coifman and Y. Meyer. "Au délà des opérateurs pseudodifférentielles." *Astérisque* **57** (1973).

[11] ———. "Nonlinear harmonic analysis, operator theory and P.D.E." In *Beijing Lectures in Harmonic Analysis*, edited by E. M. Stein. Princeton University Press, 1986.

[12] R. R. Coifman, Y. Meyer, and E. M. Stein. "Some new function spaces and their applications to harmonic analysis." *J. Funct. Anal.* **62** (1985), 304–335.

[13] P. Constantin and J. C. Saut. "Local smoothing properties of dispersive equations." *J. Amer. Math. Soc.* **1** (1988), 413–446.

[14] W. Craig, T. Kappeler, and W. A. Strauss. "Gain of regularity for equations of KdV type." Preprint.

[15] B. Dahlberg and C. E. Kenig. "A note on the almost everywhere behavior of solutions to the Schrödinger equation, Harmonic Analysis." Lecture Notes in Mathematics, no. 908. Springer Verlag, 1982.

[16] C. Fefferman, and E. M. Stein. "Some maximal inequalities." *Amer. J. Math.* **13** (1971), 107–115.

[17] ———. "H^p spaces of several variables," *Acta Math.* **129** (1972), 137–193.

[18] C. S. Gardner, J. M. Greene, M. D. Kruskal, and R. M. Miura. "A method for solving the Korteweg-de Vries equation." *Phys. Rev. Letters* **19** (1967), 1095–1097.

[19] ———. "The Korteweg-de Vries equation and generalizations. VI. Method for exact solutions." *Comm. Pure Appl. Math.* **27** (1974), 97–133.

[20] J. Ginibre, and G. Velo. "Smoothing properties and existence of solutions for the generalized Benjamin-Ono equations." Preprint.

[21] ———. "Scattering theory in the energy space for a class of nonlinear Schrödinger equations." *J. Math. Pure Appl.* **64** (1985), 363–401.

[22] ———. "Smoothing properties and retarded estimates for some dispersive evolution equations." Preprint.

[23] J. Ginibre and Y. Tsutsumi. "Uniqueness for the generalized Korteweg-de Vries equations." *SIAM J. Math. Anal.* **20** (1989), 1388–1425.

[24] N. Hayashi. "Global existence of small analytic solutions to nonlinear Schrödinger equations." *Duke Math. J.* **62** (1991), 575–592.

[25] R. J. Iorio. "On the Cauchy problem for the Benjamin-Ono equation." *Comm. Part. Diff. Eq.* **11** (1986), 1031–1081.

[26] T. Kato. "Quasilinear equations of evolution, with applications to partial differential equations." Lecture Notes in Mathematics, no. 448. Springer Verlag, 1975.

[27] ———. "On the Korteweg-de Vries equation," *Manuscripta Math.* **19** (1979), 89–99.

[28] ———. "On the Cauchy problem for the (generalized) Korteweg-de Vries equation." *Adv. Math. Suppl. Stud. Appl. Math.* **8** (1983), 93–128.

[29] T. Kato and G. Ponce. "Commutator estimates and the Euler and Navier-Stokes equations." *Comm. Pure Appl. Math.* **41** (1988), 891–907.

[30] D. J. Kaup and A. C. Newell. "An exact solution for a derivative non-linear Schrödinger equation." *J. Math. Phys.* **19** (1978), 798–801.

[31] C. E. Kenig, G. Ponce, and L. Vega. "On the (generalized) Korteweg-de Vries equation." *Duke Math. J.* **59** (1989), 585–610.

[32] ———. "Oscillatory integrals and regularity of dispersive equations." *Indiana Univ. Math. J.* **49** 1991, 33–69.

[33] ———. "Well-posedness of the initial value problem for the Korteweg-de Vries." *J. Amer. Math. Soc.* **4** (1991), 323–347.

[34] _____. "On the generalized Benjamin-Ono equation." To appear, *Trans. Amer. Math. Soc.*

[35] _____. "Small solutions to nonlinear Schrödinger equations." To appear, *Ann. Inst. Poincaré, Analyse Non-Linéaire.*

[36] _____. "Well posedness and scattering results for the generalized Korteweg-de Vries equation via the contraction principle." To appear, *Comm. Pure Appl. Math.*

[37] _____. "On the hierarchy of the generalized KdV equations." To appear, *Proceedings, Nato Lyon Workshop on Singular Limits of Dispersive Waves.*

[38] C. E. Kenig and A. Ruiz. "A strong type (2, 2) estimate for the maximal function associated to the Schrödinger equation." *Trans. Amer. Math. Soc.* **230** (1983), 239–246.

[39] S. Kichenassamy and P. J. Olver. "Existence and non-existence of solitary waves solutions to higher order model evolution equations." Preprint.

[40] D. J. Korteweg and G. de Vries. "On the change of form of long waves advancing in a rectangular canal, and on a new type of long stationary waves." *Philos. Mag.* **5** (1895), 422–443.

[41] P. D. Lax. "Integrals of nonlinear equations of evolution and solitary waves." *Comm. Pure Appl. Math.* **21** (1968), 467–490.

[42] P. L. Olver. "Hamiltonian and non-Hamiltonian models for water waves." Lecture Notes in Physics, no. 195. Springer Verlag, 1984.

[43] H. Ono. "Algebraic solitary waves in stratified fluids." *J. Phys. Soc. Japan* **39** (1975), 1082–1091.

[44] G. Ponce. "On the global well posedness of the Benjamin-Ono equation." *Diff. and Int. Eqns.* **4** (1991), 527–542.

[45] _____. "Lax pairs and higher order models for water waves." To appear *J. Diff. Eqns.*

[46] J. L. Rubio de Francia, F. J. Ruiz, and J. L. Torrea. "Calderón-Zygmund theory for operator-valued kernels." *Adv. Math.* **62** (1988), 7–48.

[47] J. C. Saut. "Sur quelques généralisations de l'équations de Korteweg-de Vries, II." *J. Diff. Eqs.* **33** (1974), 320–335.

[48] _____. "Sur quelques généralisations de l'équations de Korteweg-de Vries." *J. Math. Pures Appl.* **58** (1979), 21–61.

[49] J.-C. Saut and R. Temam. "Remarks on the Korteweg-de Vries equation." *Israel J. Math.* **24** (1976), 78–87.

[50] M. Schwarz, Jr. "The initial value problem for the sequence of generalized Korteweg-de Vries equation." *Adv. Math.* **54** (1984), 22–56.

[51] P. Sjölin. "Regularity of solutions to the Schödinger equations." *Duke Math. J.* **55** (1987), 699–715.

[52] E. M. Stein. "Interpolation of linear operators." *Trans. Amer. Math. Soc.* **83** (1956), 482–492.

[53] _____. *Singular integrals and differentiability properties of functions.* Princeton University Press, 1970.

[54] _____. "Oscillatory integrals in Fourier analysis." In *Beijing Lectures in Harmonic Analysis,* edited by E. M. Stein. Princeton University Press, 1986.

[55] E. M. Stein and G. Weiss. *Introduction to Fourier Analysis in Euclidean Spaces.* Princeton University Press, 1971.

[56] R. S. Strichartz. "Multipliers in fractional Sobolev spaces." *J. Math. Mech.* **16** (1967), 1031–1060.

[57] _____. "Restriction of Fourier transforms to quadratic surface and decay of solutions of wave equations." *Duke Math. J.* **44** (1977), 705–714.

[58] M. E. Taylor. "Pseudo-differential operators and nonlinear P.D.E." Preprint.

[59] R. Temam. "Sur un problème non linéaire." *J. Math. Pures Appl.* **48** (1969), 159–172.

[60] M. M. Tom. "Smoothing properties of some weak solutions of the Benjamin-Ono equation." *Diff. and Int. Eqns.* **3** (1990), 683–694.

[61] P. Tomas. "A restriction theorem for the Fourier transform." *Bull. Amer. Math. Soc.* **81** (1975), 477–478.

[62] Y. Tsutsumi. "L^2-solutions for nonlinear Schrödinger equations and nonlinear group." *Funkcialaj Ekvacioj* **30** (1987), 115–125.

[63] M. Tsutsumi, and I. Fukuda. "On solutions of the derivative nonlinear Schrödinger equation. Existence and Uniqueness Theorem." *Funkcialaj Ekvacioj* **23** (1980), 259–277.

[64] _____. "On solutions of the derivative nonlinear Schrödinger equation, II." *Funkcialaj Ekvacioj* **24** (1981), 85–94.

[65] L. Vega. Doctoral Thesis, Universidad Autonoma de Madrid, 1987.

[66] _____. "The Schrödinger equation: pointwise convergence to the initial data." *Proc. Amer. Math. Soc.* **102** (1988), 874–878.

[67] E. Witten. "Two dimensional gravity and intersection theory on moduli space." Preprint.

14

Singular Integrals and
Fourier Integral Operators

*D. H. Phong**

I INTRODUCTION

The theory of singular integrals and pseudo-differential and Fourier integral operators [17], [39], [43], [45] constitutes one of the most impressive achievements in classical analysis of the last few decades, both in its wide ramifications and in the coherence and unity of its structure. The operators falling outside of the scope of this classical theory have usually surfaced in widely disparate areas of analysis and geometry, often to the extent that their mutual existence was hardly known. They required methods devised individually, and with more limited success in many cases. However, they have been appearing increasingly frequently in recent years, and remarkably, many common threads have begun to emerge.

In this chapter we shall focus on two themes which seem to unify many of the very disparate operators not covered by the classical theory of Fourier integral operators. In the first theme the Lagrangian manifold may fail to be locally the graph of a canonical transformation. This was the case for diffraction theory [19], [20], [41], [42], where the Lagrangian is a Whitney fold. More systematically, it is generically the case whenever we deal with Radon transforms along submanifolds of codimension higher than 1. Thus Radon transforms and maximal functions along families of curves [34], [25], [40] and X-ray transforms [10], [12] are prime examples, as are analysis and deformation theory on manifolds of geodesics [10], [14], [44]. In the second theme, the densities of the Fourier integral operators

*This research was supported, in part, by the National Science Foundation, under grant DMS-90-04062.

are allowed singularities of an either fractional or Calderón-Zygmund type. This again seems dictated by rather general situations. A phenomenon probably typical of subelliptic problems is the following. The main term of the Green's function for the $\bar{\partial}$-Neumann on the Siegel upper half-space [29], [32], [3] is smooth outside of the diagonal of the Heisenberg group. Yet it satisfies no better estimates than a kernel supported on the distribution of maximal complex hyperplanes with additional Calderón-Zygmund singularities, and must be treated as a singular Radon transform [31], [32]. Another case of a singular density is the fundamental solution of a hyperbolic partial differential equation [21], [15]. Its wave front set is contained in the union of the cotangent space at the origin with the flow out of its intersection with the characteristic variety of the equation. This occurrence of two intersecting Lagrangians is shared by singular Radon transforms, with the Lagrangians given in this case by the diagonal and the normal bundle of the distribution of hyperplanes. Recognition of this common feature has already borne fruit, as it has led to a microlocal proof of the boundedness of singular Radon transforms [11]. It is very intriguing that the two themes we have just described seem to be themselves intimately related. In fact, composition of the Fourier integral operators singular in the first sense leads under different guises to Fourier integral operators singular in the second sense. Examples of such occurrences are given in Sections II.3 and IV below.

It can be hoped that a theory with composition calculus and Sobolev and $L^p - L^q$ bounds can be constructed which encompasses these degenerate Fourier integral operators and singular Radon transforms. Here we shall discuss encouraging progress in a number of directions [34], [35]: first, in a class of operators modeling Fourier integral operators with arbitrarily high order of degeneracy, and whose bounds can be formulated in terms of the stratification of the Lagrangian; second, in the $L^p - L^q$ bounds for Radon transforms with folding canonical relations and their relation with singular Radon transforms; third, in establishing bounds even in some cases when a normal form for the Lagrangian may not be available; and finally, in the understanding of the role of the Monge-Ampère determinant in the localization procedures required for the Sobolev estimates.

This paper is organized as follows. In Section II we provide an introduction to several problems in Fourier analysis and integral geometry leading to the operators considered here. In Section III we describe bounds for the model Fourier integral operators. The key ingredients in our approach are a refined method of stationary phase with uniform bounds and sharp estimates for the distances between the roots of an algebraic equation. In Section IV we present some aspects of Radon transforms on curves in two dimensions, with special emphasis on the emergence of the Monge-Ampère determinant in the hard analysis, and of singular densities in the composition.

This paper describes joint work with E. M. Stein, some of which stretches back to the days when I was one of his graduate students. It has been an exceptional privilege to carry on a research program with him continuously since that time. On this occasion of his sixtieth birthday, I would like to express my immense gratitude to him, and also my pride and happiness at participating in an enterprise which, if I were to judge from the boundless energy and enthusiasm he brought to it, must have been especially close to his heart.

II BASIC EXAMPLES

In this section we provide a brief discussion of several areas of analysis and integral geometry leading generically to *singular* Fourier integral operators. The singularities emerge under many different guises, either in the symbols or in the projections of the Lagrangians, and it is remarkable that they all seem to be intimately related.

1 Radon Transforms along Curves

We consider first the distribution of curves $\{M_P = P + \vec{\gamma}(t), P \in \mathbf{R}^n\}$ in $\mathbf{R}^n$ given by translates of a fixed curve $\vec{\gamma}(t)$, and the corresponding integral transform

$$(1) \qquad (Tf)(P) \equiv \int_{-\infty}^{+\infty} f(P + \vec{\gamma}(t))\chi(t)dt \equiv \int_{M_P} f$$

where χ is a fixed C_0^∞ cut-off function, $\equiv 1$ and supported in a small enough neighborhood of 0. The question is to determine geometric conditions on the curve $\vec{\gamma}(t)$ which insure the optimal regularity properties for T. As a convolution operator the multiplier of T is given by

$$(2) \qquad m(\lambda) = \int_{-\infty}^{+\infty} e^{i\langle \lambda, \vec{\gamma}(t)\rangle}\chi(t)dt.$$

The most natural geometric condition, namely that $\vec{\gamma}(t)$ have non-zero torsion at $t = 0$, i.e., that the n vectors $\dot{\vec{\gamma}}(0), \ldots, \vec{\gamma}^{(n)}(0)$ be linearly independent, is also the one guaranteeing uniform bounds in λ for the phase function in (2)

$$(3) \qquad |\langle \lambda, \dot{\vec{\gamma}}\rangle(0)| + \cdots + |\langle \lambda, \vec{\gamma}^{(n)}(0)\rangle| \geq c|\lambda|.$$

The van der Corput lemma implies $|m(\lambda)| \leq c|\lambda|^{-1/n}$, and we obtain the sharp estimate

Theorem 1. *If $\vec{\gamma}(t)$ has non-vanishing torsion at 0, then the operator T of (1) is bounded from $H_{(s)}(\mathbf{R}^n)$ to $H_{(s+1/n)}(\mathbf{R}^n)$.*

This simple example already falls outside the scope of the classical theory of Fourier integral operators. Indeed it exploits an n-th order derivative condition (torsion) rather than the second order conditions of Fourier integral operators. More precisely the kernel of (1) is the Dirac δ measure along the defining relation

(4) $$C = \{(P, Q) \in \mathbf{R}^n \times \mathbf{R}^n; \, Q \in M_P\}.$$

Its wave front is contained in the Lagrangian manifold $N^*(C) \subset T^*(\mathbf{R}^n) \times T^*(\mathbf{R}^n)$. The calculus of Fourier integral operators applies and gives L^2 estimates when the projections π_L and π_R from $N^*(C)$ on the left and right factors

(5)

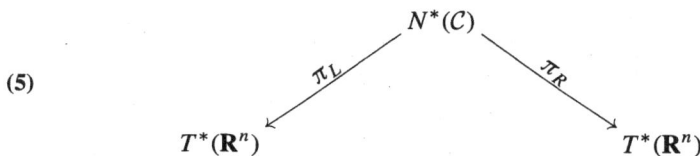

have invertible differentials. By convention the P variables are on the left and the Q variables are on the right. Now an element $(P, \xi; Q, \eta)$ is in $N^*(C)$ when $Q = P + \vec{\gamma}(t), \xi + \eta = 0$, and

(6) $$\langle \eta, \dot{\gamma} \rangle = 0.$$

An element $(\delta P, \delta \xi; \delta Q, \delta \eta)$ is then in $T_{(P,\xi;Q,\eta)}(N^*(C))$ when

(7) $\delta Q = \delta P + \dot{\gamma} \delta u,$ $\langle \delta \eta, \dot{\gamma} \rangle + \langle \eta, \ddot{\gamma} \rangle \delta u = 0,$ $\delta \xi + \delta \eta = 0$

for some δu in $\mathbf{R}$. In particular the kernel of $d\pi_L$ at $(P, \xi; Q, \eta)$ is non-trivial exactly at points where

(8) $$\langle \eta, \ddot{\gamma} \rangle = 0.$$

For $n > 2$ there are at least $n - 2$ directions η satisfying both (6) and (8), and the Lagrangian $N^*(C)$ fails there to be a local graph.

What are the optimal estimates when the condition that $N^*(C)$ be a canonical graph breaks down? A classic theorem of Hörmander [16], [17] says that when the kernels of $d\pi_L, d\pi_R$ have dimension at most k, then the corresponding Fourier integral operator satisfies bounds which lose $-k/2$ derivatives compared to when the Lagrangian is a local graph. In the formalism of Fourier integral operators, the Radon transform along curves (1) is of class $I^{-1/2}(M, M, N^*(C))$. The fact that it is of order $-1/2$ can be seen by choosing $n - 1$ defining functions $\Phi_1(P, Q) = \cdots = \Phi_{n-1}(P, Q) = 0$ for (4) and representing the Dirac measure along (4) as a Fourier integral distribution with oscillating variables $\lambda_1, \ldots, \lambda_{n-1}$, phase

$\sum_{j=1}^{n-1} \lambda_j \Phi_j(P, Q)$, and symbol 1

$$\delta_C(P, Q) = \int exp(i \sum_{j=1}^{n-1} \lambda_j \Phi_j(P, Q)) \prod_{j=1}^{n-1} d\lambda_j.$$

The order of the operator is then

$$\text{order(symbol)} + (\frac{1}{2} \text{\#oscillating variables})$$

$$(9) \qquad -\frac{1}{4}(\#P \text{ variables} + \#Q \text{ variables}) = -1/2$$

and would have been the amount of smoothing, had the projections π_L and π_R been locally invertible. In this case the kernels of $d\pi_L$ and $d\pi_R$ are one-dimensional, and the sharp estimate in the absence of any additional information as given by Hörmander's theorem is then boundedness on $H_{(s)}$. Theorem 1 shows that this can be improved by natural and generic assumptions on the geometry of π_L and π_R.

The next issue we address is whether the torsion condition, natural as it is for the family of translates of a curve, can be extended to general families of curves. This is the case, as the torsion condition for distributions of translates of a curve arises naturally from the stratification structure of the singular varieties of π_L and π_R. Let

$$(10) \qquad \Sigma = \{(P, \xi, Q, \eta) \in N^*(C) \text{ where } \pi_L, \pi_R \text{ are not locally } 1-1\}$$

be the singular variety of $d\pi_L$ and $d\pi_R$. The key geometric information is the position of the kernels of $d\pi_L, d\pi_R$ relative to the tangent space to Σ, viewed as subspaces of $T_{(P,\xi;Q,\eta)}(N^*(C))$. We discuss first, say, the left projection π_L, and make the generic assumption that the kernel of $d\pi_L$ is one-dimensional and the determinant of $d\pi_L$ vanishes only of first order along Σ. The variety Σ is then smooth, and

$$\text{Ker } d(\pi_L|_\Sigma) = \text{Ker } d\pi_L \cap T\Sigma.$$

In particular, the transversality of Ker $d\pi_L$ and $T_{(P,\xi;Q,\eta)}(N^*(C))$ is equivalent to the local injectivity of π_L when restricted to Σ. This leads us to define successively

$$(11) \qquad \Sigma_L^0 = \Sigma,$$

$$\Sigma_L^k = \{(P, \xi; Q, \eta) \in \Sigma_L^{k-1}; \text{Ker } d\pi_L(P, \xi; Q, \eta) \subset T_{(P,\xi;Q,\eta)}\Sigma_L^{k-1}\}.$$

In the language of singularity theory, $\Sigma_L^1 = \emptyset$ means for example that the projection π_L is a (Whitney) fold. For the distribution of translates of $\vec{\gamma}(t)$ it is readily seen that

$$\Sigma_L^{(n-2)} = \{(P, \xi; Q, \eta); Q = P + \vec{\gamma}(t),$$

(12) $$\langle \eta, \dot{\gamma} \rangle = \langle \eta, \ddot{\gamma} \rangle = \cdots = \langle \eta, \gamma^{(n)} \rangle = 0\}$$

and the condition that $\vec{\gamma}$ have torsion is equivalent to Σ_L^{n-2} be empty.

Finally we come to the issue of possible asymmetry between the left and right projections. To define Σ there is no need to distinguish between them, because one of them is singular if and only if both are. This is a consequence of the fact that $N^*(\mathcal{C})$ is Lagrangian and can be seen as follows. First observe that $d\pi_R(\text{Ker } d\pi_L)$ and $Image(d\pi_R)$ are orthogonal complements with respect to the symplectic form on $T^*(\mathbf{R})$. In fact $\delta \Xi$, $\delta \Pi$ are in these spaces respectively when $(0, \delta \Xi)$ and $(\delta \Psi, \delta \Pi)$ are in $T(N^*(\mathcal{C}))$ for some $\delta \Psi$. Since $N^*(\mathcal{C})$ is Lagrangian we have

$$0 = \omega_N^*(\mathcal{C})((0, \delta \Xi); (\delta \Psi, \delta \Pi)) = \omega_{T^*(\mathbf{R}^n)}(\delta \Xi, \delta \Pi).$$

This implies that $d\pi_R(\text{Ker } d\pi_L)$ and Ker $d\pi_R$ have the same dimension, and thus that Ker $d\pi_L$ and Ker $d\pi_R$ have the same dimension since $d\pi_R$ restricted to Ker $d\pi_L$ is manifestly one-to-one. For translation invariant distributions of curves, there is symmetry between left and right projections to all orders of stratification. In general, however, we also need to consider the stratification varieties Σ_R^k of π_R, as well as the mixed singular varieties defined as follows. Let $\omega = (\omega_1, \ldots, \omega_k)$ denote a sequence of k indices ω_i which can be either an L or a R index. Then

(13) $\Sigma_{\omega_1 \omega_2 \cdots \omega_k}^k \equiv \Sigma_{\omega_2 \cdots \omega_k}^{k-1}$

$$\cap \{(P, \xi; Q, \eta); \text{Ker } d\pi_{\omega_1}(P, \xi; Q, \eta) \subset T_{(P,\xi;Q,\eta)} \Sigma_{\omega_2 \cdots \omega_k}^{k-1}\}.$$

This leads us to the extension of the notion of torsion for a single curve that we were looking for. In [34] a (not necessarily translation invariant) family of curves was said to have left (respectively, right) torsion if the variety Σ_L^{n-2} (respectively, Σ_R^{n-2}) is empty. More generally it will be said to have non-vanishing "ω-torsion" at (P, Q) if for some k and ω of length k

(14) $$\Sigma_\omega^k \cap N_{(P,Q)}^*(\mathcal{C}) = \emptyset.$$

The simplest family of curves in $\mathbf{R}^n$ incorporating an asymmetric behavior is given by the parametrization

$$M_{(x,\vec{t})} = \{(y, \vec{s}); \vec{s} = \vec{t} + \vec{S}(x, y)\}$$

(15) $$\mathcal{C} = \{(x, \vec{t}; y, \vec{s}); (y, \vec{s}) \in M_{(x,\vec{t})}\}$$

with x, y in $\mathbf{R}$, and $\vec{s}, \vec{t}, \vec{S}(x, y)$ vectors in $\mathbf{R}^{n-1}$. The Lagrangian becomes
(16)
$$N^*(\mathcal{C}) = \{((x, \langle \lambda, \vec{S}_x'(x, y) \rangle), (\vec{t}, \lambda); (y, \langle \lambda, \vec{S}_y'(x, y) \rangle), (\vec{s}, -\lambda)); \lambda \in \mathbf{R}^{n-1}\}.$$

In particular it can be parametrized by $(x, y, \vec{t}, \vec{\lambda})$, and its tangent space by $(\delta x, \delta y, \delta \vec{t}, \delta \lambda)$. Linearizing the defining equations for $N^*(\mathcal{C})$, we see that the singular variety Σ and the kernels of $d\pi_L, d\pi_R$ at Σ are given by

$$\Sigma = N^*(\mathcal{C}) \cap \{\langle \lambda, S''_{xy}(x, y)\rangle = 0\}$$

$$\mathrm{Ker}\, d\pi_L = \{(\delta x = \delta t = \delta \lambda = 0, \delta y \in \mathbf{R})\}$$

$$\textbf{(17)} \qquad \mathrm{Ker}\, d\pi_R = \{(\delta x \in \mathbf{R}, \delta t = \delta \lambda = \delta y = 0)\}.$$

To each ω we associate now a set I_ω of indices $(\alpha, \beta) \in \mathbf{N}^2$ by the following inductive process reflecting that of (13)

$$I_{\omega_1\omega_2\cdots\omega_k} = I_{\omega_2\cdots\omega_k} \cup \{(\alpha + 1, \beta); (\alpha, \beta) \in I_{\omega_2\cdots\omega_k}\}, \qquad \text{if} \quad \omega_1 = R$$

$$\textbf{(18)}\ I_{\omega_1\omega_2\cdots\omega_k} = I_{\omega_2\cdots\omega_k} \cup \{(\alpha, \beta + 1); (\alpha, \beta) \in I_{\omega_2\cdots\omega_k}\}, \qquad \text{if} \quad \omega_1 = L.$$

Then we have
(19)
$$\Sigma^k_{\omega_1\omega_2\cdots\omega_k} = \{(x, y, \vec{t}, \lambda); \langle \lambda, \partial_x^{\alpha+1}\partial_y^{\beta+1}\vec{S}(x, y)\rangle = 0, \quad (\alpha, \beta) \in I_{\omega_1\omega_2\cdots\omega_k}\}.$$

We can finally formulate the first main goal of a theory of Fourier integral operators with degenerate Lagrangian manifolds: to establish optimal estimates, or, more precisely, identify the nature of TT^* and T^*T, for each possible stratification structure of the projections π_L and π_R, as described by the subvarieties Σ^k_ω. In Section III we shall provide a systematic study of a class of models which shows how the L^2 estimates and smoothing properties are indeed dictated by the stratification of the Lagrangian.

2 Green's Functions for Subelliptic Problems and Singular Radon Transforms

It is perhaps a surprising fact that Radon transforms arise in Green's functions for subelliptic problems. Indeed they are normally associated rather with propagation of singularities phenomena inherent to hyperbolic problems. More precisely the singular support of the Green's function of a subelliptic problem is contained in the diagonal, whereas the singular support of Radon transforms is usually given by a manifold of higher dimension.

Let $\mathbf{H}^n = \mathbf{C}^n \times \mathbf{R}$ be the Heisenberg group, with the group operation $(z, t)(z', t') = (z + z', t + t' + 2Imz \cdot \bar{z}')$. The vector fields

$$\textbf{(20)} \qquad Z_j = \frac{\partial}{\partial z_j} + i\bar{z}_j\frac{\partial}{\partial t}, \bar{Z}_j = \frac{\partial}{\partial \bar{z}_j} - iz_j\frac{\partial}{\partial t}, T = -i\frac{\partial}{\partial t}$$

are left invariant and form a basis for the Lie algebra of $\mathbf{H}^n$. The Laplacian is given by

$$(21) \qquad \Delta = \frac{1}{2}T^2 - \frac{1}{2}\sum_{j=1}^{n}(Z_j\bar{Z}_j + \bar{Z}_j Z_j) - (n-2)T,$$

and our task is to study the Green's function for the following operator

$$(22) \qquad \Box_+ = \Delta^{1/2} + \frac{1}{\sqrt{2}}T.$$

This problem is closely related to the $\bar{\partial}$-Neumann problem on the domain $\mathbf{U} = \mathbf{H} \times \mathbf{R}_+$. Let (z, t, ρ) denote points in $\mathbf{U}$, and set

$$(23) \qquad \Box = -\frac{1}{2}\frac{\partial^2}{\partial\rho^2} + \Delta.$$

Then the $\bar{\partial}$-Neumann problem is the following boundary value problem

$$(24) \qquad \Box u = f, \qquad (\frac{\partial}{\partial t} + i\frac{\partial}{\partial\rho})u|_{\rho=0} = 0$$

where $f(z, t, \rho)$ is a given C_0^∞ function on $\bar{\mathbf{U}}$. More precisely $\mathbf{U}$ can be identi-
fied with the Siegel upper half space $\{(z, z_{n+1}); Imz_{n+1} > |z|^2\}$ by $(z, t, \rho) \to$
$(z, z_{n+1} = t + i(\rho + |z|^2))$. The Levi metric is the metric which makes
$\omega^j = dz^j, j = 1\ldots, n$ and $\omega^{n+1} = -2^{1/2}\sum_{j=1}^{n}\bar{z}_j dz_j - i2^{-1/2}dz_{n+1}$ into
an orthonormal basis of $(1,0)$-forms. The corresponding Laplacian on $(0,1)$-forms
$\sum_{j=1}^{n+1}u_j\bar{\omega}^j$ is diagonal. The boundary condition on u_{n+1} is the Dirichlet condi-
tion $u_{n+1}|_{\rho=0} = 0$, while the Laplacian and boundary conditions on the remaining
components $u_j, j = 1, \ldots, n$ are given by (24).

The L^2[7], L^p and Schauder estimates [8], [13] for both problems (22) and (24)
are by now well-known. Here we would like to discuss instead explicit formulas
for their Green's functions (more accurately, their parametrices) and bounds for
the corresponding classes of integral operators. Translation invariance shows that
the Green's function for $\Box_+^{-1}$ is a convolution operator T on the Heisenberg group

$$(25) \qquad (T\phi)(z, t) = \iint G_b((z, t)(w, s)^{-1})\phi(w, s)\,dw\,ds.$$

On the other hand, the parametrix for the boundary value problem (24) must be of
the form

$$(Nf)(z, t, \rho) = \int_0^\infty \iint G_0((z, t)(w, s)^{-1}; |\rho - \mu|)f(w, s; \mu)\,dw\,ds\,d\rho$$

$$(26) \qquad \Big/ \quad + \int_0^\infty \iint G((z, t)(w, s)^{-1}; \rho + \mu)f(w, s; \mu)\,dw\,ds\,d\rho.$$

The first term on the right hand side of (26) is pseudolocal and is the same as for, say, the Dirichlet problem. The second term is the key one insuring the $\bar{\partial}$-Neumann condition. Such terms are called Hilbert integral operators. There is a remarkable relation [30] linking directly the Green's function of $\Box_+$ to that of (24)

(27) $$\Box_+^{-1}\phi = -(N(\phi \otimes \delta(\mu))|_{\rho=0}.$$

It is now well known [29], [32], [3] that $G(z, t; \rho)$ is given by an asymptotic expansion in kernels of the form $E_k(z, t; \rho)H_l(z, t; \rho)$, where $E_k(z, t; \rho)$ and $H_l(z, t; \rho)$ are homogeneous of order $-k$ and $-l$ with respect to the two different notions of dilations

(28) $\quad E_k(\lambda z, \lambda t; \lambda \rho) = \lambda^{-k}E_k(z, t, \rho), \qquad H_l(\lambda z, \lambda^2 t; \lambda^2 \rho) = \lambda^{-l}H_l(z, t; \rho)$

respectively. Similar notions of homogeneity are introduced for kernels which are functions of (z, t) alone. The sharp estimates for the $\bar{\partial}$-Neumann problem and for $\Box_+^{-1}$ are given by bounds on $Z_j Z_k N(f)$, $Z_j \bar{Z}_k N(f)$, $\bar{Z}_j \bar{Z}_k N(f)$, and $Z_j \Box_+^{-1}\phi$, $\bar{Z}_j \Box_+^{-1}\phi$. The theory of singular integrals with homogeneous kernels on the Heisenberg group has been developed in [8]. The above expressions lead, however, to convolutions on the Heisenberg group and Hilbert integral operators with products of kernels of two different homogeneities. The following theorems provide bounds for such operators [30], [31], [32]:

Theorem 2. *Let $G_b(z, t) = E_k(z, t)H_l(z, t)$ be a kernel of mixed homogeneity, and let T be the corresponding convolution operator (25). Then*
- *If $k + l < 2n + 2$, $k < 2n$ or $k + (l/2) < 2n + 1$, $l < 2$, T is bounded from $L_{loc}^p(\mathbf{H}^n)$ into itself for $1 \leq p \leq \infty$;*
- *If $k + l = 2n + 2$, $k < 2n$, and the kernel $E_k(z, 0)H_l(z, t)$ has mean value 0 on the parabolic (z, t)-sphere $|z|^4 + t^2 = 1$, then T is bounded on $L^p(\mathbf{H}^n)$ for $1 < p < \infty$;*
- *If $k + (l/2) = 2n + 1$, $l < 2$, and the kernel $E_k(z, t)H_l(z, 0)$ has mean value 0 on the euclidian (z, t)-sphere $|z|^2 + t^2 = 1$, then T is again bounded on $L^p(\mathbf{H}^n)$ for $1 < p < \infty$.*

Theorem 3. *Let $G(z, t; \rho) = E_k(z, t; \rho)H_l(z, t; \rho)$ be a kernel of mixed homogeneity, and let H be the corresponding Hilbert integral operator defined by the second terms on the right hand side of (26). Then*
- *If $k + l < 2n + 4$, $k < 2n$ or $k + (l/2) < 2n + 2$, $l < 4$, H is bounded from $L_{loc}^p(\mathbf{H}^n \times \mathbf{R}_+)$ into itself for $1 \leq p \leq \infty$;*
- *If $k + l = 2n + 4$, $k < 2n$, or $k + (l/2) = 2n + 2$, $l < 4$, then H is bounded on $L^p(\mathbf{H}^n \times \mathbf{R}_+)$ for $1 < p < \infty$;*

• *If $k + l = 2n + 4$, $k = 2n$, and the kernel $E_k(z, t; \rho)H_l(z, t; \rho)$ has mean value 0 on the z-sphere $|z| = 1$ for any (t, ρ), i.e.,*

(29)
$$\int_{\epsilon < |z| < \delta} E_k(z, t; \rho)H_l(z, t; \rho)dz = 0$$

for any $\epsilon, \delta, 0 < \epsilon < \delta$, then H is bounded on $L^p(\mathbf{H}^n \times \mathbf{R}_+)$ for $1 < p < \infty$.

The conditions $k + l < 2n + 2$ and $k + (l/2) < 2n + 1$ of Theorem 2 define a region in the (k, l) plane, inside of which the kernel is locally integrable. On the edges $k + l = 2n + 2$, $k < 2n$, away from the intersection point $k = 2n, l = 2$, the kernel behaves as the parabolically homogeneous kernel $E_k(z, 0)H_l(z, t)$ which has singularities along the plane $z = 0$. These singularities are, however, integrable in view of the condition $k < 2n$, and the condition that the mean value on the (z, t)-sphere of $E_k(z, 0)H_l(z, t)$ is 0 leads to L^p bounds just as in the case of smooth kernels homogeneous of degree $2n + 2$. The case of $k + (l/2) = 2n + 1$, $l < 2$ is similar, with the kernel $E_k(z, t)H_l(0, t)$ homogeneous of the (euclidian) critical degree $2n + 1$ and having integrable singularities on the (euclidian) (z, t)-sphere. The estimates for $Z_j \Box_+^{-1} \phi$, $\bar{Z}_j \Box_+^{-1}$ fall in these cases. The case $k = 2n$, $l = 2$ is not needed here, but may be of independent interest. It is likely that it can be treated by the method of singular Radon transforms used for Theorem 3, to be described next, but this has not been established.

We turn now to Hilbert integral operators. Inside the region $k + l < 2n + 4$, $k + (l/2) < 2n + 2$ the kernel $E_k(z, t; \rho)H_l(z, t; \rho)$ is locally integrable, and the first statement of Theorem 3 follows at once. On each of the edges $k + l = 2n + 4$, $k < 2n$ and $k + (l/2) = 2n + 2$, $l < 4$ the kernel behaves again as a homogeneous kernel with integrable singularities on the sphere, the homogeneities being parabolic and euclidian, respectively, in their critical degrees $2n + 4$ and $2n + 2$. Such kernels lead to operators bounded on $L^p(\mathbf{H}^n \times \mathbf{R}_+)$, by arguments similar to the one-dimensional case of $K(x, y) = (x + y)^{-1}$ on functions on $L^p(\mathbf{R}_+)$. Unlike in the case of $\Box_+^{-1}$, sharp bounds for Nf lead to Hilbert integral operators at the *doubly* critical degree of homogeneity $k = 2n, l = 4$, requiring singular Radon transforms on the Heisenberg group.

The key geometric ingredient of singular Radon transforms on $\mathbf{H}^n$ is a distribution of hypersurfaces M_P passing through each point $P \in \mathbf{H}^n$. At $P = (0, 0)$, M_P is the maximal complex space $M_P = \{Q = (w, 0); w \in \mathbf{C}^n\}$. For general $P = (z, t)$ it is the image of $M_{(0,0)}$ under the group multiplication by (z, t)

(30)
$$M_{(z,t)} = \{Q = (z + w, t + 2Imz \cdot \bar{w})\}.$$

Let $L_0(w)dw$ be a density on the hyperplane $M_{(0,0)}$, which we can transport to a density $d\mu_P(Q)$ on each $M_{(z,t)}$ again by group multiplication. Typically $L_0(w)$

is a homogeneous kernel on $\mathbb{C}^n$, and, in particular, singular at $w = 0$. The singular Radon transform associated to the distribution of hyperplanes M_P with their densities L_P is given by averaging over each M_P with respect to the density $d\mu_P(Q)$

(31) $R\phi(P) = \int_{M_P} \phi(Q)d\mu_P(Q) = \int_{\mathbb{C}^n} \phi(z + w, t + 2Imz \cdot \bar{w})L_0(w)dw.$

When $L_0(w)$ is a Calderón-Zygmund kernel on $\mathbb{C}^n$ the operator R extends as a bounded operator on L^p

(32) $\|R\phi\|_{L^p(\mathbf{H}^n)} \leq C\|\phi\|_{L^p(\mathbf{H}^n)}, \quad 1 < p < \infty.$

The key observation is that Hilbert integral operators are effectively averages of singular Radon transforms. For fixed s, ρ, μ we can construct the singular Radon transform $R_{s,\rho,\mu}$ with density $L_0(w) = K(w, s, \rho + \mu)$ and the function $\phi_{s,\mu}$ on the Heisenberg group defined by $\phi_{s,\mu}(z, t) = f(z, t - s, \mu)$. Then

(33) $Hf(z, t, \rho) = \int_0^\infty \int_{-\infty}^{+\infty} (R_{s,\rho,\mu}\phi_s)(z, t)dsd\mu.$

Now the densities $K(w, s, \rho + \mu)$ are smooth in w, but their bounds as Calderón-Zygmund kernels are only $O(s^2 + \rho^2 + \mu^2)^{-1}$. Thus (32) implies

$$\|R_{s,\rho,\mu}\| \leq C(s^2 + \rho^2 + \mu^2)^{-1}.$$

It follows that

$$\|(Hf)(\cdot, \cdot, \rho)\|_{L^p(\mathbf{H}^n)} \leq C \int_0^\infty (\int_{-\infty}^{+\infty} \frac{1}{s^2 + \rho^2 + \mu^2} \|\phi_s\|_{L^p(\mathbf{H}^n)}ds)d\mu$$

(34) $\leq C \int_0^\infty \frac{1}{\rho + \mu} \|f(\cdot, \cdot, \mu)\|_{L^p(\mathbf{H}^n)}d\mu.$

The L^p boundedness of the one-dimensional Hilbert integral operator gives the desired result, establishing Theorem 3.

   In [9] the estimate (32) was proved using the group Fourier transform on the Heisenberg group. The natural setup for singular Radon transforms is, however, in integral geometry, and the following general theorem holds [31], [32]:

Theorem 4. *Let M_P be a smooth distribution of hypersurfaces passing through P in a C^∞ compact manifold M, and let $d\mu_P(Q)$ be a density on each M_P with a Calderón-Zygmund singularity at $Q = P$. Define the corresponding singular Radon transform and maximal operator by*

$$(R\phi)(P) = \int_{M_P} \phi(Q)d\mu_P(Q)$$

(35) $$(M\phi)(P) = sup_\delta \frac{1}{|B(P, \delta)|} \int_{B(P,\delta)} \phi(Q) d\mu_P(Q)$$

where we have chosen a metric on M and denoted by $B(P, \delta)$ the geodesic ball centered at P and of radius δ. Assume that the distribution of hypersurfaces M_P has non-vanishing rotational curvature along the diagonal, i.e., if $\Phi(P, Q) = 0$, $\nabla_{P,Q}\Phi \neq 0$ is a defining equation for the relation manifold $C = \{(P, Q); Q \in M_P\}$, then

(36) $$det(J) \equiv \begin{pmatrix} \Phi(P, Q) & \nabla_P\Phi(P, Q) \\ \nabla_Q\Phi(P, Q) & \nabla_P\nabla_Q\Phi(P, Q) \end{pmatrix} \neq 0.$$

Then the operators R, M are bounded on $L^p(M)$ for $1 < p < \infty$.

It should be pointed out that (36) is exactly the Monge-Ampère determinant $J(\Phi)$ introduced by Fefferman [6] in his study of the Bergman kernel and parabolic invariants. It will also play a major role in the study of Lagrangians with folds, as discussed in Section IV below.

It is instructive to study models for the singular Radon transforms (35) in their own right. If we consider the family of hypersurfaces in $\mathbf{R}^{n+1}$ given by $M_{(x,t)} = \{(x + y, s = t + S(x, y)\}$ with $x, y \in \mathbf{R}^n$, $t \in \mathbf{R}$, and the density $d\mu_{(x,t)}(y, s)$ which is the image under $y \to (x + y, t + S(x, y))$ of a fixed homogeneous density $K(y)dy$ on $\mathbf{R}^n$, then the singular Radon transform of (35) can be viewed as a vector-valued pseudo-differential operator with symbol

(37) $$(a(\lambda)\psi)(x) = \int e^{i\lambda S(x,y)} K(x - y)\psi(y)dy.$$

Since rotational curvature is a second derivative condition, we assume that the phase $S(x, y)$ is a quadratic form. We also set $\lambda = 1$. The analogue of Theorem 4 is [32]:

Theorem 5. Let the rank of the quadratic form $S(x, y)$ be k, and let $-\mu$ be the degree of homogeneity of the kernel $K(x)$. When $\mu = n$ we also assume that $K(x)$ has mean value 0 on the unit sphere. Assume also that $k > 0$, otherwise (37) reduces to a familiar fractional integral. Then the operator a is bounded on $L^p(\mathbf{R}^n)$ for $|1/2 - 1/p| < (\mu - n + k)/(2k)$, $1 < p < \infty$, and for $|1/2 - 1/p| \leq \mu/(2n)$ when $k = n$.

More recently progress has been made in several different directions. First it has been shown in [27] that the singular Radon transforms associated to the distribution of hyperplanes (30) in the Heisenberg group and a density $L_0(w)dw$ homogeneous of degree 0 are smoothing of order $-n$ on Sobolev spaces. This is a generalization of the similar estimate for *smooth* densities which would follow from the classical theory of Fourier integral operators. Although we may expect a similar result to hold in the more general setup of Theorem 4, this has not been established to date. Second, Theorem 5 suggests that the condition that the rotational curvature form be non-degenerate can be weakened considerably if we restrict our attention to L^p bounds for singular Radon transforms with Calderón-Zygmund singularities or maximal functions. In the context of nilpotent groups, such improvements can be found in [4], [23], [36]. This has now been done in full generality by Christ, Nagel, Stein, and Wainger [5], who showed that the L^p bounds hold as long as the rotational curvature form does not vanish to infinite order. At the other extreme, when the densities are smooth, singular Radon transforms become Fourier integral operators, and non-vanishing rotational curvature is essential to insure the optimal bounds since it reduces to the condition that the Lagrangian be a local graph. It may be interesting to determine exactly how much rotational curvature is needed for a fractional singular Radon transform.

3 Integral Geometry

In a remarkable parallel development, operators sharing many of the features of the previous examples arise in integral geometry, particularly in the problems of inverting the X-ray transform and constructing Lorentz metrics, all of whose light rays are closed. More precisely, they are certain Fourier integral operators $X_{\mathcal{F}}$ whose Lagrangian exhibits an asymmetric type of singularity as discussed in Section I.1, and whose composition $X_{\mathcal{F}}^* X_{\mathcal{F}}$ is a Fourier integral operator with a singular density, analogous to the singular Radon transforms of Section II.2. This important phenomenon was discovered independently by Guillemin [14] and Greenleaf and Uhlmann [10], [12], and we discuss it briefly here.

The main example is the X-ray transform on the family of light rays in $\mathbf{R}^3$. Let a light ray l in $\mathbf{R}^3$ be any line making an angle of $\pi/4$ radians with the x_3 axis, and let $\mathcal{F}$ be the 3-dimensional family of all light rays. The corresponding X-ray transform is the operator sending a function $f \in C_0^\infty(\mathbf{R}^3)$ to the function $X_{\mathcal{F}} f$ on $\mathcal{F}$, defined by averaging f on each $l \in \mathcal{F}$. If we parametrize a light ray l by its intersection with the (x_1, x_2) plane, and the angle θ that its projection on the (x_1, x_2) plane makes with the x-axis, then

$$(38) \quad (X_{\mathcal{F}} f)(x_1, x_2, \theta) = \int_{-\infty}^{\infty} f(x_1 + s \cos\theta, x_2 + s \sin\theta, s) \chi(s) ds$$

where we have restricted the support of f to a fixed compact subset K of $\mathbf{R}^3$, and chosen a cut-off function $\chi(s)$ which is even, and $\equiv 1$ in a neighborhood of K.

As usual we define $C = \{(l, y); y \in l\}$, and the Lagrangian of $X_{\mathcal{F}}$ is

$$N^*(C) = \{((x_1, x_2, \theta), \Xi);$$

$$((y_1 = x_1 + s \cos\theta, y_2 = x_2 + s \sin\theta, y_3 = s), \Psi)\}$$

$$\Xi = (-\mu_1, -\mu_2, y_3(\mu_1 \sin\theta - \mu_2 \cos\theta))$$

(39) $\qquad \Psi = (\mu_1, \mu_2, -(\mu_1 \cos\theta + \mu_2 \sin\theta)).$

It follows that $N^*(C)$ can be parametrized by $(x_1, x_2, \theta, y_3, \mu_1, \mu_2)$ and its tangent space by infinitesimal variations $(\delta x_1, \delta x_2, \delta\theta, \delta y_3, \delta\mu_1, \delta\mu_2)$. In this instance π_R is the projection on the y, Ψ variable, so that its kernel requires $\delta\mu_1 = \delta\mu_2 = \delta y_3 = 0, \delta x_1 = y_3 sin\theta\delta\theta, \delta x_2 = -y_3 cos\theta\delta\theta$, and $(-\mu_1 sin\theta + \mu_2 cos\theta)\delta\theta = 0$. Thus Ker $d\pi_R$ is non-trivial exactly on the variety

(40) $\qquad \Sigma = N^*(C) \cup \{-\mu_1 sin\theta + \mu_2 cos\theta = 0\}$

and is generated by a non-vanishing $\delta\theta$. In particular it is one-dimensional and transversal to Σ, and the Lagrangian is a fold with respect to π_R. On the other hand, it is easy to see that $d\pi_L$ is singular along Σ, but its kernel is one-dimensional and generated rather by a non-vanishing δy_3. This implies that Ker $d\pi_L$ is tangent to Σ, which is an asymmetric behavior. In view of later generalizations, it is helpful to note that the description of Ψ in (39) is equivalent to its being perpendicular to the line l. Furthermore, the condition (40) defining Σ just means that Ψ is also orthogonal to the vector $\vec{e}_2 = (-sin\theta, cos\theta, 0)$ which, together with l, generates the tangent plane to the light cone. Thus Σ is actually characterized by requiring that Ψ be perpendicular to the light cone. The statement about $X_{\mathcal{F}}^* X_{\mathcal{F}}$ can be verified by an explicit formula which we give below.

The family of light rays is only a special case of a rather general setup, as shown by Greenleaf and Uhlmann. Let $\mathcal{L}$ be the family of all lines in $\mathbf{R}^n$. We shall see shortly that it is a smooth manifold of dimension $2n - 2$. Given a function $f \in C_0^\infty(\mathbf{R}^n)$, its X-ray transform is the function on $\mathcal{L}$ defined by

$$(Xf)(l) = \int_l f, \quad l \in \mathcal{L}.$$

Recapturing f from Xf is an overdetermined problem. Gelfand raised the problem of finding sets of uniqueness for X, i.e., find n-dimensional families $\mathcal{F}$ of lines l for which

(41) $\qquad X_{\mathcal{F}}f = Xf|_{\mathcal{F}}$

suffices to characterize f. This problem was solved in the case of complex lines in $\mathbf{C}^n$, the sets of uniqueness being the ones satisfying the following *cone condition*. For each x let Γ_x be the cone generated by the lines passing through x. Then Γ_x

is tangent to Γ_y along the line joining x and y for any y in Γ_x. Greenleaf and Uhlmann addressed the analogous problem for real lines, namely to reconstruct explicitly as much of f as possible from $X_{\mathcal{F}}f$ when $\mathcal{F}$ satisfies the cone condition. Although not generic, this condition is satisfied by many families of lines, such as the family of light rays discussed above, and the family of all lines passing through a given curve. It turns out to imply that $X_{\mathcal{F}}$ is a Fourier integral operator with a fold on one side and a blow-down on the other. Furthermore $X_{\mathcal{F}}^*X_{\mathcal{F}}$ is again essentially a singular Radon transform. We provide now some details in the simplified setting of families of lines in $\mathbf{R}^3$.

Asymmetric Behavior of the Lagrangian of $X_{\mathcal{F}}$

The first task is to obtain an explicit parametrization of the Lagrangian of $X_{\mathcal{F}}$. A line l can be described by a reference point P and a unit vector $\vec{\gamma}$. Locally we may view then the space of all lines $\mathcal{L}$ as a submanifold in $T(\mathbf{R}^3)$. As a consequence $T^*(\mathcal{L})$ can be identified with the restriction to $T(\mathcal{L})$ of $T^*(T(\mathbf{R}^3))$, viewed as linear functionals on $T(T(\mathbf{R}^3))$. Similarly $T^*(\mathcal{F})$ is the restriction to $T(\mathcal{F})$ of $T^*(T(\mathbf{R}^3))$, and in particular

$$T_l^*(\mathcal{F}) = (T_P^*(\mathbf{R}^3) \times T_P^*(\mathbf{R}^3))|_{T_l(\mathcal{F})}.$$

More explicitly the defining relation for $X_{\mathcal{F}}$ is

$$\mathcal{C} = \{(l, Q); Q \in l\}$$

(42)
$$= \{((P, \vec{\gamma}), Q); \epsilon_{ijk}(Q - P)^j \vec{\gamma}^k = 0\}$$

where ϵ_{ijk} is the antisymmetric symbol. In the above formalism the Lagrangian $N^*(\mathcal{C})$ becomes

(43)
$$N^*(\mathcal{C}) = \{((P, \vec{\gamma}), (-\epsilon_{ijk}\lambda^j \gamma^k), -\epsilon_{ijk}\lambda^j (Q - P)^k)); (Q, \epsilon_{ijk}\lambda^j \gamma^k))|_{T_l(\mathcal{F})}\}.$$

At this point we need a more concrete characterization of $T_l(\mathcal{L})$ and $T_l(\mathcal{F})$. Let s be the parametrization of l by arc length, starting from P, and let $\vec{e}_0 \equiv \vec{\gamma}$, $\vec{e}_1$, $\vec{e}_2$ be an orthonormal basis of vectors in $\mathbf{R}^3$. The line l can be deformed to another line by

(44)
$$P + s\vec{\gamma} \rightarrow (P + s\vec{\gamma}) + \sum_{i=1}^{2}(a_i s + b_i)\vec{e}_i$$

where a_i, b_i are any constants. This means that the vector fields along l given by

(45)
$$X(s) = \vec{e}_i, \qquad \text{or} \qquad s\vec{e}_i, \quad i = 1, 2$$

can be viewed as a basis for $T_l(\mathcal{L})$. They are called Jacobi fields. We note that this same discussion is valid in $\mathbf{R}^n$, in which case there would be $n - 1$ vectors $\vec{e}_i$

orthogonal to $\vec{\gamma}$, and $\mathcal{L}$ would be $(2n - 2)$-dimensional, as asserted earlier. When l is viewed as a pair $(P, \vec{\gamma}) \in T(\mathbf{R}^3)$, the deformation (44) is expressed as

(46) $$(P, \vec{\gamma}) \rightarrow (P + X(0), \vec{\gamma} + \dot{X})$$

so that a tangent vector in $T_l(\mathcal{L})$ is identified with the pair

(47) $$(X(0), \dot{X}) \in T_P(M) \times T_P(M).$$

Since $T_l(\mathcal{F})$ is a subspace of $T_l(\mathcal{L})$, it can also be described by Jacobi fields as in (45) and (47), and the covectors $(\epsilon_{ijk}\lambda^j \vec{\gamma}^k, \epsilon_{ijk}\lambda^j (Q - P)^k)$ appearing in (43) pair off naturally with vectors in (46).

The next task is to introduce a basis Φ_1, Φ_2, Φ_3 of 1-forms on $T_l^*(\mathcal{F})$, in which the covectors of (43) take a succinct form. Here we exploit the cone condition satisfied by the family $\mathcal{F}$. Now $\mathcal{F}$ is a codimension 1 submanifold of $\mathcal{L}$, so its normal space is a one-dimensional subspace of $T^*(\mathcal{L})$. An important ingredient of the geometry of $\mathcal{L}$ is that it is a symplectic manifold, with symplectic form

$$\omega_{\mathcal{L}}(\sum_{i=1}^{2}(a_i s + b_i)\vec{e}_i, \sum_{j=1}^{2}(c_j s + d_j)\vec{e}_j) = \sum_{i,j=1}^{2}(b_i c_j - a_i d_j).$$

This allows identifying $T(\mathcal{L})$ with $T^*(\mathcal{L})$ via $V_I \rightarrow \omega_{\mathcal{L}}^{I,J} V_I$. In particular the normal space to $\mathcal{F}$ corresponds to a Jacobi field $X_4(s)$ along l. The cone condition asserts that $X_4(s)$ is proportional to a fixed vector as s varies. If we consider the cone Γ_Q generated by the lines in $\mathcal{F}$ passing through Q

(48) $$\Gamma_Q \equiv \cup_{Q \in l, l \in \mathcal{F}} l$$

this means that the cones with vertices along l all share this vector as normal vector. We choose the vector $\vec{e}_1$ used earlier in the construction of Jacobi fields to be this common normal, and the vector $\vec{e}_2$ is determined accordingly. We note that $\vec{e}_2$ is tangent to Γ_Q. It is now not difficult to obtain a basis for $T_l(\mathcal{F})$. It consists of

(49) $$X_1(s) = (as + b)\vec{e}_1, \quad X_2(s) = \vec{e}_2, \quad X_3(s) = s\vec{e}_2.$$

The constant coefficients a, b are specified up to a constant multiple by the requirement that $\omega_{\mathcal{L}}(X_1, X_4) = 0$. We can now define Φ_i, $i = 1, 2, 3$ to be the dual basis to X_i, $i = 1, 2, 3$.

Returning to the expression (43) for $N^*(\mathcal{C})$, we let $Q = P + s\vec{\gamma}$ and write

(50) $$(\epsilon_{ijk}\lambda^j \vec{\gamma}^k, \epsilon_{ijk}\lambda^j (Q - P)^k) \equiv (\lambda_1 \vec{e}_1 + \vec{e}_2, s(\lambda_1 \vec{e}_1 + \lambda_2 \vec{e}_2)).$$

In terms of the $(P, \vec{\gamma})$ parametrization, the Jacobi fields X_1, X_2, X_3 of (49) become $(b\vec{e}_1, a\vec{e}_1)$, $(\vec{e}_2, 0)$, and $(0, \vec{e}_2)$, respectively. Evaluating their pairings with (50) gives

(51) $N^*(\mathcal{C}) = \{(((P, \vec{\gamma}), \Xi \equiv -[(2as + b)\lambda_1\Phi_1 + \lambda_2\Phi_2 + s\lambda_2\Phi_3)];$

$(Q, \lambda_1\vec{e}_1 + \lambda_2\vec{e}_2))\}.$

Finally we can evaluate $d\pi_L$ and $d\pi_R$. First we note that $N^*(\mathcal{C})$ can be parametrized by $(P, \vec{\gamma}) \in \mathcal{F}$, s, and λ_1, λ_2. Let the variations of the parameters be

(52) $$\sum_{i=1}^{3} \delta\alpha_i X_i, \; \delta s, \; \delta\lambda_1, \; \delta\lambda_2.$$

The corresponding infinitesimal variations $d\pi_L(T(N^*(\mathcal{C})))$ are given by

$$\delta P = \sum_{i=1}^{3} \delta\alpha_i X_i(0)$$

$$\delta\vec{\gamma} = \sum_{i=1}^{3} \delta\alpha_i \dot{X}$$

(53) $\delta\Xi = \delta s(2a\lambda_1\Phi_1 + \lambda_2\Phi_3) + \delta\lambda_1(2as + b)\Phi_1 + \delta\lambda_2(\Phi_2 + s\Phi_3).$

For (50) to give an element in $\mathrm{Ker}(d\pi_L)$ we need that $\delta\alpha_i = 0$, $i = 1, 2, 3$. Furthermore for Q generic, $2as + b$ is different from 0, and $\mathrm{Ker}(d\pi_L)$ is nontrivial (in fact, one-dimensional) at exactly the subvariety of $N^*(\mathcal{C})$ defined by $\lambda_2 = 0$

(54) $\Sigma = \{(((P, \vec{\gamma}), \Xi = (2as + b)\lambda_1\Phi_1); (Q, \Psi = \lambda_1\vec{e}_1))\}.$

Since $T(\Sigma) = T(N^*(\mathcal{C})) \cap \{\delta\lambda_2 = 0\}$, and $\mathrm{Ker}(d\pi_L)$ is generated by (52) with $\delta\alpha_i = \delta\lambda_2 = 0$, it follows that $\mathrm{Ker}(d\pi_L)$ is tangent to Σ. On the other hand we show now that if the cones Γ_Q satisfy a generic curvature condition, the kernel of $d\pi_R$ is transversal to Σ. In fact, the infinitesimal variations $d\pi_R(N^*(\mathcal{C}))$ under (52) are

$$\delta Q = \sum_{i=1}^{3} \delta\alpha_i X_i(s) + \delta s\vec{e}_0$$

(55) $$\delta\Psi = \sum_{i,j=1}^{2} \lambda_j\delta\vec{e}_i + \sum_{j=1}^{2} \delta\lambda_j\vec{e}_j.$$

For (55) to vanish, we must have $\delta s = \delta\alpha_1 = 0$ and $\delta\alpha_2 + s\delta\alpha_3 = 0$. This means that we need evaluate $\delta\vec{e}_i$ only when the line l is deformed infinitesimally along the tangent plane to Γ_Q, keeping the point Q fixed. At Σ, $\lambda_2 = 0$ and we need only evaluate $\delta\vec{e}_1$. Now the variation of $\vec{e}_1$ as we vary l along $\vec{e}_2$ is proportional to $\vec{e}_2$, with the coefficient of proportionality given by the geodesic curvature of the curve of intersection between Γ_Q and the unit sphere. If we assume that the cones Γ_Q have non-vanishing geodesic curvature, it follows that the elements (52) corresponding to $\text{Ker}(d\pi_R)$ must have $\delta\lambda_2 \neq 0$. This means that the kernel of $d\pi_R$ is transversal to the singular variety Σ. This is the asymmetric behavior of the Lagrangian $N^*(\mathcal{C})$ that we wanted to establish.

$X_{\mathcal{F}}^* X_{\mathcal{F}}$ and Singular Radon Transforms

Formally the operator $X_{\mathcal{F}}^* X_{\mathcal{F}}$ is given by

$$(56) \qquad (X_{\mathcal{F}}^* X_{\mathcal{F}} f)(Q) = \int_{\Gamma_Q} f(P) d\mu_Q(P)$$

where $d\mu_Q$ is a density on the cone Γ_Q (48), which is singular at Q. For example, when $\mathcal{F}$ is the family of light rays in $\mathbf{R}^3$ we have

$$(57) \qquad (X_{\mathcal{F}}^* g)(y_1, y_2, y_3) = \chi(y_3) \int_0^{2\pi} g(y_1 - y_3 cos\theta, y_1 - y_3 sin\theta, \theta) d\theta$$

and the operator $X_{\mathcal{F}}^* X_{\mathcal{F}}$ can be written explicitly as

$$(58) \quad (X_{\mathcal{F}}^* X_{\mathcal{F}})(f)(x_1, x_2, x_3)$$

$$= \iint f(x_1 - u, x_2 - v, x_3 \pm \sqrt{u^2 + v^2}) \chi(z \pm \sqrt{u^2 + v^2}) \frac{du\, dv}{\sqrt{u^2 + v^2}} .$$

Thus $X_{\mathcal{F}}^* X_{\mathcal{F}}$ is of the form (35), with the difference that the submanifolds passing through each Q are cones with vertices at Q and thus singular at Q, and the densities $d\mu_Q$ are of fractional degree rather than critical degree. Also as the dimension n of the ambient space increases, Γ_Q remains of dimension 2 instead of staying a hypersurface. From the microlocal viewpoint

$$(59) \qquad WF'(X_{\mathcal{F}}^* X_{\mathcal{F}}) \subset \Delta_{T^*(\mathbf{R}^3)} \cup N^*(\mathcal{G})'$$

where $\Delta_{T^*(\mathbf{R}^3)} \subset T^*(\mathbf{R}^3) \times T^*(\mathbf{R}^3)$ is the diagonal and $\mathcal{G} = \{(Q, P); P \in \Gamma_Q\}$ is the defining relation of $X_{\mathcal{F}}^* X_{\mathcal{F}}$. Operators whose wave fronts are contained in a pair of intersecting Lagrangians have appeared in a number of other contexts, and have been extensively studied [20], [15], [1]. Nevertheless a calculus powerful enough to yield Sobolev estimates is only available in certain special cases, notably when the Lagrangian $N^*(\mathcal{G})'$ is a flow-out [1]. A crucial observation of Greenleaf and Uhlmann is that this turns out to be the case for $N^*(\mathcal{G})'$ when the family $\mathcal{F}$ satisfies the cone condition. For families of lines in $\mathbf{R}^3$ we can illustrate directly the underlying geometry as follows.

For each Q in $\mathbf{R}^n$, we parametrize the unit vectors generating the cone Γ_Q by $\vec{\gamma}(Q, \theta)$. The orthonormal frame $\vec{e}_0, \vec{e}_1, \vec{e}_2$ introduced earlier is now given by

$$\vec{e}_0 = \vec{\gamma}, \qquad \vec{e}_2 = \partial_\theta \vec{\gamma} / \|\partial_\theta \vec{\gamma}\|, \qquad \vec{e}_1 = \vec{e}_2 \times \vec{e}_0.$$

The dual cone to Γ_Q is the cone Γ_Q^* of all covectors orthogonal to Γ_Q at some line in Γ_Q. It is generated by the vectors $\vec{e}_1$. We set $\Gamma^* = \cup_Q \Gamma_Q^*$. Then

$$N^*(\mathcal{G}) = \{((Q, \xi), (P = Q - s\vec{\gamma}(Q, \theta), \xi\|\vec{e}_1))\}$$

(60) $$\eta_i = \xi_i - s\partial_{x^i}\gamma^k \xi_k.$$

Under the condition that each Γ_Q have geodesic curvature, the closure $\overline{N^*(\mathcal{G})}'$ extends smoothly to $\Delta_{T^*(\mathbf{R}^3)}$, and

(61) $$\overline{N^*(\mathcal{G})}' \cap \Delta_{T^*(\mathbf{R}^3)} = \Delta_{\Gamma^*}.$$

Next we derive two key consequences of the cone condition. The first is that η in (60) is actually parallel to ξ, and the second is a simple characterization of the symplectic complement of $T(\Gamma^*)$. In fact, the cone condition says that the tangent planes to Γ_P and Γ_Q are the same along the P, Q line, i.e.,

$$\vec{\gamma}(Q - s\vec{\gamma}(Q, \theta), \theta) = \vec{\gamma}(Q, \theta)$$

(62) $$(\partial_\theta \vec{\gamma})(Q - s\vec{\gamma}(Q, \theta), \theta) \times (\partial_\theta \vec{\gamma})(Q, \theta) = 0.$$

Differentiating the first equation with respect to s shows that

(63) $$\gamma^i \partial_{x^i}\gamma^j = 0$$

so that $\partial_{x^i}\gamma^k \xi_k$ is orthogonal to $\vec{\gamma}$. Differentiating the first equation with respect to θ and using the second one shows next that $(\partial_\theta \gamma^j)(Q, \theta)\partial_{x^j}\vec{\gamma}(Q - s\vec{\gamma}(Q, \theta), \theta)$ is parallel to $\partial_\theta \vec{\gamma}(Q, \theta)$ for any $s \neq 0$. Letting s tend to 0 gives

(64) $$(\partial_\theta \gamma^j)(Q, \theta)\partial_{x^j}\vec{\gamma}(Q, \theta)\|\partial_\theta \vec{\gamma}(Q, \theta).$$

Since ξ in (60) is parallel to $\vec{\gamma} \times \partial_\theta \vec{\gamma}$, it follows at once that

$$\langle \eta, \vec{\gamma} \rangle = \langle \eta, \partial_\theta \vec{\gamma} \rangle = 0$$

and η is parallel to ξ, as claimed. Next $T_{(Q,\eta)}(\Gamma^*)$ is generated by infinitesimals $(\delta Q, \delta \eta)$ of the form

(65) $\delta \eta = -s(\vec{\gamma} \times \partial_{Q\theta}^2\vec{\gamma} + \partial_Q \vec{\gamma} \times \partial_\theta \vec{\gamma})\delta Q - (\vec{\gamma} \times \partial_\theta \vec{\gamma})\delta s - s(\vec{\gamma} \times \partial_{\theta\theta}^2\vec{\gamma})\delta \theta.$

Thus $\omega((\alpha, \beta), \delta Q, \delta \eta)) = 0$ implies that α is proportional to $\vec{\gamma}$, while

$$\beta_j = \epsilon_{lpq}\partial_{x^j}\gamma^p \partial_\theta \gamma^q \gamma^l.$$

In view of (63), (64) β is orthogonal to both $\vec{\gamma}$ and $\partial_\theta \vec{\gamma}$, and hence must be parallel to $\vec{\gamma} \times \partial_\theta \vec{\gamma}$.

We can identify now the bicharacteristic through (Q, η). By the curvature condition on each cone, the covector η corresponds to a unique ray $\vec{\gamma}(Q, \omega)$. Points (P, ξ) are on the bicharacteristic when P flows along $\vec{\gamma}(Q, \omega)$, and ξ

along η. This identifies the Lagrangian $N^*(\mathcal{G})$ as the flow-out of Γ^*, which is the desired statement.

The asymmetric behavior of the Lagrangian of $X_{\mathcal{F}}$ can lead to very different operators for $X_{\mathcal{F}}^* X_{\mathcal{F}}$ and $X_{\mathcal{F}} X_{\mathcal{F}}^*$. Again returning to the example of light rays in $\mathbf{R}^3$ and the operator (38) we find

$$(66) \quad (X_{\mathcal{F}} X_{\mathcal{F}}^* g)(l) = \iint g(x+u, y+v, \theta - \psi) \chi \left(\frac{u^2 + v^2}{2(u \cos \theta + v \sin \theta)} \right)$$

$$\frac{du\, dv}{|u \cos \theta + v \sin \theta|}$$

where ψ is the angle defined by $\cos \psi = 2uv(u^2+v^2)^{-1}$, $\sin \psi = (v^2 - u^2)(u^2 + v^2)^{-1}$. We note that the kernel of the integral operator (66) has singularities along a line, but its support is narrow enough to make it integrable. Operators of type $X_{\mathcal{F}} X_{\mathcal{F}}^*$ are central to the problem of deforming the Minkowski metric on $\mathbf{R}^4$ to metrics all of whose light rays are closed. We refer to [14] where a detailed microlocal analysis for them is provided. It would be interesting to explore these operators also from the local viewpoint, with an aim towards $L^p - L^q$ estimates.

III MODELS OF DEGENERATE
FOURIER INTEGRAL OPERATORS

A full theory of Fourier integral operators encompassing the stratification (11), (13) of the Lagrangian and singular densities is still far ahead at this point. It is important first to construct and study simple models which reflect the high order of degeneracy of the projections π_L and π_R. Returning to the family of curves (15) in $\mathbf{R}^n$, the corresponding Radon transforms can be expressed as

$$Rf(x, \vec{t}\,) = \int_{M_{(x, \vec{t}\,)}} f$$

$$(67) \qquad = \int e^{i\langle \lambda, \vec{t} - \vec{s} \rangle} \left[\int e^{i \langle \lambda, \vec{S}(x, y) \rangle} \chi(y) \hat{f}(y, \lambda) dy \right] \prod_{i=1}^{n-1} d\lambda_i$$

where $f(y, \vec{s}\,)$ is a C^∞ function with support in a fixed compact set, $\hat{f}(y, \lambda)$ is the partial Fourier transform of f with respect to $\vec{s}$, and $\chi(y)$ is a cutoff function which is identically 1 on the support of f. As in (37) the operator (67) can be viewed as a pseudo-differential operator whose symbol $a(\lambda)$ is the operator from C^∞ functions of y to C^∞ functions of x given by the expression between brackets. For each fixed λ this is an oscillatory integral with phase $S_\lambda(x, y) = \langle \lambda/|\lambda|, \vec{S}(x, y) \rangle$.

Non-vanishing right torsion at (x, y) means that for any λ

(68)
$$\sum_{\alpha=1}^{n-2} |\partial_x^\alpha \partial_y S_\lambda(x, y)| = 0 \implies \partial_x^{n-1} \partial_y S_\lambda(x, y) \neq 0.$$

Similarly the condition that any of the varieties Σ_ω^{n-2} be empty translates into the non-vanishing of an $n - th$ derivative of $S_\lambda(x, y)$ when certain derivatives of order $\leq n - 1$ vanish. Fixing λ we are led then to the following models for degenerate Fourier integral operators

(69)
$$(T\phi)(x) = \int e^{i\lambda S(x,y)} \chi(y)\phi(y)dy$$

where $S(x, y)$ is a homogeneous polynomial in (x, y) of degree n

(70)
$$S(x, y) = \sum_{i=1}^{n-1} \alpha_{i-1} x^i y^{n-i}.$$

We assume that at least one of the coefficients α_i, $i = 0, \ldots, n - 1$ is not 0. Evidently the norm $\|T\|$ is unaffected by the inclusion of terms proportional to either x^n or y^n, so we can ignore them. The models (70) generalize oscillatory integrals with quadratic phases which are models for Fourier integral operators associated to canonical graphs [18], as well as the well-known Airy operator

(71)
$$(Ai\phi)(x) = \int_{-\infty}^{+\infty} e^{i\lambda(x-y)^3} \chi(y)\phi(y)dy$$

which is the prototype of Fourier integral operators where both π_L and π_R are Whitney folds [19], [41]. It is remarkable that no systematic study of the operators (69) has been undertaken to date, and we shall address this issue here [35].

1 Low degrees of degeneracy $n = 2, 3, 4$

For low degree of degeneracy we have a complete understanding for the L^2 bounds for (69)

Theorem 6. *For each $n \leq 4$ the following bounds hold and are sharp*

(72)
$$\|T\| \leq C|\lambda|^{-1/n}$$

except when $S(x, y)$ is proportional either xy^{n-1} or $x^{n-1}y$, and n is strictly greater than 2. In these cases T satisfies the weaker estimate

(73)
$$\|T\| \leq C|\lambda|^{-1/(2(n-1))}.$$

Proof. The case $n = 2$ is trivial, since the operator T reduces to the Fourier transform (acting on $\chi\phi$) after a rescaling. For $n = 3$ either $S(x, y) = \alpha_0 xy^2 +$

$\alpha_1 x^2 y$ with $\alpha_0 \alpha_1 \neq 0$, or $S(x, y)$ is proportional to either xy^2 or $x^2 y$. In the first case, the operator T is equivalent to a similar operator with phase

$$S(x, y) = ((\frac{\alpha_0^2}{3\alpha_1})^{1/3} x + (\frac{\alpha_1^2}{3\alpha_0})^{1/3} y)^3.$$

After a linear change of variables, T reduces to a convolution operator whose multiplier is readily seen to be uniformly bounded by $|\lambda|^{-1/3}$. In the second case assume, say, that $S(x, y) = xy^2$. The kernel $K_\lambda(x, y)$ of TT^* is then given by

$$(74) \qquad |K_\lambda(x, y)| = \left| \int e^{i\lambda(x-y)z^2} \chi(z)^2 dz \right| \sim \left| \frac{i\pi}{\lambda(x-y)} \right|^{1/2}$$

which shows that $\|T\| = O(|\lambda|^{-1/4})$, as stated in (73). We turn next to $n = 4$. When $S(x, y)$ is proportional to either xy^3 or $x^3 y$, the kernel $|K_\lambda(x, y)|$ of either TT^* or T^*T can be evaluated to be proportional to $|(\lambda)^{-1/3}(x - y)^{-1/3}|$. It follows that $\|T\| \leq O(|\lambda|^{-1/6})$, which is the estimate (73) for $n = 4$. We now show that the better estimate $\|T\| \leq O(|\lambda|^{-1/4})$ holds for any other phase function which is a homogeneous polynomial of degree 4. If $S(x, y) = x^2 y^2$, the kernel $K_\lambda(x, y)$ of TT^* is $|K_\lambda(x, y)| \sim |\lambda^{-1/2}(x - y)^{-1/2}(x + y)^{-1/2}|$. Thus TT^* is bounded on $L^2(\mathbf{R})$ with norm $O(|\lambda|^{-1/2})$ and $\|T\| = O(|\lambda|^{-1/4})$ as was to be shown. Otherwise we may assume that the phase $S(x, y)$ is given by

$$(75) \qquad S(x, y) = xy^3 + \alpha_1 x^2 y^2 + \alpha_2 x^3 y$$

with at least one of the coefficients α_1, α_2 different from 0. The kernel $K_\lambda(x, y)$ is given by

$$(76) \, K_\lambda(x, y) = \int e^{i\lambda(S(x,z) - S(y,z))} \chi^2(z) dz$$

$$= e^{i\lambda(x-y)[\frac{2}{27} \alpha_1^3 (x+y)^3 - \frac{1}{3} \alpha_1 \alpha_2 (x+y)(x^2 + xy + y^2)]}$$

$$\times \int e^{i\lambda(x-y)(z^3 + [\alpha_2(x^2 + xy + y^2) - \frac{1}{3}\alpha_1^2(x+y)^2]z)} \chi^2(z + \frac{1}{3}(x + y)) dz.$$

The second integral in (76) is of the form

$$(77) \qquad I(\lambda) = \int_{-\infty}^{+\infty} e^{i\lambda(Qz - \frac{1}{3}z^3)} \chi(z) dz$$

with $\lambda(x - y)$ replaced by $-\lambda/3$. An explicit analysis, say, using Airy functions, yields

$$(78) \qquad |I(\lambda)| \leq C \min\{|\lambda|^{-1/3}, (|\lambda| |Q|^{1/2})^{-1/2}\}$$

$$(79) \qquad \text{If } A < 0, \text{ then } |I(\lambda)| \leq C(|\lambda| |Q|)^{-1}$$

$$\text{where } C \text{ is a constant independent of } A \text{ and } \lambda.$$

In particular, we obtain the following bound for the kernel $K_\lambda(x, y)$

$$(80) \quad |K_\lambda(x, y)| \le C|\lambda(x - y)|^{-1/2}|\alpha_2(x^2 + xy + y^2) - \frac{1}{3}\alpha_1^2(x + y)^2|^{-1/4}.$$

The right hand side of (80) is homogeneous of degree -1. Lemma 1 below asserts that the corresponding kernel is bounded on $L^2(\mathbf{R})$ with norm $O(|\lambda|^{-1/2})$ as long as it has at most a finite number of singularities θ_i on the circle $x^2 + y^2 = 1$, near which it is bounded by

$$|Arctan(y/x) - \theta_i|^{-1+\delta_i}$$

for some $\delta_i > 0$ in general, and $\delta_i > 1/2$ if the singularity θ_i is on one of the x, y axes. These conditions can only be violated if the quadratic form in (80) has double roots at either $x = 0$, $y = 0$, or $x = y$. It is readily seen that the first two possibilities cannot occur, while the third one occurs exactly when $9\alpha_2 - 4\alpha_1^2 = 0$. In particular, we have proved the desired estimate for all phase functions of the form (75), except when

$$S(x, y) = xy^3 + 3ax^2y^2 + 4a^2x^3y$$

for some $a \ne 0$. In this case, however, the phase in (76) becomes $\lambda(x - y)[z^3 + a^2(x - y)^2z]$. Since $a^2(x - y)^2 \ge 0$ we can apply (79) and obtain

$$(81) \quad |K_\lambda(x, y)| \le Cmin\{|\lambda(x - y)|^{-1/3}, |\lambda|^{-1}|x - y|^{-3}\}.$$

As a consequence,

$$(82) \quad \int_{-\infty}^{\infty} |K_\lambda(x, y)|dx = \int_{|x-y|<|\lambda|^{-1/4}} |K_\lambda(x, y)|dx$$

$$+ \int_{|x-y|>|\lambda|^{-1/4}} |K_\lambda(x, y)|dx$$

$$\le C|\lambda|^{-1/3} \int_{|x-y|<|\lambda|^{-1/4}} |x - y|^{-1/3}dx$$

$$+ C|\lambda|^{-1} \int_{|x-y|>|\lambda|^{-1/4}} |x - y|^{-3}dx$$

$$\le C|\lambda|^{-1/2}.$$

Similarly, the L^1 norm in y of $K_\lambda(x, y)$ is $O(|\lambda|^{-1/2})$. Thus, we still have $\|T\| \le O(|\lambda|^{-1/4})$ in this case, and the proof of Theorem 6 is complete.

General Order of Degeneracy n

In the formalism of Fourier integral operators, the non-vanishing of a coefficient α_i in $S(x, y)$ in (70) corresponds to the condition that some variety Σ_ω^{n-2} be empty.

Theorem 6 suggests that the sharp decrease $O(|\lambda|^{-1/n})$ for the operator norm $\|T\|$ holds when at least two varieties Σ_ω^{n-2}, $\Sigma_{\omega'}^{n-2}$ with suitable ω and ω' are empty, or when n is even and Σ_ω^{n-2} is empty when ω contains as many L as R indices. Strong evidence for these phenomena is provided by the following theorem which extends Theorem 6 to all orders n of degeneracy [35]:

Theorem 7. *Let* $S(x, y)$ *be as in (70), T the corresponding operator (69). Then T extends as a bounded operator on* $L^2(\mathbf{R})$ *with norm* $\|T\| \le C|\lambda|^{-1/n}$ *when*
- $\alpha_{n-2} \ne 0$, n *is even, and* $|\alpha_{(n/2)-1}| + \cdots + |\alpha_0| > 0;$
- $\alpha_{n-2} \ne 0$, n *is odd, and* $|\alpha_{(n-1)/2-1}| + \cdots + |\alpha_0| > 0;$
- n *is even,* $\alpha_{(n/2)-1} \ne 0$, *and* $\alpha_i = 0$ *either for all* $i \ge n/2$ *or for all* $i \le (n/2) - 2$.

Evidently the roles of x and y can be interchanged, and an analogous statement holds starting from the assumption $\alpha_0 \ne 0$. Perhaps the most encouraging feature of Theorem 7 is that sharp bounds depend only on the non-vanishing of *individual* coefficients in the phase function $S(x, y)$, instead of, say, the rank and signature of some quadratic form. This is certainly a necessary condition for any eventual generalization to a full-fledged theory of Fourier integral operators based on stratification of the Lagrangian by the singular varieties of π_L and π_R.

The key estimates required for the proof of Theorem 7 are contained in the following Theorem 8, which can be viewed as a composition theorem for these models of degenerate Fourier integral operators [35]:

Theorem 8. *Assume that one of the sets of conditions in Theorem 7 holds. Then the kernel* $K_\lambda(x, y)$ *of the operator* TT^* *satisfies the following bounds:*
(a) *(83)* $|K_\lambda(x, y)| \le C|\lambda(x - y)|^{-1/(n-1)}$
(b) *There exists a finite number of pairs* (a_i, b_i), $i \le N$, $a_i^2 + b_i^2 = 1$, *such that*

$$(84) \qquad |K_\lambda(x, y)| \le C|\lambda|^{-2/n-\nu} \sum_{i=1}^{N} |a_i x + b_i y|^{-1+\delta_i} (|x| + |y|)^{-\delta_i - \nu n}$$

for some $\nu \ge 0$, *and* (x, y) *outside of a conic neighborhood of the line* $x = y$. *Here the exponents* δ_i *are strictly positive, and strictly greater than* $1/2$ *if either* a_i *or* b_i *vanishes;*
(c) *In the conic neighborhood of* $x = y$ *omitted above, either an estimate of the form (84) holds, i.e.,*

$$(85) \qquad |K_\lambda(x, y)| \le C|\lambda|^{-2/n-\nu}|x - y|^{-1+\delta}(|x| + |y|)^{-\delta - \nu n}$$

for some $\delta > 0$, or we have an estimate of the form

(86)
$$|K_\lambda(x, y)| \le C|\lambda|^{-\nu-2/n}|x - y|^{-1-\nu n}$$

with $\nu \ge 0$.

Theorem 7 follows from Theorem 8, according to the following lemma:

Lemma 1. *Let $K_\lambda(x, y)$ be any kernel on $\mathbf{R} \times \mathbf{R}$ satisfying either the conditions (84) and (85), or the conditions (83) and (86). Then the corresponding operator is bounded on $L^2(\mathbf{R})$ with norm $O(|\lambda|^{-2/n})$.*

When the kernel satisfies (84) and (85), Lemma 1 is an easy generalization of the standard Hilbert integral lemma which establishes the $L^2(\mathbf{R}_+)$ boundedness of the operator with kernel $(x + y)^{-1}$. When it satisfies (83) and (86), the desired estimate can be established as in (82) assuming (81).

We discuss now the main ideas in the proof of Theorem 8. The kernel $K_\lambda(x, y)$ of TT^* can be expressed as an oscillatory integral

$$K_\lambda(x, y) = \int e^{i\lambda(x-y)P(x,y,z)}\chi(z)^2 dz$$

(87)
$$P(x, y, z) \equiv (x - y)^{-1}(S(x, z) - S(y, z)).$$

In principle, the asymptotics in λ of such integrals are given by the method of stationary phase or the van der Corput lemma, except that here it is crucial to keep track of the dependence on x, y. For example, the estimate $|\lambda Q^{1/2}|^{-1/2}$ in (78) and (79) can be worse than the estimate $\lambda^{-1/3}$ if $|Q|$ is small enough. What is needed here is a version of the method of stationary phase which keeps track of the dependence on external parameters. Since the coefficient $|Q|^{1/2}$ in (78) can be interpreted as twice the distance separating the critical points $\pm Q^{1/2}$, we are led to the following theorem which may be of intrinsic interest.

Let K_λ be an oscillatory integral of the form

(88)
$$K_\lambda = \int_a^b e^{i\lambda P(z)}\chi(z)dz$$

where $P(z)$ is a C^2 phase function on $\mathbf{R}$, $\chi(z)$ is a C^1 function which is taken to be compactly supported if either one or both of the bounds a and b are infinite. Assume that

(89)
$$|P'(z)| \ge A \prod_{j=1}^N |z - \zeta_j|$$

where A is a positive constant. The ζ_j's are called "critical points," and are allowed to be complex. If ζ_j has an imaginary part, then its conjugate also occurs in (89). We set

$$(90) \qquad \zeta_j \equiv a_j + ib_j, \quad j = 1, \ldots, N;$$

$P''(z)$ changes sign only a finite number M of times.

By a cluster L around a given critical point ζ_k we designate any subset L of the ζ_j's which contains ζ_k. L^c will denote $L \setminus \{\zeta_k\}$ and $|L^c|$ will denote the number of critical points in L^c. For each integer $m \geq 2$ and each k, $k = 1, \ldots, N$, we define $K_\lambda(k, m)$ to be

$$(91) \qquad K_\lambda(k, m) \equiv min_{L^c, |L^c|=m-2}(|A\lambda| \prod_{j \notin L} |a_k - \zeta_j|)^{-1/m}$$

when ζ_k is real, and

$$K_\lambda(k, m) \equiv min\{min_{L^c, |L^c|=m-3}(|A\lambda| \prod_{j \notin L} |a_k - \zeta_j|)^{-1/m},$$

$$min_{L^c, |L^c|=m-2}(|A\lambda| \, |b_k| \prod_{j \notin L} |a_k - \zeta_j|)^{-1/m},$$

$$(92) \qquad min_{L^c, |L^c|=m-1}(|A\lambda| \, |b_k|^2 \prod_{j \notin L} |a_k - \zeta_j|)^{-1/m}\}$$

when $\zeta_k = a_k + ib_k$ has a non-vanishing imaginary part. Then

Theorem 9. *The oscillatory integral* (88) *can be estimated by*

$$(93) \qquad |K_\lambda| \leq C_{M,N} max_{1 \leq k \leq N} min_{m \geq 2}[K_\lambda(k, m)]$$

where the constant $C_{M,N}$ depends only on M, N, and the C^1 norm of χ.

In Theorem 9 we can evidently choose to deal only with real critical points, simply by dropping the imaginary part b_j of ζ_j from the bounds (89), (91), and (92). This will lead to slower decays for K_λ. The proof of Theorem 8 actually requires the full Theorem 9 with complex critical points. To illustrate the estimate (93), we take $A = 1$, $\zeta_j = a_j$ to be real, and choose m in (91) to be 2. Then Theorem 9 implies the following bound for K_λ:

$$|K_\lambda| \leq C_{M,N} |\lambda|^{-1/2} \sum_{k=1}^{N} \prod_{j \neq k} |a_k - a_j|^{-1/2}.$$

This is the estimate predicted by the method of stationary phase, with the key improvement that the constants are uniform in all phases with bounded number of critical points and bounded number of intervals of monotonicity for $P'(z)$.

Returning to the kernel $K_\lambda(x, y)$ of TT^* (87), we replace λ by $\lambda(x - y)$, the phase is given by $P(x, y, z)$, and thus the critical points ζ_j of (90) are the (possibly complex) roots $\zeta_j(x, y)$, $j = 1, \ldots, n - 2$ of the polynomial in $z(x - y)^{-1}(S_z'(x, z) - S_z'(y, z))$. They are homogeneous functions of (x, y) of order 1, so we may assume that (x, y) is on the circle $x^2 + y^2 = 1$. For each fixed phase $S(x, y)$ there are at most a finite number of points $Arctan(y/x) = \theta_k$, $k = 1, \ldots, N$ on the circle at which two or more of the roots ζ_j coincide. Theorem 6 shows that estimates for the kernel $K_\lambda(x, y)$ reduce to estimates for $|\zeta_i(x, y) - \zeta_j(x, y)|$ as (x, y) is close to the line of slope θ_k. Exploiting the fact that these roots are algebraic functions, we can show that under the hypotheses of the Theorem not more than $(n/2) - 1$ when n is even, and $(n - 1)/2$ when n is odd can get too close together in a precise sense. This combined with Theorem 9 leads to the bounds in Theorem 8 for $K_\lambda(x, y)$, and Theorem 7 follows.

IV THE MONGE-AMPÈRE DETERMINANT AND COVARIANCE

The models we discussed in Section III correspond to Fourier integral operators with "constant coefficients." For example the symbol $a(\lambda)$ of (67) does not depend on $\vec{t}$. The task of constructing a full theory defined on smooth manifolds even when these models are already well understood turns out to be surprisingly difficult. The only reasonably general results are for $n = 3$, that is, when at least one of the projections π_L or π_R is a fold. There it is known that

- Fourier integral operators whose Lagrangians are folds with respect to both π_L and π_R can be microlocally conjugated to the Airy operator [19], [20] and are smoothing of order $-1/3$;
- If π_L is a fold and π_R is a blow-down, then the FIO is smoothing of order $-1/4$ [11], [12];
- Sobolev and $L^p - L^q$ estimates are known for Radon transforms along general variable distributions of curves in any 2-dimensional manifold [34].

The methods for the first two classes of results are microlocal in nature. Here we shall discuss the local approach of [34], which makes the role of the Monge-Ampère equation (36) transparent and also confirms the intimate relation between degenerate Fourier integral operators and singular Radon transforms.

Let M be a C^∞ two-dimensional manifold, and let M_P be a curve passing through P for each point P in M. If $d\mu_P(Q)$ is a C^∞ density on M_P supported in a small neighborhood of P we set

$$(94) \qquad Tf(P) = \int_{M_P} f(Q)d\mu_P(Q).$$

Let π_L, π_R be the projections from the Lagrangian manifold $N^*(\mathcal{C})$ on the left and right factors as in (5). Then [34]

Theorem 10.
- *If either π_L or π_R is a fold along the singular variety Σ, then T is bounded from $H_{(s)}(M)$ to $H_{(s+1/4)}(M)$;*
- *If both π_L and π_R are folds, then T is bounded from*
  (a) *$H_{(s)}(M)$ to $H_{(s+1/3)}(M)$;*
  (b) *$L^p(M)$ to $L^q(M)$ for $(1/p, 1/q)$ in the intersection of the closed triangle with vertices at $(0, 0)$, $(1, 1)$, $(2/3, 1/3)$ and the closed half-space of the equation*

(95)
$$\frac{1}{p} - \frac{1}{q} - \frac{1}{4} \le 0.$$

As mentioned above, Part (a) of the second statement is the well-known estimate of Melrose and Taylor [19], [21] in diffraction theory. The estimate in the first statement has been obtained by Greenleaf and Uhlmann [12] for any dimension, under the additional hypothesis that the other projection is completely degenerate, or, more precisely, a blow-down.

We discuss in some detail two ingredients of the proof. We focus on the case when both π_L and π_R are folds since its treatment also contains many of the ingredients required for the case when only one projection is assumed to be a fold. It suffices to prove the theorem for $M = \mathbf{R}^2$ and f is supported in a fixed small neighborhood of the origin. We can choose coordinates (x, t) on $\mathbf{R}^2$ so that locally the curve $M_{(x,t)}$ is the graph of a function

(96)
$$M_{(x,t)} = \{(y, s); s = t - S(x, t, y)\}$$

with

(97)
$$S(x, t, x) \equiv 0, \qquad S(0, t, y) \equiv 0.$$

The operator T can be written as

(98)
$$(Tf)(x, t) = \iiint e^{i\lambda(s-t+S(x,t,y))} \psi(y) f(y, s) \, dy \, ds \, d\lambda$$

where $\psi(y)$ is a suitable cutoff function in y. The standard almost-orthogonality argument reduces the Sobolev bounds to showing that the localized operators in λ

(99)
$$(T_m f)(x, t) = \iiint e^{i\lambda(s-t+S(x,t,y))} \chi(2^{-m}\lambda) \psi(y) f(y, s) \, dy \, ds \, d\lambda$$

satisfy the uniform estimates

(100) $$\|T_m\| \leq C2^{-m/3}.$$

Let $J(x, t, y)$ be the Monge-Ampère determinant (36) with $\Phi(x, t; y, s) = s - t + S(x, t, y)$

(101) $$J(x, t, y) = (1 - S_t')S_{xy}'' - S_x'S_{ty}''.$$

The condition that π_L (respectively π_R) be a fold along its singular variety translates into the fact that the zero set of $J(x, t, y)$ can be parametrized as the graph of a smooth function of $y = \bar{x}(x, t)$ (respectively $x = \bar{y}(t, y)$). In particular the map $x \to \bar{x}(x, t)$ is invertible for each t if both projections are folds. We introduce a dyadic decomposition $T_m = \sum_{l=0}^{\infty} T_{l,m}$ away from the zero set of $J(x, t, y)$

(102) $(T_{l,m}f)(x, t)$

$$= \iiint e^{i\lambda(s-t+S(x,t,y))}\chi(2^{-m}\lambda)\chi(2^l J(x, t, y))\psi(y)f(y, s)\,dy\,ds\,d\lambda.$$

The Sobolev bounds for T are consequences of the following two bounds for $T_{l,m}$

$$\|T_{l,m}\| \leq C2^{-l}$$

(103) $$\|T_{l,m}\| \leq C(2^{-m}2^l)^{1/2}.$$

The index m is fixed from now on, so we drop it from our notation. Now the fact that both π_L and π_R are folds implies that the support of the kernel of T_l in each variable is of length $\sim 2^{-l}$ when the other variable is fixed. The first estimate follows at once. We note that we would obtain only the weaker estimate $\|T_l\| \leq C2^{-l/2}$ if only one projection is assumed to be a fold. The key to the second estimate is a further localization

(104) $(T_{l,j}f)(x, t) = \chi(\bar{c}2^l(\bar{x}(x, t) - x_j))$

$$\times \iiint e^{i\lambda(s-t+S(x,t,y))}\psi(2^{-m}\lambda)\chi(2^l J(x, t, y))f(y, s)\,dy\,ds\,d\lambda$$

where $\bar{c}$ is a large constant, and x_j are points regularly spaced at a distance $(\bar{c})^{-1}2^{-l}$. Since $(x, t) \to (\bar{x}(x, t), t)$ is a smooth change of variables, we may define

(105) $$(\bar{T}_{l,j}f)(\bar{x}, t) \equiv (T_{l,j}f)(x, t).$$

The second estimate for T_l can be verified to follow from a similar estimate for $\bar{T}_{l,m}$. In terms of (y, s) and $(\bar{x}, t)$ the operator $\bar{T}_{l,m}$ is of the same form as (104) but with a different phase function

(106) $$\bar{S}(\bar{x}, t, y) \equiv S(x(\bar{x}, t), t, y).$$

Lemma 2. *In a neighborhood of the origin, the phase function* $\bar{S}(x, t, y)$ *satisfies*

(107) $$|\bar{S}_t'| = o(1), \qquad |\bar{S}_{ty}''| = o(1).$$

Lemma 3. *Let* $\bar{S}(\bar{x}, t, y)$ *be a phase function satisfying the condition* (107), *together with fold conditions on both sides. Then*

- *In a sufficiently small neighborhood of the form* $|t - s| < c|\bar{x} - y|$, *the equation*

(108) $$\bar{S}_z'(\bar{x}, t, z) - \bar{S}_z'(y, s, z) = 0$$

admits a unique solution $z_c(\bar{x}, t, y, s)$;

- *Let* $y = \tilde{x}(\bar{x}, t)$ *denote the zeroes of the Monge-Ampère determinant corresponding to* $\bar{S}(\bar{x}, t, y)$, *and set*

(109) $$X(\bar{x}, t, y, s) = \bar{S}(\bar{x}, t, \tilde{x}(\bar{x}, t)) - \bar{S}(y, s, \tilde{x}(\bar{x}, t)).$$

Then in the region where $|z - \tilde{x}(\bar{x}, t)| \leq 2^{-\ell}$, $|t - s - X(\bar{x}, t, y, s)| \geq \bar{c}2^{-l}|\bar{x} - y|$ *we have*

(110) $$|t - s - X(\bar{x}, t, y, s)| \sim |t - s - \bar{S}(\bar{x}, t, z) + \bar{S}(y, s, z)|$$

while in the region

(111) $$|t - s - X(\bar{x}, t, y, s)| \leq 2^{-l}|\bar{x} - y|$$

the critical point z_c *can be approximated by the zero of the Monge-Ampère determinant*

(112) $$|z_c(\bar{x}, t, y, s) - \tilde{x}(\bar{x}, t)| \leq C(|\bar{x} - y| + (\bar{c})^{-1}2^{-l}).$$

Lemma 4. *Let* $\bar{J}(\bar{x}, t, y)$ *be the Monge-Ampère determinant determined by the phase function* $\bar{\Phi}(\bar{x}, t; y, s) \equiv s - t + \bar{S}(\bar{x}, t, y)$. *Then we have the following transformation law*

(113) $$\bar{J}(\bar{x}, t, y) = \left(\frac{\partial x}{\partial \bar{x}}\right) J(x, t, y).$$

We sketch the proof of the second statement in Lemma 3, as it is one of the key steps in the argument. First let $\tilde{x}(\bar{x}, t)$ be *any* function of $(\bar{x}, t)$ and define X as in (109). We note that X can be expanded in terms of $(\bar{x} - y)$ and $t - s - X$ with smooth coefficients

(114) $$X \equiv (\bar{x} - y)\tilde{\beta}(\bar{x}, t, y, s) + (t - s - X)\tilde{\alpha}(\bar{x}, t, y, s).$$

In fact, a Taylor expansion gives $X \equiv (x - y)\beta_0 + (t - s)\alpha_0$, and $|\alpha_0| = o(1)$ since $|\bar{S}_t'| = o(1)$. The equation (114) follows with

(115) $$\tilde{\beta} = \beta_0(1 - \alpha_0)^{-1}, \tilde{\alpha} = \alpha_0(1 - \alpha_0)^{-1}.$$

Next we rewrite the equation for the critical points in more explicit form. Setting $\bar{S}(\bar{x}, t, z) - \bar{S}(y, s, z) \equiv (\bar{x} - y)\gamma + (t - s)\delta$ we find

$$\bar{S}(\bar{x}, t, z) - \bar{S}(y, s, z) = (\bar{x} - y)\Psi(\bar{x}, t, y, s, z)$$

$$\Psi(\bar{x}, t, y, s, z) = [\gamma + \tilde{\beta}\delta] + \frac{t - s - X}{\bar{x} - y}(\delta + \tilde{\alpha}\delta)$$

(116) $\qquad \partial_z \Psi(\bar{x}, t, y, s, z) = 0 \Longleftrightarrow z = z_c(\bar{x}, t, y, s).$

We can obtain approximations of z_c up to errors bounded by the right hand side of (112) by adding to the equation $\partial_z \Psi = 0$ any term bounded by $O(|\bar{x} - y|)$ or $O(\bar{c}^{-1}2^{-l})$. In particular in $\partial_z \Psi$ we can drop $(t - s - X)(\bar{x} - y)^{-1}(\delta + \tilde{\alpha}\delta)$, replace y by $\bar{x}$, and s by t. The approximating equation is

(117) $\qquad\qquad\qquad \partial_z(\gamma + \tilde{\beta}\delta) = 0$

with up to negligible terms

(118) $\qquad \gamma = \bar{S}'_{\bar{x}}(\bar{x}, t, z), \qquad \delta = \bar{S}'_t(\bar{x}, t, z), \qquad \tilde{\beta} = \frac{\beta_0}{1 - \alpha_0}.$

Finally, we derive explicit formulas for α_0, β_0, up to similar negligible terms

(119) $\qquad\qquad \alpha_0 = \bar{S}'_t(\bar{x}, t, \tilde{x}(\bar{x}, t)), \qquad \beta_0 = \bar{S}'_{\bar{x}}(\bar{x}, t, \tilde{x}(\bar{x}, t)).$

Altogether the equation (116) for $z_c(\bar{x}, t, y, s)$ becomes

(120) $\quad \bar{S}''_{\bar{x}z}(\bar{x}, t, z_c)(1 - \bar{S}'_t(\bar{x}, t, \tilde{x}(\bar{x}, t))) + \bar{S}'_x(\bar{x}, t, \tilde{x}(\bar{x}, t))\bar{S}''_{tz}(\bar{x}, t, z_c) = 0.$

Thus for the critical point z_c to coincide with $\tilde{x}(\bar{x}, t)$, we must choose $\tilde{x}(\bar{x}, t)$ to be the zero of the Monge-Ampère determinant! This establishes the key estimate (112) in Lemma 3.

We return to the proof of L^2 bounds for $\bar{T}_{l,j}$. Forming $\bar{T}_{l,j}\bar{T}^*_{l,j}$, we obtain an operator whose kernel is an oscillatory integral in z and in λ with phase given by

$$\lambda(\bar{S}(\bar{x}, t, z) - \bar{S}(y, s, z))$$

and a cutoff given by

(121) $\qquad \chi(\bar{c}2^l(\bar{x} - x_j))\chi(\bar{c}2^l(y - x_j))\chi(2^\ell \bar{J}(\bar{x}, t, z))\chi(2^\ell \bar{J}(y, s, z)).$

The main contribution to the oscillatory integral comes from its critical point $z_c(\bar{x}, t, y, s)$. Now Lemma 4 implies in particular that the zero set of $\bar{J}$ is given by $y = \bar{x}$. In view of Lemma 3 we have then

(122) $\qquad\qquad |z_c(\bar{x}, t, y, s) - \bar{x}| \le C(|\bar{x} - y| + (\bar{c})^{-1}2^{-l})$

in the region defined by (111). Since $|x - y| \leq (\bar{c})^{-1}2^{-l}$ on the support of the kernel of $\bar{T}\bar{T}^*$

(123)
$$|z - z_c(\bar{x}, t, y, s)| \sim |z - \bar{x}| \sim 2^{-l}$$

and we obtain the following lower bound for the phase variation

(124)
$$|\bar{S}'_z(\bar{x}, t, z) - \bar{S}'_z(y, s, z)| \sim |x - y|2^{-l}.$$

This allows to exploit integration by parts with respect to z in the region (111). Outside of this region, the desired estimates follow from a simpler integration by parts in λ using (110).

We would like to point out an interesting feature of the above arguments. For classical Fourier integral operators we can choose coordinate systems so that the non-vanishing of the Monge-Ampère determinant in small neighborhoods is equivalent to the non-vanishing of the Hessian in x, y of $S(x, t, y)$. This is done from the outset and the Monge-Ampère determinant itself no longer enters the analysis [32], [33]. In the above argument on the other hand it is crucial to introduce dyadic partitions away from the zeroes of the Monge-Ampère determinant itself, rather than from the zeroes of say the Hessian, since these approximate better the critical points z_c in (111). Now the critical points z_c are zeroes of a linear expression in S, while the zeroes of J are of course zeroes of a non-linear expression in S. The appearance of this non-linearity is due to the fact that the approximation (112) does not hold everywhere, but only in a region determined by the cutoff itself. It is very intriguing that the localization procedures required by linear analysis inevitably introduce some non-linearity and lead us exactly to a covariant expression.

We conclude by sketching the relation between the Radon transforms of (94) and singular Radon transforms. For the sake of simplicity we shall assume that the family M_P is semi translation-invariant in the sense that the defining function $S(x, t, y)$ of (96) is independent of t. The operators T and T^*T can then be expressed as

$$(Tf)(x, t) = \int f(y, t - S(x, y))dy$$

(125)
$$(T^*Tf)(x, t) = \iint f(y, t + S(x, z) - S(y, z))dzdy.$$

Fixing (x, t) we make the change of variables $(y, z) \to (u, v)$

$$u = x - y$$

(126)
$$v = S(y, z) - S(x, z) - (S(y, z_c(x, y)) - S(x, z_c(x, y)))$$

where $z_c(x, y)$ is the critical point $S'_z(x, z_c) - S'_z(y, z_c) = 0$. Observe that

$$|Jacobian| = |S'_z(x, z) - S'_z(y, z)|$$

$$\sim |x - y|\,|z - z_c(x, y)|$$

(127)
$$|v| \sim |x - y|\,|z - z_c(x, y)|^2$$

so that the operator T^*T corresponds to the case $s = 1/2$ of the following operators

(128) $(U_s f)(x, t)$

$$= \iint f(x - u, t - v + S(x, z_c(x, x - u)) - S(x - u, z_c(x, x - u)))$$

$$\frac{du\,dv}{|u|^s |v|^s}.$$

This relation reduces the $L^p - L^q$ bounds for T to L^2 bounds for singular Radon transforms. In fact, the estimates along the segment $(1/p) - (1/q) - (1/4) \le 0$ of Theorem 10 follow from the boundedness of T^*T from $L^{4/3}(\mathbf{R}^2)$ to $L^4(\mathbf{R}^2)$. By complex interpolation this is a consequence of the endpoint bounds

- U_s is bounded from L^1 to L^∞ for Re $s = 0$;
- U_s is bounded from L^2 to L^2 for Re $s = 1$.

The first statement is trivial. As for the second, the operator U_s can be factorized as $U_s = L_s M_s$ where

(129)
$$(M_s f)(x, t) = \int f(x, t - v)\,\frac{dv}{|v|^s}$$

$$(L_s f)(x, t) = \int f\big(x - u, t + S(x, z_c(x, x - u))$$

$$- S(x - u, z_c(x, x - u))\big)\,\frac{du}{|u|^s}.$$

The boundedness of M_s on L^2 is the standard one-dimensional theorem on singular integrals. As for L_s we recognize that it is a singular Radon transform with phase function

(130)
$$\Phi(x, y) \equiv S(x, z_c(x, y)) - S(y, z_c(x, y)).$$

In general, this phase function is degenerate. However we have

Lemma 5. *Let $\bar{z}(x)$ be the zero of the Monge-Ampère determinant $S''_{xy}(x, \bar{z}(x)) = 0$. Then*

(131)
$$\Phi'''_{xyy} = \frac{1}{4}\,S'''_{xyy}(y, \bar{z}(y))\Big(\frac{d\bar{z}(y)}{dy}\Big)^2 + O(|x - y|).$$

This shows that for two sided folds $|\Phi'''_{xyy}| \ge c > 0$. Thus the Hessian of the phase (130) does not vanish to infinite order, and the refined boundedness Christ-

Nagel-Stein-Wainger theorem [5] (see also the discussion at the end of Section II.2) for singular Radon transforms applies, giving the desired estimate.

Columbia University

REFERENCES

[1] J. Antoniono and G. Uhlmann. "A functional calculus for a class of pseudo-differential operators with singular symbols." *Proc. Symp. Pure Math.* **43** (1985) 5–16.

[2] V. Arnold, A. Varchenko, and S. Goussein-Zade. "Singularités des applications différentiables." Editions Mir, 1986.

[3] M. Beals, C. Fefferman, and R. Grossman. "Strongly pseudo-convex domains." *Bull. Amer. Math. Soc.* **8** (1983), 125–322.

[4] M. Christ. "Hilbert transforms along curves I." *Ann. of Math.* **122** (1985), 575–596.

[5] M. Christ, A. Nagel, E. M. Stein, and S. Wainger. "Singular and maximal Radon transforms." In preparation.

[6] C. Fefferman. "Monge-Ampère equations, the Bergman kernel, and geometry of pseudo convex domains." *Ann. of Math.* **103** (1976), 395–416.

[7] G. Folland and J. J. Kohn. "The Neumann problem for the Cauchy-Riemann complex." Annals of Mathematics Studies **75**, Princeton University Press, 1972.

[8] G. Folland and E. M. Stein. "Estimates for the $\bar{\partial}_b$-complex and analysis on the Heisenberg group." *Comm. Pure Appl. Math.* **27** (1974), 429–522.

[9] D. Geller and E. M. Stein. "Estimates for singular convolution operators on the Heisenberg group." *Math. Ann.* **267** (1984), 1–15.

[10] A. Greenleaf and G. Uhlmann. "Non-local inversion formulas for the X-ray transform." *Duke Math. J.* **58** (1989), 205–240.

[11] _____. "Estimates for singular Radon transforms and pseudo-differential operators with singular symbols." *J. Funct. Anal.* **89** (1990), 202–232.

[12] _____. "Composition of some singular Fourier integral operators and estimates for the X-ray transform I." *Ann. Inst. Fourier* **40** (1990), II, 1991 preprint.

[13] P. Greiner and E. M. Stein. "A parametrix for the $\bar{\partial}$-Neumann problem." In *Estimates of the Neumann Problem*, edited by P. C. Greiner. Mathematical Notes 19. Princeton University Press, 1977.

[14] V. Guillemin. "Cosmology in $(2 + 1)$ dimensions, cyclic models, and deformations of $M_{2,1}$." Annals of Mathematics Studies **121**. Princeton University Press, 1989.

[15] V. Guillemin and G. Uhlmann. "Oscillatory integrals with singular symbols." *Duke Math. J.* **48** (1981), 251–267.

[16] L. Hörmander. "Fourier integral operators I." *Acta Math.* **127** (1971), 79–183.

[17] _____. *The Analysis of Linear Partial Differential Operators I-IV*. Springer Verlag, 1985.

[18] A. Melin. "Lower bounds for pseudo-differential operators." *Arkiv Math.* **9** (1971), 117–140.

[19] R. Melrose. "Equivalence of glancing hypersurfaces." *Inv. Math.* **37** (1976), 165–191.

[20] R. Melrose and M. Taylor. "Near peak scattering and the correct Kirchhoff approximation for a convex obstacle." *Adv. Math.* **55** (1985), 242–315.

[21] R. Melrose and G. Uhlmann. "Lagrangian intersection and the Cauchy problem." *Comm. Pure Appl. Math.* **32** (1979), 483–519.

[22] G. Mockenhaupt, A. Seeger, and C. Sogge. "Wave front sets, local smoothing, and Bourgain's circular maximal theorem." *Ann. of Math.* **136** (1992), 207–218.

[23] D. Müller. "Singular kernels supported by homogeneous submanifolds." *J. Reine Angew. Math.* **356** (1985), 90–118.

[24] D. Muller and F. Ricci. "Analysis of second order differential operators on the Heisenberg group I." *Inv. Math.* **101** (1990), 545–582.

[25] A. Nagel, N. Riviere, and S. Wainger. "Hilbert transforms associated with plane curves." *Bull. Amer. Math. Soc.* **223** (1974), 235–252.

[26] D. Oberlin. "Convolution estimates for some measures on curves." *Proc. Amer. Math. Soc.* **99** (1987), 56–60.

[27] Y. Pan. Ph.D. diss., Princeton University, 1989.

[28] Y. Pan and C. Sogge. "Oscillatory integrals associated to folding canonical relations." *Colloq. Math.* **60** (1990), 413–419.

[29] D. H. Phong. "On integral representations for the Neumann operator." *Proc. Nat. Acad. Sci.* **76** (1979), 1554–1558.

[30] D. H. Phong and E. M. Stein. "Some further classes of pseudo-differential and singular integral operators arising in boundary value problems I." *Amer. J. Math.* **104** (1982), 141–172.

[31] _____. "Singular integrals related to the Radon transform and boundary value problems." *Proc. Nat. Acad. Sci.* **80** (1983), 7697–7701.

[32] _____. "Hilbert integrals, singular integrals, and Radon transforms I and II." *Acta Math.* **157** (1986), 99–157 and *Inv. Math.* **86** (1986), 75–113.

[33] _____. "Singular Radon transforms." *Duke Math. J.* **58** (1989), 349–369.

[34] _____. "Radon transforms and torsion." *Int. Math. Res. Notices* **4** (1991), 49–60.

[35] _____. "Oscillatory integrals with polynomial phases." *Inv. Math.* **110** (1992), 39–62.

[36] F. Ricci and E. M. Stein. "Harmonic Analysis on nilpotent groups and singular integrals, I. Oscillatory integrals." *J. Funct. Anal.* **73** (1987), 179–194; "II. Singular kernels supported on submanifolds." *J. Funct. Anal.* **78** (1988), 56–94.

[37] C. Sogge and E. M. Stein. "Averages of functions over hypersurfaces: smoothness of generalized Radon transforms." *J. Analyse Math.* **54** (1990), 165–188.

[38] E. M. Stein. "Maximal functions: spherical means." *Proc. Nat. Acad. Sci.* **73** (1976).

[39] _____. *Harmonic Analysis: Real-Variable Methods, Orthogonality, and Oscillatory Integrals.* Princeton Mathematical Series 43. Princeton University Press, 1993.

[40] E. M. Stein and S. Wainger. "Problems in harmonic analysis related to curvature." *Bull. Amer. Math. Soc.* **84** (1978), 1239–1295.

[41] M. Taylor. "Propagation, reflection, and diffraction of singularities of solutions to wave equations." *Bull. Amer. Math. Soc.* **84** (1978), 589–611.

[42] _____. "Fefferman-Phong inequalities in diffraction theory." *Proc. Symp. Pure Math.* **43** (1985), 261–300.

[43] _____. *Pseudo-differential operators.* Princeton Mathematical Series 34. Princeton University Press, 1981.

[44] A. Thompson. "A singular Radon transform in $\mathbf{RP}^3$." *Comm. Part. Diff. Eq.* **14** (1989), 1461–1470.

[45] F. Treves. *Introduction to pseudo-differential and Fourier integral operators.* Plenum, 1980.

15

Counterexamples with Harmonic Gradients in $\mathbb{R}^3$

*Thomas H. Wolff**

INTRODUCTION

Many results about harmonic functions in $\mathbb{R}^2$ are proved using complex analysis explicitly or implicitly, so it is natural that some of them should fail in higher dimensions, or should at least need to be reformulated. The purpose of this paper is to work out three such counterexamples. Let $\alpha > 0$ be small enough.

Theorem 1. *There is a nonconstant harmonic function on the upper half space $\mathbb{R}^3_+ \subset \mathbb{R}^3$ which is $C^{1+\alpha}$ up to the boundary, and whose gradient vanishes on a boundary set of positive measure.*

Theorem 2. *There is a harmonic function u on $\mathbb{R}^3_+$ such that*

$$(0.1) \qquad \sup_{t>0} \int_{\mathbb{R}^2} |\nabla u(x, t)|^\alpha dx_1 dx_2 < \infty,$$

and such that ∇u does not have nontangential limits almost everywhere.

Theorem 3. *There is a bounded domain in $\mathbb{R}^3$ whose harmonic measure puts no mass on any set with Hausdorff dimension less than $2 + \alpha$.*

These results extend easily to $\mathbb{R}^n$, $n \geq 4$. What they have in common, in addition to being counterexamples to the obvious generalizations of two-dimensional

*This research was supported by the National Science Foundation, under grant DMS 87-03456.

theorems, is that the two-dimensional results may all be proved using the sub-harmonicity of $\log |\nabla u|$ when u is a harmonic function on a domain in $\mathbb{R}^2$. For Theorems 1 and 2 this is well known, e.g., if $u : \mathbb{R}^2_+ \to \mathbb{R}$ and ∇u vanishes on a boundary set E with positive measure, then $\log |\nabla u|$ is $-\infty$ on E and, being subharmonic, is identically $-\infty$. I will now give a brief discussion of Theorem 3, which motivated this paper.

In [12], Jones and I showed that harmonic measure for any domain in $\mathbb{R}^2$ puts full mass on some set with Hausdorff dimension 1. This completed a line of work due to Oksendal [19], [20] (who also made the conjecture), Kaufman-Wu [13], Carleson [6], and Makarov [15]. In particular, Makarov had proved the same result for simply connected domains in [15, Theorem 3]. It is sharp for any smooth domain, so the apparent generalization to $\mathbb{R}^n$ would be that harmonic measure always lives on a set with dimension $n - 1$.

The relevance of subharmonicity of $\log |\nabla u|$ for Oksendal's problem is due to Carleson [6] who used the following argument. Consider a scale invariant boundary such as the von Koch snowflake. Harmonic measure at a fixed interior point is then mutually absolutely continuous to an invariant measure for the scaling transformation on the boundary. In ergodic theory it is known that the dimension of such an invariant measure and its entropy are equivalent quantities. Computing the entropy turns out to be a matter of evaluating

$$(0.2) \qquad \lim_{n \to \infty} \frac{1}{n} \int_{\partial \Omega_n} |\nabla g_n| \log |\nabla g_n| d\sigma,$$

where Ω_n is the nth stage in the construction of the snowflake and g_n is the Green's function, and σ is surface measure. Subharmonicity of $\log |\nabla g_n|$ away from the pole implies the integrals in (0.2) are bounded below in terms of an upper bound for diam Ω_n. So (0.2) is certainly nonnegative, which is what is needed to conclude that harmonic measure is at most 1-dimensional.

This argument is reversible, so the main step in proving Theorem 3 is to find a snowflake-type construction in $\mathbb{R}^3$ for which (0.2) is negative. Theorems 1 and 2 are similar. Instead of needing to construct a domain and its Green's function, we have a fixed domain (the upper half-space) and we need a sequence of harmonic functions u_n with suitable normalization, such that

$$(0.3) \qquad \int_{Q(1)} \log |\nabla u_n| dx_1 dx_2 \to -\infty.$$

Here $Q(1)$ is the unit square in $\mathbb{R}^2$ and $\mathbb{R}^2$ is the boundary of the upper half-space.

The constructions for (0.2) and (0.3) are done by the same method. This method seems to be the canonical one for counterexamples involving boundary behavior; e.g., it has been used by Aleksandrov, Hakim-Sibony, and Low (see [1]) to construct inner functions on the ball in $\mathbb{C}^n$ and by Lewis [14] and myself [31] to construct p-harmonic functions without almost everywhere boundary values. The method itself goes back to Men'shov—the Men'shov correction theorem and the fact that every measurable function is a.e. the sum of a trigonometric series. Consider (0.3). One shows first that there is no true obstruction to (0.3) such as the subharmonicity in the 2-dimensional case. This comes down to proving that if e is a unit vector in $\mathbb{R}^3$, then there is a harmonic gradient ∇u with suitable decay at ∞ such that

(0.4) $$\int_{\mathbb{R}^2} \log |e + \nabla u| dx_1 dx_2 < 0.$$

Once (0.4) is known, one sets up a recursive construction as follows. Let ∇g_1 be any smooth harmonic gradient. On a small enough scale ∇g_1 is approximately constant, so by adding on dilates of the functions ∇u one obtains ∇g_2 with

$$\int_{\mathbb{R}^2} \log |\nabla g_2| dx_1 dx_2 < \int_{\mathbb{R}^2} \log |\nabla g_1| dx_1 dx_2 - \eta, \quad \eta > 0 \text{ fixed.}$$

One then changes to a still smaller scale and repeats the procedure. (0.2) is treated the same except that the modifications are made in the domain and the initial step is to find a perturbation, Ω, of the upper half-space with the property that

$$\int_{\partial \Omega} |\nabla g_\infty| \log |\nabla g_\infty| d\sigma < 0,$$

where g_∞ (Green's function with pole at ∞) satisfies $g_\infty(x) = 0$ when $x \in \partial \Omega$ and $\lim_{x \to \infty} g_\infty(x) - x_3 = 0$.

Theorems 1, 2, and 3 leave open a number of questions having to do with the size of the number α. Consider first Theorem 2. Stein and Weiss [25] found the analogue of subharmonicity of $\log |\nabla u|$ in higher dimensions, namely that $|\nabla u|^{(n-2)/(n-1)}$ is subharmonic if u is harmonic on a domain in $\mathbb{R}^n$. They introduced the space of functions satisfying (0.1) (now called the Stein-Weiss H^α space) as a generalization of H^p theory to $\mathbb{R}^n$, and proved that functions in H^α have a.e. boundary values if $\alpha \geq \frac{n-2}{n-1}$. One can ask whether Theorem 2 can be pushed up to their critical exponent. The main obstacle to doing this by our method is that one would need the analogue of (0.4), i.e., for each $p < \frac{n-2}{n-1}$ and unit vector $e \in \mathbb{R}^n$ a harmonic gradient ∇u on $\mathbb{R}^n_+$ with rapid decay at ∞ such that

(0.5) $$\int_{\mathbb{R}^{n-1}} \left(|e + \nabla u|^p - 1 \right) dx_1 \ldots dx_{n-1} < 0.$$

Such functions probably exist, although my argument for (0.4), which consists of explicitly writing down u and computing the relevant integral, does not show how to construct them. Similar remarks apply for questions like Theorem 2 where Riesz transforms are replaced by double Riesz transforms, etc., as in [5].

One can also ask whether C^2 or C^∞ can replace $C^{1+\alpha}$ in Theorem 1. This has to do with the rate at which (0.3) can diverge and could be a delicate question. A C^2 counterexample would lead (by taking one derivative) to a harmonic function u with $u = \frac{du}{dx_3} = 0$ on a boundary set of positive measure, which is how the problem has sometimes been stated.

Regarding Theorem 3, Bourgain [3] showed that any harmonic measure in $\mathbb{R}^n$ puts full mass on some set with dimension $n - \epsilon_n$ where $\epsilon_n > 0$ depends only on n. It would be interesting if one could compute the best value of ϵ_n. Both Bourgain's argument and the argument of this paper are too crude for that.

I want to mention some previous work related to Theorems 1 and 2. Uchiyama [26], [27], [28] gave a new method for studying singular integrals on L^p, $p \le 1$. His method does not use subharmonicity at all and applies to more general situations than harmonic gradients, but a special case of the result of [28] is another proof that Stein-Weiss H^p functions have boundary values for $p > 1 - \epsilon$. In [27], he considered the small p case and gave counterexamples like Theorem 2 for related problems on discrete martingales. The proof of Theorems 1 and 2 uses some of the ideas of his method for the positive results in [26]; in particular, the functions satisfying (0.4) play the same role as his Lemma 2.3.

The question of whether there are uniqueness theorems for boundary sets of positive measure in higher dimensions was posed by Bers. The previous work that has been done on it appears to be concerned mainly with local properties near one boundary point; cf. Mergelyan [18] and for the strongest known result, Rao [21].

Acknowledgments. I am grateful to J. Goodman for verifying Lemma 1.2 numerically. This work was done while I was at the Courant Institute.

Added September, 1991

Almost all current research in Euclidean harmonic analysis owes something to the work of E. M. Stein and this paper is quite clearly not an exception. I dedicate it to him with appreciation and respect.

Since it was first circulated in October 1987 there has been some further progress which I will now summarize. A. B. Aleksandrov and P. Kargaev [2] have given an alternate, less computational approach to the existence of the building block functions (0.4), (0.5). For given $n \ge 3$, let $\mathbb{R}^n_+ = \{x \in \mathbb{R}^n : x_n > 0\}$ and denote by $S_H(\mathbb{R}^n_+)$ the harmonic gradients $\nabla u : \mathbb{R}^n_+ \to \mathbb{R}^n$ vanishing at infinity, smooth up to the boundary and with boundary values in the Schwarz space. Then Aleksandrov and Kargaev show that for any $p < \frac{n-2}{n}$ and unit vector $e \in \mathbb{R}^n$ there is a harmonic gradient $\nabla u \in S_H(\mathbb{R}^n_+)$ such that (0.5) holds. They also give several additional applications of this construction including among others the following: (1) given any continuous function $f : \mathbb{R}^{n-1} \to \mathbb{R}^n$, $n \geq 3$, there is a harmonic function u on $\mathbb{R}^n_+ C^1$ up to the boundary and such that ∇u coincides with f on $\mathbb{R}^n$ outside a set with arbitrarily small measure; (2) for $p < \frac{n-2}{n}$, any L^p function $f : \mathbb{R}^{n-1} \to \mathbb{R}^n$ coincides with the a.e. boundary values of some harmonic gradient ∇u belonging to the Stein-Weiss H^p space; (3) for $p < \frac{n-1}{n}$ there is a function in the Stein-Weiss H^p space whose a.e. boundary values belong to a ray $\{e : r > 0\}$, for some $e \in \mathbb{R}^n$.

   J. Bourgain and the author [3] used the construction of Aleksandrov and Kargaev to show also that there are harmonic functions u on $\mathbb{R}^n_+ C^1$ up to the boundary and such that u and $\frac{du}{dx_n}$ vanish on a common boundary set of positive measure. This answers one of the questions mentioned above. W. S. Wang [30] has proved further related results, in particular, that examples analogous to those of Bourgain and the author exist on any domain with a $C^{1+\epsilon}$ boundary.

   On the other hand the question of higher smoothness in Theorem 1 remains open and no further bounds have been proved on the dimension of harmonic measure. It is also still unknown whether (0.5) can be pushed all the way up to the Stein-Weiss critical exponent $\frac{n-2}{n-1}$.

   I have made some changes in the preceding introduction mainly reflecting the fact that in 1987 I was unaware of relevant previous work—Men'shov's classical work as well as [17], [21]. I have also corrected some misprints and minor inaccuracies in the subsequent sections but have not attempted a serious revision. It should therefore be pointed out that several arguments can be done more simply, and in particular, the appendix is now definitely unnecessary because of the work of Aleksandrov and Kargaev.

   I would like to take this opportunity to present the following observations which show that functions proving (0.5) will tend to depend strongly on p as $p \to \frac{n-2}{n-1}$. First we have the following "rigidity" result for the Stein-Weiss subharmonicity inequality.

Proposition A. *Suppose u is harmonic on a domain $\Omega \subset \mathbb{R}^n$, $n \geq 3$, and $|\nabla u|^{\frac{n-2}{n-1}}$ is also harmonic. Then u is either affine, or of the form $b|x - e|^{2-n} + c$ for some $b, c \in \mathbb{R}$ and $e \in \mathbb{R}^n$.*

Proof. Since u is real analytic we may shrink Ω and may therefore assume that ∇u never vanishes. If we examine the proof of the Stein-Weiss inequality [24], [25] we see that strict inequality must hold at a given point unless ∇u is an eigenvector of the Hessian H_u and, furthermore, all eigenvalues other than the one corresponding to ∇u are equal. Then in the present situation we conclude that there is a function $\lambda : \Omega \to \mathbb{R}$, such that

(i) $H_u(x)(\nabla u(x)) = \lambda(x)\nabla u(x)$

(ii) $H_u(x) \upharpoonright (\nabla u(x))^\perp = -\frac{1}{n-1}\lambda(x) \cdot \text{id}.$

where id is the $(n-1) \times (n-1)$ identity matrix. Now (i) implies that $|\nabla u|$ is constant on level sets of u, while (ii) implies that the level sets are umbilical (all principal curvatures at any given point are equal). By [23] Theorem 26, the level sets are contained in planes or spheres. The proposition now follows by applying the uniqueness statement in the Cauchy-Kowalewski theorem on one of the level sets. ∎

A somewhat deeper application of the same general principle is given in Theorem 2 of [8]. We now apply Proposition A to prove

Proposition B. *Suppose that* $n \geq 3$, e *is a unit vector in* $\mathbb{R}^n$, $\{\nabla u_k\} \subset S_H(\mathbb{R}^n_+)\{p_k\} \to \frac{n-2}{n-1}$ *and* $\int_{\mathbb{R}^{n-1}}(|e + \nabla u_k|^{p_k} - 1)dx < 0$. *Then the only possible* $L^{\frac{n-2}{n-1}} + L^1$ *limit point of the sequence* $\{\nabla u_k\}$ *is zero.*

Proof. Let $p = \frac{n-2}{n-1}$. We have

$$\int_{\mathbb{R}^{n-1}} \big(|e + \nabla u_k|^p - 1\big)dx = \int_{\mathbb{R}^{n-1}} \big(|\nabla(x \cdot e + u_k)|^p - 1\big)dx$$

$$\geq \int_{\mathbb{R}^n_+} x_n \Delta\big(|\nabla(x \cdot e + u_k)|^p\big)dx.$$

The Laplacian is taken in the pointwise sense. It is well defined except at critical points and therefore a.e. The inequality is easily justified by a limiting argument starting from the smooth subharmonic functions $\frac{(|\nabla(x \cdot e + u_k)|^2 + \epsilon^2)^{p/2}}{(1+\epsilon^2)^{p/2}} - 1$ and using Fatou's lemma.

Suppose now that $\{\nabla u_k\}$ converges in $L^p + L^1$. Then the maximal function estimates in [25] imply $\{\nabla u_k\}$ also converges uniformly on compacts of $\mathbb{R}^n_+$ and the limit function is a harmonic gradient ∇u. Then using the $L^p + L^1$ convergence and the fact that $||e + \nabla u|^q - 1| \lesssim \min(|\nabla u|, |\nabla u|^p)$ for $q \leq p$, we get

$$\lim_{k \to \infty} \int_{\mathbb{R}^{n-1}} (|e + \nabla u_k|^{p_k} - 1) = \int_{\mathbb{R}^{n-1}} (|e + \nabla u|^p - 1)$$

$$= \lim_{k \to \infty} \int_{\mathbb{R}^{n-1}} (|e + \nabla u_k|^p - 1).$$

In particular, $\lim_{k \to \infty} \int |e + \nabla u_k|^p - 1) \leq 0$. On the other hand $\Delta(|\nabla(x \cdot e + u_k)|^p) \to \Delta(|\nabla(x \cdot e + u)|^p)$ on $\mathbb{R}^n_+$ except at critical points of $x \cdot e + u$, and therefore a.e. By Fatou's lemma

$$\int_{\mathbb{R}^n_+} x_n \Delta(|\nabla(x \cdot e + u)|^p) \leq \liminf_{k \to \infty} \int_{\mathbb{R}^n_+} x_n \Delta\big(|\nabla(x \cdot e + u_k)|^p\big)$$

$$\leq \liminf_{k \to \infty} \int_{\mathbb{R}^{n-1}} \left(|e + \nabla u_k|^p - 1 \right)$$

$$\leq 0.$$

It follows that $\triangle(|\nabla(x \cdot e + u)|^p)$ must vanish identically, so Proposition A applies to $x \cdot e + u$. Since ∇u cannot be bounded below at infinity, the only possibility is that $x \cdot e + u = x \cdot e + $ const and the proposition is proved.   ∎

We remark that convergence to zero can take place only in a fairly weak sense, if at all. Variational calculations like those in Section 1 below imply that $\{\nabla u_n\}$ cannot converge to zero in L^∞.

1   BOUNDARY PROBLEMS FOR RIESZ SYSTEMS

We work in $\mathbb{R}^3$ and often write points of $\mathbb{R}^3$ as $x = (\bar{x}, x_3)$ with $\bar{x} \in \mathbb{R}^2$. The upper half space is $\mathbb{R}^3_+ = (x \in \mathbb{R}^3 : x_3 > 0)$. We write $\overline{\mathbb{R}^3_+}$ for its closure and identify $\mathbb{R}^2$ with $\overline{\mathbb{R}^3_+} \backslash \mathbb{R}^3_+$. Squares in $\mathbb{R}^2$ with sides parallel to the axes are denoted Q, and $Q(N) = [-\frac{1}{2}N, \frac{1}{2}N] \times [-\frac{1}{2}N, \frac{1}{2}N]$. The standard basis is e_1, e_2, e_3.

A harmonic gradient is a function $f : \mathbb{R}^3_+ \to \mathbb{R}^3$ which is the gradient of some harmonic function u, or else the boundary values of such an f. Both f and u will always be assumed to vanish at ∞. If X is a function space on $\mathbb{R}^2$, then $X_H = \{f : \mathbb{R}^2 \to \mathbb{R}^3 : f$ is a harmonic gradient and the components of f belong to $X\}$. Thus S_H is the harmonic gradients whose components belong to the Schwarz space S, etc.

Before doing the constructions we summarize some properties of Schwarz space harmonic gradients.

(i)   Let $f : \mathbb{R}^2 \to \mathbb{R}^3$ be a harmonic gradient such that $f_3 \in S$. Then $f \in S_H$ if and only if $\int x^\alpha f_3 dx = 0$ for all multiindices $\alpha = (\alpha_1, \alpha_2)$ of length ≥ 0. This follows by taking Fourier transforms: $\hat{f}_1 = -i\xi_1|\xi|^{-1}\hat{f}_3$ and $\hat{f}_2 = -i\xi_2|\xi|^{-1}\hat{f}_3$ are smooth at zero if $\hat{f}_3$ vanishes to infinite order there.

(ii)   (a) S_H is dense in L^p_H, $1 < p < \infty$.

      (b)   For any $k \in \mathbb{Z}^+$ there is $\ell \in \mathbb{Z}^+$ making the following true: if f is a harmonic gradient such that f_3 is a smooth function with compact support and with $\int x^\alpha f(x)dx = 0$ when $|\alpha| \leq \ell$, then for any $\epsilon > 0$, there is $g \in S_H$ such that $\sup_{x \in \mathbb{R}^2}(1 + |x|)^k|f(x) - g(x)| < \epsilon$.

I believe these facts to be well known but will include a proof. The proof in my 1987 manuscript was incorrect; I thank W. S. Wang, who very tactfully pointed this out.

Let ψ be a Schwarz function on $\mathbb{R}^2$ such that $\hat{\psi}$ has compact support and $\hat{\psi} = 1$ in a neighborhood of 0. Let $\psi_t(x) = t^{-2}\psi(t^{-1}x)$. To prove

(a) it suffices to approximate functions $f \in L_H^p$ such that $f_3 \in S$, since such functions are dense in L_H^p by boundedness of the Riesz transforms. However, if $f_3 \in S$, then by (i) and the fact that $\hat{\psi} = 1$ near 0, we will have $f - \psi_t * f \in S_H$. Moreover $\psi_t * f \to 0$ as $t \to \infty$ by Young's inequality since $f \in L^q$ for $1 < q < p$.

To prove (b) note first that the moment assumption implies $|f(x)| \lesssim (1 + |x|)^{-(\ell+2)}$. It therefore suffices to show that if $\int x^\alpha f(x)dx = 0(|\alpha| \le \ell)$ and $|f(x)| \lesssim (1 + |x|)^{-(\ell+2)}$ then $\sup_{x \in \mathbb{R}^2}(1 + |x|)^k|\psi_t * f(x)| \to 0$ as $t \to \infty$. This is done as follows: Fix x and $t \ge 1$. On the one hand, since $\psi \in S$

$$|\psi_t * f(x)| \lesssim t^{-2} \int (1 + |y|)^{-(\ell+2)}\left(1 + \frac{|x - y|}{t}\right)^{-(\ell+2)} dx$$

$$\lesssim \left(1 + \frac{|x|}{t}\right)^{-(\ell+2)}.$$

On the other hand let p be the degree $\ell - 2$ Taylor polynomial of ψ at $\frac{x}{t}$; then

$$|\psi_t * f(x)| = t^{-2} \left| \int f(y)\left(\psi\left(\frac{x - y}{t}\right) - p\left(\frac{x - y}{t}\right)\right)dy \right|$$

$$\lesssim t^{-2} \int |f(y)|\left|\frac{y}{t}\right|^{\ell-1}dy$$

$$\lesssim t^{-(\ell+1)}.$$

Taking $\ell = 2k$ we obtain

$$|\psi_t * f(x)| \le \left(\left(1 + \frac{|x|}{t}\right)^{-(2k+2)} t^{-(2k+1)}\right)^{1/2}$$

$$\le t^{-1/2}(t + |x|)^{-k},$$

which suffices.
(iii) If $g \in S_H$ then g is rapidly decreasing on the upper half space, i.e., $\forall k \exists C_k : |g(x)| \le C_k(1 + |x|)^{-k}, x \in \mathbb{R}_+^3$.

One expands the Poisson kernel P_1 as a Taylor series at 0 with remainder, scales to obtain an expansion of P_t for t large, and uses (i).
(iv) Harmonic gradients $g \in S_H$ such that $g \ne 0$ on $Q(2)$ are dense in S_H.

Choose finitely many $\phi_1 \ldots \phi_n \in S_H$ such that $\{\phi_j(x)\}$ spans $\mathbb{R}^3$ for each $x \in Q(2)$. This can be done by compactness since $\{x : \phi_j(x)$ do not span$\}$ is closed and clearly $S_H(x)$ spans for each fixed x. If $f \in S_H$ is given, then let δ be small and consider

$$\int_{Q(2)} \int_{\frac{1}{2}\delta<\delta_j<\delta} |f + \Sigma\delta_j\phi_j|^{-2}d\delta_1 \ldots d\delta_n dx_1 dx_2.$$

The inner integral is finite for each fixed x since the zero set of $f(x) + \Sigma \delta_j \phi_j(x)$ has codimension 3, and is bounded by a compactness argument on the minors of the matrices $\{\phi_j(x)\}$. On the other hand, if $f + \Sigma \delta_j \phi_j$ has a zero for a certain $\delta_1, \ldots, \delta_n$ then $\int_{Q(2)} |f + \Sigma \delta_j \phi_j|^{-2} dx_1 dx_2 = \infty$. If this happened for all choices of the δ's, Fubini's theorem would be contradicted.

The first step in proving Theorems 1 and 2 is to distinguish the 3-dimensional case from the 2-dimensional case by means of the following fact.

Lemma 1.1. *For each unit vector $e \in \mathbb{R}^3$ there is $q \in S_H$ such that $\int_{\mathbb{R}^2} \log |e + q(x)| dx < 0$. We can choose q so that $e + q$ never vanishes on $\mathbb{R}^2$.*

It is natural to try to prove Lemma 1.1 by perturbing off the case $q = 0$, and we now look where this leads. Fix $q \in S_H$ and let $I(t) = \int_{\mathbb{R}^2} \log |e + tq| dx$. By calculations which are justified (for t small) by the rapid decay at ∞,

$$\frac{dI}{dt} = \int_{\mathbb{R}^2} \frac{\langle e + tq, q \rangle}{|e + tq|^2} dx$$

$$\frac{d^2I}{dt^2} = \int_{\mathbb{R}^2} \frac{|q|^2 |e + tq|^2 - 2\langle q, e + tq \rangle^2}{|e + tq|^4} dx.$$

Taking $t = 0$,

$$\left. \frac{dI}{dt} \right|_{t=0} = \int_{\mathbb{R}^2} \langle e, q \rangle dx = 0,$$

$$\left. \frac{d^2I}{dt^2} \right|_{t=0} = \int_{\mathbb{R}^2} \left(|q|^2 - 2\langle e, q \rangle^2 \right) dx.$$

If $e = e_3$ (or $-e_3$) it follows, e.g., using the Fourier transform that $\left. \frac{d^2I}{dt^2} \right|_{t=0} = 0$ for any $q \in S_H$. But if e has a nonzero tangential component, then for similar reasons $\left. \frac{d^2I}{dt^2} \right|_{t=0} > 0$ for any $q \in S_H$, i.e., 0 is a local minimum.

Proof of Lemma 1.1 when $e = \pm e_3$ We can assume $e = e_3$. Differentiating once more,

$$\left. \frac{d^3I}{dt^3} \right|_{t=0} = \int_{\mathbb{R}^2} (8q_3^3 - 6q_3 |q|^2) dx$$

$$= 2 \int_{\mathbb{R}^2} \left(q_3^3 - 3q_3 (q_1^2 + q_2^2) \right) dx.$$

Temporarily drop the requirement that $q \in S_H$ and try $q = \nabla \phi$, $\phi(x) = -|x + e_3|^{-1}$. Then

$$q_3 = |x + e_3|^{-3}, \qquad q_1^2 + q_2^2 = \frac{x_1^2 + x_2^2}{|x + e_3|^6}$$

$$\int_{\mathbb{R}^2} (q_3^3 - 3q_3(q_1^2 + q_2^2))dx = 2 \int_0^\infty \left((r^2 + 1)^{-9/2} - 3r^2(r^2 + 1)^{-9/2} \right) r\,dr$$

$$= -\frac{2\pi}{35} < 0.$$

Approximating this q in L^3 norm by one in S_H we obtain $q \in S_H$ with $\frac{d^3 I}{dt^3}\big|_{t=0} < 0$. Then for small positive t, $\int_{\mathbb{R}^2} \log |e + tq| < 0$ and $e + tq$ has no zeros.

Proof of Lemma 1.1 when e has nonzero tangential component We can assume $e = (\alpha, 0, \sqrt{1 - \alpha^2})$, $0 < \alpha \leq 1$. Again we temporarily drop the requirement that $q \in S$. To emphasize the $+e_1$-direction it is natural to consider $q = \nabla f$ where in polar coordinates on $\mathbb{R}^2$, $f(x_1, x_2, 0) = \phi(r) \cos \theta$. Next observe that if $\phi(r) = -\alpha r$ for $r < 1$, then $(e + q)_1 = (e + q)_2 = 0$ when $r < 1$. Having defined ϕ this way for $r < 1$ we decide to extend it to $\mathbb{R}^2$ keeping the L^2 norm of q as small as possible. That leads to the choice $q = \alpha \nabla g$, where g is the harmonic function with boundary values

$$g(x_1, x_2, 0) = \begin{cases} -r \cos \theta, & \text{if } r < 1 \\ -r^{-1} \cos \theta, & \text{if } r > 1 \end{cases}.$$

We then need to show

Lemma 1.2. $J(\alpha) \overset{\text{def}}{=} \frac{1}{2\pi} \int_{\mathbb{R}^2} (\log |e + q| - \langle e, q \rangle) dx < 0$ *for each* $\alpha \in (0, 1]$.

The integral is proved absolutely convergent at ∞ by expanding the logarithm to second order and using that $q \in L^2$. It is easily seen to be absolutely convergent locally, e.g., this follows from formulas (A1) and (A2) and statements A3.1 and A3.5 in the appendix to this paper.

Lemma 1.2 could in principle have been proved by computer work and error analysis, but I decided to treat it as a calculus exercise instead. This argument is messy and is therefore postponed to the appendix. It turns out that if one uses polar coordinates the θ-integral can be evaluated explicitly in terms of certain elliptic integrals (cf. Lemma A1.1 and the subsequent remark). Jonathan Goodman offered to integrate the resulting function of r numerically, and found that J is decreasing in α with $J \to 0$ as $\alpha \to 0$ and $J(1) \approx -0.28$, so the reader can believe the lemma without going through the technicalities in the last three sections of the appendix.

We now show how to obtain Lemma 1.1 from Lemma 1.2. If $e = (\alpha, 0, \sqrt{1 - \alpha^2})$, and q is as in Lemma 1, then $q \in L^2$ and $(e + q)_1$ is everywhere nonnegative. So by first convolving q with a strictly positive approximate identity and then approximating the resulting harmonic gradient by one in the Schwarz space, we obtain $q_n \in S_H$ such that $q_n \to q$ in L^2_H and $e + q_n$ never vanishes. By (i), $\int q_n = 0$. Also

$$\log |e + q_n| - \langle e, q_n \rangle \leq C|q_n|^2.$$

Since $q_n \to q$ in L^2 and $\log |e + q| - \langle e, q \rangle \in L^1$, we can pass to a subsequence and assume that

$$\log |e + q_n| - \langle e, q_n \rangle - (\log |e + q| - \langle e, q \rangle) \leq f,$$

with $f \in Ł^1$. By Fatou's lemma,

$$\limsup_{n \to \infty} \int_{\mathbb{R}^2} \left(\log |e + q_n| - \langle e, q_n \rangle \right) dx \leq \int_{\mathbb{R}^2} \left(\log |e + q| - \langle e, q \rangle \right) dx < 0.$$

Since $\int \langle e, q_n \rangle = 0$, q_n will satisfy the conditions of Lemma 1 when n is large enough. ■

Lemma 1.1 leads immediately to a refinement of itself where instead of a constant vector e we consider a constant vector plus a small error.

Lemma 1.3. *There are finitely many harmonic gradients* $q_j \in S_H$ *and numbers* $\kappa > 0, \rho > 0, \eta > 0$ *making the following true. If N is sufficiently large, then for each* $v \in \mathbb{R}^3$ *there is j such that if* $g : Q(N) \to \mathbb{R}^3$ *and* $\|g - v\|_\infty < \kappa N^{-2}|v|$, *then*

$$\rho|g| \leq |g + |v|q_j| \leq \rho^{-1}|g| \qquad on \quad Q(N),$$

$$\int_{Q(N)} \log |g + |v|q_j| \leq -\eta + \int_{Q(N)} \log |g|.$$

Proof. Lemma 1.1 and a compactness argument show there are finitely many q_j and numbers ρ, η such that if e is any unit vector then for some j,

$$2\rho \leq |e + q_j| \leq (2\rho)^{-1}$$

(1.1) $$\int_{\mathbb{R}^2} \log |e + q_j| \leq -3\eta.$$

Replacing 3η by 2η, the integral in (1.1) may be taken over $Q(N)$ provided N is large enough. When v is a unit vector, the lemma follows by easy estimates with absolutely convergent integrals. The general case follows by properties of the logarithm function. ■

We now set up a recursive procedure with the basic step provided by Lemma 1.3, which will prove Theorem 1 and the following result:

Lemma 1.4. *Suppose $p > 0$ is small enough, and let $k = k_p < \infty$ be large enough. Then for given $\epsilon > 0$ there is a harmonic gradient g which is continuous on $\overline{\mathbb{R}}_+^3$ and satisfies: $|g(x)| < \epsilon |x|^{-k}$ when $|x| > 1$, $\int_{\mathbb{R}^2} g(x) x^\alpha dx_1 dx_2 = 0$ for all $\alpha = (\alpha_1, \alpha_2)$ with $|\alpha| \le k$, $\sup_t \int_{\mathbb{R}^2} |g(\overline{x}, t)|^p dx_1 dx_2 \le C$, $(\int_{\mathbb{R}^2} |g(\overline{x}, 0)|^p dx_1 dx_2)^{1/p} < \epsilon$, and $|\{\overline{x} \in \mathbb{R}^2 : |g(\overline{x}, a)| \ge C^{-1}\}| \ge C^{-1}$ for a certain $a = a_\epsilon > 0$.*

Here C is independent of ϵ. The last two properties are the significant ones. Theorem 2 follows easily from Lemma 1.4 as we explain at the end of the section.

Choose three large enough constants A, B, N in that order. They depend on properties of the functions q_j and will be kept fixed throughout the proofs. Next choose an initial harmonic gradient $g_1 \in S_H$ with $|g_1| \ne 0$ on $Q(2)$, and then a large number λ and, finally, a small number $\delta > 0$ with $\delta^{-1} \in \mathbb{Z}$. Constants will initially be independent of $A, B, N, g_1, \lambda, \delta$.

Let G_n be the grid on $Q(1)$ consisting of squares with side δ^n. For each $Q \in G_n$, let a_Q be its center point. For each j, let

$$q_j^Q(x) = q_j(N\delta^{-n}(x - a_Q))$$

with q_j as in Lemma 1.3.

At stage n ($n \ge 2$) we will have constructed a harmonic gradient $g_{n-1} \in S_H$ obeying the following estimates when $x \in Q(2)$, $y \in Q(2) \times [0, 1]$.

(1.2) $|g_{n-1}(y)| \le A \max(1, \delta^{-(n-1)}|x - y|)|g_{n-1}(x)|.$

If $|x - y| < \delta^{n-1}$, then

(1.3) $|g_{n-1}(y) - g_{n-1}(x)| \le BN\delta^{-(n-1)}|g_{n-1}(x)| \, |x - y|.$

We will also have a subset $A_{n-1} \subset G_{n-1}$, called nonstopped squares at stage $n - 1$. Take g_1 as above and take A_1 to be those squares in G_1 which do not touch the boundary of $Q(1)$. Then (1.2), (1.3) are satisfied when $n = 2$, provided δ is small enough.

We now define A_n and g_n. A square $Q \in G_n$ belongs to A_n, provided every $Q' \in G_{n-1}$ which contains or touches Q belongs to A_{n-1} and, furthermore,

(1.4) $\max_Q \log(|g_{n-1}|/|g_1|) \le -\frac{\eta}{2} N^{-2} n + \lambda(n \log n)^{1/2},$

with η as in Lemma 1.3. And

$$g_n = g_{n-1} + \sum_{Q \in A_n} |v_Q| q_{j(v_Q)}^Q$$

where $v_Q = g_{n-1}(a_Q)$ and $j(v_Q)$ is the index provided by Lemma 1.3. We now make some estimates and show in particular that (1.2), (1.3) are again satisfied.

Claim. For any $d \geq \frac{1}{2}\delta^n$ and $y \in Q(2) \times [0, 1]$,

$$(1.5) \quad \sum_{\substack{Q \in A_n \\ |y-a_Q| \geq d}} |v_Q| \, |q^Q_{j(v_Q)}(y)| \leq CAN^{-4}\delta^{2n}|g_{n-1}(\bar{y})|d^{-1} \max(d^{-1}, \delta^{-(n-1)})$$

$$(1.6) \quad \sum_{\substack{Q \in A_n \\ |y-a_Q| \geq d}} |v_Q| \, |\nabla q^Q_{j(v_Q)}(y)| \leq CAN^{-4}\delta^{2n}|g_{n-1}(\bar{y})|d^{-2} \max(d^{-1}, \delta^{-(n-1)})$$

Proof of (1.5) Since $q_j \in S_H$ we have $|q^Q_{j(v_Q)}(y)| \leq CN^{-4}\delta^{4n}|y - a_Q|^{-4}$. Using this fact and property (1.2) of g_{n-1}, the sum in (1.5) is

$$\leq CA \sum_{\substack{Q \in A_n \\ |y-a_Q| \geq d}} \max(1, \delta^{-(n-1)}|\bar{y} - a_Q|)|g_{n-1}(\bar{y})| \, |y - a_Q|^{-4}N^{-4}\delta^{4n}$$

$$\leq CAN^{-4}\delta^{4n}|g_{n-1}(\bar{y})|\left(\sum_{\delta^{n-1} > |y-a_Q| > d} |y - a_Q|^{-4} + \sum_{|y-a_Q| \geq \max(d,\delta^{n-1})} \delta^{-(n-1)}|y - a_Q|^{-3} \right).$$

Since the sums are comparable to integrals, the expression in brackets is

$$\leq \begin{cases} C\delta^{-2n}d^{-2}, & \text{if } d < \delta^{n-1} \\ C\delta^{-2n}\delta^{-(n-1)}d^{-1}, & \text{if } d \geq \delta^{n-1} \end{cases}$$

and (1.5) follows. For (1.6), use $|\nabla q^Q_{j(v_Q)}(y)| \leq CN^{-4}\delta^{-4n}|y - a_Q|^{-5}$ and argue the same way. ∎

Let

$$\tilde{g}_n(y) = g_{n-1}(y) + \sum_{\substack{\bar{y} \notin Q \\ Q \in A_n}} |v_Q|q^Q_{j(v_Q)}(y).$$

Taking $d = \frac{1}{2}\delta^n$ and using (1.5),

$$(1.7) \qquad |\tilde{g}_n(y) - g_{n-1}(y)| \leq CAN^{-4}|g_{n-1}(\bar{y})|,$$

when $y \in Q(2) \times [0, 1]$. We make N large enough that $CAN^{-4} < \frac{1}{3}\kappa N^{-2}$ with C as in (1.7) and κ as in Lemma 1.3. Now restrict temporarily to $x = \bar{x} \in \mathbb{R}^2$. If $x \in Q \in A_n$ then

$$|g_{n-1}(x) - g_{n-1}(a_Q)| \leq BN\delta|g_{n-1}(a_Q)|$$

by (1.3), and we take δ small enough that $BN\delta < \frac{1}{3}\kappa N^{-2}$. Then $x \in Q \in A_n$ implies

$$|\tilde{g}_n(x) - g_{n-1}(a_Q)| \leq |\tilde{g}_n(x) - g_{n-1}(x)| + |g_{n-1}(x) - g_{n-1}(a_Q)|$$

$$\leq \frac{1}{3}\kappa N^{-2}(|g_{n-1}(x)| + |g_{n-1}(a_Q)|)$$

$$\leq \frac{1}{3}\kappa N^{-2}(2|g_{n-1}(a_Q)| + |g_{n-1}(x) - g_{n-1}(a_Q)|)$$

$$< \kappa N^{-2}|g_{n-1}(a_Q)|.$$

Therefore, Lemma 1.3 applies to $g = \tilde{g}_n$ and $q_j = q^Q_{j(v_Q)}$, after scaling by $N\delta^{-n}$, and we get $\rho^{-1}|\tilde{g}_n| \geq |g_n| \geq \rho|\tilde{g}_n|$ on $\cup(Q : Q \in A_n)$. Using that $\tilde{g}_n = g_n$ on $Q(2) \backslash \cup \{Q : Q \in A_n\}$, (1.7) implies

(1.8) $$2\rho^{-1}|g_{n-1}| \geq |g_n| \geq \frac{1}{2}\rho|g_{n-1}|,$$

on $Q(2)$. Moreover if $Q \in A_n$ then

$$\int_Q \log|g_n| \leq -\eta\delta^{2n}N^{-2} + \int_Q \log|\tilde{g}_n|$$

$$\leq -\eta\delta^{2n}N^{-2} + \int_Q \log|g_{n-1}| + CAN^{-6}\delta^{2n}$$

by (1.7). So

(1.9) $$\int_Q \log|g_n| \leq -\frac{1}{2}\eta N^{-2}\delta^{2n} + \int_Q \log|g_{n-1}|,$$

provided N is large enough.

Next we prove (1.2) for g_n. Observe that (1.3) for g_{n-1} implies

(1.10) $$|g_{n-1}(y)| \leq (1 + BN\delta^{1/2})|g_{n-1}(x)| \quad \text{if} \quad |x - y| < \delta^{n-1/2}.$$

Now if $\overline{y} \in Q \in A_n$ (the case where $\overline{y} \notin \cup\{Q : Q \in A_n\}$ is slightly easier),

$$|g_n(y) - g_{n-1}(y)| \leq |g_n(y) - \tilde{g}_n(y)| + |\tilde{g}_n(y) - g_{n-1}(y)|$$

$$\leq C|g_{n-1}(a_Q)| + |\tilde{g}_n(y) - g_{n-1}(y)|$$

$$\leq C(1 + BN\delta^{1/2})|g_{n-1}(\overline{y})| + CAN^{-4}|g_{n-1}(\overline{y})|,$$

using (1.7), (1.10), so if N is large and δ small,

$$|g_n(y)| \leq C(|g_{n-1}(\overline{y})| + |g_{n-1}(y)|)$$

and by (1.8),

$$\frac{|g_n(y)|}{|g_n(x)|} \leq C_0 \frac{|g_{n-1}(y)| + |g_{n-1}(\overline{y})|}{|g_{n-1}(x)|}.$$

If $|x - y| > \delta^{n-1/2}$, (1.2) for g_n now follows from (1.2) for g_{n-1} provided δ is small enough. If $|x - y| \leq \delta^{n-1/2}$, (1.2) for g_n follows from (1.10) by choosing δ small enough, provided $A > 2C_0$.

To prove (1.3) for g_n, write

$$|g_n(x) - g_n(y)| \leq |g_{n-1}(x) - g_{n-1}(y)| + |(g_n - g_{n-1})(x) - (g_n - g_{n-1})(y)|.$$

The first term is $\leq BN\delta^{-(n-1)}|x - y|\,|g_{n-1}(x)|$. To bound the second suppose first that x and $\overline{y}$ belong to the same $Q \in A_n$. Then

$$|(g_n - g_{n-1})(x) - (g_n - g_{n-1})(y)| \leq |g_{n-1}(a_Q)|\,|q^Q_{j(v_Q)}(x) - q^Q_{j(v_Q)}(y)|$$

$$+ |(\tilde{g}_n - g_{n-1})(x) - (\tilde{g}_n - g_{n-1})(y)|.$$

The first term is $\leq C|g_{n-1}(a_Q)|N\delta^{-n}|x - y|$ where C depends on Lipschitz bounds for the q_j. The second term is $\leq CAN^{-4}|x - y|\delta^{-n}\sup_Q|g_{n-1}|$ by (1.6) (with $d = \frac{1}{2}\delta^n$) at the points of a straight line path from x to y and the mean value theorem. By (1.2) for g_{n-1},

$$|(g_n - g_{n-1})(x) - (g_n - g_{n-1})(y)| \leq CAN\delta^{-n}|x - y|\,|g_{n-1}(x)|$$

if N is large. If neither x nor $\overline{y}$ belongs to any $Q \in A_n$, the same estimate is valid and slightly easier. For general x, y with $|x - y| < \delta^n$, draw a straight line from x to $\overline{y}$ and then from $\overline{y}$ to y. That gives

$$|(g_n - g_{n-1})(x) - (g_n - g_{n-1})(y)| \leq CAN\delta^{-n}|x - y| \max_{t \in [0,1]} |g_{n-1}(tx + (1-t)\overline{y})|,$$

and the right side is $\leq CA^2N\delta^{-n}|x - y|\,|g_{n-1}(x))|$ by (1.2). So

$$|g_n(x) - g_n(y)| \leq (BN\delta^{-(n-1)} + CA^2N\delta^{-n})|x - y|\,|g_{n-1}(x)|$$

$$\leq 2\rho^{-1}(BN\delta^{-(n-1)} + CA^2N\delta^{-n})|x - y|\,|g_n(x)|,$$

which proves (1.3) for g_n provided $B > 2C\rho^{-1}A^2$ and $\delta < \frac{1}{2}\rho B^{-1}(B - 2C\rho^{-1}A^2)$.

That means that the construction makes sense, and now we show that it converges. From here on, A, B, N are fixed. Constants can depend on them. Let $E = \bigcap_n \cup\{Q : Q \in A_n\}$. Let $\tau : Q(2) \to \mathbb{Z}^+ \cup (\infty)$ be the stopping time in the construction: $\tau(x) = \infty$ if $x \in E$ and otherwise $\tau(x) = \min(n : x \notin Q \in A_n)$. We use the usual notation $g_\tau(x) = g_{\tau(x)}(x)$.

Lemma 1.5. *The g_k converge uniformly on $\overline{\mathbb{R}}^3_+$ to a harmonic gradient g. This g is Holder continuous with $g = 0$ on E and $C^{-1}|g_\tau| \leq |g| \leq C|g_\tau|$ on $Q(2)\backslash E$. If $a > 0, \epsilon > 0, k < \infty$ are given then the following estimates hold if δ is small enough: $|g(x) - g_1(x)| < \epsilon|x|^{-k}$ when $|x| > 1$ and $|g(x) - g_1(x)| < \epsilon$ when $x \in Q(2) \times [a, 1]$.*

Proof. Fix $x \in Q(2)\backslash E$. If $k \geq \tau(x) + 2$ and $Q \in A_k$ then $|x - a_Q| \geq \delta^{\tau(x)+1}$ by the rule about touching squares in the definition of A_k. By (1.5), with $d = \delta^{\tau+1} \geq \delta^{k-1}$,

$$|g_k(x) - g_{k-1}(x)| \leq C|g_{k-1}(x)|\delta^{k-\tau}.$$

Calculus then shows $|g_{k-1}(x)| \leq 2|g_{\tau+1}(x)|$ for $k \geq \tau + 2$ if δ is small. By (1.8), $|g_k - g_{k-1}| \leq C\delta^{k-\tau}|g_\tau|$ when $k \geq \tau + 1$. This implies $C^{-1}|g_\tau| \leq |g_k| \leq C|g_\tau|$ when $k \geq \tau$, and that $\{g_k(x)\}$ converges uniformly on $\{x \in Q(2) : \tau(x) \leq n\}$ for any given n. However if $k \leq \tau(x)$ then the stopping rule (1.4) implies $|g_k(x)| \leq Ce^{-\alpha k}$ (here α and C depend on g_1 and λ), so it follows that $\{g_k\}$ converge uniformly on $Q(2)$, and clearly the limit satisfies $g = 0$ on E and $C^{-1}|g_\tau| \leq |g| \leq C|g_\tau|$ on $Q(2)\backslash E$. To estimate when $|x| > 1$ note the preceding argument gives a bound $\|g_j\|_{L^\infty(Q(2))} \leq M$ where $M = M(g_1, \lambda)$ is independent of δ for δ small enough. Using that $q_j \in S_H$ and that there are at most δ^{-2j} squares in A_j,

(1.11) $$|g_j(x) - g_{j-1}(x)| \leq C_k\delta^{-2j}(\delta^{-j}|x|)^{-k}M,$$

for any given k if $|x| > 1$. This implies uniform convergence on $\mathbb{R}^2\backslash Q(2)$ (hence on $\mathbb{R}^3_+$) and also the estimate $|g - g_1| < \epsilon|x|^{-k}$ when $|x| > 1$. The estimate $|g - g_1| < \epsilon$ on $Q(2) \times [a, 1]$ also follows in this way; replace $|x|$ by a on the right side of (1.11). We prove the Holder continuity on $Q(2)$ only since on $\mathbb{R}^2\backslash Q(2)$ it follows easily from (1.6). In this argument, constants depend on all parameters including δ. By the stopping rule (1.4) we know $|g_k| \leq C\eta^k$ for some $\eta < 1$, when $k < \tau$, and therefore $|g| \leq C\eta^\tau$. Also (1.3) together with boundedness of the g_k implies $|\nabla g_k| \leq C\delta^{-k}$. If $j \geq \tau(x) + 2$, then using (1.6) and the rule about touching squares, $|\nabla(g_j - g_{j-1})(x)| \leq C\delta^{j-2\tau(x)}$. Taking $k = \tau(x) + 1$ and summing a series gives $|\nabla g| \leq C\delta^{-\tau}$, i.e., $|\nabla g| \leq C|g|^{-\omega}$ when $g \neq 0$, where $\omega = (\log \delta)(\log \eta)^{-1}$. This estimate implies g is Holder with exponent $(1 + \omega)^{-1}$. ∎

Lemma 1.6. *Fix $t \in (0, 1]$. If $t > \delta$, then $|g(\overline{x}, t)| \leq C(|g_1(\overline{x}, t)| + |g_1(\overline{x})|)$ for all $\overline{x} \in Q(2)$. If $t < \delta$, then $|g(\overline{x}, t)| \leq C|g_k(\overline{x})|$ where k is the largest number with $\delta^k \geq t$.*

Proof. We do the $t < \delta$ case, leaving the other to the reader. By (1.2), $|g_k(\bar{x}, t)| \leq C|g_k(\bar{x})|$. By (1.5), (1.8), if $j \geq k + 1$ then

$$|g_j(\bar{x}, t) - g_{j-1}(\bar{x}, t)| \leq C|g_{j-1}(\bar{x})||\delta^{2j-k-1-\max(k+1, j-1)}$$

$$\leq C|g_{j-1}(\bar{x})||\delta^{j-k-1}$$

$$\leq C\left(\frac{2\delta}{\rho}\right)^{j-k-1}|g_k(\bar{x})|,$$

and the result follows if $\delta < \frac{1}{2}\rho$.   ∎

We still have to show why g can be taken to have the special properties in Theorem 1 and Lemma 1.4. Basically, this is because of (1.9). Let $\tau \wedge k = \min(\tau, k)$, and if Q is a square, ϕ a function, let $Q(\phi) = |Q|^{-1} \int_Q \phi$.

Lemma 1.7. *On* $Q(2)$, $\log|g_k| \leq \log|g_1| - \frac{1}{2}N^{-2}\eta\tau \wedge k + h_k$ *where* h_k *satisfies*

$$\int_{x \in Q(2) \, : \, \tau \wedge k = n} e^{ph_k} \leq C_1 e^{C_1 np^2}$$

for all $n \in \mathbb{Z}^+$, $p \in (0, \infty)$.

Proof. Let $\phi_n = \log|g_n| - \log|g_{n-1}|$. By (1.3), $\log|g_n|$ and, therefore, ϕ_n are Lipschitz with Lipschitz norm $\leq C\delta^{-n}$. Also $Q(\phi_n) \leq -\frac{1}{2}N^{-2}\eta$ for each $Q \in A_n$ by (1.9). Using a partition of unity, write $\phi_n = \psi_n + \xi_n$ where $\xi_n \leq -\frac{1}{2}N^{-2}\eta$ on $\cup\{Q : Q \in A_n\}$ and $Q(\psi_n) = 0$ for each $Q \in A_n$, and $\|\psi_n\|_{\text{Lip}(1)} \leq C\delta^{-n}$. For $n \geq 2$ define $f_n : \cup\{Q : Q \in A_n\} \to \mathbb{R}$ by $f_n = \sum_{j=1}^n \psi_j$. The ψ_j are weakly dependent and we can estimate f_n by a standard argument from elementary probability. (See the proof of Khinchin's inequality in the appendix of [24].) Fix $Q \in A_n$. If $Q(\psi) = 0$ then $Q(e^{p\psi}) \leq e^{Cp^2}$ where C depends on $\|\psi\|_\infty$. Since $\psi_n + f_{n-1} - Q(f_{n-1})$ has mean zero on Q and Lipschitz norm $\leq C\delta^n$, hence sup norm $\leq C$,

$$Q(e^{pf_n}) = e^{pQ(f_{n-1})}Q(e^{p(\psi_n + f_{n-1} - Q(f_{n-1}))})$$

$$\leq e^{pQ(f_{n-1})}e^{Cp^2}$$

$$\leq e^{Cp^2}Q(e^{pf_{n-1}})$$

$$\int_{\cup\{Q \, : \, Q \in A_n\}} e^{pf_n} \leq e^{Cp^2}\int_{\cup\{Q \, : \, Q \in A_n\}} e^{pf_{n-1}}$$

$$\leq e^{Cp^2}\int_{\cup\{Q \, : \, Q \in A_{n-1}\}} e^{pf_{n-1}}.$$

Iterating $n - 1$ times,

(1.12)
$$\int_{\cup\{Q:\, Q \in A_n\}} e^{pf_n} \le e^{C(n-1)p^2}.$$

Now we have, using Lemma 1.5,

$$\log |g_k| \le \log |g_{\tau \wedge k}| + C_2$$
$$= \log |g_1| + \sum_{2 \le j \le \tau \wedge k} \phi_j + C_2$$
$$\le \log |g_1| - \frac{1}{2} N^{-2} \eta(\tau \wedge k - 1) + f_{\tau \wedge k} + C_2.$$

The set $\{x : \tau \wedge k = n\}$ is contained in $\cup\{Q : Q \in A_{n-1}\}$, so the lemma follows from (1.12) if we take $h_k = f_{\tau \wedge k} + \frac{1}{2} N^{-2} \eta + C_2$. ∎

Proof of Theorem 1 Fix $g_1 \in S_H$ with $g_1 \ne 0$ on $Q(2)$. We claim that $|E| > 0$ if λ is large and δ small. For $n \ge 2$, let F_n be the union of all squares such that $Q \subset Q' \in A_{n-1}$ and (1.4) fails on Q at stage n. Then $\tau \wedge n = n$ on F_n. Also, if Q is one of the squares comprising F_n then $h_{n-1} \ge \lambda(n \log n)^{1/2}$ somewhere on Q by the fact that (1.4) fails. So $h_{n-1} \ge \lambda(n \log n)^{1/2} - C$ everywhere on Q by (1.3), and then by (1.8), $h_n \ge \lambda(n \log n)^{1/2} - C_1$ on Q. So $h_n \ge \frac{1}{2}\lambda(n \log n)^{1/2}$ on F_n provided $\lambda > 2C_1(2 \log 2)^{-1/2}$. Lemma 1.7 implies

$$|F_n| \le C_2 e^{C_2 p^2 n - \frac{1}{2}\lambda p(n \log n)^{1/2}},$$

for any p. Taking $p = (4C_1)^{-1}\lambda n^{-1/2}(\log n)^{1/2}$ gives

$$|F_n| \le C_1 n^{-\frac{1}{16C_1}\lambda^2}.$$

Accordingly $\Sigma|F_n|$ may be made arbitrarily small by taking λ large. If $x \in Q(1)\backslash E$, then x belongs either to the double of one of the squares used to define one of the F_n, or to the double of one of the squares in G_1/A_1, so

$$|Q_1\backslash E| \le 4\Big(\sum_{n \ge 2} |F_n| + 4\delta\Big),$$

and the proof is finished. ∎

Proof of Lemma 1.4 We need $p < (2C_1)^{-1}N^{-2}\eta$ with C_1 as in Lemma 1.7. Let k be large enough and fix $\epsilon > 0$. We need g_1 to satisfy $|g_1| < \epsilon(4|x|)^{-k}$ when $|x| > \frac{1}{4}$, $\int_{\mathbb{R}^2} x^\alpha g_1(x) = 0$ when $|\alpha| \le k$, $|g_1| \le C$ and $|\{x \in Q(1): |g_1(x)| > C^{-1}\}| > C^{-1}$. C as well as the constants appearing below may depend on k, but not on ϵ. Such a g_1 is easily constructed using "atoms": let ℓ be large and fix a smooth harmonic gradient ϕ with ϕ_3 supported in $|x| < \frac{1}{4}$ and $\int x^\alpha \phi dx = 0$ when

$|\alpha| \leq \ell$, and with $|\{x : |\phi_3(x)| > C_1^{-1}\}| > C_1^{-1}$. This is simply the existence of compact support functions with many vanishing moments. Let ψ be a Schwarz space approximation to ϕ obtained by convolution as in (ii). We now modify ψ so as to obtain the first mentioned property of g_1. For a given ϵ, let t be small enough and $h(x) = \sum_z \psi(t^{-1}x - z)$ where the sum is over lattice points $z \in \mathbb{Z}^2$ with $|z| < (8t)^{-1}$. The rapid decay of ψ implies that $|h| \leq C$, and $(1 + |x|)^k|\psi_3(x)|$ may be made arbitrarily small when $|x| > \frac{1}{4}$ as described in (ii). So there is no significant cancellation among the different functions $\psi_3(t^{-1}x - z)$ and it follows that $|\{x : |h(x)| \geq \frac{1}{2}C_1^{-1}\}| \geq \frac{1}{2}C_1^{-1}$. If $|x| \geq \frac{1}{4}$, then $|t^{-1}x - z| > \frac{1}{2}t^{-1}|x|$ for all z, so $|h(x)| \leq C_3 t^{-2}(\frac{1}{2}t^{-1}|x|)^{-(k+2)}$ which is $< \epsilon(4|x|)^{-k}$ if t is small enough. To obtain g_1, approximate h by a harmonic gradient without zeros on $Q(2)$ using (iv).

Let n_0 be a large enough integer. By taking λ very large we can guarantee that if $\tau(x) \leq n_0$ then $x \in Q(2)\backslash Q(1)$ or x belongs to the double of one of the squares in $G_1\backslash A_1$. In either case, $|x| \geq \frac{1}{4}$. Let $a > 0$ be such that $|\{\overline{x} \in Q(1) : |g_1(\overline{x}, a)| > C^{-1}\}| > C^{-1}$ and make δ small enough that the last statement of Lemma 1.5 is valid with this a. If g is as produced by the construction, then $|\{\overline{x} : |g(\overline{x}, a)| > \frac{1}{2}C^{-1}\}| > \frac{1}{2}C^{-1}$. Moreover,

$$\int_{Q(2)} |g|^p = \sum_{n \leq n_0} \int_{\tau(x)=n} |g_1|^p \frac{|g|^p}{|g_1|^p} + \sum_{n > n_0} \int_{\tau(x)=n} |g_1|^p \frac{|g|^p}{|g_1|^p}$$

$$\leq \epsilon^p \sum_{n \leq n_0} \int_{\tau(x)=n} \frac{|g|^p}{|g_1|^p} + C \sum_{n > n_0} \int_{\tau(x)=n} \frac{|g|^p}{|g_1|^p}.$$

By Lemma 1.7,

$$\int_{\tau(x)=n} \frac{|g|^p}{|g_1|^p} \leq C_1 e^{-pn\eta/2N^2} e^{C_1 p^2 n}$$

$$= C_1 e^{-\beta n},$$

where $\beta > 0$ since $p < \eta/2C_1N^2$. So $\int_{Q(2)} |g|^p < C\epsilon^p$ if n_0 is large enough. If $0 < t \leq 1$, then choosing k as in Lemma 1.6,

$$\int_{Q(2)} |g(\overline{x}, t)|^p \leq C + C \int_{Q(2)} |g_k(\overline{x})|^p$$

$$\leq C + C \int_{Q(2)} \frac{|g_k(\overline{x})|^p}{|g_1(\overline{x})|^p}$$

$$\leq C + C + C \sum_n e^{-\beta n}$$

$$\leq C.$$

The next to last line here follows from Lemma 1.7 by conditioning on $\tau \wedge k$.

Estimates when $|x| > 1$ follow from Lemma 1.5 if δ is small enough, and the vanishing moment condition is built into the construction.　■

Proof of Theorem 2 This is a very standard argument; probably it is well known that Lemma 1.4 implies Theorem 2.

If g is a harmonic gradient, then we call the function whose gradient is g the primitive of g and denote it u_g. The example will have the following additional properties which will be used in extending to higher dimensions: u_g is continuous on $\overline{\mathbb{R}}_+^3$ and C^1 for large $|x|$, and for suitable ℓ, $\int_{\mathbb{R}^2} x^\alpha u_g(x) = 0$ when $|\alpha| \le \ell$ and g and u_g vanish at ∞ faster than $|x|^{-\ell}$.

Fix k large enough. Let g_ϵ be the function in Lemma 1.4. Note that $M_\epsilon \stackrel{\text{def}}{=} \|u_{g_\epsilon}\|_\infty$ and $|u_{g_\epsilon}| \le C\epsilon|x|^{1-k}$ for large $|x|$, and

$$\left(\int_{\mathbb{R}^2} |g_\epsilon(\overline{x}, t)|^p dx_1 dx_2 \right)^{1/p} < \epsilon \qquad \text{unless} \quad t \in (\tau_\epsilon, 2),$$

with $\tau_\epsilon > 0$. Let

$$g_{\epsilon,N}(x) = \sum_m g_\epsilon(Nx - m)$$

where m runs over all points $(2z, 0)$ with $z \in \mathbb{Z}^2$ and $|z| \le \frac{1}{2} N$.

Consider $\sum_m g_\epsilon(Nx - m)$ where we now restrict to points m as above with $|Nx - m| > 1$. In view of the estimate $|g_\epsilon(x)| \le \epsilon|x|^{-k}$ when $|x| > 1$, this sum will be $\le C\epsilon$ for all x and in fact $\le C\epsilon d^{-(k-2)}$, where $d = d(x) = \min_m(|Nx - m| : |Nx - m| > 1)$. It now follows that

(1.13) $$\left| \{\overline{x} : |g_{\epsilon,N}(\overline{x}, \frac{a_\epsilon}{N})| > C_3^{-1}\} \right| > C_3^{-1}.$$

with $a_\epsilon = a$ as in Lemma 1.4. Also,

$$\int |g_{\epsilon,N}(\overline{x}, t)|^p dx_1 dx_2 \le \sum_m \int |g_\epsilon(N\overline{x} - m, Nt)|^p dx_1 dx_2$$

which is $\le C$ for all t and $\le C\epsilon^p$ if $t \notin \left(\frac{\tau_\epsilon}{N}, \frac{2}{N} \right)$.

Let $u_{\epsilon,N}$ be the primitive of $g_{\epsilon,N}$. Since u_ϵ vanishes rapidly at ∞, an argument like the one for (1.12) bounds $\|u_{\epsilon,N}\|_\infty$; due to the different scaling we obtain

$$\|u_{\epsilon,N}\|_\infty \le CM_\epsilon/N.$$

Now let $\{\epsilon_j\}_1^\infty$ and $\{N_j\}_1^\infty$ converge rapidly enough to 0 and ∞ respectively. Specifically, let B be a large constant. We require

$$\epsilon_j < B^{-j}.$$

The intervals $\left(\frac{\tau_{\epsilon_j}}{N_j}, \frac{2}{N_j} \right)$ are disjoint

(1.14) $$\sum \frac{M_{\epsilon_j}}{N_j} < \infty.$$

These are obtained recursively by choosing $\epsilon_j < B^{-j}$ and then $N_j > \max(2^j M_{\epsilon_j}, 2N_{j-1}\tau_{\epsilon_{j-1}}^{-1})$. Let $g = \sum_j g_{\epsilon_j, N_j}$. Property (1.14) implies that the series converges uniformly on compact sets in $\mathbb{R}^3_+$ and the limit has a primitive which extends continuously to $\overline{R}^3_+$. For any t,

$$\int |g(\overline{x}, t)|^p \leq \sum \int |g_{\epsilon_j, N_j}(\overline{x}, t)|^p$$

$$\leq C,$$

since at most one term in the sum is $\geq B^{-jp}$. Moreover, if $t \notin \cup_j (\tau_{\epsilon_j} N_j^{-1}, 2N_j^{-1})$, then

$$\int |g(\overline{x}, t)|^p < \sum B^{-jp},$$

which can be made arbitrarily small by taking B large. In particular, we can guarantee that for such t

$$|\{\overline{x} : |g(\overline{x}, t)| > \frac{1}{3}C_3^{-1}\}| < \frac{1}{3}C_3^{-1},$$

with C_3 as in (1.13). In the same way, we guarantee that for $t \in (\tau_{\epsilon_j} N_j^{-1}, 2N_j^{-1})$,

$$|\{\overline{x} : |g(\overline{x}, t) - g_{\epsilon_j, N_j}(\overline{x}, t)| > \frac{1}{3}C_3^{-1}\}| < \frac{1}{3}C_3^{-1},$$

and, therefore,

$$|\{\overline{x} : |g(\overline{x}, N_j^{-1} a_{\epsilon_j})| > \frac{2}{3}C_3^{-1}\}| > \frac{2}{3}C_3^{-1}.$$

It follows that there is a set of $\overline{x}$ with measure at least $\frac{1}{3}C_3^{-1}$ on which $\lim \sup_{t \to 0} |g(\overline{x}, t)| > \frac{2}{3}C_3^{-1}$ and $\lim \inf_{t \to 0} |g(\overline{x}, t)| < \frac{1}{3}C_3^{-1}$, so Theorem 2 is proved. From the construction, u_g has as many zero moments as we like, and it is easily checked that it is C^1 for large $|x|$ and that g and u_g die faster than a given power of $|x|^{-1}$. ∎

The extension to higher dimensions is achieved by adding dummy variables. For Theorem 1 this may be done in a trivial way, but for Theorem 2 an argument is required. Denote variables in $\mathbb{R}^n_+ (n \geq 4)$ by $z = (x, y, t)$, $x \in \mathbb{R}^2$, $y \in \mathbb{R}^{n-3}$, $t \geq 0$. Let $\psi : \mathbb{R}^{n-3} \to \mathbb{R}$ be a smooth function with $\psi(y) = 1$ when $|y| \leq 1$, $\psi(y) = 0$ when $|y| \geq 2$, and with sufficiently many zero moments.

Let $u = u_g$ where g is as in Theorem 2, and let F be the harmonic extension of $\psi(y)u(x, 0)$ to $\mathbb{R}^n_+$. By the reflection principle, $F(x, y, t) - u(x, t)$ extends smoothly across the set $\{(x, y, 0) : |y| \leq 1\} \subset \mathbb{R}^{n-1}$, so ∇F doesn't have radial limits almost everywhere. We will show that ∇F is in the Stein-Weiss H^p space. Since $\phi(y)u(x, 0)$ is C^1 for large $|x| + |y|$ and rapidly vanishing at ∞ and has lots of zero moments, it is enough to show

$$\sup_{0 < t < 1} \int_{\max(|x|, |y|) < R} |\nabla F(x, y, t)|^p dx dy < \infty$$

for fixed R. Fix y_0 and write

$$\psi(y)u(x, 0) = q(y)u(x, 0) + (\psi(y) - q(y))u(x, 0)$$

where $q(y) = \psi(y_0) + \langle \nabla\psi(y_0), y - y_0 \rangle$. The function $G_{y_0}(x, y, t) = q(y)u(x, t)$ is harmonic (compute its Laplacian) and

$$\int_{|x| < R} |\nabla G_{y_0}(x, y_0, t)|^p dx \leq \frac{4}{3}\pi R^3 \|u\|^p_\infty |\nabla\psi(y_0)|^p$$

$$+ |\psi(y_0)|^p \int_{|x| < R} |u(x, t)|^p dx \leq C$$

uniformly in y_0. Let $H_{y_0}(x, y, t)$ be the harmonic extension of $(\psi(y) - q(y))u(x, 0)$. Since $|(\psi(y) - q(y))u(x, 0)| \leq C|y - y_0|^2$ locally and $\leq C|y| |x|^{-k}$ at ∞, the formula for the derivatives of the Poisson kernel applies and shows that $|\nabla H_{y_0}(x, y_0, t)|$ is bounded. We conclude

$$\int_{\max(|x|, |y|) < R} |\nabla F(x, y, t)|^p dx dy$$

$$\leq \int \left(\int [|\nabla G_y(x, y, t)|^p + |\nabla H_y(x, y, t)|^p] dx \right) dy$$

$$\leq \int C dy$$

$$< \infty.$$

2 DIMENSION OF HARMONIC MEASURES

The purpose of this section is to prove Theorem 3. We will do the main construction in somewhat greater generality than is needed for this since it doesn't require any extra effort and tends to show that there is no reason at all for a harmonic measure in $\mathbb{R}^3$ to be 2-dimensional. As explained in the introduction, ideas of Carleson [6] show that for scale invariant boundaries the dimension of the harmonic measure should be determined by the quantity

$$(2.1) \qquad \lim_{n \to \infty} \frac{1}{n} \int_{\partial \Omega^n} |\nabla g_n| \log |\nabla g_n| d\sigma$$

where Ω^n is a natural smooth (enough) approximation to our domain Ω and g_n is its Green's function with fixed pole p. (We normalize Green's function to be positive and satisfy $|\nabla g| = \frac{d\omega}{d\sigma}$ where σ is surface measure and ω is harmonic measure at p.) In particular, in $\mathbb{R}^3$ the sign of (2.1) determines whether ω is > 2- or < 2-dimensional. In contrast to the planar case considered in [6], the sign of (2.1) may be chosen freely by adjusting the parameters of our construction.

The situation is similar to section 1 in that once we know how to "improve a constant" in a suitable sense, we can set up an iterative procedure to construct fractal boundaries with control over (2.1). Now the role of a constant is played by the upper half space. Suppose $\phi : \mathbb{R}^2 \to \mathbb{R}$ is a Lipschitz function with $\mathrm{supp}\, \phi \subset Q(1)$. Let $U = \{x \in \mathbb{R}^3 : x_3 > \phi(\bar{x})\}$. Then there is a unique harmonic function $g_\infty : U \to \mathbb{R}^+$ ("Green's function with pole at ∞") such that g_∞ vanishes on ∂U and $\lim_{x \to \infty}(g_\infty(x) - x_3) = 0$. One shows that for $x \in \mathbb{R}^2$, $|\nabla g(x)| = 1 + \mathcal{O}(|x|^{-3})$ as $|x| \to \infty$. This is done as follows: let ω be harmonic measure for the upper half-space of a disc centered at the origin and containing $\mathrm{supp}\, \phi$. Then using the maximum principle,

$$x_3 - C\omega \leq g_\infty \leq x_3 + C\omega, \quad x \in U \cap \mathbb{R}^3_+$$

for suitable C. And $|\nabla \omega|$ is $\mathcal{O}(|x|^{-3})$ at ∞ when $x \in \mathbb{R}^2$ by the Poisson integral formula, so the statement follows from l'Hôpital's rule. The integral

$$I(\phi) = \int_{\partial U} |\nabla g_\infty| \log |\nabla g_\infty| d\sigma$$

is then easily (given [7]) seen to be absolutely convergent. At the end of the section we will find piecewise linear functions ϕ with support in $Q(1)$ and arbitrarily small Lipschitz norm, such that $I(\phi)$ has either sign.

For now, fix any piecewise linear ϕ with $\mathrm{supp}\, \phi \subset Q(1)$. Next fix $b > 0$ and small. If Q is a square (i.e., a square in some plane in $\mathbb{R}^3$) with center point a_Q and given unit normal e, define "pyramids" P_Q and $\tilde{P}_Q$ by

$$P_Q = \mathrm{cch}(Q \cup \{a_Q + b\ell(Q)e\})$$

$$\tilde{P}_Q = \mathrm{int}\, \mathrm{cch}(Q \cup \{a_Q - b\ell(Q)e\})$$

where cch denotes the closed convex hull.

Let N be a large number. In particular, N should be large enough that $\{x : x_3 = N^{-1}\phi(N\bar{x})\}$ is well inside $P_{Q(1)} \cup \tilde{P}_{Q(1)}$. Constants will be independent of N

provided N is large enough. Define (with $e = -e_3$)

$$\Lambda = \{x \in P_{Q(1)} \cup \tilde{P}_{Q(1)} : x_3 > N^{-1}\phi(N\bar{x})\}$$

$$\partial = \{x \in \mathbb{R}^3 : \bar{x} \in Q(1),\ x_3 = N^{-1}\phi(N\bar{x})\}$$

Then $\partial \subset P_{Q(1)} \cup \tilde{P}_{Q(1)}$. ∂ is a polygon with lots of faces. Fix a Whitney decomposition W of each face into squares Q with side $8^{-k}, k = 1, 2, \ldots$. Choose a distinguished edge for each square. These choices may be made in an arbitrary way, but once made, they are kept fixed. The following is easy to see if N is large enough.

Lemma 2.1. *There are at most $C8^k$ squares $Q \in W$ with $\ell(Q) = 8^{-k}$. And for each $k \geq 1$,*

$$\bigcup\{Q \in W : \ell(Q) \leq 8^{-k}\} \subset \bigcup\{D(a_Q, C8^{-k}) : Q \in W, \ell(Q) = 8^{-k}\}.$$

Suppose next that Ω is a domain and $Q \subset \partial\Omega$ is a square with $P_Q \cap \Omega = \emptyset$, $\tilde{P}_Q \subset \Omega$, and with a distinguished edge γ. Let e be the normal into P_Q. Form a new domain $\tilde{\Omega}$ as follows: Let T be the affine map with $T(Q(1)) = Q$, $T(0) = a_Q$, $T(-e_3) = e$, and $T(\{\frac{1}{2}\} \times [-\frac{1}{2}, \frac{1}{2}] \times \{0\}) = \gamma$. Let $\Lambda_Q = T(\Lambda)$ and $\partial_Q = T(\partial)$. Then $\tilde{\Omega} \cap (P_Q \cup \tilde{P}_Q) = \Lambda_Q$ and $\tilde{\Omega}\backslash(P_Q \cup \tilde{P}_Q) = \Omega\backslash(P_Q \cup \tilde{P}_Q)$.

We call this construction "adding a blip to Ω along Q." Transferring W by T, $\partial\tilde{\Omega} \cap (P_Q \cup \tilde{P}_Q) = \partial_Q$ has a natural decomposition into squares with distinguished edges together with a set of σ-finite length.

The domains we want are obtained by iterating this procedure. Let Ω_0 be the unit cube. If Ω_{n-1} has been constructed, then its boundary will be given as $\partial\Omega_{n-1} = E_{n-1} \cup (\bigcup_{\underline{G}_{n-1}} Q)$ where E_{n-1} has σ-finite length and each $Q \in \underline{G}_{n-1}$ is a square with a distinguished edge and, with $P_Q \cap \Omega_{n-1} = \emptyset$, $\tilde{P}_Q \subset \Omega_{n-1}$, and, moreover, if $Q_1, Q_2 \in \underline{G}_{n-1}$, then $\mathrm{int}\, P_{Q_1} \cap \mathrm{int}\, P_{Q_2} = \emptyset$ and $\tilde{P}_{Q_1} \cap \tilde{P}_{Q_2} = \emptyset$. To form Ω_n, add a blip along each $Q \in \underline{G}_{n-1}$.

Then $\partial\Omega_n = E_{n-1} \cup (\bigcup_{Q \in \underline{G}_{n-1}} \partial_Q)$ is decomposed as $E_n \cup (\bigcup_{Q \in \underline{G}_{n-1}} Q')$ where the Q' are obtained from the preceding decomposition of the ∂_Q. If $Q \in \underline{G}_{n-1}$, $Q' \in \underline{G}_n$ and $Q' \subset \partial_Q$ then we say Q' is directly descended from Q.

If N is large enough then it is clear that for Q' directly descended from Q, $P_{Q'} \cup \tilde{P}_{Q'} \subset P_Q \cup \tilde{P}_Q$ and in fact by properties of the Whitney decomposition, $\mathrm{dist}(P_{Q'} \cup \tilde{P}_{Q'}, (P_Q \cup \tilde{P}_Q)^c) \geq C^{-1}\ell(Q')$. Moreover, $P_{Q'} \subset \Lambda_Q^c$, $\tilde{P}_{Q'} \subset \Lambda_Q$, and if Q_1', Q_2' are directly descended from the same Q then $P_{Q_1'} \cap P_{Q_2'} = \emptyset$, $\tilde{P}_{Q_1'} \cap \tilde{P}_{Q_2'} = \emptyset$. So the induction hypothesis is satisfied and the construction can continue.

Let $\Omega = \lim_{n \to \infty} \Omega_n$. Also let $\underline{G} = \bigcup_{n=1}^{\infty} \underline{G}_n$. If $Q, Q' \in \underline{G}$ we say Q' is descended from Q in n stages if there are $Q_0 = Q', Q_1, \ldots, Q_n = Q$ with Q_j

directly descended from Q_{j+1} for $j = 0, \ldots, n - 1$. If Q' is descended from Q we write $Q' \prec Q$.

We will write $\omega(U, Y, z)$ for the harmonic measure of the set Y (i.e., $Y \cap \partial U$), relative to the domain U, evaluated at $z \in U$. Let $\omega = \omega(\Omega, \cdot, 0)$.

Main Lemma. *Suppose ϕ has sufficiently small Lipschitz norm and* $\mathrm{I}(\phi) \neq 0$. *Then if N is large enough there is a number d such that*

$$\lim_{r \to 0} \frac{\log \omega(D(x, r))}{\log r} = d \qquad \text{for a.e.} \quad (d\omega) \, x \in \partial\Omega$$

where $D(x, r)$ is the intersection of $\partial\Omega$ with a ball of radius r, centered at x. Thus ω puts full mass on some set with dimension d and no mass on any set with dimension less than d. If $\mathrm{I}(\phi) < 0$, then $d > 2$, and if $\mathrm{I}(\phi) > 0$, then $d < 2$.

We will prove only the $\mathrm{I}(\phi) < 0$ case explicitly but we have set up the lemmas so that the $\mathrm{I}(\phi) > 0$ case follows by changing the notation.

Certain estimates of harmonic measure will have to be made. Fortunately the whole construction takes place within the class of NTA (nontangentially accessible) domains of Jerison and Kenig [9] and most of the necessary estimates are proved in [9]. Let Ω be a bounded domain. If diam $\Omega = 1$, then Ω is an NTA domain with NTA constant A if it has the following properties:

- For each $x \in \partial\Omega$, $r < A^{-1}$, there are points $A_r(x) \in \Omega$, $B_r(x) \in \Omega^c$ with $|A_r(x) - x| \leq Ar$, $|B_r(x) - x| \leq Ar$, and dist$(A_r(x), \partial\Omega) \geq A^{-1}r$, dist$(B_r(x), \partial\Omega) \geq A^{-1}r$. Existence of the points $A_r(x)$ (resp. $B_r(x)$) is called the interior (exterior) corkscrew condition.
- If $x, y \in \Omega$ then there is a path γ from x to y with length$(\gamma) \leq A|x - y|$ and dist$(\gamma(t), \partial\Omega) \geq A^{-1} \min(|\gamma(t) - x|, |\gamma(t) - y|)$. This is called the Harnack chain condition.

In general Ω is an NTA domain with NTA constant A if a dilation of it with diameter 1 is such. This definition is easily seen to be equivalent to (say) the one in [11].

If Ω is an NTA domain, $t > 0$, then we say Ω is Lipschitz on scale t with Lipschitz constant M if $D(x, t) \cap \partial\Omega$ is the graph of a function with Lipschitz norm $\leq M$ for each $x \in \partial\Omega$.

Now suppose $\Gamma \subset \underline{G}$ is closed under the descent relation, i.e., $Q \in \Gamma$ and $Q \prec Q'$ imply $Q' \in \Gamma$. Then we can form a domain Ω_Γ by adding blips only along squares in Γ: $\Omega_\Gamma^0 = \Omega_0$, Ω_Γ^{n+1} is obtained from Ω_Γ^n by adding blips along squares $Q \in \Gamma \cap \underline{G}_n$—it is clear by induction that such squares are contained in $\partial\Omega_\Gamma^n$ and satisfy $\tilde{P}_Q \subset \Omega_\Gamma^n$, $P_Q \cap \Omega_\Gamma^n = \emptyset$—and $\Omega_\Gamma = \lim_{n \to \infty} \Omega_\Gamma^n$.

Lemma 2.2. Ω_Γ *is an NTA domain with NTA constant independent of N and Γ.* *If Γ contains no squares with length $< t$ then Ω_Γ is Lipschitz on scale $C^{-1}t$ with C and the Lipschitz constant independent of N, t, and Γ.*

Proof. We can assume Γ is finite. Squares denoted Q, Q', $\overline{Q}$, etc., will be understood to belong to Γ.

For each Q, define $V_Q = \Lambda_Q \setminus \cup \{\tilde{P}_{Q'} : Q'$ directly descended from $Q\}$ and $W_Q = (P_Q \cup \tilde{P}_Q) \setminus (\Lambda_Q \cup \bigcup \{P_{Q'} : Q'$ directly descended from $Q\})$. Then V_Q and W_Q are Lipschitz domains provided b is small enough, and $V_Q \subset \Omega_\Gamma$, $W_Q \cap \Omega_\Gamma = \emptyset$. Moreover, $V_Q \cap V_{Q'} = \emptyset$ if $Q \neq Q'$, and $\Omega_\Gamma = S \cup (\bigcup_Q V_Q)$ where $S = \Omega_0 \setminus \cup \{\tilde{P}_Q : Q \in \underline{G}_0 \cap \Gamma\}$ is a "trivial" set. For each Q, let b_Q be a point of V_Q with $\text{dist}(b_Q, \partial V_Q) \approx \text{diam } V_Q \approx \ell(Q)$. There are three things to check in order to know that Ω_Γ is an NTA.

Exterior corkscrew condition. Fix $x \in \partial \Omega_\Gamma$, $0 < r < 1$. We will ignore the easy case where $x \in \Sigma \in \underline{G}_0 \setminus \Gamma$. Jiggling x slightly to avoid edges, it follows that $x \in \partial_Q$ for some $Q \in \Gamma$. Let Q_x be a smallest such square. Let Q be a smallest square such that $Q_x \prec Q$ and $\ell(Q) \geq r$. Since $x \in P_Q \cup \tilde{P}_Q$ it follows that $\text{dist}(x, W_Q) \leq Cr$. Let z be a closest point to x in ∂W_Q. Since W_Q is Lipschitz with diameter $\geq r$ there is $B_r(x) \in W_Q$ with $|B_r(x) - z| \leq Cr$ and $\text{dist}(B_r(x), \partial W_Q) \geq C^{-1}r$. Then $|B_r(x) - x| \leq Cr$ and $\text{dist}(B_r(x), \overline{\Omega}_\Gamma) \geq \text{dist}(B_r(x), \partial W_Q) \geq C^{-1}r$.

Interior corkscrew condition. Choose Q_x and Q as before, let z be a closest point to x in ∂V_Q, and let $A_r(x)$ satisfy $|A_r(x) - z| \leq Cr$ and $\text{dist}(A_r(x), V_Q^c) \geq C^{-1}r$.

Harnack chain condition. It is here that we use that $\|\phi\|_{\text{Lip}(1)}$ is sufficiently small. We use this assumption only to assert that domains (2.2) below are Lipschitz and therefore NTA.

We show first that if Q, $Q_j^{(i)}$, $i = 1, 2$, $j = 1, 2$, are squares with $Q_j^{(1)}$ directly descended from Q and $Q_j^{(2)}$ directly descended from $Q_j^{(1)}$ then

$$(2.2) \qquad V_Q \cup V_{Q_j^{(1)}} \cup V_{Q_j^{(2)}} \qquad \text{and} \qquad V_Q \cup \bigcup_{j=1}^{2} \left(V_{Q_j^{(1)}} \cup V_{Q_j^{(2)}} \right)$$

are Lipschitz domains. Consider, for example, the first domain. It is obtained from $\tilde{P}_Q$ by adding a blip, adding two more blips, then deleting certain pyramids all of whose bases make a small angle with the plane containing Q (provided $\|\phi\|_{\text{Lip}(1)}$ is small), resulting (if b is small) in a figure whose boundary is $\partial \tilde{P}_Q \setminus Q$ together with a Lipschitz graph over Q.

Suppose $x \in \partial\Omega_\Gamma \backslash S$, Q_x is the square with $x \in V_{Q_x}$, and $Q_x \prec Q$.

Claim. There is a path γ from x to b_Q with length$(\gamma) \leq C\ell(Q)$ and dist$(\gamma(t), \partial\Omega_\Gamma) \geq C^{-1}|\gamma(t) - x|$.

Proof. Let $Q_x = Q_0 \prec Q_1 \prec \cdots \prec Q_n = Q$ with Q_j directly descended from Q_{j+1}. Using the Lipschitz domains V_{Q_0} and $V_{Q_{j-1}} \cup V_{Q_j}$ we can find paths γ_0 from x to b_{Q_0} and γ_j from $b_{Q_{j-1}}$ to b_{Q_j} $(j = 1, \ldots, n)$ such that

$$\text{length}(\gamma_0) \leq C\ell(Q_0),$$

$$\text{dist}(\gamma_0(t), \partial\Omega_\Gamma) \geq C^{-1}\min(|\gamma_0(t) - x|, |\gamma_0(t) - b_{Q_0}|)$$

$$\text{length}(\gamma_j) \leq C\ell(Q_j),$$

$$\text{dist}(\gamma_j(t), \partial\Omega_\Gamma) \geq C^{-1}\min(|\gamma_j(t) - b_{Q_{j-1}}|, |\gamma_j(t) - b_{Q_j}|).$$

Now $|b_{Q_j} - x| \leq \text{diam}(P_{Q_j} \cup \tilde{P}_{Q_j}) \leq C \, \text{dist}(b_{Q_j}, \partial V_{Q_j}) \leq C \, \text{dist}(b_{Q_j}, \partial\Omega_\Gamma)$ for any j. It follows that $|y - x| \leq |y - b_{Q_j}| + C \, \text{dist}(b_{Q_j}, \partial\Omega_\Gamma)$ for all $y \in \Omega_\Gamma$, and, therefore, that dist$(\gamma_j(t), \partial\Omega_\Gamma) \geq C^{-1}|\gamma_j(t) - x|$. Let γ be γ_0 followed by γ_1 followed by ... followed by γ_n. Then dist$(\gamma(t), \partial\Omega_\Gamma) \geq C^{-1}|\gamma(t) - x|$ and length$(\gamma) \leq C \sum_j \ell(Q_j) \leq C\ell(Q)$, since $\ell(Q_j) \leq \frac{1}{8}\ell(Q_{j+1})$. ∎

Now fix $x, y \in \Omega_\Gamma$. We assume neither of them belongs to S. So there are Q_x, Q_y with $x \in V_{Q_x}$, $y \in V_{Q_y}$. Let $\overline{Q}$ be the minimal square such that $Q_x \preceq \overline{Q}$ and $Q_y \preceq \overline{Q}$.

Case 1. Each of Q_x, Q_y is either equal to $\overline{Q}$ or descended from $\overline{Q}$ in at most two stages.

Then we simply use the Lipschitz domains (2.2).

Case 2. One of Q_x or Q_y is descended from $\overline{Q}$ in more than two stages but not the other.

We suppose Q_x is descended from $\overline{Q}$ in more than two stages. Let Q satisfy $Q_x \prec Q$ and Q is descended from $\overline{Q}$ in exactly two stages. Then $|x - b_Q| \leq C|x - y|$. To see this, let Q' satisfy $Q \prec Q' \prec \overline{Q}$. Then $|x - y| \geq \text{dist}(x, (P_{Q'} \cup \tilde{P}_Q)^c) \geq \text{dist}(P_Q \cup \tilde{P}_Q, (P_{Q'} \cup \tilde{P}_{Q'})^c) \geq C^{-1}\ell(Q) \geq C^{-1}|x - b_Q|$, proving the assertion. Now connect x to b_Q by a path γ_1 as in the claim. The length of this path is $\leq C|x - y|$. Connect b_Q to y by a path γ_2 with length$(\gamma_2) \leq C|b_Q - y|$ and dist$(\gamma_2(t), \partial\Omega_\Gamma) \geq C^{-1}\min(|\gamma_2(t) - b_Q|, |\gamma_2(t) - y|)$ using the Lipschitz domains (2.2) associated to $\overline{Q}$. Since $|b_Q - y| \leq |x - y| + |b_Q - x| \leq C|x - y|$, we have $\mathrm{dist}(\gamma_2(t), \partial\Omega_\Gamma) \geq C^{-1}\min(|\gamma_j(t) - y|, |\gamma_j(t) - x|)$ and $\mathrm{length}(\gamma_j) \leq C|x - y|$. Now let γ be γ_1 followed by γ_2.

Case 3. Both Q_x and Q_y are descended from $\overline{Q}$ in more than two stages.

Choose $Q_x^{(1)}$, $Q_y^{(1)}$ so that $Q_x^{(1)}$, $Q_y^{(1)}$ are descended from $\overline{Q}$ in exactly two stages and $Q_x \prec Q_x^{(1)}$, $Q_y \prec Q_y^{(1)}$ Let $b_x = b_{Q_x^{(1)}}$, $b_y = b_{Q_y^{(1)}}$. It follows as before that $|x - b_x| \leq C|x - y|$, $|y - b_y| \leq C|x - y|$. Use the claim to connect x to b_x by γ_1 and y to b_y by γ_3, and (2.2) to connect b_x to b_y by γ_2. Then γ_1 followed by γ_2 followed by γ_3 is the desired path.

To prove the last statement of the lemma, let $\Sigma_Q = \Lambda_Q \triangle \tilde{P}_Q$ ($\triangle =$ symmetric difference)—the change in Ω_Γ made in adding the blip along Q. Note $\mathrm{dist}(\Sigma_Q, (P_Q \cup \tilde{P}_Q)^c) \geq C^{-1}\ell(Q)$. If $x \in \partial\Omega_\Gamma$, let $Q \in \Gamma$ be minimal subject to $D(x, C^{-1}t) \cap \Sigma_Q \neq \emptyset$. Then $D(x, C^{-1}t) \subset P_Q \cup \tilde{P}_Q$. By minimality it follows that $D(x, C^{-1}t) \cap \partial\Omega_\Gamma = D(x, C^{-1}t) \cap \partial_Q$, which is a Lipschitz graph. ∎

We now give some general estimates for harmonic measures. We will assume that $\Omega \subset \mathbb{R}^n$ is NTA and a basepoint $p \in \Omega$ has been fixed; a lower bound for $(\mathrm{diam}\,\Omega)^{-1}\mathrm{dist}(p, \partial\Omega)$ is incorporated into the NTA constant A. All constants C, α, etc., depend only on A. The basic result proved in [9], Lemma 4.10, and Theorem 7.9 is the "rate theorem": if u and v are positive harmonic functions on Ω, $x \in \partial\Omega$, $r > 0$ and u and v vanish continuously on $D(x, r) \cap \partial\Omega$, then on $D(x, \frac{r}{2})$, $\frac{u}{v} \leq C\frac{u(A_r(x))}{v(A_r(x))}$. Here $D(x, r)$ denotes a ball of radius r, centered at x. Moreover $\frac{u}{v}$ extends to a Hölder continuous function on $\overline{\Omega} \cap D(x, \frac{r}{2})$ and satisfies the following estimate when $y, z \in \partial\Omega \cap D(x, \frac{r}{2})$

$$(2.3) \qquad \left|\frac{u(y)}{v(y)} - \frac{u(z)}{v(z)}\right| \leq C\left(\frac{|y - z|}{r}\right)^\beta \frac{u(A_r(x))}{v(A_r(x))}, \qquad \beta > 0.$$

Other results that will be used, such as the doubling property of harmonic measure, are in Section 4 of [9]. We also need some further estimates which are not in [9] but follow easily from the techniques used there.

Lemma 2.3. *There is $\alpha > n - 2$ such that if $0 < r < R$, $x \in \Omega$ then* $\omega(D(x, r)) \leq C(\frac{r}{R})^\alpha \omega(D(x, R))$.

Proof. If we required only $\alpha > 0$ this would follow from the doubling property, as does a corresponding inequality in the opposite direction. To get $\alpha > n - 2$, normalize Ω so that $\mathrm{diam}\,\Omega = 1$. By "localization" ([9], Lemma 4.11; see also [11]) we can assume $R = 1$. Lemma 4.1 of [9] applied to the Green's function g of Ω with pole at p on the domain $\Omega/D(0, C^{-1})$ gives $g(A_r(x)) \leq Cr^\beta$ with $\beta > 0$. Lemma 4.8 of [9] then implies $\omega(D(x, r)) \leq Cr^{\beta+n-2} = Cr^\alpha$. ∎

Lemma 2.4. *Let Ω_1 and Ω_2 be two NTA's with base points p_1, p_2, and $\omega_i = \omega(\Omega_i, \cdot, p_i)$. Suppose $E \subset \partial\Omega_1 \cap \partial\Omega_2$ satisfies $D(x, C_1^{-1}r) \cap \partial\Omega_1 \subset E$, diam $E \le r$ and $\mathrm{dist}(E, \partial\Omega_1 \triangle \partial\Omega_2) \ge C_1^{-1}r$. Suppose $Y \subset E$. Then $\frac{\omega_2(Y)}{\omega_1(Y)} \le C \frac{\omega_2(E)}{\omega_1(E)}$.*

Proof. The assumptions imply that E is contained in the union of a bounded number of discs of radius $\frac{1}{2}C_1^{-1}r$ whose doubles do not intersect $\partial\Omega_1 \triangle \partial\Omega_2$. The statement now follows from the rate theorem, Lemma 4.8 of [9], and the doubling property.   ■

Lemma 2.5. *Let Ω_1 and Ω_2 be two NTA's with a common base point p. Suppose the Hausdorff distance between $\partial\Omega_1$ and $\partial\Omega_2$ is $\le C_1 t$, and that $x \in \partial\Omega_1 \cap \partial\Omega_2$ with $D(x, t) \cap \partial\Omega_1 = D(x, t) \cap \partial\Omega_2$. Then $\omega_2(D(x, t)) \approx \omega_1(D(x, t))$ where the constants depend on A and C_1.*

Proof. We will show $\omega_2(D(x, t)) \le C\omega_1(D(x, t))$. By the doubling property we can assume $D(x, 2t) \cap \partial\Omega_1 = D(x, 2t) \cap \partial\Omega_2$. Let $D = D(x, t)$, and let $D_j = D(x, 2^j t)$, $j = 0, 1, 2, \ldots$. We have

(2.4) $$\omega_2(D) - \omega_1(D) \le \int_{\Omega_2 \cap \partial\Omega_1} \omega(\Omega_2, D, \zeta) d\omega_1(\zeta).$$

To see (2.4), regard the function $\omega(\Omega_2, D, z)$ as extended to $\Omega_1 \cup \Omega_2$ by setting it to zero on $\Omega_1 \setminus \Omega_2$. Then $\omega(\Omega_2, D, z) - \omega(\Omega_1, D, z)$ is subharmonic on Ω_1 and zero on $\partial\Omega_1 \setminus \Omega_2$ and (2.4) follows.

Let k_ζ be the kernel function $\frac{d\omega(\Omega_2, \cdot, \zeta)}{d\omega_2}$. If $\zeta \in (D_{j+1} \setminus D_j) \cap \Omega_2 \cap \partial\Omega_1$, then, estimating k_ζ by [9], Lemma 4.14, we have

$$\omega(\Omega_2, D, \zeta) \le \omega_2(D) \max_{Y \in D \cap \partial\Omega_2} k_\zeta(Y)$$

$$\le C \frac{\omega_2(D)}{\omega_2(D_{j+1})} \left(\frac{\mathrm{dist}(\zeta, \partial\Omega_2)}{2^j t} \right)^\alpha, \quad \alpha > 0$$

$$\le C 2^{-j\alpha} \frac{\omega_2(D)}{\omega_2(D_{j+1})}$$

by the assumption about the Hausdorff distance. So,

$$\omega_2(D) - \omega_1(D) \le C \sum_j \frac{\omega_1(D_{j+1} \cap \Omega_2)}{\omega_2(D_{j+1})} 2^{-j\alpha} \omega_2(D).$$

We make two estimates on $\omega_1(D_{j+1} \cap \Omega_2)$. By [9], Lemma 4.2 we know $\omega(\Omega_2, D_{j+2}, \zeta) \ge C^{-1}$ if $\zeta \in D_{j+1} \cap \partial\Omega_1 \cap \Omega_2$. It follows by the maximum principle and doubling property that

$$\omega_1(D_{j+1} \cap \Omega_2) \le C\omega_2(D_{j+2}) \le C\omega_2(D_{j+1}).$$

On the other hand $\omega_1(D_{j+1}) \leq C^{j+1}\omega_2(D_{j+1}) \frac{\omega_1(D)}{\omega_2(D)}$ by the doubling property. So for any $j_0 < \infty$,

$$\omega_2(D) - \omega_1(D) \leq C \sum_{j > j_0} 2^{-j\alpha}\omega_2(D) + C \sum_{j \leq j_0} C^j 2^{-j\alpha}\omega_1(D).$$

The lemma follows by taking j_0 large enough that $C \sum_{j > j_0} 2^{-j\alpha} < 1$. ∎

Suppose next that Ω is also Lipschitz on scale t. Then ([7]; see also [10]) harmonic measure (with basepoint p) is absolutely continuous to surface measure and the density $\frac{d\omega}{d\sigma}$ is the normal derivative of Green's function. Moreover $|\nabla g|$ satisfies an A_2 condition with respect to ω on discs of radius $\leq Mt$ for any $M < \infty$, i.e.,

$$(2.5) \qquad \frac{1}{r^{n-1}} \int_{D(x,r)} |\nabla g|^2 d\sigma \leq C \left(\frac{\omega(D(x,r))}{r^{n-1}} \right)^2$$

if $r \leq Mt$, where C depends on the NTA constant A, the Lipschitz constant and M. Actually only the case $t = \text{diam}\,\Omega$ is stated in [7], [10] but the localized version follows using Lemma 4.11 of [9].

Lemma 2.6. *Let Ω be NTA and Lipschitz on scale t. Let $x \in \partial\Omega$, $r \leq Mt$, and let $\lambda > 0$ satisfy*

$$C_1^{-1} \frac{\omega(D(x,r))}{r^{n-1}} \leq \lambda \leq C_1 \frac{\omega(D(x,r))}{r^{n-1}}.$$

Then

$$\int_{D(x,r) \cap \partial\Omega} |\log(\lambda^{-1}|\nabla g|)| \frac{|\nabla g|}{\omega(D(x,r))} d\sigma \leq C.$$

Proof. This is a formal consequence of (2.5). One can quote the fact that the logarithm of an A_2 weight is BMO, or prove it using Jensen's inequality and boundedness of the function $x \log_- x$. ∎

The next lemma is one of the main steps in the proof. We are assuming $I(\phi) < 0$, i.e., adding a blip to a half-space decreases the value of $\int |\nabla g_\infty| \log |\nabla g_\infty|$, and we want to conclude that adding a blip to an NTA-domain decreases the value of $\int |\nabla g| \log |\nabla g|$. This will follow in a standard, if slightly complicated, way using the rate theorem to compare with the half-space case, provided N is large enough that the blip can be regarded as far from the nonflat part of the boundary.

Lemma 2.7. *Suppose Ω is NTA and Lipschitz on scale t. Let $Q \subset \partial\Omega$ be a square with $\ell(Q) \leq Mt$, and with $P_Q \cap \Omega = \emptyset$, $\tilde{P}_Q \subset \Omega$. Let $\tilde{\Omega}$ be the result of adding* *a blip along Q, and let g and $\tilde{g}$ be the Green's functions with pole at p, ω and $\tilde{\omega}$ the harmonic measures at p. Then*

$$\int_{\partial\tilde{\Omega}} |\nabla\tilde{g}| \log |\nabla\tilde{g}| d\sigma \leq \int_{\partial\Omega} |\nabla g| \log |\nabla g| d\sigma - \eta N^{-2}\omega(Q)$$

provided $N > N_0$ is large enough. N_0 and η depend on the Lipschitz and NTA constants and on M, and of course on ϕ.

Proof. We will use [9] as a reference for estimates although often only the smooth or Lipschitz case is needed. We scale so that $Q = Q(1)$ and $\tilde{P}_Q \subset \mathbb{R}^3_+$. Let $q = \frac{1}{2} be_3$ and let $\tilde{g}_\infty$ be the Green's function of $\{x : x_3 > N^{-1}\phi(N\bar{x})\}$ with pole at ∞.

It is easy to see that $\tilde{\omega}(D(0, t)) \approx \omega(D(0, t))$ for all $t > N^{-1}$. (If we want, we can use Lemma 2.5 and the doubling property for this.) Likewise $\int_{D(0,t)} |\nabla\tilde{g}_\infty| d\sigma \approx t^2$ for all $t > N^{-1}$.

Also $g(q) \approx \omega(Q(1))$ by [9], Lemma 4.8, so by [9], Lemma 4.4, $g \leq \omega(Q(1))$ on $\tilde{P}_{Q(1)}$. By the reflection principle and Hopf lemma,

$$|\nabla g| \approx \omega(Q(1)) \qquad \text{on} \quad Q\left(\frac{1}{2}\right)$$

(2.6) $\qquad |\nabla g(x) - \nabla g(y)| \leq C\omega(Q(1))|x - y| \qquad \text{if} \quad x, y \in Q\left(\frac{1}{2}\right).$

Similarly $g_\infty(q) \approx 1$. Let $m = (C_1/N)e_3$ where C_1 is a constant which is $> 2\|\phi\|_{\text{Lip}(1)}$, so that $\text{dist}(m, \partial\tilde{\Omega}) \approx \frac{1}{N}$. Then $\tilde{g}(m) - g(m) \leq CN\tilde{\omega}((D(0, \frac{1}{N}))$ by [9], Lemma 4.8. So by [9] Lemma 4.4, we have $\tilde{g} - g \leq C\tilde{\omega}(D(0, \frac{1}{N}))g_m$ on $\Omega\setminus D(0, \frac{2}{N})$ where g_m is the Green's function of Ω with pole at m. So $|\nabla(\tilde{g} - g)| \leq C\tilde{\omega}(D(0, \frac{1}{N}))|\nabla g_m|$ on $\partial\Omega/D(0, \frac{2}{N})$ and by [9] Lemma 4.14, it follows that (for some $\beta > 0$)

$$|\nabla(\tilde{g} - g)| \leq C\tilde{\omega}\left(D\left(0, \frac{1}{N}\right)\right)(N|z|)^{-\beta}\omega(D(0, |z|))^{-1}|\nabla g|$$

on $\Omega\setminus D(0, \frac{2}{N})$. Using the preceding bounds,

(2.7) $\qquad |\nabla(\tilde{g} - g)(z)| \leq CN^{-2}\omega(Q(1))\omega(D(0, |z|))^{-1}(N|z|)^{-\beta}|\nabla g(z)|.$

If in addition $z \in Q(\frac{1}{2})$, this implies by the formula preceding (2.6) that

(2.8) $\qquad |\nabla(\tilde{g} - g)(z)| \leq C(N|z|)^{-(2+\beta)}\omega(Q(1)).$

In particular, $|\nabla\tilde{g}(z)| \approx \omega(Q(1))$ when $z \in Q(\frac{1}{2})$ and $N|z|$ is large enough. Likewise, comparing $\tilde{g}_\infty$ with x_3, we get

$$|\nabla\tilde{g}_\infty(z) - e_3| \leq C(N|z|)^{-(2+\beta)},$$

if $z \in \mathbb{R}^2$ and $N|z|$ is large enough (actually, we could take $\beta = 1$ here). By the rate theorem,

$$\frac{|\nabla \tilde{g}(z)|}{|\nabla \tilde{g}_\infty(z)|} \approx \frac{\tilde{\omega}(Q(1))}{\tilde{\omega}_\infty(Q(1))} \approx \omega(Q(1)).$$

Moreover, by (2.3) and L'Hôpital's rule,

$$(2.9) \qquad \left| \frac{|\nabla \tilde{g}(y)|}{|\nabla \tilde{g}_\infty(y)|} - \frac{|\nabla \tilde{g}(z)|}{|\nabla \tilde{g}_\infty(z)|} \right| \le C|y - z|^\beta \omega(Q(1))$$

if $y, z \in$ graph ϕ with $\bar{y}, \bar{z} \in Q(\frac{1}{2})$. Now let $\lambda = |\nabla g(0)|$ and fix $\alpha < 1$ with $\alpha(2 + \beta) > 2$ and (for notational convenience) $\alpha > (1 - \alpha)(2 + \beta)$.

Claim.

$$\int_{\partial\tilde{\Omega} \cap D(0, N^{-\alpha})} \lambda^{-1} |\nabla \tilde{g}| \log(\lambda^{-1} |\nabla \tilde{g}|) d\sigma - \int_{\partial\Omega \cap D(0, N^{-\alpha})} \lambda^{-1} |\nabla g| \log(\lambda^{-1} |\nabla g|) d\sigma$$

$$\le -\eta N^{-2}$$

for N large enough.

To prove this, fix $z \in \mathbb{R}^2$ with $|z| = N^{-\alpha}$. Let $\tilde{\lambda} = \frac{|\nabla \tilde{g}(z)|}{|\nabla \tilde{g}_\infty(z)|}$. Then $\tilde{\lambda} \approx \omega(Q(1))$ and

$$|\tilde{\lambda} - \lambda| \le \left| |\nabla g(0)| - |\nabla g(z)| \right| + \left| |\nabla g(z)| - |\nabla \tilde{g}(z)| \right|$$

$$+ \frac{|\nabla \tilde{g}(z)|}{|\nabla \tilde{g}_\infty(z)|} \left| |\nabla \tilde{g}_\infty(z)| - 1 \right|$$

$$\le C(N^{-\alpha} + N^{-(1-\alpha)(2+\beta)} + N^{-(1-\alpha)(2+\beta)})\omega(Q(1))$$

$$(2.10) \qquad |\tilde{\lambda} - \lambda| \le C N^{-(1-\alpha)(2+\beta)} \omega(Q(1))$$

where we used (2.6), (2.8). Also if $|x| < N^{-\alpha}$, $x \in \partial\tilde{\Omega}$,

$$\left| \tilde{\lambda}^{-1} |\nabla \tilde{g}(x)| - |\nabla \tilde{g}_\infty(x)| \right| = \tilde{\lambda}^{-1} \left| \frac{|\nabla \tilde{g}(x)|}{|\nabla \tilde{g}_\infty(x)|} - \frac{|\nabla \tilde{g}(z)|}{|\nabla \tilde{g}_\infty(z)|} \right| |\nabla \tilde{g}_\infty(x)|$$

$$(2.11) \qquad\qquad\qquad\qquad \le N^{-\alpha\beta} |\nabla \tilde{g}_\infty(x)|$$

using (2.9) and $\tilde{\lambda} \approx \omega(Q(1))$. The expression in the claim is I + II + III + IV where

$$
\mathrm{I} = \int_{\partial\tilde{\Omega}\cap D(0,N^{-\alpha})} \left[\lambda^{-1}|\nabla\tilde{g}|\log(\lambda^{-1}|\nabla\tilde{g}|) - \tilde{\lambda}^{-1}|\nabla\tilde{g}|\log(\tilde{\lambda}^{-1}|\nabla\tilde{g}|)\right]d\sigma
$$

$$
\mathrm{II} = \int_{\partial\tilde{\Omega}\cap D(0,N^{-\alpha})} \left[\tilde{\lambda}^{-1}|\nabla\tilde{g}|\log(\tilde{\lambda}^{-1}|\nabla\tilde{g}|) - |\nabla\tilde{g}_\infty|\log|\nabla\tilde{g}_\infty|\right]d\sigma
$$

$$
\mathrm{III} = \int_{\partial\tilde{\Omega}\cap D(0,N^{-\alpha})} |\nabla\tilde{g}_\infty|\log|\nabla\tilde{g}_\infty|d\sigma
$$

$$
\mathrm{IV} = -\int_{\partial\Omega\cap D(0,N^{-\alpha})} \lambda^{-1}|\nabla g|\log(\lambda^{-1}|\nabla g|)d\sigma.
$$

If N is large enough, then expression III is $\leq -\eta N^{-2}$ by a change of scale and the fact that $\mathrm{I}(\phi) < 0$. We will show the others are error terms.

$$
|\mathrm{I}| \leq \lambda^{-1}|\log(\lambda\tilde{\lambda}^{-1})|\int_{\partial\tilde{\Omega}\cap D(0,N^{-\alpha})} |\nabla\tilde{g}|d\sigma + |\tilde{\lambda}^{-1} - \lambda^{-1}|
$$

$$
\int_{\partial\tilde{\Omega}\cap D(0,N^{-\alpha})} |\nabla\tilde{g}||\log(\tilde{\lambda}^{-1}|\nabla\tilde{g}|)|\,d\sigma.
$$

The first integral is $\leq CN^{-2\alpha}\omega(Q(1))$. Lemma 2.6 applies to the second integral showing that it is $\leq C\tilde{\omega}(D(0, N^{-\alpha})) \leq CN^{-2\alpha}\omega(Q(1))$. Since (2.10) implies the constants in front of the integrals are $\leq CN^{-(1-\alpha)(2+\beta)}\omega(Q(1))^{-1}$, we get

$$
|\mathrm{I}| \leq CN^{-(1-\alpha)(2+\beta)-2\alpha} \leq CN^{-2-\beta(1-\alpha)}.
$$

Expression II is

$$
\leq \int_{\partial\tilde{\Omega}\cap D(0,N^{-\alpha})} \tilde{\lambda}^{-1}|\nabla\tilde{g}|\left|\log\left(\frac{\tilde{\lambda}^{-1}|\nabla\tilde{g}|}{|\nabla\tilde{g}_\infty|}\right)\right|d\sigma
$$

$$
+ \int_{\partial\tilde{\Omega}\cap D(0,N^{-\alpha})} \left|\tilde{\lambda}^{-1}|\nabla\tilde{g}| - |\nabla\tilde{g}_\infty|\right||\log|\nabla\tilde{g}_\infty||\,d\sigma
$$

$$
\leq C\tilde{\lambda}^{-1}N^{-\alpha\beta}\int_{\partial\tilde{\Omega}\cap D(0,N^{-\alpha})} |\nabla\tilde{g}|d\sigma + N^{-\alpha\beta}\int_{\partial\tilde{\Omega}\cap D(0,N^{-\alpha})} |\nabla\tilde{g}_\infty||\log|\nabla\tilde{g}_\infty||\,d\sigma.
$$

Lemma 2.6 also holds for the Green's function at ∞ of a Lipschitz domain, so the second integral is $\leq CN^{-2\alpha}$. Thus

$$
|\mathrm{II}| \leq CN^{-\alpha(2+\beta)}.
$$

Finally, $|\mathrm{IV}| \leq CN^{-3\alpha}$ by (2.6); the claim follows.

Next we estimate

(2.12) $\left| \int_{\partial\Omega \setminus D(0,N^{-\alpha})} [\lambda^{-1}|\nabla\tilde{g}|\log(\lambda^{-1}|\nabla\tilde{g}|) - \lambda^{-1}|\nabla g|\log(\lambda^{-1}|\nabla g|)]d\sigma \right|$

$$\leq \int_{\partial\Omega \setminus D(0,N^{-\alpha})} \lambda^{-1}|\nabla\tilde{g}|\left|\log\left(\frac{|\nabla\tilde{g}|}{|\nabla g|}\right)\right|d\sigma$$

$$+ \int_{\partial\Omega \setminus D(0,N^{-\alpha})} \lambda^{-1}|\nabla(\tilde{g}-g)| \, |\log(\lambda^{-1}|\nabla g|)|d\sigma.$$

We have $|\nabla\tilde{g}| \approx |\nabla g|$ by (2.7). So the first term is

$$\leq \lambda^{-1} \int_{\partial\Omega \setminus D(0,N^{-\alpha})} |\nabla(\tilde{g}-g)|d\sigma.$$

Let $D_j = D(0, 2^j N^{-\alpha})$. Using (2.7),

$$\int_{\partial\Omega \cap (D_{j+1}\setminus D_j)} |\nabla(\tilde{g}-g)|d\sigma \leq CN^{-2}\omega(Q(1))(2^j N^{1-\alpha})^{-\beta}.$$

Summing over j and using $\lambda \approx \omega(Q(1))$, the first term in (2.12) is $\leq$ $CN^{-2-\beta(1-\alpha)}$. Once again splitting in annuli and using (2.7), the second term is

$$\leq CN^{-2}\sum_{j=0}^{\infty}(2^j N^{1-\alpha})^{-\beta}\omega(D_j)^{-1}\int_{(D_{j+1}\setminus D_j)\cap\partial\Omega} |\nabla g|\log(\lambda^{-1}|\nabla g|)d\sigma.$$

Cover $D_{j+1}\setminus D_j$ by discs of radius $\approx N^{-\alpha}$ with each point belonging to a bounded number of them. Let D be one of these discs. Then

$$\int_D |\nabla g|\log(\lambda^{-1}|\nabla g|)d\sigma = \omega(D)\int_D \omega(D)^{-1}|\nabla g|\log(N^{-2\alpha}\omega(D)^{-1}|\nabla g|)d\sigma$$

$$+ \omega(D)\log(N^{2\alpha}\lambda^{-1}\omega(D)).$$

The first term is $\leq C\omega(D)$ by Lemma 2.6. In the second term,

$$|\log(N^{2\alpha}\lambda^{-1}\omega(D))| \leq \left|\log\frac{\omega(D)}{\omega(D(0, N^{-\alpha)})}\right| + C$$

$$\leq C(j+1)$$

where the second inequality follows from the doubling property. So the second term in (2.12) is

$$\leq CN^{-2}\sum_{j=0}^{\infty}(2^j N^{1-\alpha})^{-\beta}(j+1)\omega(D_j)^{-1}\sum_D \omega(D)$$

$$\leq CN^{-2}\sum_{j=0}^{\infty}(j+1)(2^j N^{1-\alpha})^{-\beta}$$

$$\leq CN^{-2-\beta(1-\alpha)}.$$

We conclude that

$$\int_{\partial\tilde{\Omega}} \lambda^{-1} |\nabla \tilde{g}| \log(\lambda^{-1} |\nabla \tilde{g}|) d\sigma \leq \int_{\partial\Omega} \lambda^{-1} |\nabla g| \log(\lambda^{-1} |\nabla g|) d\sigma - \eta N^{-2}.$$

The lemma follows since

$$\int_{\partial\tilde{\Omega}} |\nabla \tilde{g}| \log(\lambda^{-1}) d\sigma = \log(\lambda^{-1}) = \int_{\partial\Omega} |\nabla g| \log(\lambda^{-1}) d\sigma$$

and $\lambda \approx \omega(Q(1))$. ∎

At this point we fix N large enough that Lemma 2.7 is true with the NTA constant, etc., from Lemma 2.2, and allow constants to depend on N. In the construction, define $\Omega^k = \Omega_\Gamma$ where $\Gamma = \{Q \in \underline{G} : \ell(Q) \geq 8^{-k}\}$ and let g^k, ω^k be its Green's function and harmonic measure. All Green's functions and harmonic measures will be taken based at 0 unless specified otherwise.

Lemma 2.8. $\int |\nabla g^k| \log |\nabla g^k| d\sigma \leq -\rho k$ *if* k *is large;* $\rho > 0$ *is independent of* k.

Proof. It follows by induction that all squares $Q_1, \ldots, Q_r \in \underline{G}$ with $\ell(Q_j) = 8^{-k}$ are contained in $\partial\Omega^{k-1}$. To pass from Ω^{k-1} to Ω^k, we add a blip along each Q_i. Let Ω_j^k be the domain obtained by adding blips along $Q_1 \ldots Q_{j-1}$. Then Ω_j^k is Lipschitz on scale $C^{-1}8^{-k}$, so, by Lemma 2.7,

$$\int_{\partial\Omega_{j+1}^k} |\nabla \tilde{g}| \log |\nabla \tilde{g}| d\sigma \leq \int_{\partial\Omega_j^k} |\nabla g| \log |\nabla g| d\sigma - \rho_1 \omega(\Omega_j^k, Q_j, 0)$$

where g and $\tilde{g}$ are the relevant Green's functions. By Lemma 2.5, $\omega(\Omega_j^k, Q_j, 0) \geq C^{-1}\omega^{k-1}(Q_j)$, so, iterating on j,

$$\int_{\partial\Omega^k} |\nabla g^k| \log |\nabla g^k| d\sigma \leq \int_{\partial\Omega^{k-1}} |\nabla g^{k-1}| \log |\nabla g^{k-1}| d\sigma - \rho_2 \sum_{j=1}^{r} \omega^{k-1}(Q_j).$$

We now show that

$$(2.13) \qquad \partial\Omega^{k-1} \subset \bigcup_{j=1}^{r} D(a_{Q_j}, C8^{-k}).$$

Let $x \in \partial\Omega^{k-1}$. Then $x \in Q_x$ for some square $Q_x \subset \partial\Omega^{k-1}$, and clearly $\ell(Q_x) \leq 8^{-k}$. Let Q be a smallest square with $\ell(Q) \geq 8^{-k}$ and $Q_x \prec Q$. By Lemma 2.1, there is $Q_j \preceq Q$ with $\ell(Q_j) = 8^{-k}$ and $Q_x \subset D(a_{Q_j}, C8^{-k})$.

The doubling property implies

$$\int_{\partial\Omega^k} |\nabla g^k| \log |\nabla g^k| d\sigma \le \int_{\partial\Omega^{k-1}} |\nabla g^{k-1}| \log |\nabla g^{k-1}| d\sigma - \rho_3,$$

and the lemma follows by iterating on k. ∎

For each $Q \in \underline{G}$, let $\Gamma_Q = \partial\Omega \cap (P_Q \cup \tilde{P}_Q)$. Then $Q' \prec Q$ implies $\Gamma_{Q'} \subset \Gamma_Q$ and the Γ_Q are otherwise disjoint.

Lemma 2.9. *If $\eta > 0$ is small enough, then for each sufficiently large k there are squares $Q_1 \ldots Q_s$ with $\ell(Q_j) = 8^{-k}$, such that $\sum_{j=1}^s \omega(\Gamma_{Q_j}) \ge \eta$ and $\log \frac{\omega(\Gamma_{Q_j})}{\ell(Q_j)^2} \le -\eta k$ for all j.*

Proof. Fix $\eta > 0$ sufficiently small. We first find such squares with $\omega^{k-1}(Q_j)$ replacing $\omega(\Gamma_{Q_j})$. Let $Q_1 \ldots Q_r$ be all squares of length 8^{-k}, with $Q_1 \ldots Q_s$ being those for which $\log \frac{\omega^{k-1}(Q_j)}{\ell(Q_j)^2} \le -\eta k$. Let $D_j = D(a_{Q_j}, C8^{-k})$. By (2.13), there are pairwise disjoint regions R_j with $Q_j \subset R_j \subset D_j$ such that $\partial\Omega^{k-1} = \bigcup_j R_j$. Take $\lambda = \frac{\omega_{k-1}(D_j)}{\ell(Q_j)^2}$ and apply Lemma 2.6 on D_j, restricting the integral to R_j. This gives

$$\sum_{j=1}^r \omega^{k-1}(R_j) \log \frac{\omega^{k-1}(D_j)}{\ell(Q_j)^2} \le C \sum_j \omega_{k-1}(D_j)$$

$$+ \sum_j \int_{R_j} |\nabla g^{k-1}| \log |\nabla g^{k-1}| d\sigma.$$

By Lemma 2.8 and the doubling property

$$\sum_{j=1}^r \omega^{k-1}(R_j) \log \frac{\omega^{k-1}(Q_j)}{\ell(Q_j)^2} \le -\rho k - C \le -\rho_2 k.$$

If $\sum_{j=1}^s \omega^{k-1}(Q_j) \le \eta$, then using $\omega^{k-1}(R_j) \le C\omega^{k-1}(Q_j)$ and $\omega^{k-1}(Q_j) \ge C^{-k}$, we get

$$-C\eta k(\log C - \log 64) - \eta k \le -\rho_2 k,$$

which is a contradiction if η is small enough.

Now, for each $j \in \{1, \ldots, s\}$ let $\Omega_{Q_j} = \Omega_\Gamma$ where $\Gamma = \{Q \in \underline{G} : Q \not\preceq Q_j\}$. Since Ω_{Q_j} is obtained from Ω^{k-1} by adding blips along squares of side $\le 8^{-k}$, the Hausdorff distance from $\partial\Omega_{Q_j}$ to $\partial\Omega^{k-1}$ is $\le C8^{-k}$ and Lemma 2.5 shows $\omega(\Omega_{Q_j}, Q_j, 0) \approx \omega^{k-1}(Q_j)$. A symmetry argument (or [9], Lemma 4.2, and the doubling property) and the maximum principle imply $\omega(\Gamma_{Q_j}) \approx \omega(\Omega_{Q_j}, Q_j, 0)$. So $\omega(\Gamma_{Q_j}) \approx \omega^{k-1}(Q_j)$. The lemma follows. ∎

We now follow [6] and derive the main lemma from Lemma 2.9 using elementary results in ergodic theory. Fix a square $\Sigma \in \underline{G}_0$. The shift T on infinitely many symbols acts on Γ_Σ. Namely, for each $Q \in \underline{G}_1$ with $Q \prec \Sigma$, the restriction of T to Γ_Q is the conformal affine map which takes Q to Σ and preserves distinguished edges. This defines T except on a set with σ-finite length and therefore a.e. $(d\omega)$. Letting ξ be the partition $\{\Gamma_Q : Q \in \underline{G}_1\}$, the nth partition $\bigvee_{j=0}^{n-1} T^{-j}\xi$ is $\{\Gamma_Q : Q \in \underline{G}_n\}$. For $x \in \Gamma_\Sigma$, let $Q_n(x)$ satisfy $x \in \Gamma_{Q_n(x)}$ and $Q_n(x) \in \underline{G}_n$. Define $\sigma_n(x) = \ell(Q_n(x))^2$ and $h_n(x) = \omega(\Gamma_{Q_n(x)})$.

Lemma 2.10. *ω is mutually absolutely continuous to a certain invariant measure μ for T. T is ergodic with respect to μ and ξ is a generating partition. The limits $\lim_{n\to\infty} \frac{1}{n} \log \frac{1}{\sigma_n(x)}$ and $\lim_{n\to\infty} \frac{1}{n} \log \frac{1}{h_n(x)}$ exist a.e. / $(d\mu)$ and are constant.*

Proof. Let $\mu^{(n)}(Y) = \frac{1}{n} \sum_{j=0}^{n-1} \omega(T^{-j}Y)$. Let μ be a weak* limit point of $\{\mu^{(n)}\}$. Then μ is an invariant measure [22]. Let $E_\epsilon = \{x \in \Gamma_\Sigma : \text{dist}(x, B) < \epsilon\}$ where B is the boundary of the square Σ. Then

(2.14)   If $Y \cap E_\epsilon = \emptyset$, then $C_\epsilon^{-1}\omega(Y) \le \mu(Y) \le C_\epsilon\omega(Y)$.

(2.15) $\mu^{(n)}(E_\epsilon) \le C\epsilon^\beta$ with $\beta > 0$, independently of n.

For let Y be as in (2.14). Fix $Q \in \underline{G}_n$. Let S be the affine map with $S \upharpoonright \Gamma_Q = T^n$. By Lemma 2.4 with $\Omega_1 = \Omega$, $p_1 = 0$, $\Omega_2 = S\Omega$, $p_2 = S(0)$, $E = \Gamma_\Sigma \backslash E_\epsilon$, we have

(2.16) $C_\epsilon^{-1}\omega(Y)\omega(\Gamma_Q) \le \omega(T^{-n}Y \cap \Gamma_Q) \le C_\epsilon\omega(Y)\omega(\Gamma_Q)$.

So $C_\epsilon^{-1}\omega(Y) \le \omega(T^{-n}Y) \le C_\epsilon\omega(Y)$ and (2.14) follows. To do (2.15) fix again $Q \in \underline{G}_n$. Then $T^{-n}E_\epsilon \cap \Gamma_Q$ can be covered with $C\epsilon^{-1}$ discs of radius $\epsilon\ell(Q)$. By Lemma 2.3 and the doubling property,

$$\omega(T^{-n}E_\epsilon \cap \Gamma_Q) \le C\epsilon^\beta\omega(\Gamma_Q),$$

with $\beta > 0$. (2.15) follows.

By (2.14) μ and ω are equivalent on each $\Gamma_\Sigma \backslash E_\epsilon$ and since (2.15) shows $\lim_{\epsilon\to 0} \mu(E_\epsilon) = \lim_{\epsilon\to 0} \omega(E_\epsilon) = 0$, they are equivalent. Given (2.16), the fact that T is ergodic and ξ is a generating partition follows by a standard argument as in [6]. Next, define T' by $T'(x) = \ell(Q)^{-1}$ if $x \in \Gamma_Q$ and $Q \in \underline{G}_1$.

Claim. $\|(\log T') \circ T^n\|_{L^1(d\omega)} \le C$ independently of n.

Fix $Q \in \underline{G}_n$. The set $\{x \in \Gamma_Q : (\log T') \circ T^n(x) > \lambda\}$ is identical with $\cup\{\Gamma_{Q'} : Q'$ directly descended from Q and $\ell(Q') < e^{-\lambda}\ell(Q)\}$. By Lemma 2.1, the latter set can be covered by Ce^λ discs of radius $e^{-\lambda}$. By Lemma 2.3,

$$\omega(\{x \in \Gamma_Q : (\log T') \circ T^n(x) > \lambda\}) \leq Ce^{-\beta\lambda}\omega(\Gamma_Q),$$

with $\beta > 0$. The claim follows. From the claim we conclude that $\log T' \in L^1(d\mu)$. Since $\frac{1}{n} \log \frac{1}{\sigma_n} = \frac{2}{n} \sum_{j=0}^{n-1} \log T' \circ T^j$, the ergodic theorem shows $\lim_{n\to\infty} \frac{1}{n} \log \frac{1}{\sigma_n}$ exists a.e. $d\mu$ and is constant. Now fix $Q_1 \in \underline{G}_1$. By the doubling property, there is a constant r such that

$$\omega(T^{-n}\Gamma_{Q_1} \cap \Gamma_Q) \geq C\ell(Q_1)^r\omega(\Gamma_Q),$$

for all $Q \in \underline{G}_n$. It follows that $\mu(\Gamma_{Q_1}) \geq C\ell(Q_1)^r$ and, since $\log T' \in L^1(d\mu)$, that

$$\sum_{Q_1 \in \underline{G}_1} \mu(\Gamma_{Q_1}) \log \frac{1}{\mu(\Gamma_{Q_1})} < \infty.$$

Since ξ is a generating partition, this implies that T has finite entropy with respect to μ. Let $\overline{h}_n(x) = \mu(Q_n(x))$. Then $\lim_{n\to\infty} \frac{1}{n} \log \frac{1}{\overline{h}_n}$ exists a.e. $(d\mu)$ and is constant by the Shannon-McMillan-Breiman theorem (the version we need for infinite partitions with finite entropy is in [16]). By (2.14), $\lim_{n\to\infty} \frac{1}{n} \log \frac{1}{h_n}$ exists and is constant. ∎

Proof of Main Lemma Note that any squares $Q \in \underline{G}$ with $\ell(Q) = 8^{-k}$ must belong to $\underline{G}_n$ for some $n \leq k$. By Lemma 2.9, we can choose Σ so that

$$\omega\left(\left\{x \in \Gamma_\Sigma : \frac{1}{k} \inf_{n \leq k} \log \frac{h_n(x)}{\sigma_n(x)} \leq -\eta\right\}\right) \geq \frac{\eta}{6}$$

for a sequence of k tending to ∞. It follows that $\lim_{n\to\infty} \frac{1}{n} \log \frac{h_n}{\sigma_n} < 0$ and therefore that

$$(2.17) \qquad \lim_{n\to\infty} \frac{\log h_n}{\log \sigma_n} = 1 + \frac{\lim_{n\to\infty} \frac{1}{n} \log \frac{h_n}{\sigma_n}}{\lim_{n\to\infty} \frac{1}{n} \log \sigma_n} = \beta$$

exists a.e. and is > 1. Suppose x is a point where (2.17) and $\lim_{n\to\infty} \frac{1}{n} \log \sigma_n(x)$ exist. For $r > 0$, let n be the largest number such that $\ell(Q_n(x)) > r$. By the doubling property

$$\frac{\log \omega(D(x,r))}{\log r} \geq 2 \frac{\log h_n(x)}{\log \sigma_{n+1}(x)} + 0\left(\left(\log \frac{1}{\sigma_{n+1}(x)}\right)^{-1}\right)$$

$$\geq 2 \frac{\log h_n(x)}{\log \sigma_n(x)} + o(1)$$

as $r \to 0$, since $\lim_{n\to\infty} \frac{\log \sigma_{n+1}(x)}{\log \sigma_n(x)} = 1$. So

$$\liminf_{r\to 0} \frac{\log \omega(D(x,r))}{\log r} \geq 2\beta > 2.$$

Likewise,

$$\limsup_{r \to 0} \frac{\log \omega(D(x, r))}{\log r} \leq 2\beta$$

and the proof is complete.   ∎

We still have to construct suitable variations ϕ. First we get rid of the piecewise linearity requirement.

Lemma 2.11. *Suppose* $\{\phi_n\}$ *is a sequence of Lipschitz functions on* $\mathbb{R}^2$ *with common compact support and* $\phi_n \to \phi$ *in Lipschitz norm. Then* $I(\phi_n) \to I(\phi)$.

Proof. Let g_n, g be Green's function at ∞ for $\Omega_n = \{x : x_3 > \phi_n(\overline{x})\}$ and $\Lambda = \{x : x_3 > \phi(\overline{x})\}$. We can assume $\operatorname{supp} \phi_n \subset Q(1)$.

There is no difficulty at ∞ since $|\nabla g_n(x)| = 1 + 0(|x|^{-3})$ uniformly in n as $x \in \mathbb{R}^2 \to \infty$ and similarly with $|\nabla g|$. It follows that it is enough to prove

$$(2.18) \quad \int_{Q(2)} |\nabla g(\overline{x}, \phi(\overline{x})) - \nabla g_n(\overline{x}, \phi_n(\overline{x}))|^2 dx_1 dx_2 \to 0 \quad \text{as} \quad n \to \infty.$$

In proving (2.18) we may assume $\phi_n \geq \phi$. To see that, let $\epsilon_n = \|\phi_n - \phi\|_\infty$. Let ψ_n satisfy $\psi_n = \phi_n$ on $Q(3)$ and $-\epsilon_n \leq \psi_n - \phi_n \leq 0$ outside $Q(3)$ with $\psi_n - \phi_n = -\epsilon_n$ outside $Q(4)$. Let $\tilde{g}_n$ be Green's function at ∞ of $x_3 > \psi_n(\overline{x})$, i.e., $\tilde{g}_n = 0$ when $x_3 = \psi_n(\overline{x})$ and $\tilde{g}_n - (x_3 + \epsilon_n) \to 0$ as $x \to \infty$. The rate theorem applied to $\tilde{g}_n - g_n$ and g_n implies that $|\nabla g_n - \nabla \tilde{g}_n| \leq \delta(\epsilon_n)|\nabla g_n|$ on $\{x \in \partial \Omega_n : \overline{x} \in Q(2)\}$, where $\delta(\epsilon)$ depends only on ϵ and a bound for $\|\phi_n\|_{\operatorname{Lip}(1)}$, and $\delta(\epsilon) \to 0$ as $\epsilon \to 0$. So

$$(2.19) \quad \int_{Q(2)} |\nabla g_n(\overline{x}, \phi_n(\overline{x})) - \nabla \tilde{g}_n(\overline{x}, \phi_n(\overline{x}))|^2 dx_1 dx_2 \to 0,$$

as $n \to \infty$. We can replace $Q(2)$ by $Q(8)$ in (2.19) by standard theorems on continuous dependence of solutions of the Dirichlet problem, since both ϕ_n and ψ_n are smooth outside $Q(1)$.

If we now replace $Q(1)$ by $Q(4)$ and ϕ_n by $\psi_n + \epsilon_n$ we have a reduction of (2.18) to the case $\phi_n \geq \phi$. This case will follow from the regularity for the Dirichlet problem on Lipschitz domains due to Verchota [29]. That theorem implies

$$\int_{Q(2)} |\nabla g(x, \phi(x)) - \nabla g(x, \phi_n(x))|^2 dx_1 dx_2 \to 0$$

as $n \to \infty$. In particular, by the Lip(1) convergence,

$$\int_{Q(2)} |\nabla_\tau g(x, \phi_n(x))|^2 dx_1 dx_2 \to 0,$$

where ∇_r is differentiation along $\partial\Omega_n$. Using regularity for the Dirichlet problem again, this time on the domain Ω_n, we obtain

$$\int_{Q(2)} |\nabla g_n(x, \phi_n(x)) - \nabla g(x, \phi_n(x))|^2 dx_1 dx_2$$

$$\leq C \int_{\partial\Omega_n} |\nabla_r (g_n - g)|^2 dx_1 dx_2$$

$$\to 0 \quad \text{as} \quad n \to \infty$$

and the lemma follows.   ∎

Now fix $\phi \in C_0^\infty(\mathbb{R}^2)$. Let $\Omega_t = \{x \in \mathbb{R}^3 : x_3 > t\phi(\overline{x})\}$. Let $g(t, x)$ $(t \in \mathbb{R}, x \in \overline{\Omega}_t)$ be the Green's function of Ω_t with pole at ∞. Then $g(t, x)$ is smooth in t and x jointly. Denote $I(t\phi)$ by $I(t)$. We claim that I vanishes precisely to third order at zero for suitable ϕ, as in the first case of Lemma 1.1.

Lemma 2.12.

$$\frac{dI}{dt}\Big|_{t=0} = \frac{d^2I}{dt^2}\Big|_{t=0} = 0 \quad and \quad \frac{d^3I}{dt^3} = \int_{\mathbb{R}^2}\left[\left(\frac{d\hat{\phi}}{dx_3}\right)^3 - 3|\overline{\nabla}\phi|^2\frac{d\hat{\phi}}{dx_3}\right]dx_1 dx_2$$

Here $\hat{\phi}$ is the harmonic extension into the upper half space and $\overline{\nabla}$ is the $\mathbb{R}^2$ gradient.

Proof. We also let $\overline{\triangle}$ be the $\mathbb{R}^2$ Laplacian. Let g', g'', g''' be the first three t-derivatives of g with x fixed. Then $g'(t, \cdot)$, $g''(t, \cdot)$, $g'''(t, \cdot)$ are harmonic on Ω_t, vanish at ∞, and vanish on $\partial\Omega_t$ outside the support of ϕ. Moreover, $g'(0, \cdot) = -\hat{\phi}$ ("Hadamard variational formula"); to prove it on $\mathbb{R}^2$, which suffices, differentiate the equation $g(t, \overline{x}, t\phi(\overline{x})) = 0$ with respect to t.

We do the calculations using coordinates $(t, u) \in \mathbb{R} \times \overline{\mathbb{R}}_+^3$ and the map $(t, u) \to (t, x)$; with $x \in \Omega_t$ defined by $x = (\overline{u}, u_3 + t\phi(\overline{u}))$. We write $g(t, u)$ for $g(t, \overline{u}, u_3 + t\phi(\overline{u}))$, $n(t, \overline{u}, 0)$ for the normal into Ω_t at $(\overline{u}, t\phi(\overline{u}))$, etc. Then

(2.20) $\dfrac{d}{dt}\Big|_{t=0} d\sigma = 0, \qquad \dfrac{d^2}{dt^2}\Big|_{t=0} d\sigma = |\overline{\nabla}\phi|^2 du_1 du_2$

(2.21) $\left\langle\dfrac{dn}{dt}, e_3\right\rangle\Big|_{t=0} = 0, \quad \left\langle\dfrac{d^2n}{dt^2}, e_3\right\rangle\Big|_{t=0} = -|\overline{\nabla}\phi|^2, \quad \left\langle\dfrac{d^3n}{dt^3}, e_3\right\rangle\Big|_{t=0} = 0$

$$\frac{dg}{dt} = g' + \phi(\overline{u})\frac{dg}{du_3}$$

$$\frac{d}{dt}\frac{dg}{du_3} = \frac{d}{du_3}\frac{dg}{dt} = \frac{d}{du_3}\left(g' + \phi(\overline{u})\frac{dg}{du_3}\right) = \frac{dg'}{du_3} + \phi(\overline{u})\frac{d^2g}{du_3^2},$$

$$\frac{d^2}{dt^2}\frac{dg}{du_3} = \frac{d}{du_3}\frac{d^2g}{dt^2} = \frac{d}{du_3}\frac{d}{dt}\left(g' + \phi(\overline{u})\frac{dg}{du_3}\right)$$

$$= \frac{d}{du_3}\left(g'' + \phi\frac{dg'}{du_3} + \phi\frac{d}{du_3}\frac{dg}{dt}\right) = \frac{dg''}{du_3} + 2\phi\frac{d^2g'}{du_3^2} + \phi^2\frac{d^3g}{du_3^3},$$

$$\frac{d^3}{dt^3}\frac{dg}{du_3} = \frac{dg'''}{du_3} + 3\phi\frac{d^2g''}{du_3^2} + 3\phi^2\frac{d^3g'}{du_3^3} + \phi^3\frac{d^4g}{du_3^4}.$$

When $t = 0$, g is linear and the last term in each of the last three expressions drops out. If $u_3 = 0$, then $|\nabla g| = n_3^{-1}\frac{dg}{du_3}$, so when $u_3 = t = 0$ (2.21) gives

$$\frac{d|\nabla g|}{dt} = \frac{dg'}{du_3},$$

(2.22) $\quad\quad \dfrac{d^2|\nabla g|}{dt^2} = \dfrac{dg''}{du_3} + 2\phi\dfrac{d^2g'}{du_3^2} + |\overline{\nabla}\phi|^2,$

$$\frac{d^3|\nabla g|}{dt^3} = \frac{dg'''}{du_3} + 3\phi\frac{d^2g''}{du_3^2} + 3\phi^2\frac{d^3g'}{du_3^3} + 3\frac{dg'}{du_3}|\overline{\nabla}\phi|^2.$$

Also, for t arbitrary but $u_3 = 0$ and $\overline{u}$ outside the support of ϕ we have

$$\frac{d|\nabla g|}{dt} = \frac{dg'}{du_3},$$

(2.23) $\quad\quad \dfrac{d^2|\nabla g|}{dt^2} = \dfrac{dg''}{du_3},$

$$\frac{d^3|\nabla g|}{dt^3} = \frac{dg'''}{du_3}.$$

In computing I_t, I_{tt}, I_{ttt} we can differentiate freely under the integral sign. For there is no problem locally since all functions are smooth, while if $\overline{u}$ is outside the support of ϕ, then the first three t-derivatives of $|\nabla g|\log|\nabla g|d\sigma$ are sums of terms each of which (by (2.23)) involves either $\log|\nabla g|$, $\frac{dg'}{du_3}$, $\frac{dg''}{du_3}$, or $\frac{dg'''}{du_3}$. Since g', g'', g''' and $g(x) - x_3$ are harmonic functions vanishing at ∞ and with boundary values of compact support, their u_3-derivatives die like $|u|^{-3}$ as $u \to \infty$ along $\mathbb{R}^2$, uniformly in any finite t-interval. For $g - x_3$ this was proved at the beginning of the section and the same proof works in general. So we have uniform integrability.

Keeping in mind that $|\nabla g| = 1$ when $t = 0$, and using (2.20), (2.22),

$$\frac{dI}{dt}\bigg|_{t=0} = \int_{\mathbb{R}^2}\frac{dg'}{du_3}\,du_1du_2$$

(2.24) $\dfrac{d^2I}{dt^2}\bigg|_{t=0} = \displaystyle\int_{\mathbb{R}^2}\left[\frac{dg''}{du_3} + 2\phi\frac{d^2g'}{du_3^2} + \left(\frac{dg'}{du_3}\right)^2 + |\overline{\nabla}\phi|^2\right]du_1du_2,$

(2.25) $\dfrac{d^3I}{dt^3}\bigg|_{t=0} = \displaystyle\int_{\mathbb{R}^2}\left[9|\overline{\nabla}\phi|^2\frac{dg'}{du_3} + 6\phi\frac{dg'}{du_3}\frac{d^2g}{du_3^2} + 3\phi^2\frac{d^3g'}{du_3^3} - \left(\frac{dg'}{du_3}\right)^3\right.$

$$+ 3\phi \frac{d^2 g''}{du_3^2} + 3\frac{dg'}{du_3}\frac{dg''}{du_3} + \frac{dg'''}{du_3}\Big]du_1 du_2.$$

Now we use that if $f : \mathbb{R}^3_+ \to \mathbb{R}$ is harmonic and vanishes at ∞ then $\int_{\mathbb{R}^2} \frac{df}{du_3} = 0$, $\int_{\mathbb{R}^2} \left(\frac{df}{du_3}\right)^2 = \int_{\mathbb{R}^2} |\overline{\nabla} f|^2$. Therefore, $I_t\big|_{t=0} = 0$. Also, the first term on the right side of (2.24) is zero, and, using $g' = -\hat\phi$,

$$\frac{d^2 I}{dt^2}\Big|_{t=0} = \int_{\mathbb{R}^2}\Big[-2\phi\frac{d^2\hat\phi}{du_3^2} + \Big(\frac{d\hat\phi}{du_3}\Big)^2 + |\overline{\nabla}\phi|^2\Big]du_1 du_2$$

$$= \int_{\mathbb{R}^2}\Big[2\phi\overline{\Delta}\phi + \Big(\frac{d\hat\phi}{du_3}\Big)^2 + |\overline{\nabla}\phi|^2\Big]du_1 du_2$$

$$= \int_{\mathbb{R}^2}\Big[\Big(\frac{d\hat\phi}{du_3}\Big)^2 - |\overline{\nabla}\phi|^2\Big]du_1 du_2$$

$$= 0.$$

The last term in (2.25) is zero. The preceding two terms are

$$3\int_{\mathbb{R}^2}\Big[\phi\frac{d^2 g''}{du_3^2} - \frac{d\hat\phi}{du_3}\frac{dg''}{du_3}\Big]du_1 du_2$$

which is zero by Green's identity for the harmonic functions $\hat\phi$ and $\frac{dg''}{du_3}$. The remaining terms in (2.25) are

$$\int_{\mathbb{R}^2}\Big[-9|\overline{\nabla}\phi|^2\frac{d\hat\phi}{du_3} + 6\phi\frac{d\hat\phi}{du_3}\frac{d^2\hat\phi}{du_3^2} - 3\phi^2\frac{d^3\hat\phi}{du_3^3} + \Big(\frac{d\hat\phi}{du_3}\Big)^3\Big]du_1 du_2$$

$$= \int_{\mathbb{R}^2}\Big[-9|\overline{\nabla}\phi|^2\frac{d\hat\phi}{du_3} - 6\phi\frac{d\hat\phi}{du_3}\overline{\Delta}\phi + 3\phi^2\overline{\Delta}\frac{d\hat\phi}{du_3} + \Big(\frac{d\hat\phi}{du_3}\Big)^3\Big]du_1 du_2$$

$$= \int_{\mathbb{R}^2}\Big[-3|\overline{\nabla}\phi|^2\frac{d\hat\phi}{du_3} - 3\frac{d\hat\phi}{du_3}\overline{\Delta}(\phi^2) + 3\phi^2\overline{\Delta}\frac{d\hat\phi}{du_3} + \Big(\frac{d\hat\phi}{du_3}\Big)^3\Big]du_1 du_2.$$

The middle terms vanish by Green's identity for $\overline{\Delta}$ since ϕ has compact support.∎

Remark. Similar calculations may be done with any initial domain in place of the upper half space and for most domains one does not have to compute so many derivatives since curvature terms enter into the first variation. However, this is useless for proving Theorem 3. The curvature effects scale out as $N \to \infty$.

The expression $\int\big[\big(\frac{d\hat\phi}{du_3}\big)^3 - 3|\overline{\nabla}\phi|^2\frac{d\hat\phi}{du_3}\big]$ also came up in the proof of Lemma 1.1 and was shown to be negative if $\phi = \phi_0 = -|x + e_3|^{-1}$. Approximating ϕ_0 in the $W^{1,3}$ Sobolev norm by a C_0^∞ function ψ, and then approximating $t\psi$ with

$t > 0$ small in Lipschitz norm by a piecewise linear function, we have a proof of Theorem 3.

Taking $t < 0$ instead of $t > 0$ we obtain a domain whose harmonic measure is less than two-dimensional. Domains homeomorphic to a ball with this property (in fact, with 1-dimensional harmonic measure) may be found much more easily by other means (a closely related result is in [32]), but it may be new that there is a scale invariant and/or NTA example. It is also amusing to note the following: the proof of the main lemma works just as well if Ω is replaced by the complement of $\overline{\Omega}$, except that $I(\phi)$ must be replaced by $\int |\nabla g| \log |\nabla g|$ where g is the Green's function at ∞ on the domain *below* the graph of ϕ. It follows that if Ω is the domain just used to prove Theorem 3, then harmonic measure for $\overline{\Omega}^c$ is < 2-dimensional. It is probably possible to make similar examples where the interior and exterior harmonic measures are both > 2-, or both < 2-dimensional. This would involve working with a function ϕ as in Lemma 2.12 for which $\frac{d^3I}{dt^3}\big|_{t=0} = 0$ and $\frac{d^4I}{dt^4}\big|_{t=0}$ has the appropriate sign. The calculations for the fourth derivative are rather complicated and I have not worked this out. Another remark is that some of the technicalities in the proof of the main lemma could be avoided by working in a special situation where it is possible to use a finite partition. However, one would then have to compute sgn $I(\phi)$ for special ϕ, which (unless it were done numerically) would appear to be a difficult problem.

Finally, we show how to generalize to $\mathbb{R}^n$, $n \geq 4$. An unbounded domain with dim $\omega \geq n - 1 + \epsilon$ may be obtained by crossing the domain in Theorem 3 with $\mathbb{R}^{n-3}$. To get a bounded domain, intersect domains $U_1 \ldots U_{n-2}$ of this type with different choices of the dummy variables. If U is the resulting domain and $E \subset \partial U$ with dim $E < n - 1 + \epsilon$, then

$$\omega(U, E) \leq \sum_i \omega(U, E \cap \partial U_i) \leq \sum_i \omega(U_i, E \cap \partial U_i) = 0.$$

APPENDIX: PROOF OF LEMMA 1.2

This argument uses a lot of high school math; the details of some of the calculations will be omitted. The last part is a proof-by-cases and a hand calculator is used to evaluate logarithms and the like. The reader who wants to check through it all will need a calculator and scratch pad.

While I was working out the proof, it often seemed to be a huge waste of time. It would be desirable to have a conceptual proof of Lemma 1.1.

Added September, 1991

See the remarks after the introduction, and reference [2].

A.1 The setup

In this section we reduce the problem to estimating an explicit integral. We will use cylindrical coordinates r, θ, z as well as rectangular coordinates x, y, z on $\overline{\mathbb{R}}^3_+$.

Lemma A.1.1.
(a) $\int_{-\pi}^{\pi} \langle e, q(r, \theta) \rangle \frac{d\theta}{2\pi} = 0$ for all r.
(b) Let $\psi(r) = \frac{1}{\pi} \int_{-\pi}^{\pi} \frac{\cos \phi}{(1+r^2 - 2r \cos \phi)^{1/2}} \, d\phi$.
Then for $r < 1$,

$$\int_{-\pi}^{\pi} \log |e + q(r, \theta)| \frac{d\theta}{2\pi}$$

$$= \begin{cases} \log\left(\frac{1}{2}\left(\sqrt{1 - \alpha^2} + \sqrt{1 - \alpha^2 - \alpha^2 \psi(r)^2}\right)\right), & \psi(r) \leq \dfrac{\sqrt{1 - \alpha^2}}{\alpha} \\[2ex] \log \dfrac{\alpha \psi(r)}{2}, & \psi(r) \geq \dfrac{\sqrt{1 - \alpha^2}}{\alpha}. \end{cases}$$

(c) Let $a = \alpha^2 \left(\frac{4}{r^2} + \psi(r)^2\right)$, $b = 2\alpha\sqrt{1 - \alpha^2}\psi(r)$, $c = 1 - \frac{2\alpha^2}{r^2} + \frac{\alpha^2}{r^4}$.
Then for $r > 1$,

$$\int_{-\pi}^{\pi} \log |e + q(r, \theta)| \frac{d\theta}{2\pi} \leq \log\left(\frac{1}{2}\left(\sqrt{c} + \sqrt[4]{(a + c)^2 - b^2}\right)\right).$$

Remarks.
(1) It is easy to check that $4ac - b^2$; therefore, $(a + c)^2 - b^2$, and also $\psi(r)$, are ≥ 0 for all r.
(2) It will be clear from the argument for (c) that we could also obtain a formula for the integral appearing there. This formula is too messy to be useful in the proof, but was used by J. Goodman to evaluate $J(\alpha)$ numerically.

Proof. Recall $e = (\alpha, 0, \sqrt{1 - \alpha^2})$, g is the harmonic function on $\mathbb{R}^3_+$ with boundary values

$$g(r, \theta) = \begin{cases} -r \cos \theta, & \text{if } r < 1 \\ -r^{-1} \cos \theta, & \text{if } r > 1 \end{cases}$$

and $q = \alpha \nabla g$. We write $\nabla_T = \left(\frac{d}{dx}, \frac{d}{dy}\right)$, $\Delta_T = \frac{d^2}{dx^2} + \frac{d^2}{dy^2}$. When $z = 0$ we have

(A.1)
$$\nabla_T g = \begin{cases} (-1, 0) & \text{if } r < 1 \\ \left(\frac{1}{r^2} \cos 2\theta, \frac{1}{r^2} \sin 2\theta\right) & \text{if } r > 1 \end{cases}$$

and $\Delta_T g = 0$ when $r \neq 1$. Thus as distributions on $\mathbb{R}^2 = \partial \mathbb{R}^3_+$,

$$\frac{d^2 g}{dz^2} = -\Delta_T g = -\left(\frac{dg}{dr_+} + \frac{dg}{dr_-}\right) d\theta \Big|_{r=1} = -2\cos\theta \, d\theta \Big|_{r=1}.$$

If $(x, y, z) \in \mathbb{R}^3_+$ then by the Poisson integral formula

$$\frac{d^2 g}{dz^2} = -\frac{1}{2\pi} \int_{-\pi}^{\pi} \frac{2z \cos\phi \, d\phi}{((x - \cos\phi)^2 + (y - \sin\phi)^2 + z^2)^{3/2}}.$$

When $z = 0$, the fundamental theorem of calculus gives

(A.2)
$$\frac{dg}{dz} = \frac{1}{\pi} \int_{-\pi}^{\pi} \frac{\cos\phi \, d\phi}{((x - \cos\phi)^2 + (y - \sin\phi)^2)^{1/2}}$$
$$= \psi(r)\cos\theta,$$

with ψ as in the lemma. Part (a) of the lemma is now obvious. Also

$$|e + q|^2 = 1 + \alpha^2|\nabla g|^2 + 2(\alpha^2 g_x + \alpha\sqrt{1 - \alpha^2} g_z)$$
$$= \begin{cases} (\sqrt{1 - \alpha^2} + \alpha\psi(r)\cos\theta)^2, & r < 1 \\ a\cos^2\theta + b\cos\theta + c, & r > 1. \end{cases}$$

We record the formula

(A.3)
$$\int_{-\pi}^{\pi} \log|\cos\theta - p| \frac{d\theta}{2\pi} = \log|p + \sqrt{p^2 - 1}| - \log 2, \quad p \in \mathbb{C},$$

where the sign of the radical is chosen so that $|p + \sqrt{p^2 - 1}| \geq 1$. When $p_1, p_2 \in \mathbb{R}$ we obtain

$$\int_{-\pi}^{\pi} \log|p_2\cos\theta - p_1| \frac{d\theta}{2\pi} = \begin{cases} \log|p_2| - \log 2, & |p_1| \leq |p_2| \\ \log(|p_1| + \sqrt{p_1^2 - p_2^2}) - \log 2, & |p_1| \geq |p_2|. \end{cases}$$

Part (b) of the lemma follows from this. For (c) we use

Lemma A.1.2. *Suppose* $p(t) = at^2 + bt + c$ *is nonnegative on the real axis. Then*

$$\int_{-\pi}^{\pi} \log p(\cos\theta) \frac{d\theta}{2\pi} \leq 2\log\left(\frac{1}{2}\left(\sqrt{c} + \sqrt[4]{(a + c)^2 - b^2}\right)\right).$$

Proof. Let $D = 4ac - b^2 \geq 0$. Write $p(t) = a(t - \zeta)(t - \bar\zeta)$, $\zeta = \frac{1}{2a}(-b + i\sqrt{D})$. Then

$$\int_{-\pi}^{\pi} \log p(\cos\theta) \frac{d\theta}{2\pi} = \log a - 2\log 2 + 2\log|\zeta + \sqrt{\zeta^2 - 1}|,$$

with the sign of the radical as in (A.3). Now

$$|\zeta + \sqrt{\zeta^2 - 1}|^2 = \frac{c}{a} + \frac{1}{a}\sqrt{(a + c)^2 - b^2} + \frac{1}{a} r e \sqrt{E},$$

where $E = 4c^2 - 2b^2 + 4ac + 2ib\sqrt{D}$ and $\sqrt{E}$ is taken to have nonnegative real part. Estimating $er\sqrt{E}$ by $\sqrt{|E|}$, we obtain

$$|\zeta + \sqrt{\zeta^2 - 1}|^2 \leq \frac{c}{a} + \frac{1}{a}\sqrt{(a + c)^2 - b^2} + \frac{2}{a}\sqrt{c}\sqrt[4]{(a + c)^2 - b^2}$$

$$= \frac{1}{a}\left(\sqrt{c} + \sqrt[4]{(a + c)^2 - b^2}\right)^2,$$

and the lemma follows. ∎

Part (c) of Lemma A.1.1 is immediate from this. ∎

A.2 Simplifying the formulas

We want to get rid of the roots and logarithms. Define $J_{\text{in}}(\alpha) = \int_{r<1} \log |e + q(r, \theta)| \frac{d\theta}{2\pi} r \, dr$, $J_{\text{out}}(\alpha) = \int_{r>1} \log |e + q(r, \theta)| \frac{d\theta}{2\pi} r \, dr$. Also define

$$d = \int_{r<1} \psi^2 r \, dr, \qquad x = \int_{\substack{r<1 \\ \psi \leq \alpha^{-1}(1-\alpha^2)^{1/2}}} \psi^2 r \, dr, \qquad Q = \int_{\substack{r<1 \\ \psi \leq \alpha^{-1}(1-\alpha^2)^{1/2}}} r \, dr.$$

Lemma A.2.1. $J_{\text{in}} \leq \frac{Q}{2} \log(1 - \alpha^2) - \frac{1}{4}\frac{\alpha^2}{1-\alpha^2} x + \frac{1}{2}\left(\frac{1}{2} - Q\right) \log\left[\frac{\alpha^2(d-x)}{4(\frac{1}{2}-Q)}\right]$.

Lemma A.2.2. $J_{\text{out}} \leq \frac{1}{8}(1 - \alpha^2) \log\left(\frac{1+\alpha^2}{1-\alpha^2}\right) + V$ *where V satisfies the following two estimates. Let*

$$\Gamma(r) = \min\left(\psi(r)^2, \left[\psi(r)^2 + \frac{2}{\alpha^2}\left(2\alpha^2 - 1 + \alpha^2\left(\frac{2}{r^2} + \frac{1}{r^4}\right)\right)\right]_+\right).$$

Then

$$V(r) \leq \frac{1}{4}\alpha^2 \int_1^\infty \left(1 - \frac{\alpha^2}{(1 + \alpha^2)r^2}\right)\Gamma(r) r \, dr,$$

$$V(r) \leq 12^{-3/4}(4\alpha^2(1 - \alpha^2))^{-1/4}\alpha^2 \int_1^\infty \left(1 + \frac{1}{r^2}\right)^{-1/2}\Gamma(r) r \, dr$$

Proof of Lemma A.2.1 It is understood that $r < 1$ in all formulas.

$$\int_{\psi \leq \sqrt{1-\alpha^2}/\alpha} \log |e + q| \frac{d\theta}{2\pi} r \, dr$$

$$= \int_{\psi \leq \sqrt{1-\alpha^2}/\alpha} \log\left[\frac{1}{2}\left(\sqrt{1-\alpha^2} + \sqrt{1-\alpha^2 - \alpha^2\psi^2} \right) \right] r \, dr$$

$$= \int_{\psi \leq \sqrt{1-\alpha^2}/\alpha} \left(\frac{1}{2}\log(1-\alpha^2) + \log\left[\frac{1}{2}\left(1 + \sqrt{1 - \frac{\alpha^2}{1-\alpha^2}\psi^2} \right) \right] \right) r \, dr$$

$$\leq \frac{Q}{2}\log(1-\alpha^2) - \int_{\psi \leq \sqrt{1-\alpha^2}/\alpha} \frac{1}{2}\left(1 - \sqrt{1 - \frac{\alpha^2}{1-\alpha^2}\psi^2} \right) r \, dr$$

$$\leq \frac{Q}{2}\log(1-\alpha^2) - \frac{1}{4}\frac{\alpha^2}{1-\alpha^2} x,$$

where we used $\log(1-t) \leq -t,\ 1 - \sqrt{1-t} \geq \frac{1}{2}t$. Also

$$\int_{\psi > \sqrt{1-\alpha^2}/\alpha} \log |e + q| \frac{d\theta}{2\pi} r \, dr = \int_{\psi > \sqrt{1-\alpha^2}/\alpha} \log\left|\left(\frac{\alpha\psi}{2} \right)\right| r \, dr$$

$$= \frac{1}{2}\left(\frac{1}{2} - Q \right) \int_{\psi > \sqrt{1-\alpha^2}/\alpha} \log\left(\frac{\alpha^2\psi^2}{4} \right) \frac{r \, dr}{\frac{1}{2} - Q}$$

$$\leq \frac{1}{2}\left(\frac{1}{2} - Q \right) \log\left[\frac{\alpha^2(d-x)}{4(\frac{1}{2} - Q)} \right]$$

by Jensen's inequality. ∎

Proof of Lemma A.2.2 Here it is understood that $r > 1$. We use the following fact.

Lemma A.2.3. *Let* $e \in [-1, 1]$ *and let* $\gamma = \min(\frac{1}{2}, 2^{-1/2}3^{-3/4}(1 - e^2)^{-1/4})$. *Then* $(1 + 2et + t^2)^{1/4} \leq 1 + \gamma \min(t, (t + 2e)_+)$ *for* $t \in [0, \infty)$.

Proof. Let $f(t) = (1 + 2et + t^2)^{1/4} - 1$. Then $f(0) = f(-2e) = 0$, and $f < 0$ on $(0, -2e)$ if $e < 0$, so it is enough to show $f'(t) \leq \gamma$ when $t > \max(0, -2e)$. We have $f'(t) = \frac{1}{2}(t + e)(1 + 2et + t^2)^{-3/4}$, $f''(t) = (-\frac{1}{4}t^2 - \frac{e}{2}t + \frac{1}{2} - \frac{3}{4}e^2)(1 + 2et + t^2)^{-7/4}$, so the critical points of f' are $-e \pm \sqrt{2 - 2e^2}$.

If $e > \sqrt{2/3}$, then f' is decreasing on $[0, \infty)$; hence $f' \leq f'(0) = \frac{e}{2}$. If $0 \leq e < \sqrt{2/3}$, then f' has a maximum at $-e + \sqrt{2 - 2e^2}$; its value there is $2^{-1/2}3^{-3/4}(1 - e^2)^{-1/4}$. One computes $\frac{e}{2} \leq \gamma$ for all e while $2^{-1/2}3^{-3/4}(1 - e^2)^{-1/4} = \gamma$ when $e < \sqrt{2/3}$. The $e < 0$ case follows the same way: if

$e \leq -\sqrt{2/3}$, then f' is decreasing on $[-2e, \infty)$ and if $-\sqrt{2/3} < e < 0$ then f' has a maximum at $-e + \sqrt{2 - 2e^2}$. ∎

Now, working from A.1.1 (c), let

$$t = \frac{\alpha^2 \psi^2}{1 + \frac{2\alpha^2}{r^2} + \frac{\alpha^2}{r^4}}, \qquad e = \frac{2\alpha^2 - 1 + \alpha^2(\frac{2}{r^2} + \frac{1}{r^4})}{1 + \frac{2\alpha^2}{r^2} + \frac{\alpha^2}{r^4}} \in [-1, 1],$$

and define γ relative to this e. Then

$$\frac{1}{2}(\sqrt{c} + \sqrt[4]{(a+c)^2 - b^2})$$

$$= \frac{1}{2}\left(\sqrt{1 - \frac{2\alpha^2}{r^2} + \frac{\alpha^2}{r^4}} + \sqrt{1 + \frac{2\alpha^2}{r^2} + \frac{\alpha^2}{r^4}}\sqrt[4]{t^2 + 2et + 1}\right)$$

$$\leq \frac{1}{2}\left(\sqrt{1 - \frac{2\alpha^2}{r^2} + \frac{\alpha^2}{r^4}} + \sqrt{1 + \frac{2\alpha^2}{r^2} + \frac{\alpha^2}{r^4}}\right)$$

$$+ \frac{\gamma}{2}\sqrt{1 + \frac{2\alpha^2}{r^2} + \frac{\alpha^2}{r^4}} \, \min(t, (t + 2e)_+)$$

where we used A.2.3. So, since $\log x \leq x - 1$,

$$\int \log|e + q| \frac{d\theta}{2\pi} \leq \left[\frac{1}{2}\sqrt{1 - \frac{2\alpha^2}{r^2} + \frac{\alpha^2}{r^4}} + \frac{1}{2}\sqrt{1 + \frac{2\alpha^2}{r^2} + \frac{\alpha^2}{r^4}} - 1\right]$$

$$+ \frac{\gamma}{2}\sqrt{1 + \frac{2\alpha^2}{r^2} + \frac{\alpha^2}{r^4}} \, \min(t, (t + 2e)_+)$$

$$= \mathrm{I}(r) + \mathrm{II}(r).$$

Now

$$\mathrm{I}(r) = \frac{1}{2}\left[\left(1 - \frac{\alpha^2}{r^2}\right)\sqrt{1 + \frac{\alpha^2 - \alpha^4}{(r^2 - \alpha^2)^2}} + \left(1 + \frac{\alpha^2}{r^2}\right)\sqrt{1 + \frac{\alpha^2 - \alpha^4}{(r^2 + \alpha^2)^2}}\right] - 1$$

$$\leq \frac{1}{2}\frac{\alpha^2 - \alpha^4}{r^4 - \alpha^4}$$

since $\sqrt{1 + x} \leq 1 + \frac{1}{2}x$. It follows that

$$\int_1^\infty \mathrm{I}(r)r\,dr \leq \frac{1}{8}(1 - \alpha^2)\log\left(\frac{1 + \alpha^2}{1 - \alpha^2}\right).$$

We set $V = \int_1^\infty \text{II}(r)r\,dr$ and make two estimates corresponding to $\gamma \leq \frac{1}{2}$ and $\gamma \leq 2^{-1/2}3^{-3/4}(1 - e^2)^{-1/4}$. Using $\gamma \leq \frac{1}{2}$, we get

$$\text{II}(r) \leq \frac{1}{4}\sqrt{1 + \frac{2\alpha^2}{r^2} + \frac{\alpha^2}{r^4}}\, \min(t, (t + 2e)_+)$$

$$= \frac{1}{4}\left(1 + \frac{2\alpha^2}{r^2} + \frac{\alpha^2}{r^4}\right)^{-1/2}\Gamma(r)\alpha^2$$

and the first stated estimate for V follows using

$$\left(1 + \frac{2\alpha^2}{r^2} + \frac{\alpha^2}{r^4}\right)^{-1/2} \leq \left(1 + \frac{\alpha^2}{r^2}\right)^{-1} \leq 1 - \frac{\alpha^2}{(1 + \alpha^2)r^2}, \quad r > 1.$$

The other estimate comes from

$$\text{II}(r) \leq \frac{1}{2}2^{-1/2}3^{-3/4}(1 - e^2)^{-1/4}\sqrt{1 + \frac{2\alpha^2}{r^2} + \frac{\alpha^2}{r^4}}\, \min(t, (t + 2e)_+)$$

$$= 12^{-3/4}(1 - e^2)^{-1/4}\left(1 + \frac{2\alpha^2}{r^2} + \frac{\alpha^2}{r^4}\right)^{-1/2}\Gamma(r)\alpha^2$$

because $(1 - e^2)\left(1 + \frac{2\alpha^2}{r^2} + \frac{\alpha^2}{r^4}\right)^2 = 4\alpha^2(1 - \alpha^2)(1 + \frac{1}{r^2})^2.$ ∎

A.3 Estimates for ψ

We give some formulas for ψ and then some estimates.

(A3.1) $$\psi(r^{-1}) = r\psi(r)$$

(A3.2) $$\int_0^\infty \psi(r)^2 r\,dr = 2$$

Proof. On the one hand $\int g_z^2 \frac{d\theta}{2\pi} r\,dr = \frac{1}{2}\int \psi(r)^2 r\,dr$, and on the other hand $\int g_z^2 \frac{d\theta}{2\pi} r\,dr = \int |\nabla_r g|^2 \frac{d\theta}{2\pi} r\,dr = \int_0^1 r\,dr + \int_1^\infty r^{-3}\,dr = 1.$ ∎

(A3.3) $$\psi(r) = \frac{2}{\pi}r\int_0^1 \frac{t^{1/2}dt}{(1 - t)^{1/2}(1 - r^2t)^{1/2}} \qquad \text{if} \quad r < 1.$$

Proof. The function $\left(\frac{z}{(1+r^2)z - r(1+z^2)}\right)^{1/2}$ has a single valued branch on the unit disc slit from 0 to r. Accordingly,

$$\psi(r) = \text{e}\left(\frac{1}{\pi}\int_{|z|=1} \frac{z}{(1 + r^2 - r(z + \bar{z}))^{1/2}} \frac{dz}{iz}\right)$$

$$= \text{e}\left(\frac{1}{\pi i}\int_{|z|=1}\left(\frac{z}{(1 + r^2)z - r(z^2 + 1)}\right)^{1/2}dz\right)$$

$$= e\left(\frac{-2}{\pi i} \int_0^r \left(\frac{x}{(1+r^2)x - r(x^2+1)}\right)^{1/2} dx\right),$$

and now let $x = rt$.   ∎

(A3.4) $\dfrac{\psi(r)}{r}$ is increasing on $[0, 1)$.

This follows from A3.3. Note it implies ψ is also increasing.

(A3.5) $\psi(r) \geq r$ and (with x, Q as in A2) $x \geq Q^2$.

The first statement follows from A3.4 since $\psi'(0) = 1$. The second statement
follows from the first.

(A3.6) $\dfrac{2}{3} \leq d \leq \dfrac{7}{8}$ (d as in A2).

Proof. Let $v(r) = \frac{\psi(r)}{r}$. Then $d = \int_0^1 v^2 r^3 dr$ and by A3.1, A3.2, $2 - d = \int_0^1 v^2 r\, dr$. v^2 is increasing, so

$$\int_0^1 v^2(r - 2r^3)dr = \frac{1}{2}(r^2 - r^4)v^2\Big|_0^1 - \int_0^1 \frac{1}{2}(r^2 - r^4)\frac{dv^2}{dr}dr$$

$$\leq 0.$$

It follows that $d \geq \frac{2}{3}$. Also,

$$\int_0^1 v^2(r - r^3)dr \geq \int_0^1 (r - r^3)dr = \frac{1}{4},$$

by A3.5, and this implies $d \leq \frac{7}{8}$.   ∎

(A3.7) $\displaystyle\int_1^\infty (1 + \frac{1}{r^2})^{-1/2}\psi(r)^2 r\, dr \leq 2(\sqrt{2} - 1)(2 - d).$

Proof. Let $p = 2(\sqrt{2} - 1)$. Let v be as in A3.6. Let $f(r) = (1 + r^2)^{1/2} - \frac{1}{2}pr^2 - 1$. Then $f(0) = f(1) = 0$ and $f > 0$ on $(0, 1)$ (this follows because f is concave down at the critical point $r = (p^{-2} - 1)^{1/2}$) so

$$\int_0^1 \left(\frac{r}{\sqrt{1+r^2}} - pr\right)v(r)^2 dr = fv^2\Big|_0^1 - \int_0^1 f\frac{dv^2}{dr}dr$$

$$< 0.$$

By A3.1, $\int_1^\infty (1 + \frac{1}{r^2})^{-1/2}\psi(r)^2 r\, dr = \int_0^1 (1 + r^2)^{-1/2} r v(r)^2 dr$, and the
statement follows by A3.1, A3.2 as above.   ∎

(A3.8)   For $r < 1$, $\dfrac{d}{dr}((1 + r^2)^{1/2}\psi(r)) \leq (1 - r^2)^{-1}(1 + 4r^2 - r^4)^{1/2}$.

Proof. Let $t = \frac{2r}{1+r^2}$. Then $(1 + r^2)^{1/2}\psi(r) = \frac{1}{\pi}\int_{-\infty}^{\infty}\frac{\cos\theta\,d\theta}{(1-t\cos\theta)^{1/2}} \overset{\text{def}}{=} f(t)$.

$$\frac{df}{dt} = \frac{1}{2\pi}\int_{-\pi}^{\pi}\frac{\cos^2\theta\,d\theta}{(1 - t\cos\theta)^{3/2}}$$

$$\leq A(t)^{1/2}B(t)^{1/2},$$

where

$$A(t) = \frac{1}{2\pi}\int_{-\pi}^{\pi}\frac{\cos^2\theta}{1 - t\cos\theta}\,d\theta$$

$$B(t) = \frac{1}{2\pi}\int_{-\pi}^{\pi}\frac{\cos^2\theta}{(1 - t\cos\theta)^2}\,d\theta$$

may be evaluated using residues to be

$$A(t) = t^{-2}((1 - t^2)^{-1/2} - 1) = \frac{(1 + r^2)^2}{2(1 - r^2)}$$

$$B(t) = t^{-2}(1 + (1 - t^2)^{-3/2} - 2(1 - t^2)^{-1/2}) = \frac{(1 + r^2)^2}{2(1 - r^2)^3}(1 + 4r^2 - r^4).$$

The statement follows from these formulas and the fact that $\frac{dt}{dr} = \frac{2(1-r^2)}{(1+r^2)^2}$. ∎

(A3.9)
$$\psi(r) \leq (1 + r^2)^{-1/2}\left(\log\frac{1}{1 - r} + \log(1 + r) - \arctan r\right) \qquad \text{when} \quad r < 1.$$

Proof. A3.8 implies

$$\frac{d}{dr}((1 + r^2)^{1/2}\psi(r)) \leq (1 - r^2)^{-1}((1 + r^2)^2 + 2(r^2 - r^4))^{1/2}$$

$$\leq (1 - r^2)^{-1}(1 + r^2)\left(1 + \frac{r^2 - r^4}{(1 + r^2)^2}\right)$$

$$= \frac{1}{1 - r} + \frac{1}{1 + r} - \frac{1}{1 + r^2}. \qquad ∎$$

(A3.10) $\qquad \dfrac{d\psi}{dr} \leq \dfrac{4}{\pi}r^{-1/2}(1 - r^2)^{-1} \qquad$ when $\quad r < 1$.

Proof. Let $f(r) = r^{1/2}(1 - r^2)\frac{d\psi}{dr}$. We claim $f(r) \le \limsup_{t \to 1} f(t)$. This will follow if $f(r) > f(r^2)$. From A3.3,

$$f(r) = \frac{2}{\pi} \int_0^1 \frac{r^{1/2}(1 - r^2)t^{1/2}}{(1 - t)^{1/2}(1 - r^2 t)^{3/2}} \, dt$$

$$f(r^2) = \frac{2}{\pi} \int_0^1 \frac{r(1 - r^4)t^{1/2}}{(1 - t)^{1/2}(1 - r^4 t)^{3/2}} \, dt$$

$$= \frac{2}{\pi} \int_0^1 \frac{\phi(r, u)r^{1/2}(1 - r^2)u^{1/2}}{(1 - u)^{1/2}(1 - r^2 u)^{3/2}} \, du \quad (u^2 = t),$$

with

$$\phi(r, u) = \frac{2u^{3/2}r^{1/2}(1 + r^2)}{(1 + u)^{1/2}(1 + r^2 u)^{3/2}}.$$

One computes $\frac{d\phi}{du} > 0$, so $\phi(r, u) < \phi(r, 1) = \left(\frac{2r}{1+r^2}\right)^{1/2} < 1$, and the claim follows. To evaluate the limit, write

$$\frac{d\psi}{dr} = \frac{1}{\pi} \int_{-\pi}^{\pi} \frac{(\cos\theta - r)\cos\theta}{((1 - r)^2 + 2r(1 - \cos\theta))^{3/2}} \, d\theta.$$

We show this is $\frac{2}{\pi}(1 - r)^{-1}$ + lower order as $r \to 1$. The denominator is $\gtrsim \max((1 - r)^3, \theta^3)$, so we can ignore $O(\theta^2)$ terms in the numerator and $O(\theta^4)$ terms in the denominator. This leads to

$$\frac{d\psi}{dr} = \frac{1}{\pi} \int_{-\pi}^{\pi} \frac{1 - r}{((1 - r)^2 + r\theta^2)^{3/2}} \, d\theta + \text{lower order terms}$$

$$= \frac{1}{\pi(1 - r)} \int_{-\infty}^{\infty} (1 + t^2)^{-3/2} dt + \text{lower order terms} \left(t = \frac{r^{1/2}\theta}{1 - r}\right)$$

$$= \frac{2}{\pi(1 - r)} + \text{lower order terms},$$

and A3.10 follows. ∎

(A3.11) $$\int_{s < r < 1} r\psi(r)^2 dr \le \frac{1}{2}(1 - s^2)(\psi(s)^2 + J(s)\psi(s) + K(s))$$

where $J(s) = \frac{16}{\pi}\frac{1 - s^{1/2}}{1 - s^2}$, $K(s) = \frac{128}{\pi^2}\frac{(1+s^{1/2})^{-2} - \frac{1}{4}}{1 - s^2}$.

Remark. J and K are decreasing on $(0, 1)$. For K, this may be seen by letting $t = s^{1/2}$—then $\frac{4\pi^2}{128K(s)} = (1 + t^2)\frac{(1+t)^3}{3+t}$, and calculus shows that $\frac{(1+t)^3}{3+t}$ is increasing.

Proof. $\int_s^1 r\psi(r)^2 dr = \frac{1}{2}(1-s^2)\psi(s)^2 + 2\psi(s)\int_s^1(\psi(r)-\psi(s))r\,dr + \int_s^1(\psi(r)-\psi(s))^2 r\,dr$ and

$$2\int_s^1(\psi(r)-\psi(s))r\,dr = 2\int_s^1\int_s^r \psi'(t)dt\,r\,dr$$

$$= \int_s^1(1-t^2)\psi'(t)dt$$

$$\leq \frac{4}{\pi}\int_s^1 t^{-1/2}dt$$

$$= J(s)\frac{1}{2}(1-s^2)$$

$$\int_s^1(\psi(r)-\psi(s))^2 r\,dr = \int_s^1\int_s^1\frac{1}{2}(1-\max(t,u)^2)\psi'(t)\psi'(u)dt\,du$$

$$= \int_s^1\int_s^t \psi'(u)\psi'(t)(1-t^2)du\,dt$$

$$\leq \frac{16}{\pi^2}\int_s^1\int_s^t t^{-1/2}u^{-1/2}(1-u^2)^{-1}du\,dt$$

$$= \frac{64}{\pi^2}\int_{\sqrt{s}}^1(1+v)^{-1}(1+v^2)^{-1}dv$$

$$\leq \frac{128}{\pi^2}\int_{\sqrt{s}}^1(1+v)^{-3}dv$$

$$= K(s)\frac{1}{2}(1-s^2). \qquad\blacksquare$$

A.4 Final calculations

The argument will proceed roughly as follows. Suppose a function f, involving logarithms or whatever, can be shown by sections A2, A3 to be an upper bound for $J(\alpha)$ on the interval $[0.2, 0.27]$. Suppose f is increasing. If $f(0.27) < 0$, then Lemma 1.2 is proved when $0.2 \leq \alpha \leq 0.27$. The fact that $f(0.27) < 0$ will be proved by evaluating $f(0.27)$ on a calculator.

With V as in A2.2, we can use $\Gamma \leq \psi^2$ and A3.1, A3.2, A3.7 to get

$$V \leq \min\left(\frac{1}{4}\alpha^2\left(2 - \frac{1+2\alpha^2}{1+\alpha^2}d\right), 2(2^{1/2}-1)12^{-3/4}(4\alpha^2(1-\alpha^2))^{-1/4}\alpha^2(2-d)\right).$$

Another frequently used fact will be that all Taylor coefficients at 0 of $\phi(\alpha) = \frac{1}{8}(1-\alpha^2)\log\left(\frac{1+\alpha^2}{1-\alpha^2}\right) + \frac{1}{4}\log(1-\alpha^2) = \frac{1}{8}(1-\alpha^2)\log(1+\alpha^2) + \frac{1}{8}(1+$

$\alpha^2) \log(1 - \alpha^2)$ are nonpositive, and in particular (look at the first nonzero coefficient) $\phi(\alpha) \leq -\frac{3}{8}\alpha^4$.

Case 1. $0.85 \leq \alpha^2 \leq 1$.

$$J_{\text{out}} \leq \frac{1}{8}(1 - \alpha^2) \log\left(\frac{1 + \alpha^2}{1 - \alpha^2}\right) + \frac{1}{4}\alpha^2\left(2 - \frac{1 + 2\alpha^2}{1 + \alpha^2}d\right)$$

$$J_{\text{in}} \leq \frac{Q}{2} \log(1 - \alpha^2) + \frac{1}{2}\left(\frac{1}{2} - Q\right) \log\left(\frac{\alpha^2 d}{4(\frac{1}{2} - Q)}\right)$$

where we have overestimated in Lemma A2.1 using $x \geq 0$. Adding the right sides, then maximizing over $d \in (0, \infty)$ gives

$$J \leq \frac{1}{8}(1 - \alpha^2) \log\left(\frac{1 + \alpha^2}{1 - \alpha^2}\right) + \frac{1}{2}\alpha^2 + \frac{Q}{2} \log(1 - \alpha^2)$$

$$\text{(A.4)} \qquad + \frac{1}{2}\left(\frac{1}{2} - Q\right) \log\left[\frac{1 + \alpha^2}{2e(1 + 2\alpha^2)}\right].$$

This is linear in Q so we need only check the endpoints, 0 or $\frac{1}{2}$.

$Q = \frac{1}{2}$:

(A.5) $\text{(A.4)} = \frac{1}{8}(1 - \alpha^2) \log\left(\frac{1 + \alpha^2}{1 - \alpha^2}\right) + \frac{1}{4} \log(1 - \alpha^2) + \frac{1}{2}\alpha^2$.

$Q = 0$:

(A.6) $\text{(A.4)} = \frac{1}{8}(1 - \alpha^2) \log\left(\frac{1 + \alpha^2}{1 - \alpha^2}\right) + \frac{1}{2}\alpha^2 + \frac{1}{4} \log\left[\frac{1 + \alpha^2}{2e(1 + 2\alpha^2)}\right]$.

(A.5)$/\alpha^2$ is decreasing in α by the remark about negative coefficients. The calculator shows (A.5) < 0 when $\alpha^2 = 0.85$, so it remains to consider (A.6). Letting $x = 1 - \alpha^2$, we get

$$\text{(A.6)} = \frac{1}{4}\left(\log \frac{e}{3} - 2x + \frac{1}{2}x \log(2 - x) + \frac{1}{2}x \log \frac{1}{x} + \log\left[\frac{3(2 - x)}{2(3 - 2x)}\right]\right).$$

By concavity, if $a > 0$ then $\frac{1}{2}t \log \frac{1}{t} \leq \frac{1}{2}a \log \frac{1}{a} + \frac{1}{2}(t - a)(\log \frac{1}{a} - 1) = \frac{1}{2}t \log \frac{1}{ae} + \frac{1}{2}a$. Taking $a = 2 \log \frac{3}{e}$ and evaluating $\log \frac{1}{ae}$ on the calculator, we obtain $\frac{1}{2}t \log \frac{1}{t} \leq 0.32t + \log \frac{3}{e}$; hence, using $\log x \leq x - 1$,

$$\text{(A.6)} \leq \frac{1}{4}\left(-2x + \frac{1}{2}x \log 2 + 0.32x + \frac{x}{6 - 4x}\right) \leq -0.2x.$$

Case 2. $0.75 \leq \alpha^2 \leq 0.85$.

Use the same bound for J_{in} and maximize over Q for fixed d. The worst Q is $\frac{1}{2} - \frac{\alpha^2 d}{4e(1-\alpha^2)}$ leading to $J_{\text{in}} \le \frac{1}{4} \log(1 - \alpha^2) + \frac{\alpha^2 d}{8e(1-\alpha^2)}$. The second bound for V leads to

$$J \le \frac{1}{8}(1 - \alpha^2) \log\left(\frac{1 + \alpha^2}{1 - \alpha^2}\right) + \frac{1}{4}\log(1 - \alpha^2) + \frac{\alpha^2 d}{8e(1 - \alpha^2)}$$

$$+ 2(2^{1/2} - 1)12^{-3/4}(2 - d)(4\alpha^2(1 - \alpha^2))^{-1/4}\alpha^2$$

$$\le \frac{1}{8}(1 - \alpha^2) \log\left(\frac{1 + \alpha^2}{1 - \alpha^2}\right) + \frac{1}{4}\log(1 - \alpha^2) + \frac{\alpha^2 d}{8e(1 - \alpha^2)}$$

$$+ 0.153\alpha^2(2 - d)$$

using that $2(2^{1/2} - 1)12^{-3/4}(4\alpha^2(1 - \alpha^2))^{-1/4}$ is increasing and a calculator for the value when $\alpha^2 = 0.85$. The bound is increasing in d when $\alpha^2 \ge \frac{3}{4}$, so by A3.6 we can take $d = 7/8$ getting

$$J \le \frac{1}{8}(1 - \alpha^2) \log \frac{1 + \alpha^2}{1 - \alpha^2} + \frac{1}{4}\log(1 - \alpha^2) + \frac{7}{64e}\frac{\alpha^2}{1 - \alpha^2} + 0.173\alpha^2.$$

The calculator shows this is negative, as follows: Evaluate $\frac{7}{64e(1-\alpha^2)} + 0.173$ at $\alpha_3^2 = 0.85, \alpha_2^2 = 0.82, \alpha_1^2 = 0.79$, obtaining values $\le 0.442 = t_3, \le 0.397 = t_2$, $\le 0.365 = t_1$. Let $\alpha_0^2 = 0.75$. By calculator, $\frac{1}{8}(1 - \alpha^2) \log\left(\frac{1+\alpha^2}{1-\alpha^2}\right) + \frac{1}{4}\log(1 - \alpha^2) + t_j\alpha^2$ is $\le -0.01 < 0$ when $\alpha = \alpha_{j-1}$ ($j = 1, 2, 3$). The statement follows using monotonicity, as with (A5) in Case 1.

Case 3. $0.5 \le \alpha^2 \le 0.75$.

Subcase A. $Q \ge \frac{1}{4}$. Let s satisfy $\psi(s) = \frac{\sqrt{1-\alpha^2}}{\alpha}$. Then $s \ge 2^{-1/2}$ and in A3.11, $J(s) \le J(2^{-1/2}) \le 1.7$, $K(s) \le K(2^{-1/2}) \le 1.2$. In the notation of A2,

$$d - x \le \left(\frac{1 - \alpha^2}{\alpha^2} + 1.7\frac{\sqrt{1 - \alpha^2}}{\alpha} + 1.2\right)\left(\frac{1}{2} - Q\right)$$

$$\le 2\left(\frac{1 - \alpha^2}{\alpha^2} + 1\right)\left(\frac{1}{2} - Q\right).$$

Using this to bound the last term in A2.1 and A3.5 for the second term, we get

$$J_{\text{in}} \le \frac{Q}{2}\log(1 - \alpha^2) - \frac{1}{64}\frac{\alpha^2}{1 - \alpha^2} - \frac{1}{2}\left(\frac{1}{2} - Q\right)\log 2$$

$$= \frac{1}{4}\log(1 - \alpha^2) - \frac{1}{64}\frac{\alpha^2}{1 - \alpha^2} - \frac{1}{2}\left(\frac{1}{2} - Q\right)\log(2(1 - \alpha^2))$$

(A.7) $\qquad \le \frac{1}{4}\log(1 - \alpha^2) - \frac{1}{64}\frac{\alpha^2}{1 - \alpha^2} + \frac{\log 2}{4}(2\alpha^2 - 1)$

where the last line follows by concavity of the logarithm function on the interval $[\frac{1}{2}, 1]$. Also

$$J_{\text{out}} \leq \frac{1}{8}(1 - \alpha^2) \log\left(\frac{1 + \alpha^2}{1 - \alpha^2}\right) + 2(\sqrt{2} - 1)12^{-3/4}(2 - d)(4\alpha^2(1 - \alpha^2))^{-1/4}\alpha^2.$$

By A3.6, $2 - d \leq \frac{4}{3}$. By calculator $\frac{8}{3}(\sqrt{2} - 1)12^{-3/4}(4\alpha^2(1 - \alpha^2))^{-1/4} \leq \frac{3}{16}$ when $\alpha^2 = 0.75$, so

$$J_{\text{out}} \leq \frac{1}{8}(1 - \alpha^2) \log\left(\frac{1 + \alpha^2}{1 - \alpha^2}\right) + \frac{3}{16}\alpha^2$$

$$J \leq \frac{1}{8}(1 - \alpha^2) \log\left(\frac{1 + \alpha^2}{1 - \alpha^2}\right) + \frac{1}{4}\log(1 - \alpha^2) - \frac{1}{64}\frac{\alpha^2}{1 - \alpha^2} + \frac{3}{16}\alpha^2$$

$$+ (2\alpha^2 - 1)\frac{\log 2}{4}$$

$$\leq -\frac{3}{8}\alpha^4 - \frac{1}{32}\alpha^2 + \frac{3}{16}\alpha^2 + (2\alpha^2 - 1)\frac{\log 2}{4}$$

$$< 0$$

where the last line is easily checked since the preceding line is a quadratic polynomial in α^2.

Subcase B. $Q \leq \frac{1}{4}$. Estimating J_{in} as in case 1, then rearranging some terms, we find that

$$J_{\text{in}} \leq \frac{1}{4}\log(1 - \alpha^2) + \frac{1}{2}\left(\frac{1}{2} - Q\right)\log\left(\frac{\alpha^2 d}{4(1 - \alpha^2)(\frac{1}{2} - Q)}\right).$$

This is increasing in Q since $d \leq \frac{7}{8}, Q \leq \frac{1}{4}, \alpha^2 \leq \frac{3}{4}$, so

$$J_{\text{in}} \leq \frac{1}{4}\log(1 - \alpha^2) + \frac{1}{8}\log\left(\frac{\alpha^2 d}{1 - \alpha^2}\right)$$

$$J \leq \frac{1}{8}\log(1 - \alpha^2)\log\left(\frac{1 + \alpha^2}{1 - \alpha^2}\right) + \frac{1}{4}\log(1 - \alpha^2) + \frac{1}{8}\log\left(\frac{\alpha^2 d}{1 - \alpha^2}\right)$$

$$+ 2(2^{1/2} - 1)12^{-3/4}(2 - d)(4\alpha^2(1 - \alpha^2))^{-1/4}\alpha^2.$$

Using that $(4\alpha^2(1 - \alpha^2))^{-1/4}$ is largest at $\alpha^2 = \frac{3}{4}$ (and a calculator for the coefficient of d in the last term when $\alpha^2 = \frac{3}{4}$) this is seen to be increasing in d for $d \leq \frac{7}{8}$, so, evaluating $2(2^{1/2} - 1)12^{-3/4}(2 - \frac{7}{8})(4 \cdot \frac{3}{4} \cdot \frac{1}{4})^{-1/4} \leq 0.16$, we get

$$J \leq \frac{1}{8}(1 - \alpha^2) \log\frac{1 + \alpha^2}{1 - \alpha^2} + \frac{1}{4}\log(1 - \alpha^2) + \frac{1}{8}\log\left[\frac{7\alpha^2}{8(1 - \alpha^2)}\right] + 0.16\alpha^2$$

$$= \frac{1}{8}(1 - \alpha^2)\log\left(\frac{1 + \alpha^2}{1 - \alpha^2}\right) + \frac{1}{8}\log(4\alpha^2(1 - \alpha^2)) + \frac{1}{8}\log\left(\frac{7}{32}\right) + 0.16\alpha^2$$

$$\leq \frac{1}{8}(1 - \alpha^2)\log\left(\frac{1 + \alpha^2}{1 - \alpha^2}\right) - \frac{1}{8}(2\alpha^2 - 1)^2 - 0.18 + 0.16\alpha^2$$

using $\log x \leq x - 1$ and evaluating $\frac{1}{8}\log\frac{7}{32}$. The function $\frac{1}{8}(1 - \alpha^2)\log\left(\frac{1+\alpha^2}{1-\alpha^2}\right)$ is concave down in α^2 and so has a unique maximum. Checking values at $\alpha^2 = 0.55, 0.56, 0.57$, we see that the maximum occurs between 0.55 and 0.57. At the maximum $0.43 < 1 - \alpha^2 < 0.45, 1 + \alpha^2 < 1.57$, so the value is $< 0.073 < 0.08$, and

$$J \leq -\frac{1}{10} - \frac{1}{8}(2\alpha^2 - 1)^2 + 0.16\alpha^2$$

$$< 0.$$

Case 4. $0.36 \leq \alpha^2 \leq 0.5$.

If $\alpha^2 = 0.36$, then $\frac{8}{3}(\sqrt{2} - 1)12^{-3/4}(4\alpha^2(1 - \alpha^2))^{-1/4} \leq 0.175$, so using $2 - d \leq \frac{4}{3}$, we have as before

$$J_{\text{out}} \leq \frac{1}{8}(1 - \alpha^2)\log\frac{1 + \alpha^2}{1 - \alpha^2} + 0.175\alpha^2.$$

If $\alpha^2 \leq \frac{1}{2}$, then $\frac{\sqrt{1-\alpha^2}}{\alpha} \geq 1$, and A3.9 shows $\psi(2^{-1/2}) < 1$, so if $\psi(s) > \frac{\sqrt{1-\alpha^2}}{\alpha}$ then $s \geq 2^{-1/2}$. So we can bound J_{in} as in the argument leading to (A.7), the only difference being at the last step where $\log(2(1 - \alpha^2))$ is now positive so the worst case is $Q = \frac{1}{2}$. We obtain

$$J_{\text{in}} \leq \frac{1}{4}\log(1 - \alpha^2) - \frac{1}{64}\frac{\alpha^2}{1 - \alpha^2}$$

$$J < -\frac{3}{8}\alpha^4 - \frac{1}{64}\frac{\alpha^2}{1 - \alpha^2} + 0.175\alpha^2.$$

If $\alpha^2 \geq 0.4$, this implies $J \leq \left(-\frac{3}{8}(0.4) - \frac{1}{(64)(0.6)} + 0.175\right)\alpha^2 < 0$.

If $0.36 \leq \alpha^2 \leq 0.4$, we modify as follows: If $\alpha^2 \leq 0.4$, then $\frac{\sqrt{1-\alpha^2}}{\alpha} \geq 1.22$, and if $r = 0.8$, then $\psi(r) < 1.19$ by A3.9. So if $\alpha^2 < 0.4$, then $Q > \frac{(0.8)^2}{2} = 0.32$, $\frac{1}{4}Q^2 \geq 0.0256$, and we can improve the second term in the bound for J_{in} in obtaining

$$J < -\frac{3}{8}\alpha^4 - 0.0256\frac{\alpha^2}{1 - \alpha^2} + 0.175\alpha^2$$

$$\leq \left(-\frac{3}{8}(0.36) - \frac{0.0256}{0.64} + 0.175 \right) \alpha^2$$

$$= 0.$$

Now we make estimates that will be used for all the remaining cases.

We always assume $\alpha^2 \leq 0.36$. Then $\frac{\sqrt{1-\alpha^2}}{\alpha} \geq \frac{4}{3}$. A3.9 shows $\psi(0.83) \leq 1.30 < \frac{4}{3}$ and in A3.11, $J(0.83) \leq \frac{3}{2}$ and $K(0.83) \leq 1$. It follows that if $\alpha^2 \leq 0.36$, then

$$d - x \leq \left(\frac{1-\alpha^2}{\alpha^2} + \frac{3}{2} \frac{\sqrt{1-\alpha^2}}{\alpha} + 1 \right) \left(\frac{1}{2} - Q \right).$$

We set $A_\alpha = \frac{1-\alpha^2}{\alpha^2} + \frac{3}{2} \frac{\sqrt{1-\alpha^2}}{\alpha} + 1$ and $C_\alpha = \frac{\alpha^2 A_\alpha}{4(1-\alpha^2)}$. A2.1 then implies

(A.8) $$J_{\text{in}} - \frac{1}{4} \log(1-\alpha^2) \leq \frac{1}{2} \left(\frac{1}{2} - Q \right) \log C_\alpha - \frac{1}{4} \frac{\alpha^2}{1-\alpha^2} x,$$

where we know $x \geq \max \left(Q^2, d - A_\alpha \left(\frac{1}{2} - Q \right) \right)$.

Claim. $\frac{1}{2} \left(\frac{1}{2} - Q \right) \log C_\alpha - \frac{1}{4} \frac{\alpha^2}{1-\alpha^2} Q^2$ is increasing in Q on $[0, \frac{1}{2}]$.

Proof. The critical point is $Q = Q_{\text{critZ}}(\alpha^2) = -\frac{(1-\alpha^2)}{\alpha^2} \log C_\alpha$. Let $t = \frac{\alpha}{\sqrt{1-\alpha^2}}$. Then $Q_{\text{critZ}} = \frac{-\log\left(\frac{1}{4} + \frac{3}{8} t + \frac{1}{4} t^2 \right)}{t^2}$. This is decreasing in t as long as $\frac{1}{4} + \frac{3}{8} t + \frac{1}{4} t^2 < 1$, as will be the case since $\alpha^2 < 0.36$. So $Q_{\text{critZ}} \geq Q_{\text{critZ}}(0.36) > \frac{1}{2}$. ∎

So in using (A.8) to bound J_{in} we can restrict to the case where $Q^2 \leq d - A_\alpha \left(\frac{1}{2} - Q \right)$. Writing this as a quadratic for $\frac{1}{2} - Q$ it implies $\frac{1}{2} - Q \leq \frac{d - \frac{1}{4}}{A_\alpha - 1}$, so

$$J_{\text{in}} - \frac{1}{4} \log(1-\alpha^2) \leq \max_{0 \leq \frac{1}{2} - Q \leq (d - \frac{1}{4})/(A_\alpha - 1)} \frac{1}{2} \left(\frac{1}{2} - Q \right) \log C_\alpha$$

$$- \frac{1}{4} \frac{\alpha^2}{1-\alpha^2} \left(d - A_\alpha \left(\frac{1}{2} - Q \right) \right)$$

$$\leq \max_{0 \leq \frac{1}{2} - Q \leq (d - \frac{1}{4})/(A_\alpha - 1)} \left(\frac{1}{2} - Q \right) \left(\frac{1}{2} \log C_\alpha + C_\alpha \right)$$

$$- \frac{1}{4} \frac{\alpha^2}{1-\alpha^2} d,$$

and now we have the following:

Lemma A4.1.

(a) If $\frac{1}{2} \log C_\alpha + C_\alpha < 0$, then $J_{\text{in}} - \frac{1}{4} \log(1-\alpha^2) \leq -\frac{1}{4} \frac{\alpha^2}{1-\alpha^2} d \leq -\frac{1}{6} \frac{\alpha^2}{1-\alpha^2}$.

(b) *If* $\frac{1}{2} \log C_\alpha + C_\alpha > 0$, *then*

$$J_{\text{in}} - \frac{1}{4} \log(1 - \alpha^2) \leq \left(\frac{d - \frac{1}{4}}{A_\alpha - 1} \right) \left(\frac{1}{2} \log C_\alpha + C_\alpha \right) - \frac{1}{4} \frac{\alpha^2}{1 - \alpha^2} d$$

$$\leq \frac{5}{12(A_\alpha - 1)} \left(\frac{1}{2} \log C_\alpha + C_\alpha \right) - \frac{1}{6} \frac{\alpha^2}{1 - \alpha^2}.$$

Proof. We have everything but the second inequality in (b), and that will follow if we know that the quantity on the right side of the preceding inequality is decreasing as a function of d. For this we show

Lemma A4.2. *The function* $\phi(\alpha) = \frac{1-\alpha^2}{\alpha^2} \frac{1}{A_\alpha - 1} (C_\alpha + \frac{1}{2} \log C_\alpha)$ *is increasing in* α.

Proof. Let $y = 1 + \frac{3}{2} \frac{\alpha}{\sqrt{1-\alpha^2}}$; then

$$\phi(\alpha) = \frac{1}{9y} (y^2 + \frac{1}{4} y + 1) + \frac{1}{2y} \log\left(\frac{1}{9} \left(y^2 + \frac{1}{4} y + 1 \right) \right),$$

which is increasing in y since $y \geq 1$ and the logarithm is negative. ∎

We use the calculator to make the following table.

α^2	$\phi(\alpha)$
0.36	≤ 0.223
0.27	≤ 0.152
0.2	≤ 0.088
0.13	≥ 0
0.12	≤ 0

In particular, $\phi(\alpha) \leq 0.223 < 0.25$ for all α, which finishes Lemma A4.1. ∎

Case 5. $0.2 \leq \alpha^2 \leq 0.36$.

Subcase A. $0.27 \leq \alpha^2 \leq 0.36$. By A4.1 (b), A4.2, and $\phi(\sqrt{0.36}) \leq 0.223$,

$$J_{\text{in}} - \frac{1}{4} \log(1 - \alpha^2) \leq \left(\frac{(5)(0.223)}{12} - \frac{1}{6} \right) \frac{\alpha^2}{1 - \alpha^2}$$

$$\leq \left(\frac{(5)(0.223)}{12} - \frac{1}{6} \right) \frac{\alpha^2}{0.73}$$

$$\leq -\frac{1}{10} \alpha^2.$$

When $\alpha^2 = 0.27$, $\frac{8}{3}(\sqrt{2}-1)12^{-3/4}(4\alpha^2(1-\alpha^2))^{-1/4} \leq 0.182$, and in the usual way we get

$$J_{\text{out}} \leq \frac{1}{8}(1-\alpha^2)\log\left(\frac{1+\alpha^2}{1-\alpha^2}\right) + 0.182\alpha^2$$

$$J \leq -\frac{3}{8}\alpha^4 + 0.182\alpha^2 - \frac{1}{10}\alpha^2$$

$$< 0.$$

Subcase B. $0.2 \leq \alpha^2 \leq 0.27$. Exactly the same argument, but using $\phi(\sqrt{0.27}) \leq 0.152$ and $\frac{8}{3}(\sqrt{2}-1)12^{-3/4}(4(0.20)(0.80))^{-1/4} \leq 0.192$, gives

$$J_{\text{in}} - \frac{1}{4}\log(1-\alpha^2) \leq \left(\frac{(5)(0.152)}{12} - \frac{1}{6}\right)\frac{\alpha^2}{0.8}$$

$$\leq -\frac{1}{8}\alpha^2$$

$$J \leq -\frac{3}{8}\alpha^4 + 0.192\alpha^2 - \frac{1}{8}\alpha^2$$

$$< 0.$$

Case 6. $\alpha^2 \leq 0.2$, but $\phi(\alpha) \geq 0$.

By the table, $\alpha^2 \geq 0.12$ and $\phi(\alpha) \leq 0.088$. We have

$$J_{\text{in}} - \frac{1}{4}\log(1-\alpha^2) \leq \left(\frac{(5)(0.088)}{12} - \frac{1}{6}\right)\frac{\alpha^2}{0.88}$$

$$\leq -0.147\alpha^2$$

$$J_{\text{out}} \leq \frac{1}{8}(1-\alpha^2)\log\left(\frac{1+\alpha^2}{1-\alpha^2}\right) + V, \quad \text{where by Lemma A2.2,}$$

$$V \leq 12^{-3/4}(4\alpha^2(1-\alpha^2))^{-1/4}\alpha^2 \int_1^\infty \frac{1}{\sqrt{1+r^{-2}}}\left[\psi(r)^2\right.$$

$$+ \frac{2}{\alpha^2}\left((2\alpha^2 - 1 + \alpha^2\left(\frac{2}{r^2} + \frac{1}{r^4}\right)\right)\Big]_+ r\,dr$$

$$\leq 12^{-3/4}(4\alpha^2(1-\alpha^2))^{-1/4}\alpha^2 \int_0^1 (1+r^2)^{-1/2}\left[\psi(r)^2 - \frac{6}{r^2}\right.$$

$$+ 2r^2 + 4\Big]_+ \frac{dr}{r}.$$

The last line was obtained as follows: The integrand on the previous line was increasing in α—this uses $\alpha^2 \leq 0.2$ which implies $2\alpha^2 - 1 + \alpha^2\left(\frac{2}{r^2} + \frac{1}{r^4}\right) < 0$— so we could replace α^2 by 0.2. We then changed variables $r \to 1/r$ and used A3.1.

If the (last) integrand is nonzero, then

(A.9)
$$\psi(r)^2 \geq \frac{6}{r^2} - 2r^2 - 4.$$

Since ψ is increasing and $\frac{6}{r^2} - 2r^2 - 4$ decreasing, the set where (A.9) happens is an interval $[R, 1)$. By A3.9 and calculator, $R \geq 0.87$. So

$$V \leq 12^{-3/4}(4\alpha^2(1 - \alpha^2))^{-1/4}\alpha^2 \int_{0.87}^1 (1 + r^2)^{-1/2} \frac{1}{r} \psi(r)^2 dr$$

$$\leq 12^{-3/4}(4\alpha^2(1 - \alpha^2))^{-1/4}\alpha^2(1 + (0.87)^2)^{-1/2} \int_{0.87}^1 \frac{1}{r} \psi(r)^2 dr.$$

Now by (A3.5), (A3.1), (A3.2), (A3.6)

$$\int_{0.87}^1 \frac{1}{r} \psi(r)^2 dr \leq \int_0^1 \frac{1}{r} \psi(r)^2 dr - \int_0^{0.87} r\, dr$$

$$\leq \frac{4}{3} - \frac{(0.87)^2}{2}.$$

If we use $\alpha^2 \geq 0.12$ to estimate $(4\alpha^2(1 - \alpha^2))^{-1/4}$, the calculator now gives $V \leq 0.1386\alpha^2 \leq 0.14\alpha^2$ and, therefore,

$$J \leq -0.147\alpha^2 - \frac{3}{8}\alpha^4 + 0.14\alpha^2$$

$$< 0.$$

Case 7. $\phi(\alpha) < 0$.

Then $\alpha^2 < 0.13$. By Lemma A.4.1,

$$J_{\text{in}} \leq \frac{1}{4} \log(1 - \alpha^2) - \frac{1}{6} \frac{\alpha^2}{1 - \alpha^2}.$$

We estimate J_{out} using the first alternative in A2.2 and dropping the $\frac{\alpha^2}{(1+\alpha^2)r^2}$ term, i.e.,

$$J_{\text{out}} \leq \frac{1}{8}(1 - \alpha^2) \log\left(\frac{1 + \alpha^2}{1 - \alpha^2}\right) + V$$

$$V \leq \frac{1}{4}\alpha^2 \int_1^\infty \left[\psi(r)^2 + \frac{2}{\alpha^2}\left(2\alpha^2 - 1 + \alpha^2\left(\frac{2}{r^2} + \frac{1}{r^4}\right)\right)\right]_+ r\, dr$$

$$= \frac{1}{4}\alpha^2 \int_0^1 \left[\psi(r)^2 + \frac{2}{\alpha^2} \left(\frac{2\alpha^2 - 1}{r^2} + \alpha^2 \left(2 + r^2 \right) \right) \right]_+ \frac{1}{r} \, dr.$$

Now $\alpha^2 < 0.13$ implies that

$$\frac{2}{\alpha^2} \left(\frac{2\alpha^2 - 1}{r^2} + \alpha^2(2 + r^2) \right) < \frac{2}{\alpha^2} (5\alpha^2 - 1) < -5.38$$

and

$$V \le \frac{1}{4}\alpha^2 \int_0^1 [\psi(r)^2 - 5.38]_+ \frac{dr}{r}.$$

Let s satisfy $\psi(s)^2 = 5.38$. By A3.9, $s > 0.96$ so

$$V \le \frac{1}{4}\alpha^2 (0.96)^{-2} \int_s^1 (\psi(r)^2 - 5.38) r \, dr$$

$$\le \frac{1}{4}\alpha^2 (0.96)^{-2} \left(\frac{3}{2} (5.38)^{1/2} + 1 \right) \frac{1}{2} (1 - s^2)$$

by A3.11 and the estimates $J \le \frac{3}{2}$, $K \le 1$ that we have been using. So

$$V \le \frac{1}{4}\alpha^2 (0.96)^{-2} \left(\frac{3}{2} (5.38)^{1/2} + 1 \right) \frac{1}{2} (1 - (0.96)^2)$$

$$\le 0.05\alpha^2.$$

We conclude that

$$J \le -\frac{1}{6} \frac{\alpha^2}{1 - \alpha^2} - \frac{3}{8}\alpha^4 + 0.05\alpha^2$$

$$< 0,$$

and no more cases remain.

University of California, Berkeley

REFERENCES

[1] A. B. Aleksandrov. "Inner functions: results, methods, problems." ICM Proceedings. Berkeley, 1986.
[2] A. B. Aleksandrov, P. Kargaev. Letter to author, 1988. To appear.
[3] J. Bourgain. "On the Hausdorff dimension of harmonic measure in higher dimensions." *Inv. Math.* **87** (1987), 477–483.
[4] J. Bourgain, T. Wolff. "Note on gradients of harmonic functions in dimension ≥ 3." *Colloq. Math.* **40/41**, 1 (1990), 253–260.
[5] A. P. Calderón, A. Zygmund. "On higher gradients of harmonic functions." *Studia Math.* **24** (1964), 211–226.

[6] L. Carleson. "On the support of harmonic measures for sets of Cantor type." *Ann. Acad. Sci. Fenn.* **10** (1985), 113–123.

[7] B. Dahlberg. "Estimates of harmonic measure." *Arch. Rat. Mech. Anal.* **65** (1977), 275–288.

[8] S. Gleason, T. Wolff. "Lewy's harmonic gradient maps in higher dimension." *Comm. Part. Diff. Eq.* **16** (1991), 1925–1968.

[9] D. Jerison, C. Kenig. "Boundary behavior of harmonic functions in nontangentially accessible domains." *Adv. Math.* **46** (1982), 80–147.

[10] ———. "Boundary value problems on Lipschitz domains." In *Studies in Partial Differential Equations*, edited by W. Littman. MAA Studies in Mathematics, Vol. 23, 1982.

[11] P. Jones. "A geometric localization theorem." *Adv. Math.* **46** (1982), 71–79.

[12] P. Jones, T. Wolff. "Hausdorff dimension of harmonic measures in the plane." *Acta Math.* **161** (1988), 131–144.

[13] R. Kaufman, J.-M. G. Wu. "On the snowflake domain." *Arkiv for Math.* **23** (1985), 177–183.

[14] J. L. Lewis. "Note on a theorem of Wolff." In *Holomorphic Functions and Moduli*, edited by D. Drasin, et al. MSRI Publication, Vol. 10–11. Springer Verlag, 1988.

[15] N. G. Makarov. "Distortion of boundary sets under conformal mappings." *Proc. London Math. Soc.* **51** (1985), 369–384.

[16] N. Martin, J. England. "Mathematical Theory of Entropy." *Encyclopedia of Mathematics and its Applications*, Vol. 12. Addison-Wesley, 1981.

[17] V. G. Maz'ja, V. P. Havin. "The Cauchy problem for Laplace's equation." *Vestnik Leningrad Univ.* **23**, 7 (1968), 146–147.

[18] S. N. Mergelyan. "Harmonic approximation and approximate solution of the Cauchy problem for Laplace's equation." (In Russian.) *Uspekhi Mat. Nauk (N. S.)* **11** (1956), no. 5 (71), 3–26.

[19] B. Oksendal. "Null sets for measures orthogonal to $R(X)$." *Amer. J. Math.* **94** (1972), 331–332.

[20] ———. "Brownian motion and sets of harmonic measure zero." *Pacific J. Math.* **95** (1981), 179–192.

[21] N. V. Rao. "Uniqueness theorems for harmonic functions." *Math. Notes USSR* **3** (1968), 159–162.

[22] Ya. G. Sinai. *Introduction to Ergodic Theory.* Translated by V. Scheffer. Princeton University Press, 1976.

[23] M. Spivak. "A Comprehensive Introduction to Differential Geometry." Vol. 4, Publish or Perish, Berkeley, 1979.

[24] E. M. Stein. *Singular Integrals and Differentiability Properties of Functions.* Princeton University Press, 1970.

[25] E. M. Stein, G. Weiss. "On the theory of harmonic functions of several variables, I: the theory of H^p spaces." *Acta Math.* **103** (1960), 25–62.

[26] A. Uchiyama. "A constructive proof of the Fefferman-Stein decomposition of $BMO(R^n)$." *Acta Math.* **148** (1982), 215–241.

[27] ———. "The singular integral characterization of H^p on simple martingales." *Proc. Amer. Math. Soc.* **88** (1983), 617–621.

[28] ———. "The Fefferman-Stein decomposition of smooth functions and its applications to $H^p(R^n)$." *Pacific J. Math.* **115** (1984), 217–255.

[29] G. Verchota. "Layer potentials and regularity for the Dirichlet problem for Laplace's equation on Lipschitz domains." *J. Funct. Anal.* **59** (1984), 572–611.

[30] W. S. Wang. To appear.

[31] T. Wolff. "Gap series constructions for the p-Laplacian." Typescript, 1984.

[32] J.-M. G. Wu. "On singularity of harmonic measure in space." *Pacific J. Math.* **121** (1986), 485–496.

GPSR Authorized Representative: Easy Access System Europe - Mustamäe tee 50, 10621 Tallinn, Estonia, gpsr.requests@easproject.com